Continuous Thermal Processing of Foods

Pasteurization and UHT Sterilization

ASPEN FOOD ENGINEERING SERIES

Aspen Food Engineering Series

Jose M. Aguilera and David W. Stanley, *Microstructural Principles of Food Processing and Engineering*, Second Edition (1999)
Stella M. Alzamora, María S. Tapia, and Aurelio López-Malo, *Minimally Processed Fruits and Vegetables: Fundamental Aspects and Applications*
Gustavo Barbosa-Cánovas and Humberto Vega-Mercado, *Dehydration of Foods* (1996)
Pedro Fito, Enrique Ortega-Rodríguez, and Gustavo Barbosa-Cánovas, *Food Engineering 2000* (1997)
P. J. Fryer, D. L. Pyle, and C. D. Rielly, *Chemical Engineering for the Food Industry* (1997)
S. D. Holdsworth, *Thermal Processing of Packaged Foods* (1997)
Rosana G. Moreira, M. Elena Castell-Perez, and Maria A. Barrufet, *Deep-Fat Frying: Fundamentals and Applications* (1999)
M. Anandha Rao, *Rheology of Fluid and Semisolid Foods: Principles and Applications* (1999)

Continuous Thermal Processing of Foods

Pasteurization and UHT Sterilization

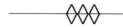

Michael Lewis, BSc, MSc, PhD
Senior Lecturer
Department of Food Science and Technology
The University of Reading
Whiteknights, Reading
United Kingdom

Neil Heppell, BSc, MSc, PhD
Senior Lecturer
School of Biological and Molecular Sciences
Oxford Brookes University
Headington, Oxford
United Kingdom

AN ASPEN PUBLICATION®
Aspen Publishers, Inc.
Gaithersburg, Maryland
2000

The author has made every effort to ensure the accuracy of the information herein. However, appropriate information sources should be consulted, especially for new or unfamiliar procedures. It is the responsibility of every practitioner to evaluate the appropriateness of a particular opinion in the context of the actual clinical situations and with due considerations to new developments. The author, editors, and the publisher cannot be held responsible for any typographical or other errors found in this book.

Aspen Publishers, Inc., is not affiliated with the American Society of Parenteral and Enteral Nutrition.

The schematic drawing on the cover shows a typical continuous-flow thermal process with steam injection and expansion cooling. The drawing is adapted with permission from H. Burton, *Ultra-High-Temperature Processing of Milk and Milk Products*, p. 108, © 1988.

Library of Congress Cataloging-in-Publication Data
Lewis, M.J. (Michael John), 1949–
Continuous thermal processing of foods: pasteurization and UHT sterilization /
Michael Lewis, Neil Heppell.
p. cm. — (Aspen food engineering series)
Includes bibliographical references and index.
ISBN 0-8342-1259-5
1. Food—Effect of heat on. 2. Food—Preservation. I. Heppell, N.J.
II. Title. III. Food engineering series (Aspen Publishers)
TP371 .L49 2000
664′.028—dc21
00-040601

Editorial Services: Kathy Litzenberg
Library of Congress Catalog Card Number: 00-040601
ISBN: 0-8342-1259-5

Printed in the United States of America

1 2 3 4 5

Table of Contents

Foreword xi

Preface xiii

Chapter 1—Introduction 1

1.1 Introduction 1
1.2 Food Composition 7
1.3 Physical Properties of Foods 10
1.4 Thermal Properties 15
1.5 Heat Transfer 19
1.6 Water Activity 25
1.7 Electrical Properties 25
1.8 Diffusion Properties 27
1.9 Surface Properties 28
1.10 Optical Properties and Sensory Characteristics 29
1.11 Other Physical Variables 30
1.12 Competing Techniques 31

Chapter 2—Kinetics for Microorganism Death and Changes in Biochemical Components 33

2.1 Death of Microorganisms at Constant Lethal Temperature 33
2.2 Effect of Temperature on Decimal Reduction Time 36
2.3 Overall Effect of a Changing Temperature 40
2.4 Thermal Treatment Criteria 41
2.5 Methods for Determination of Thermal Death Kinetics 43
2.6 Problems with Thermal Death Measurements 44
2.7 Nonlinear Thermal Death Kinetic Data 45
2.8 Log-Logistic Theory of Microbial Death 47
2.9 Changes in Biochemical Components at High Temperatures 49

2.10 Kinetics of Biochemical Changes 49
2.11 General Cooking Changes 50

Chapter 3—Continuous-Flow Thermal Processing Plant 57

3.1 Process Equipment for Continuous-Flow Thermal Processes 59
3.2 Heat Exchange Equipment 62
3.3 Process Equipment/Preprocessing Heat Treatment 62
Indirect Heating Systems 63
3.4 Plate Heat Exchangers 64
3.5 Pasteurization Process 65
3.6 Sterilization Processes for Low-Viscosity Liquid Products 66
3.7 Sterilization Processes for Low-Viscosity Liquids 68
3.8 Tubular Heat Exchangers 72
3.9 Scraped-Surface Heat Exchangers 79
3.10 Comparison of Indirect Heat Exchangers 81
Direct Heating Systems 84
3.11 The Steam Injection System 85
3.12 The Infusion System 91
3.13 Comparison of Injection with Infusion Systems 94
3.14 Control of Product Concentration or Dilution During Direct Heating Processes 95
3.15 The Supply of Culinary Steam for Direct Heating Processes 99
Comparison of Indirect and Direct Heating Systems 101
Processes Involving the Direct Use of Electricity 103
3.16 Friction Heating 103
3.17 Use of Microwaves 104
3.18 Ohmic Heating 104
Evaluation of Process Plant Performance 105
3.19 Acceptable Pack Failure Rates 105
3.20 Prediction of Process Performance Using Measured Plant Data 107
3.21 Prediction of Process Performance from Time-Temperature Profile 107
3.22 Calculation of Process Performance from Time-Temperature Data 109
Residence Time Distribution 113
3.23 Effect of Residence Time Distribution on Plant Performance 113
3.24 Velocity Profiles for Low-Viscosity Newtonian Liquids 114
3.25 Residence Time Distribution Theory 115
3.26 Measurement of Residence Time Distribution 117
3.27 Models for RTD Data 119
3.28 Further Analysis of Practical RTD Curves 123

3.29 Errors in Prediction of RTD 123
3.30 Effect of RTD on Sterilization Efficiency 124
3.31 Evaluation of Plant Performance with Temperature Distribution and RTD 127

Chapter 4—Continuous-Flow Thermal Processing of Viscous and Particulate Liquids 133

Viscous Liquid Products 133
4.1 Equipment Selection for Viscous Liquid Products 134
4.2 Characterization of Viscosity 134
4.3 Effect of Non-Newtonian Viscosity on Critical Reynolds Number 136
4.4 Effect of Non-Newtonian Viscosity on Laminar Flow Velocity Profile 137
4.5 Residence Time Distribution for Non-Newtonian Liquids in Different-Shaped Sections 139
4.6 Residence Time Distribution in Process Equipment 140
4.7 Practical Evaluation of Residence Time Distribution 141
4.8 Heat Transfer Coefficients for Viscous Liquids in Heat Exchangers 141
Liquids Containing Solid Particulates 143
4.9 Processing Equipment for Liquids Containing Particulate Solids 144
4.10 Pumping of Liquids Containing Particulate Solids 145
4.11 Transport of Particulate Solids by Liquids 148
4.12 Physical Factors Affecting the Transport of Particulate Solids ... 149
4.13 Effect of Different Parameters on the Residence Time Distribution of Particulates 152
4.14 Effect of Particulates on Heat Transfer Coefficient in Heat Exchangers 159
4.15 Effect of Surface Heat Transfer on Rate of Heating of Particulate Solid 160
4.16 Heat Transfer Coefficient between Liquids and Solid Bodies ... 164
4.17 Experimental Methods for Determination of Liquid-Particulate Heat Transfer Coefficient 166
4.18 Use of TTI Markers 168
4.19 Experimental Measurement of Liquid-Particle Heat Transfer Coefficient 170
4.20 Mathematical Modeling of the Thermal Process 174
4.21 Overall Process Validation 178
4.22 Stork Rota-Hold Process 181
4.23 Ohmic Heating of Particulate Foods 183

Chapter 5—Pasteurization **193**

5.1 Introduction 193
5.2 HTST Pasteurization 196
5.3 Factors Affecting Keeping Quality 201
5.4 Specific Products 211
5.5 Other Heat Treatments 223
5.6 Strategies for Extended Shelf Life of Refrigerated Products 224
5.7 Extended Shelf Life at Ambient Temperatures 231

Chapter 6—Sterilization **237**

6.1 Introduction: Criteria for Sterilization 237
6.2 Continuous Systems 242
6.3 Homogenization 248
6.4 Milk 250
6.5 Cream 271
6.6 Evaporated Milk and UHT Evaporated Milks 273
6.7 Ice Cream Mix 274
6.8 Soy Milk and Other Milk Analogues 275
6.9 Starch-Based Products: Soups and Custards 278
6.10 Other Drinks and Beverages: Coffees/Teas 280
6.11 Specialized Products 280

Chapter 7—Packaging Systems **285**

7.1 Connection of Packaging Equipment to Process Plant 285
7.2 Cleaning and Sterilization of Sterile Tanks 290
7.3 Use of Sterile Barriers 292
7.4 General Mechanical Principles 293
7.5 Packaging Systems for Continuous-Flow Thermally Processed Products 294
7.6 Hot Filling 294
7.7 Aseptic Packaging 295
7.8 Controls and Safeguards in Aseptic Filling Systems 325
7.9 Performance of Aseptic Fillers 326

Chapter 8—Fouling, Cleaning, and Disinfecting **331**

8.1 Introduction 331
8.2 Measurement of Fouling 335
8.3 Review of Factors Affecting Fouling 341
8.4 Fouling of Specific Products 344
8.5 Rinsing, Cleaning, Disinfecting, and Sterilizing 356
8.6 Rinsing 357

8.7 Cleaning 358
8.8 Disinfecting and Sterilizing 364

Chapter 9—Storage 369

9.1 Introduction: Raw Materials/Processing and Storage 369
9.2 Color/Browning Reaction 371
9.3 Destabilization/Deposit Formation and Gelation 374
9.4 Dissolved Oxygen 380
9.5 Flavor Changes 380
9.6 Changes to Other Components During Storage 384
9.7 Some Other Products 389
9.8 Accelerated Storage 389
9.9 Refrigerated Storage 390

Chapter 10—Quality Assurance 395

10.1 Introduction 395
10.2 Microbiological Specifications for Foods 398
10.3 Commercially Sterile Products 401
10.4 Quality Assurance/Commercially Sterile Products: The Current Approach 407
10.5 The Role of Analytical Testing 412
10.6 Sensory Characteristics 421
10.7 Some Legal Aspects 422

Appendix A—Mathematical Model for Heat Transfer to a Sphere 427

Index 435

Foreword

My own book, *Ultra-High-Temperature Processing of Milk and Milk Products*, was published by Elsevier in 1988. A few years ago it became out of print, and I was asked to consider a revised second edition. After thought and discussion with others, it seemed to me that there was insufficient new material to justify a simple revision of the same book. There did, however, seem to be a need to extend the book to cover the ultrahigh-temperature (UHT) processing of more-viscous food products and of those containing particulates. I did not feel able to do this myself; by that time I had been away from active work for too long, and my direct experience did not include these newer applications. I suggested that Michael Lewis and Neil Heppell, both of whom I had known for some time, might be interested in writing such a book. They were prepared to take over where I left off, and *Continuous Thermal Processing of Foods: Pasteurization and UHT Sterilization* is the result of their work.

In most respects this is a new and different book. Apart from an extension to viscous and particulate materials, they decided to include material on all continuous-flow heat treatment processes, which include pasteurization as well as UHT processes. Their approach has been that of chemical engineers, and much of the material specific to milk that I included has been omitted to accommodate material applicable to other products. However, some of the subjects covered in my book, particularly relating to UHT heat treatment systems and to aseptic filling methods, have changed little. Parts of my book dealing with these subjects have therefore been reused with little modification.

I hope that their work will be well received and accepted as an important contribution to food-processing literature. I am pleased to have played some part in its inception and to have contributed to it.

Harold Burton
Reading, United Kingdom
2000

Preface

This book originated from a suggestion by the publisher for a revision of Harold Burton's book on ultrahigh-temperature (UHT) processing of milk, which was published in 1988 and which is now out of print. Harold dedicated this seminal book to all those who worked on UHT processing and aseptic filling at the National Institute for Research in Dairying (NIRD) between 1948 and 1985 and particularly those in the Process Engineering Group, of which he was head for much of the time. During that period his group did much of the fundamental work on understanding the safety and quality of UHT milk, and his name was known worldwide.

Harold retired in 1985 and felt that he had been away from an active work and research environment for too long to undertake the task of revision, so he suggested to the publishers that Neil Heppell and I may wish to tackle it. Our relationship with Harold extends for more than twenty years. I was introduced to Harold by Reg Scott (the author of *Cheesemaking Practice*) shortly after my appointment as a lecturer in the Department of Food Science at University of Reading in 1973. Neil was a member of Harold's department and later registered for his PhD in the Department of Food Science. Others who followed a similar path were Sami Al-Roubaie, Geoffrey Andrews, Monika Schröder, and Paul Skudder, all of whom have contributed to the further understanding of continuous heat treatment. This helped to develop a good working relationship between NIRD and the university. Harold also worked with many other international experts on UHT during this period, some of whom spent periods of time studying at the NIRD. Meanwhile, the university obtained an APV Junior UHT plant in 1976, which has been used extensively for teaching, product development, and research since then. UHT activity at the university was summarized in a recent article (Lewis, M.J., 1995, UHT Processing, Research and Application, *European Food and Drink Review*, spring edition, 21–23, 25).

Our first reaction to the suggestion for a revised edition was that there had not been sufficient further developments within ten years to warrant a direct updating of the book (I think this coincided with Harold's view). Therefore, we proposed a more ambitious project, extending the basic format of the book to cover continuous heat treatment in its wider aspects. Three major areas were identified to provide a wider coverage.

The first was to extend the range of food products covered beyond milk products alone. This is a simple aim to state, but one that in practice is much more difficult to achieve, as the bulk of the published research work on continuous heat treatment still relates to milk and milk-based products. This probably stems from milk being widely available, cheap, and extremely nutritious. The reason it is so nutritious is its chemical complexity, which in turn leads to its being both a very difficult and fascinating product to study. However, in contrast, the commercial reality is that there is now a much wider range of aseptic products available to the consumer, although the relevant technical information on such matters as formulations and processing conditions is not so easy to find in the public domain. There would appear to be plenty of scope for further experimental work on these products. This has been particularly noticeable in the chapters on sterilization, storage, and fouling and cleaning, which are predominantly milk-based. Although this book deals predominantly with milk and milk-based products, it also covers a wider range of products than any other currently on the market.

The second aspect involved the realization that many of the products now processed are considerably more viscous than are milk and cream, and some of them contain discrete particles (deliberately added and not present as sediment). It was felt important to cover the heat treatment of more-viscous fluids, where streamline flow conditions are likely to prevail, as well as the thornier problem of heat-treating products containing particles, ideally ensuring uniform heating of the solid and liquid phases. It became necessary to cover the underlying principles and problems in more general terms without reference to specific products.

The third important aspect was to extend the coverage to incorporate pasteurization and heat treatments designed to extend further the shelf life of pasteurized products. In fact, pasteurized products are more widespread than sterilized products in many countries. It was felt that this could be most effectively done by positioning the processes of pasteurization and sterilization next to each other to emphasize further their similarities and the differences. This was also felt to be the most appropriate place to discuss the heat treatment of acidic products, fruit juices in particular, and strategies for extending the keeping quality of pasteurized products.

We have aimed to produce a book that will give a clear explanation of the principles involved in continuous heat treatment processes. The emphasis throughout is on product safety and quality. To fully understand these issues involves integrating a number of important scientific disciplines covering the physical aspects of foods: the transfer of energy and the effects of heat on the chemical, biochemical, and sensory characteristics; and the problems inherent in dealing with biological raw materials.

How best to arrange the material to achieve these objectives was a major problem. The need to integrate material led to the book's being structured into discrete packages (no pun intended). There is a detailed introductory chapter, which sets the scene and covers, in some detail, the physical properties of foods and their influence on thermal processing operations. It is our experience that these important aspects are often overlooked in courses on heat processing, despite their considerable influence on the overall process itself.

This is followed by chapters that cover the general principles of reaction kinetics, the different types of heat exchangers, and the problems encountered with viscous and particulate matter, as these will be appropriate to both pasteurization and sterilization. Following these are specific chapters on pasteurization and sterilization, where there is considerable detail, mainly on milk but also on a number of other products. To complete the book, specific chapters are devoted to aseptic packaging, storage, fouling and cleaning, and quality assurance.

This layout should help the reader who wishes to explore specific topics in depth. We have taken care to ensure that the book is well cross-referenced and indexed, which will help the reader who wishes to browse. Perhaps a novelty is the chapter on storage; although this subject is covered in Harold's book, elsewhere it is rarely given as much attention as it properly warrants.

We hope that this book will be particularly useful to undergraduate and postgraduate students of food science and technology, and to biotechnologists and engineers who need to heat and cool biological raw materials.

We believe that it will provide a useful reference source to those in the industry and provide a focus for gaining a better understanding of the factors influencing safety and quality. We believe that one of the strengths of the book is the combination of theoretical knowledge derived from the considerable research output in the subject area with our practical experience of heat processing. We have tried to make our explanations as clear as possible, especially so when interpreting results from those articles where it was not too clear what was really intended.

There is a great deal of interest (rightly so) in alternative technologies and processes for pasteurizing and sterilizing foods. Unfortunately, most articles that deal with these subjects ignore the fact that they have to compete against heat treatment, which is a very effective, convenient, and energy-efficient method of processing foods. In fact, the application of heat in higher temperatures–shorter times (HTST) pasteurization and UHT sterilization are two well-established minimum processes.

There have been many other constraints and competing pressures: meeting the publisher's deadlines, as well as other pressing priorities such as teaching quality audits (TQAs) and research assessment exercises (RAEs). One thing appreciated at an early stage in the writing of this book is that no single book on thermal processing can cover in "academic depth" such a wide and diverse subject area. In fact, many books have been written on thermal processing. We have aimed to ensure that this one will give a good balance between the engineering aspects and problems and the chemical, biochemical, and microbiological issues that have to be considered to produce foods that are both safe and of a high quality.

Finally, returning to Harold Burton, in brewing technology there is a process known as "Burtonizing" the water, which ensures water used for beer production has a composition similar to that found in Burton-on-Trent, one of the great brewing centers in the United Kingdom. It could well be argued, considering his enormous contribution to the subject area, that the term *Burtonizing the milk* would be synonymous with UHT treatment.

CHAPTER 1

Introduction

1.1 INTRODUCTION

This book is devoted to the continuous heat treatment of foods that are capable of being pumped through a heat exchanger. The aim of this introductory chapter is to supply the necessary background for understanding the principles behind the production of safe, high-quality heat-treated foods. One way of achieving these objectives is to heat and cool the food as quickly as possible. Therefore, it is important to understand the basic mechanisms of heat transfer that occur in foods as well as the interaction of that thermal energy with the components of that food.

1.1.1 Reasons for Heating Foods

Foods are heated for a number of reasons, the main ones being to inactivate pathogenic or spoilage microorganisms. Other reasons include inactivation of enzymes. For example, foods may also change and become unacceptable due to reactions catalyzed by enzymes, examples being the browning of fruit by polyphenol oxidases and flavor changes resulting from lipase and proteolytic activity. The process of heating a food will also induce physical changes or chemical reactions, such as starch gelatinization, protein denaturation, or browning, which in turn will affect the sensory characteristics, such as color, flavor, and texture, either advantageously or adversely. For example, heating pretreatments are used in the production of evaporated milk to prevent gelation and age thickening and for yogurt manufacture to achieve the required final texture in the product. However, such heating processes will also change the nutritional value of the food.

Once the food has been heated, it is held for a short period of time to inactivate the microorganisms before being cooled. Continuous processes provide scope for energy savings, whereby the hot fluid is used to heat the incoming fluid; this is known as regeneration and saves both heating and cooling costs (see Section 5.2).

Thermal processes vary considerably in their severity, ranging from mild processes such as thermization and pasteurization through to more severe processes such as in-container sterilization processes. The severity of the process will affect both the shelf life as well as other quality characteristics.

A wide range of products are heat treated, ranging from low-viscosity fluids through to highly viscous fluids. The presence of particles (up to 25 mm in diameter) further complicates the process, as it becomes necessary to ensure that both the liquid and solid phases are at least adequately and if possible equally heated. The presence of dissolved air in either of the phases becomes a problem, as it becomes less soluble as temperature increases and will come out of solution. Air is a poor heat transfer fluid.

Heating is also involved in many other operations, which are not covered in such detail in this book, such as evaporation and drying. It is also used for solids and in processing powders and other particulate materials, for example extrusion, baking, and spice sterilization.

1.1.2 Brief History

Food sterilization in sealed containers is usually attributed to the pioneering work of Appert, although recent research by Cowell (1994 and 1995) has indicated that experimental work on heating foods in sealed containers was documented much earlier than that. He also describes in some detail the early commercialization of the canning process in East London at the turn of the nineteenth century, which includes the contributions not only of Nicholas Appert, but of Peter Durand, Bryan Donkin, John Gamble, and Phillipe de Girard. It is interesting and worrying that the causative agents of food poisoning and spoilage were not understood until considerably later that century, through the work of Pasteur, who confirmed that the many food fermentations that were spoiling foods were not spontaneous but the result of microbial metabolism. He also discovered that both yeasts and *Acetobacter* species could be destroyed by relatively mild heat treatments (about 55°C). His work on the production of beer, wine, and vinegar laid the foundations for hygienic processing and the recognition of the public health implications of hygiene and heat treatment (Wilbey, 1993).

Early heat treatment processes were essentially batch in nature. The features of batch processing are described in more detail in Section 5.1. These still have an important role in food processing operations and provide the small-scale food producer with a cheap and flexible means of heat treating foods (Figure 1–1). The history of continuous sterilization process has been reviewed by Burton (1988). Continuous sterilizers had been patented and constructed to heat milk to temperatures of 130° to 140°C before the end of nineteenth century, again well before the benefits of the process were understood. Aseptically canned milk was produced in 1921, and a steam injection system was developed in 1927 by Grindrod in the United States. However, the major initiatives leading to commercialization of the ultrahigh-temperature (UHT) systems began in the late 1940s through the development of concentric-tube sterilizers and the uperization steam into milk UHT system, which was developed in conjunction with the Dole aseptic canning system. Much greater impetus came with the introduction of aseptic cartoning systems in 1961. Regulations permitting indirect UHT treatment of milk came into force in the United Kingdom in 1965, and direct steam injection was approved in 1972.

Historical developments in pasteurization are covered in Chapter 5. Now, continuous processes are widespread and include both high temperature-short time (HTST) pasteurization and UHT sterilization.

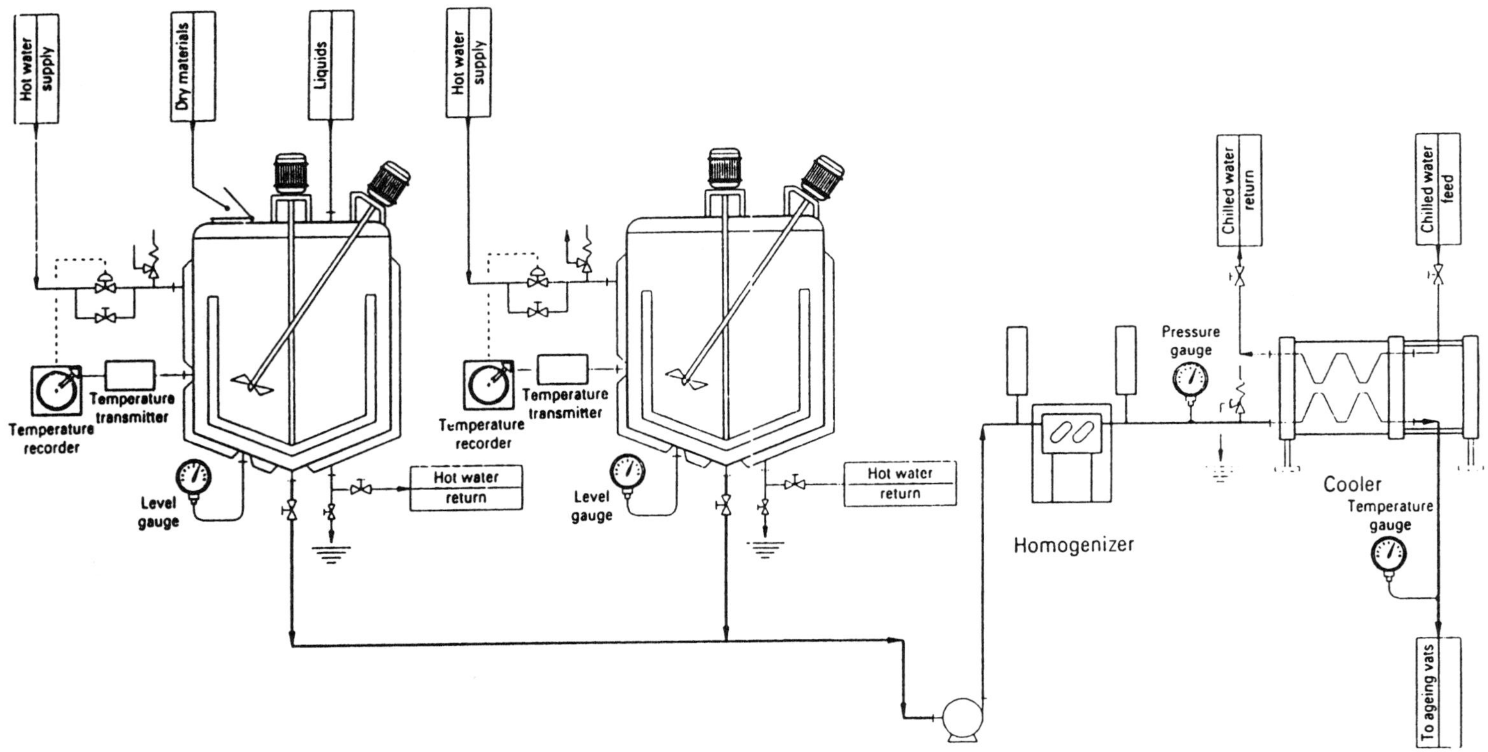

Figure 1–1 Basic flow diagram of "flip-flop" batch vat pasteurization of ice cream mix. *Source:* Reprinted with permission from M.A. Pearse, Pasteurization of Liquid Products, *Encyclopedia of Food Science and Nutrition*, p. 3443, © 1993, Academic Press.

1.1.3 Safety and Quality Issues

Some important features of continuous processing are illustrated in Figure 1–2. Note that this does not show regeneration. Some advantages of continuous processes are as follows:

- Foods can be heated and cooled more rapidly compared to in-container processes; this improves the economics of the process and the quality of the product.
- There are none of the pressure constraints that apply to heating products in sealed containers. This allows the use of higher temperatures and shorter times, which results in less damage to the nutrients and improved sensory characteristics: these being appearance, color, flavor, and texture.

The two most important issues connected with thermal processing are food safety and food quality.

Food Safety

The major safety issue involves inactivating pathogenic microorganisms that are of a public health concern. The World Health Organization estimates that there are over 100 million cases of food poisoning each year and that 1 million of these result in death. These pathogens will show considerable variation in their heat resistance; some are heat labile, such as *Campylobacter, Mycobacterium tuberculosis, Salmonella*, *Lysteria*, and of more recent concern *Escherichia coli* O157, which are inactivated by pasteurization; of greater heat resistance is *Bacillus cereus*, which may survive pasteurization and also grow at low temperatures. The most heat-resistant pathogenic bacterial spore is *Clostridium botulinum.* As well as these major foodborne pathogens, it is important to inactivate those microorganisms that will cause food spoilage, such as yeasts, molds, and gas-producing and souring bacteria. Again there is considerable variation in their heat resistance, the most heat resistant being the spores of *Bacillus stearothermophilus.* The heat resistance of any microorganism will change as the environment changes; for example, pH, water activity, or chemical composition changes and foods themselves provide such a complex and variable environment.

It is important to be aware of the type of microbial flora associated with all raw materials that are to be heat treated. After processing, it is very important to avoid reinfection of the product; such contamination is known generally as postprocessing contamination (ppc) and can cause problems in both pasteurization and sterilization. Therefore, raw materials and

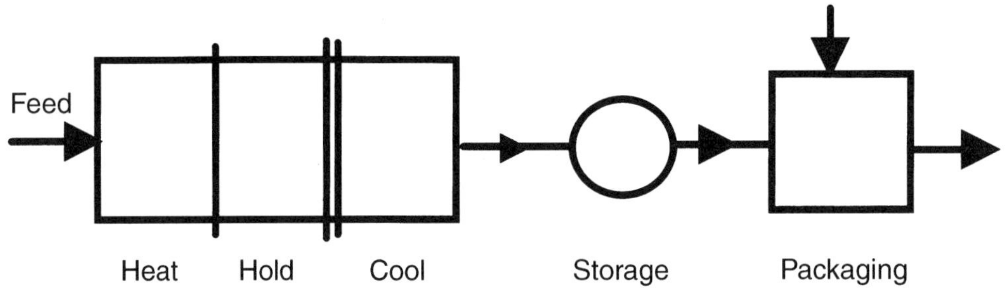

Figure 1–2 Continuous processing

finished product should not be allowed in close proximity of each other. Other safety issues are concerned with natural toxins, pesticides, herbicides, antibiotics, growth hormones, and environmental contaminants. Again, it is important to ensure that steps are taken to ensure that these do not finish up in the final product. Recently there have been some serious cases of strong allergic reactions, with some deaths, shown by some individuals to foods such as peanuts and shellfish. There is also concern and misunderstanding about bovine spongiform encephalopathy (BSE) in cattle and new variant Creutzfeldt-Jakob disease in humans (most probably caught by eating meat from cattle that have been fed contaminated animal feed).

Food Quality

Quality issues revolve around minimizing chemical reactions and loss of nutrients and ensuring that sensory characteristics (appearance, color, flavor, and texture) are acceptable to the consumer. Quality changes that may result from enzyme activity must also be considered.

There may also be conflicts between safety and quality issues. For example, microbial inactivation and food safety are increased by more severe heating conditions, but product quality will in general deteriorate.

To summarize, it is important to understand reaction kinetics and how they relate to the following:

- microbial inactivation
- chemical damage
- enzyme inactivation
- physical changes

In general, heat processing eliminates the need to use further additives to extend the shelf life, although additives may help improve the sensory characteristics or make processes less susceptible to fouling. In addition to reactions taking place during the heat treatment, chemical, enzymatic, and physical changes will continue to take place during storage. Microorganisms that survive the heat treatment may also grow if conditions are favorable. Pasteurized products are normally kept refrigerated during storage to retard microbial growth, and low temperatures must be maintained throughout the cold chain. On the other hand, sterilized products are not normally refrigerated and are stored at ambient temperature. This may vary considerably throughout the world, ranging from below 0°C to above 50°C. All the reactions mentioned above are temperature dependent, and considerable changes may take place during the storage period. One example is browning of milk and milk products, which is very significant after 4 months of storage at 40°C. Changes during storage are discussed in more detail for pasteurized products in Chapter 5 and for sterilized products in Chapters 6 and 9.

1.1.4 Product Range

The products covered in this book include all those that can be processed by passing them through a continuous heat exchanger. This includes milks and milk-based drinks, fruit and vegetable juices, purées, soups, and sauces (both sweet and savory), and a range of products containing particulate matter, up to about 25 mm diameter.

There are two distinct market sectors. The first is those products that are given a mild heat treatment and then kept refrigerated throughout storage; the second is those that are sterilized and stored at ambient temperature. The relative importance of these two sectors will vary from country to country for different products and from country to country for the same product. For example, in England and Wales, pasteurized milk, which is stored chilled, accounts for about 87% of liquid milk consumption, while UHT and sterilized milk accounts for about 10%. However, UHT milk accounts for a much greater proportion of milk consumed in other countries, for example, France (78%) and Germany (58%) (see Table 1–1). In the United Kingdom, there has also been a significant increase in semiskim milk consumption, at the expense of whole milk, as well as an increase in the sale of milk in plastic bottles (Dairy Facts and Figures, 1996). In Great Britain in 1995, household consumptions (kg/head/year) were whole milk (42.3), low-fat milks (57.5), cream (0.7), fruit juice (12.7), soft drinks (47.3), mineral water (5.7), tea (2.0), and coffee (0.6). In the United Kingdom the aseptically packaged food and drinks market is estimated to be worth £1.2 billion, with cartons accounting for greater than 90% of this. Fruit juices and fruit drinks accounted for 76% of this market, UHT milk 15%, milk drinks 6%, with other products about 2%. (James, 1996).

Table 1–1 Sales of Liquid Milk by Type of Treatment '000 Tonnes

		Type of Milk			
Country	*Year*	*Pasteurized*	*Sterilized*	*UHT*	*Total*
Germany	1983	1,942	46	1,651	3,639
	1993	2,173	24	3,057	5,254
France	1983	1,106	468	2,123	3,697
	1993	609	362	3,437	4,408
Italy	1988	1,452	246	1,072	2,770
	1993	1,796	286	1,007	3,089
Netherlands	1983	788	— 166	—	954
	1993	815	— 69	—	884
Belgium	1983	27	497	212	736
	1993	44	363	420	827
Luxembourg	1983	24	—	7	31
	1993	29	—	38	67
United Kingdom	1983/4	5,244	342	59	5,645
	1993/4	4,671	155	271	5,097
Irish Republic	1983	524	—	—	524
	1993	533	—	—	533
Denmark	1983	578	—	—	578
	1993	487	—	—	487
Greece	1988	280	59	—	339
	1993	315	19	—	334
Spain	1988	617	819	1,226	2,662
	1993	270	639	2,679	3,588
Portugal	1988	243	3	440	686
	1993	133	—	627	760

Source: Reprinted with permission from *EC Facts and Figures*, © 1994, Residency Milk Marketing Board.

1.2 FOOD COMPOSITION

Foods composition tables are useful as they provide detailed information on the chemical composition of foods that are to be heat treated. They also help to illustrate the chemical complexity of food and the large number of chemical reactions that can take place on heating. Most foods are extremely complex; for example, in milk it is possible to measure over 100 components in measurable concentrations; there may also be unwanted components arising from the environment, such as antibiotics or other veterinary residues, radionuclides, and plasticizers such as phthalates from plastic contact materials.

Some representative data from food composition tables for products encountered in this text are given in Table 1–2 (For a fuller list of components, the tables themselves should be consulted). For most fresh foods the main component is water, and one very important piece of information is its moisture content or its solid content. The tables also give information on the major components (protein, fat, carbohydrate) as well as minerals, vitamins, and trace metals.

One important limitation of these tables is that they give average values and do not account for the biological variability found for most natural products. This is not so important for using these data to estimate physical properties of foods (see later). However, it is relevant where components, such as vitamin C, are used as an index of nutritional damage during processing. This is because for most processed foods, the analyst does not have access to the raw material, and it would be misleading (incorrect) to use values in these tables as the starting value. If this is done, it will lead to inaccuracies in estimates for nutritional losses. In order to assess the severity of a heating process, it is more satisfactory to make use of "indicator" components, which are not present in the raw material but are produced after heat treatment (e.g., lactulose in milk).

Furthermore, these tables do not provide any information on the major enzymes associated with the foods or the main types of pathogenic and spoilage organisms that may be present. This is covered in more specialized literature on specific commodities. Other information of relevance will be an awareness of the presence of antinutritional compounds and environmental pollutants, which may need to be removed as part of the process. Some separation processes that might influence the properties of heat processed foods include reverse osmosis, ultrafiltration, ion exchange, electrodialysis, and fat fractionation. These processes have been reviewed recently by Grandison and Lewis (1996).

1.2.1 Acidity—pH

A very important property as far as heat treatment is concerned is the acidity of the food. This is usually expressed as pH, which is the reciprocal of the log of the hydrogen ion concentration:

$$pH = \log (1 / [H^+])$$

Some pH values for a range of foods are given in Exhibit 1–1. These are derived from a number of sources. Historically foods are classified as acid or low acid (see Section 6.1). Most bacteria, including *Clostridium botulinum*, will not be active and reproduce below a pH

Table 1–2 Food Composition Data (g/100 g)

	Milk	*Apple, Cooking, Raw*	*Orange Flesh*	*Mango Flesh*	*Tomato*	*Carrot, Young, Raw*	*Peas, Raw*	*Eggs Chicken, Raw, Whole*	*Vegetable Soup, Canned*	*Soy Milk, Plain*
Water (g)	87.8	87.7	86.1	82.4	93.1	88.8	74.6	75.1	86.4	89.7
Protein (g)	3.2	0.3	1.1	0.7	0.7	0.7	6.9	12.5	1.5	2.9
Fat (g)	3.9	0.1	0.1	0.2	0.3	0.5	1.5	10.8	0.7	1.9
Sugar (g)	4.8	8.9	8.5	13.8	3.1	5.6	2.3	tr	2.5	0
Starch (g)	0	tr	—	0.3	tr	0.2	7.0	0	4.2	0.8
Fiber (g)	0	2.2	1.8	2.9	1.3	2.6	4.7	0	1.5	tr
Potassium (mg)	140	88	150	180	250	240	330	130	140	120
Sodium (mg)	55	2	5	2	9	40	1	140	500	32
Calcium (mg)	115	4	47	12	7	34	21	57	17	13
Iron (mg)	0.06	0.1	0.1	0.7	0.5	0.4	2.8	1.9	0.6	0.4
Phosphorus (mg)	92	7	21	16	24	25	130	200	27	47
Vitamin C (mg)	1.0	14	54	37	17	4	24	0	tr	0
Vitamin B_1 (mg)	0.06	0.04	0.11	0.04	0.09	0.04	0.21	0.09	0.03	0.06
Vitamin B_6 (mg)	0.06	0.06	0.10	0.13	0.14	0.07	0.12	0.12	0.05	0.07
Vitamin D (μg)	0.03	—	0	0	0	0	0	1.75	0	0
Vitamin E (mg)	0.09	0.27	0.24	1.05	1.22	0.56	0.21	1.11	N*	0.74

**Note:* N = The nutrient is present in significant quantities but there is no reliable information on the amount.

Source: Data from McCance and Widdowsons, *Composition of Foods Tables*, 5th ed., © 1991, Royal Society of Chemistry, MAFF.

Exhibit 1–1 Mean pH Values of Selected Foods

pH Value	
2.3	Lemon Juice (2.3), Cranberry sauce (2.3)
3.0	Rhubarb (3.1)
	Applesauce (3.4), Cherries, RSP (3.4)
	Berries (3.0–3.9), Sauerkraut (3.5)
	Peaches (3.7), Orange juice (3.7)
	Apricots (3.8)
4.0	
	Cabbage, red (4.2), Pears (4.2)
	Tomatoes (4.3)
	Onions (4.4)
4.6	Ravioli (4.6)
	Pimentos (4.7)
	Spaghetti in tomato sauce (4.9)
5.0	Figs (5.0)
	Carrots (5.2)
	Green beans (5.3), Beans with pork (5.3)
	Asparagus (5.5), Potatoes (5.5)
	Lima beans (5.9), Tuna (5.9), Tamales (5.9)
6.0	Codfish (6.0), Sardines (6.0), Beef (6.0)
	Pork (6.1), Evaporated milk (6.1)
	Frankfurters (6.2), Chicken (6.2)
	Corn (6.3)
	Salmon (6.4)
	Crabmeat (6.8), Milk (6.8)
	Ripe olives (6.9)
7.0	Hominy (7.0) (cornmeal)

of 4.5, so foods that are below this value are classified as acidic foods; most fruit fall into this category, and the main concern in these foods is to inactivate yeasts and molds. Temperatures up to 100°C may be required. Low-acid foods (pH > 4.5) need to be more severely treated, with a treatment of 121°C for 3 minutes (or equivalent) being recommended as a minimum process (see Section 6.1 for further details). Vegetables, meat, fish, and milk fall into this category. Many of these products can be made acidic by direct addition of acids (e.g., vinegar, citric acid) or through acid addition by lactic acid fermentation (e.g., yogurt or sauerkraut).

The pH of the food may influence many other aspects related to quality, such as colloidal stability in milk and the rate of color development in both enzymatic and nonenzymatic browning. Microbial inactivation will also be influenced by pH.

It is worth noting that the pH of foods decreases during heating, by as much as 1 pH unit. This results from the change of the dissociation constant for water, combined with other changes to the chemical components. The pH usually returns close to its original value after cooling.

For some products such as milk, titratable acidity is also often used. However, in a complex food environment, it is more than just a measure of the amount of acid in the product. What is being measured is the amount of alkali required to bring the pH of the sample from its initial starting value to that which corresponds to the pH change of the indicator (about 8 for phenolphthalein). Therefore, as well as being influenced by the amount of acid, it will also be affected by other components that exert a buffering action, for example, protein and salts in milk. However, it is widely used in milk processing to observe the progress of lactic acid fermentations, for example, in cheeses and yogurt production.

1.3 PHYSICAL PROPERTIES OF FOODS

As described earlier, it is desirable both to heat foods and to cool them as quickly as possible. Therefore, it is important to understand the factors affecting heating and cooling of foods. In this context, the physical properties of foods play an important role.

The composition of the food will affect the physical properties, particularly its moisture content. Many models for predicting physical properties are based on food composition; the simplest consider the food to be a two-component system (water and solids), whereas the more complex will break down the solid fraction into protein, carbohydrate, fat, and minerals. Therefore, a useful starting point for many foods is the food composition tables. Also, the physical form of the food is important—whether it is solid or liquid; this will determine whether the main mechanism of heat transfer is by conduction or convection. Canned foods are often classified as either conduction or convection packs (see Section 1.4).

1.3.1 Size–Shape–Surface Area

The size and shape presented is also important. Heat transfer rates are proportional to the surface area. To maximize the heating and cooling rate, the surface area to volume (SAV) ratio should be increased. The worse possible shape is the sphere, which presents the minimum SAV ratio. As a sphere is flattened, this ratio increases. With canned foods, as the can sizes get larger the SAV ratio decreases and the heating rate decreases. New shapes have evolved, such as plastic trays or flexible pouches, which increase the SAV ratio, thereby heating more quickly, which in turn may give rise to an improvement in product quality.

In a plate heat exchanger, the plate gap width is narrow, offering a relatively large SAV ratio. This, combined with turbulence, helps promote rapid heat transfer. Tubes tend to be wider in diameter, which exposes a smaller heating surface per unit volume, which leads to less rapid heating. For more-viscous products, internal agitation may be used to improve heat transfer and reduce burn-on.

For colloidal systems the size of the dispersed phase is important; for example, in milk casein micelles range between 30 and 300 nm diameter, whereas fat globules range between about 1 and 10 μm. If air is incorporated, (e.g., whipped cream or ice-cream) the air bubbles may be an order of magnitude larger than the fat globules. The size of the fat globules, as well as their density will influence their separation (Stokes law, see Section 1.3.2 and 4.12).

A decrease in particle size, for example, after homogenization, will increase the SAV ratio. On the other hand, clumping of bacteria will lead to decrease in the SAV ratio, as will the presence of any clumps of poorly dispersed or dissolved solid matter. Heat transfer will be slower into larger clumps.

1.3.2 Density

Although it does not directly contribute to the thermal efficiency of a heat treatment process, an awareness of density can be important, and neglecting it may lead to some serious defects in product quality. In fact, density differences between components in a formulated food may well lead to separation, for example, cream plug in milk and cream, sediment formation in UHT milk, and separation of components in particulate systems.

Density (ρ) is simply defined as mass/volume, with Septìme International (SI) units of kg m^{-3}. Water has a density of 1,000 kg m^{-3}. It can also be expressed as specific gravity (SG) by reference to water (ρ_w); this is dimensionless;

$$SG = \rho/\rho_w \tag{1.1}$$

The densities (kg m^{-3}) of other main components in foods are oils and fats (900–950), sucrose (1,590), starch (1,500), cellulose (1,270–1,610), protein (1,400); salt (2,160), and citric acid (1,540).

It can be seen that most are substantially different from water, and this may lead to separation if not fully dissolved. The separation velocity can be estimated from Stokes law:

$$v = gD^2(\rho_s - \rho_f)/18\,\mu \tag{1.2}$$

where v = terminal velocity (m s^{-1}); D = particle diameter (m), and ρ_s and ρ_f are the solid and fluid densities, respectively (kg m^{-3}) ; μ = fluid viscosity (Pa.s). This allows comparisons of the effects of density differences between the phases, with changes in particle diameter and fluid viscosity.

Although most solids and liquids are considered to be incompressible (i.e., density is not affected by moderate changes in temperature and pressure), the density of liquids decreases as temperature increases. Thus a heated liquid will tend to rise and this provides the driving force for natural convection. Thus, the influence of temperature on density may become relevant in thermal processing. Some examples of temperature dependence are given in Table 1–3.

The presence of air ($\rho = 1.27$ kg m^{-3}) will substantially decrease the density of a fluid; for example, the density of apples can range between about 600 and 950 kg m^{-3}; these differences are due to the different amounts of air within the apple.

Whipping and foaming will lead to a decrease in density of a liquid. Many products that are to be heat treated are mixed beforehand; it is important to avoid too much air incorporation by excessive agitation.

Gases and vapors are described as compressible. The compressibility is best illustrated by reference to steam tables, where it is presented as specific volumes (Table 1–4). For example, at atmospheric pressure steam has a specific volume of 1.67 $m^3\,kg^{-1}$, whereas at 10°C (under vacuum) it has increased to 106.4 $m^3\,kg^{-1}$.

For processes involving direct contact with steam, it is important to check that there has been no net dilution. Freezing point depression or accurate density measurement could be used for this purpose.

For foods that contain little entrapped air, the density of a food can be estimated from a knowledge of its composition (mass fractions of the main components and their densities; see earlier values): this is useful for most liquid formulations.

$$\rho = 1/\Sigma\, m_i/\rho_i \qquad (1.3)$$

Where there is a substantial volume fraction of air in the sample, the density can be expressed in terms of the volume fraction: $\rho = \Sigma\, V_i\, \rho_i$.

There are some unusual density scales used, for example Brix, Baume, and Twaddell: these are summarized by Lewis (1990) and Hayes (1987).

One important function of blanching (for canned foods) and other heating processes is to remove air from food; this helps minimize strain during the subsequent sterilization process as well as increasing the density of the food and ensuring that it packs better; for example, some apples contain up to 40% air by volume. However, in general there is very little published information on the air content of foods.

Table 1–3 Examples of Temperature Dependence on Density (kg m^{-3})

	0°C	*20°C*	*40°C*	*60°C*
Water	999.9	998.2	992.2	983.2
Corn oil	933	920	906	893
Cottonseed oil	935	921	908	894

Table 1–4 Properties of Saturated Steam

Temperature	Pressure (Absolute)	Specific Volume (v_g)	Enthalpy (kJ/kg)			Entropy (kJ/kg K)		
°C	bar	(m^3kg^{-1})	h_f	h_{fg}	h_g	s_f	s_{fg}	s_g
0	0.006	206.30	0.0	2501.6	2501.6	0.0	9.16	9.16
2	0.007	179.92	8.4	2496.8	2505.2	0.03	9.07	9.10
10	0.012	106.43	42.0	2477.9	2519.9	0.15	8.75	8.90
20	0.023	57.84	83.9	2454.3	2538.2	0.30	8.37	8.67
30	0.042	32.93	125.7	2430.7	2556.4	0.44	8.02	8.45
40	0.073	19.55	167.5	2406.9	2574.4	0.57	7.69	8.26
50	0.123	12.05	209.3	2382.9	2592.2	0.70	7.37	8.08
60	0.199	7.68	251.1	2358.6	2609.7	0.83	7.10	7.93
70	0.311	5.05	293.0	2334.0	2626.9	0.95	6.80	7.75
80	0.474	3.41	334.9	2308.8	2643.8	1.08	6.54	7.62
90	0.701	2.36	376.9	2283.2	2660.1	1.19	6.29	7.40
100	1.013	1.67	419.1	2256.9	2676.0	1.31	6.05	7.36
110	1.433	1.21	461.3	2230.0	2691.3	1.42	5.82	7.24
120	1.985	0.89	503.7	2202.2	2706.0	1.53	5.60	7.13
130	2.701	0.67	546.3	2173.6	2719.9	1.63	5.39	7.03
140	3.610	0.51	589.1	2144.0	2733.1	1.74	5.19	6.93
150	4.760	0.39	632.1	2113.3	2745.4	1.84	4.99	6.83
175	8.924	0.22	741.1	2030.7	2771.8	2.09	4.53	6.62
200	15.549	0.13	852.4	1938.6	2790.9	2.33	4.10	6.43
225	25.501	0.08	966.9	1834.3	2801.2	2.56	3.68	6.25
250	39.776	0.05	1085.8	1714.7	2800.4	2.79	3.28	6.07
300	85.927	0.02	1345.1	1406.0	2751.0	3.26	2.45	5.71

Source: Reprinted with permission from M.J. Lewis, *Physical Properties of Foods and Food Processing Systems*, © 1990, Woodhead Publishing Ltd.

For liquids, the level of dissolved oxygen may also be important. Note that the solubility of dissolved oxygen decreases as temperature increases. Therefore, air will come out of solution during heating and may cause air bubbles or air locks (see Section 8.4.3).

1.3.3 Viscosity

The main resistance to heat transfer by convection is given by the viscosity (μ) of the fluid. Viscosity is considered to be a measure of the internal friction within a fluid. It is a measure of how easily it will flow when it is subject to shear force. One of the most common such forces is that due to gravity.

Viscosity is defined as the ratio of shear stress (τ) (Pa) to shear rate (dv/dy) (s^{-1}). It has SI units of Pa.s (Nsm^{-2}). The cgs unit (centipoise [cp]) is still widely used, the conversion factor being 1 mPa.s = 1 cp.

Water is a low-viscosity fluid with a value of about 1 mPa.s at 20°C. Milk has a value of about twice that of water. Some viscosity values for some other fluids are given in Table 1–5.

Viscosity decreases as temperature increases (Figure 1–3). This makes fluids easier to pump at high temperatures. It can also lead to a problem of the development of high viscosities in the cooling section of some heat exchangers. This is especially so for thickened products or those containing substantial amounts of fat, which will be undergoing crystallization. Not only will the viscosity be increasing, but the material will be giving out its latent heat of crystallization. More information about temperature dependence of viscosity for dairy products is given by Lewis (1993).

Another problem is that many foods are non-Newtonian. For a non-Newtonian fluid, the measured viscosity may change with both the shear rate and the time of application of the shear. Non-Newtonian behavior can be classified as time independent or time dependent. (Figure 1–4). The two types of time-independent behavior are pseudoplastic (shear thinning) and dilatant (shear thickening). Pseudoplastic or shear-thinning behavior is the most common behavior. These fluids decrease in viscosity with increasing shear rate. Many dilute solutions of macromolecules show this type of behavior.

Dilatant or shear-thickening behavior in foods is comparatively rare. One place it has been encountered is in cream products that have been rapidly cooled. It is perhaps fortunate that it is rarely encountered, as it makes heat transfer at high shear rates more difficult to achieve. Examples of models that have been used to characterize time-independent non-Newtonian liquids are given in Section 4.2. Some values for K and n for the power law equation (Equation 4.2) for different foods are given in Table 1–6.

A further complication is time dependency. Again the most common behavior is shear-thinning time dependency, known as thixotropy. When measured at a constant speed, such fluids will show a decreasing viscosity value; an equilibrium may be eventually reached. These fluids may also recover their viscosity when rested. They show hysteresis when subjected to increasing followed by decreasing shear rate. Some fluids may also exhibit some elastic characteristics, superimposed on their viscous nature; such fluids are termed *viscoelastic* and this property will make them more difficult to pump in some circumstances.

Table 1–5 Viscosity Values at 20°C

Fluid	*Viscosity (Nsm^{-2})*	*Fluid*	*Viscosity (Nsm^{-2})*
Carbon dioxide	1.48×10^{-5}	20% sucrose (g/100 g of solution)	2×10^{-3}
Water	1.002×10^{-3}	40% sucrose g/100 g of solution)	6.2×10^{-3}
Carbon tetrachloride	0.969×10^{-3}	60% sucrose (g/100 g of solution)	58.9×10^{-3}
Olive oil	84×10^{-3}	Honey (average values after mixing) (25°C)	6000×10^{-3}
Castor oil	986×10^{-3}		
Glycerol	1490×10^{-3}	Milk	2×10^{-3}
		Ethanol	1.20×10^{-3}
		n-hexane	0.326×10^{-3}

Source: Reprinted with permission from M.J. Lewis, *Physical Properties of Foods and Food Processing Systems*, p. 111, © 1990, Woodhead Publishing Ltd.

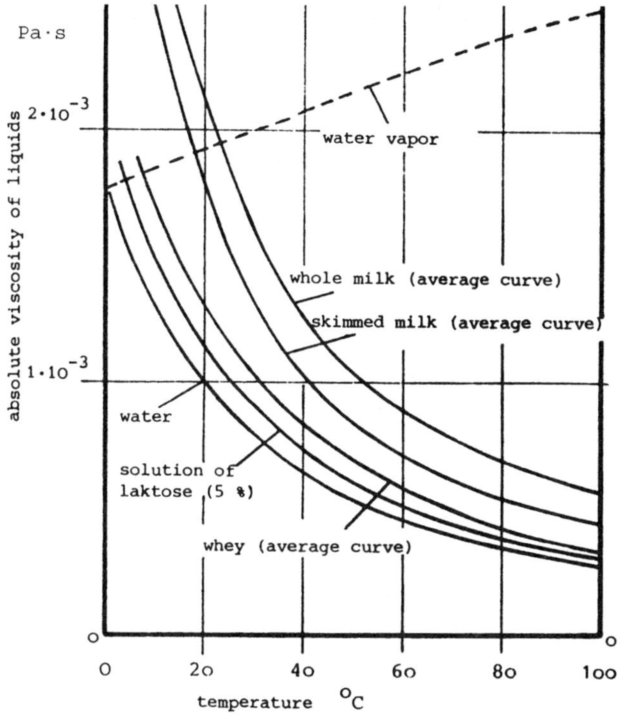

Figure 1–3 Absolute viscosity, η, of a number of liquids and gases. *Source:* Reprinted with permission from *Food Engineering and Dairy Technology*, p. 13, Copyright © Verlag Kessler.

The viscosity of a fluid can affect the choice of heat exchanger, pumps, and ancillary equipment. As a fluid becomes more viscous, more energy is required to pump it and the pressure drops become much larger. The viscosity will strongly influence the Reynolds number and determine whether the flow regime is either streamlined or turbulent. This in turn will influence the rate of heat transfer and the distribution of residence times. This is discussed in more detail in Chapters 3 and 4.

1.4 THERMAL PROPERTIES

Since heating and cooling are the central theme of this text, the thermal properties of foods are very important. These will influence both the total amount of energy and the rate of energy transfer required.

1.4.1 Specific Heat

Specific heat is defined as the amount of energy required to raise unit mass by unit temperature rise. It has SI units of kJ kg^{-1} K^{-1}. Water has a very high value (4.18 kJ kg^{-1} K^{-1}) compared to most other substances; this combined with its ubiquitous nature has led to its

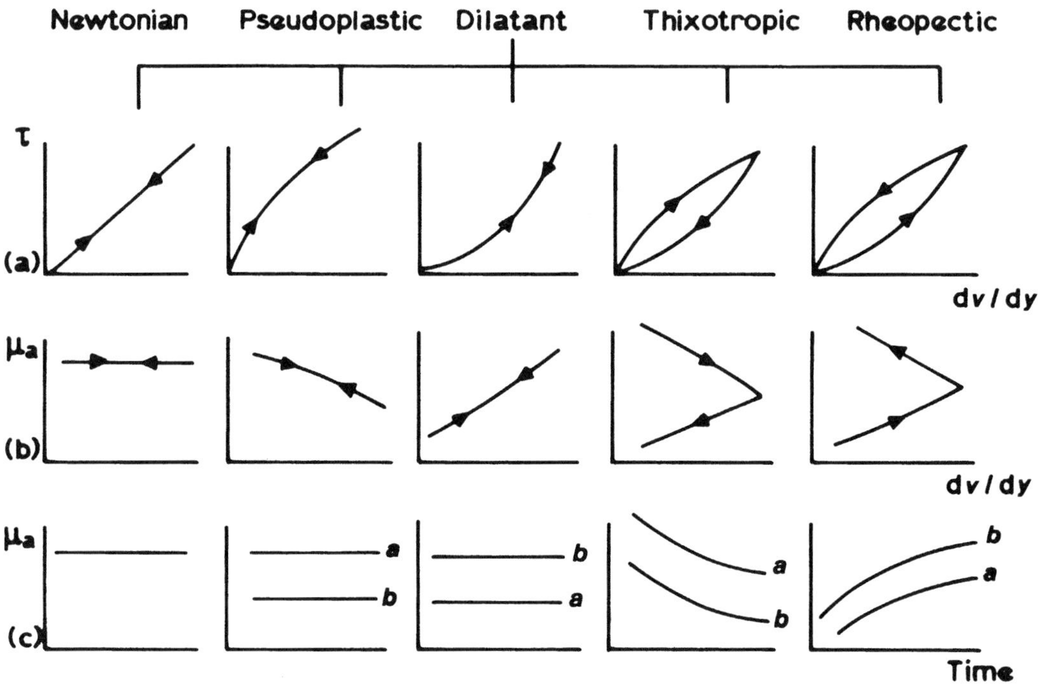

Figure 1–4 Rheograms for non-Newtonian fluids; (a) Shear stress (τ) against shear rate (dv/dy); (b) apparent viscosity (μa) against shear rate; (c) apparent viscosity against time at two different shear rates, a and b, where b>a. *Source:* Reprinted from M.J. Lewis, Physical Properties of Dairy Products, in *Modern Dairy Technology*, Vol. 2, 1st ed., R.K. Robinson, ed., pp. 331–380, © 1993, Aspen Publishers, Inc.

wide use as a heating and cooling medium. When used as a coolant, close attention should be paid to its microbial quality.

A number of models are available to predict the specific heat of a food from its composition. The simplest treats the food as a two-component system: water and solid:

$$c = m_w c_w + m_s c_s \tag{1.4}$$

Table 1–6 Power Law and Consistency Index Values: = k(dv/dy)^n

Product	*Temperature* *°C*	*K* *(Ns^nm^{-2})*	*n*	*Reference*
Custard	80	7.24	0.36	Milson (1980)
Gravy	80	2.88	0.39	Milson (1980)
Tomato juice (12%TS)	32	2.0	0.43	Toledo (1980)
Tomato juice (25%TS)	32	12.9	0.40	Toledo (1980)
Tomato juice (30%TS)	32	18.1	0.40	Toledo (1980)

where c = specific heat; m = mass fraction. For this the specific heat of solids is taken as 1.46 kJ kg^{-1} K^{-1}.

For multicomponent systems, a more accurate form of the equation becomes $c = \Sigma m c$, that is, the sum of products of the mass fraction and the specific heats for water, fat, carbohydrate, protein, and minerals. The specific heats (kJ kg^{-1} K^{-1}) for these different components are water (4.18), fat (1.7), carbohydrate (1.4), protein (1.6), minerals (0.8).

It can be seen that the specific heats of all other food components is much lower than that for water. Ice has a specific heat of about half that for liquid water, as does also water vapor. Some specific heat values for some different foods are given in Table 1–7. Where data are unavailable, food composition tables together with the models above are useful for predicting specific heat values.

Table 1–7 Specific Heat of Some Foods and Food-Processing Materials

		Specific Heat	
Material	*Temperature*	*(kJ kg^{-1} K^{-1})*	*(kcal kg^{-1} K^{-1}) (BTU/1b °F)*
Water	59°F	4.18	1.000
Ice	32°F	2.04	0.487
Water vapor	212°F	2.05	0.490
Air	−10°F–80°F	1.00	0.240
Copper	20°C	0.38	0.092
Aluminum	20°C	0.89	0.214
Stainless steel	20°C	0.46	0.110
Ethylene glycol	40°C	2.21	0.528
Alcohol	0°C	2.24	0.535
Oils	20°C	1.73	0.414
Corn	0°C	1.86	0.446
Sunflower	20°C	1.93	0.460
Apples (84.1% m.c.)	above F.pt.	3.59	0.860
	below F.pt.	1.88	0.45
Potatoes (77.8% m.c.)	above F.pt.	3.43	0.82
	below F.pt.	1.80	0.43
	dried (10.9% m.c.)	1.85	0.443
Lamb (58.0% m.c.)	above F.pt.	2.80	0.67
	below F.pt.	1.25	0.30
Cod	above F.pt.	3.76	0.90
	below F.pt.	2.05	0.49
Milk (87.5% m.c.)	above F.pt.	3.89	0.930
	below F.pt.	2.05	0.490
Soya beans, 8.7% m.c.		1.85	0.442
Wheat, 10.0% m.c.		1.46–1.80	0.35–0.43

Source: Reprinted with permission from M.J. Lewis, *Physical Properties of Foods and Food Processing Systems*, © 1990, Woodhead Publishing Ltd., with data from N.N. Mohsenin, *Thermal Properties of Food and Agricultural Materials*, © 1980; and H.G. Kessler, *Food Engineering and Dairy Technology*, © 1981.

Specific heat is a measure of sensible heat change. For substances that may be subject to some crystallization, an apparent specific heat may be used, for example, fat in milk or cream (Table 1–8). This takes into account changes brought about by crystallization (latent heat) and sensible heat changes.

1.4.2 Latent Heat

Latent heat changes assume importance when evaporating or condensing liquids or during crystallization or melting. Water has very high latent heat of vaporization and steam is widely used as a heating medium. The properties of steam are summarized in the saturated steam tables (see Table 1–4). These tables give the relationship between steam pressure and temperature, as well as information on the specific volumes enthalpy and entropy values for the liquid (f) and vapor (g) states (specific volumes of liquids are often not recorded). Special attention must be paid to the quality of steam for direct heating purposes (see Sections 3.11 and 6.22).

Thermodynamic data are also available for the common refrigerants used in heat treatment. Other fluids used for cooling include brines, glycols, and cryogenic fluids (American Society of Heating, Refrigeration and Air Conditioning Engineers, Inc., 1994).

1.4.3 Enthalpy and Specific Enthalpy

Enthalpy (H) is a thermodynamic function that is defined as $H = U + PV$, where U = internal energy, P = pressure, and V = volume.

Of special interest in any chemical or physical reaction is the enthalpy change (ΔH). It can be shown that specific enthalpy changes can be equated to heat changes when processes take place at constant pressure.

$$\Delta H = q = \int c_p \, d\theta \tag{1.5}$$

This is useful, since most processes occur at constant pressure.

The units of enthalpy are joules (J). In most compilations of thermodynamic properties, specific enthalpy is used, which is the enthalpy per unit mass (J kg^{-1}). Thus in the steam tables, the heat given out when steam condenses at a constant pressure is equal to ($h_g - h_f$). The same applies for refrigerants when they evaporate or condense. It is worth mentioning that water has a very high latent heat of vaporization (see h_{fg} column in steam tables) compared to most other substances.

Table 1–8 Variation of Apparent Specific Heat for Butterfat with Temperature

Temperature (°C)	–40	–20	–10	0	10	20	30	40	50
Apparent specific heat (kJ kg^{-1} K^{-1})	1.59	1.84	2.01	3.34	4.39	5.35	3.34	2.09	2.01

Enthalpy data are particularly useful when crystallization also takes place during cooling, for example, fat crystallization during cooling of emulsions and water crystallization during freezing. Most data also provide information on the percentage of crystalline solids (α) present at any temperature. An interesting system is ice cream mix, which is pasteurized or sterilized, homogenized, cooled, aged cooled, and frozen. During the cooling and freezing, fat crystallization takes place from about 40°C downward, and ice formation starts at about –2°C.

Enthalpy changes can be determined by differential scanning calorimetry (DSC) (Biliaderis, 1983). Enthalpy data for fruit and vegetable juices is shown in Figure 1–5. This shows how enthalpy is affected by changes in both temperature and solids content. Riedel (1955) also presents detailed enthalpy data for over 20 different oils and fats. These data are particularly useful during cooling for products containing fat, when crystallization will also take place (Table 1–9).

1.5 HEAT TRANSFER

The three main mechanisms of heat transfer are conduction, convection, and radiation. Conduction is the predominant mechanism in solids and convection in liquids. Most processes usually involve both mechanisms (see Section 1.5.4). Radiation involves heat transfer by the electromagnetic radiation that can cover the whole electromagnetic spectrum, for example microwaves, infrared, ultraviolet, X rays, and gamma rays.

1.5.1 Conduction and Thermal Conductivity

In solids, the main mechanism of heat transfer is by conduction. However, the thermal conductivity of foods is low compared to metals, and in general solid foods heat very slowly. An increase in the surface area to volume ratio caused by size reduction processes will have a very beneficial effect on the overall rate of heat transfer.

The thermal conductivity of a material is defined as the rate of heat transfer passing through a 1 m^2 cross-sectional area of the material at steady state, when a temperature gradient of 1 K (1°C) is maintained over a distance of 1 m. The SI units are $Wm^{-1} K^{-1}$. Some values for a number of materials are given in Table 1–9. It is notable that the thermal conductivity of stainless steel is much less than that of both aluminium and copper. In comparison it can be seen that foods are poor conductors of heat compared to metals, with typical values between 0.4 and 0.6 $Wm^{-1} K^{-1}$. As moisture content is reduced the thermal conductivity decreases. Air is a very poor conductor of heat, so this means that porous foods are also very poor conductors; one example is freeze-dried products.

Again there are a number of models for predicting thermal conductivity from food composition. These have been reviewed by Rahman (1995). Thermal conductivity is anisotropic; that is, it is a property that depends upon direction of measurement. For example, for fibrous foods heat transfer is better in a direction along the fibers than across the fiber bundle. Be-

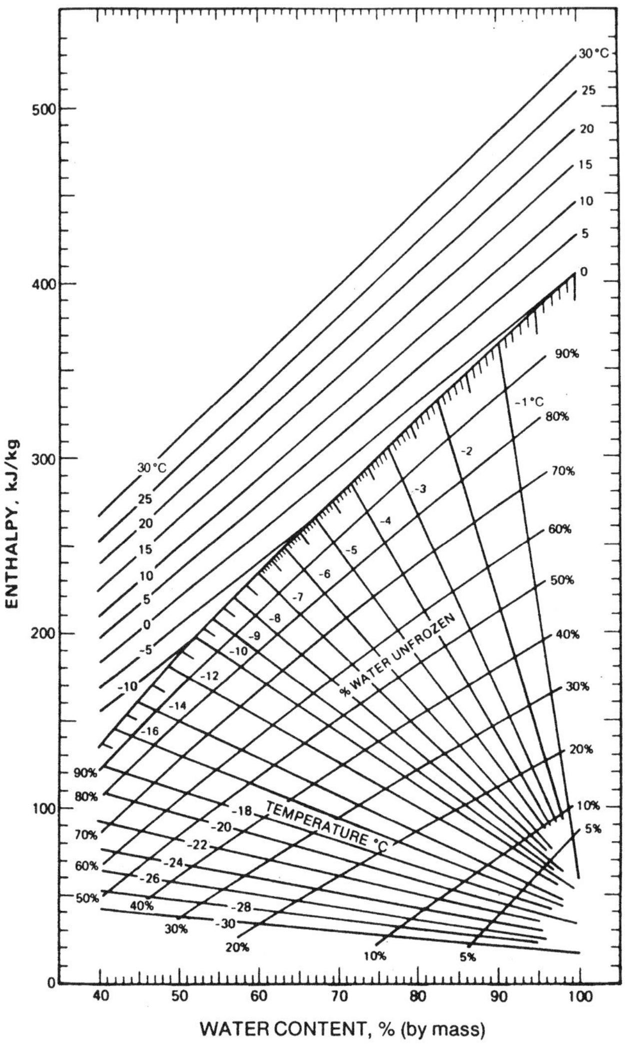

Figure 1–5 Enthalpy of fruit juices and vegetable juices. *Source:* Reprinted from Enthalpy of Fruit Juices and Vegetable Juices, *ASHRAE Handbook Fundamentals*, © 1993, ASHRAE.

cause of this, models for estimating the thermal conductivity of composite foods are based upon resistance in series or resistances in parallel.

Thermal conductivity increases slightly as temperature increases: models for a wide range of foods are listed by Rahman (1995, p. 327). It is noteworthy that values are also cited for liquids. This is done by ensuring that temperature distribution by convection is eliminated.

The resistance due to conduction can be characterized by L/k, where L = the half thickness (or radius of the food); thus resistance to heat transfer is increased by increasing the dimension of the food and by decreasing the thermal conductivity. Changing the material of con-

Table 1–9 Thermal Conductivity Values

Material	*Temperature (°C)*	*Thermal Conductivity ($W\ m^{-1}\ K^{-1}$)*
Silver	0	428
Copper	0	403
Copper	100	395
Aluminum	20	218
Stainless steel	0	8–16
Glass	0	0.1–1.0
Ice	0	2.3
Water	0	0.573
Corn oil	0	0.17
Glycerol	30	0.135
Ethyl alcohol	20	0.24
Air	0	2.42×10^{-2}
Cellular polystyrene	0	3.5×10^{-2}
Freeze-dried peach (1 atm)	0	4.18×10^{-2}
Freeze-dried peach (10^{-2} Torr)	0	1.35×10^{-2}
Whole soya beans	0	0.097–0.133
Starch (compact powders)	0	0.15
Beef, parallel to fibers	0	0.491
Frozen beef	–10	1.37
Fish		$0.0324+0.3294m_w$
Sorghum		$0.564+0.0858m_w$

m_w is the mass fraction of moisture. 1 Btu $h^{-1}\ ft^{-1}\ degF^{-1}$ = 1.731 $W\ m^{-1}\ K^{-1}$.
Source: Reprinted with permission from M.J. Lewis, *Physical Properties of Foods and Food Processing Systems*, p. 249, © 1990, Woodhead Publishing, Ltd.

struction, from tin plate to glass, usually increases the resistance very significantly as the glass has a lower thermal conductivity and a greater wall thickness.

Air has a very low thermal conductivity and is effectively a good insulator. The presence of air in steam or in foods will reduce the efficiency of heat transfer processes.

1.5.2 Thermal Diffusivity

Thermal diffusivity (α) is a complex property, defined as $k/(c\rho)$, with SI unit of $m^2\ s^{-1}$. It is an unsteady-state heat transfer property that measures how quickly temperature changes with time when a substance is heated or cooled. It can be measured directly by using a number of methods that involve unsteady-state heat transfer principles (Mohsenin, 1980; Sweat, 1995; Rahman, 1995). A metal such as copper will have a very high thermal diffusivity ($m^2\ s^{-1}$) of 1.1×10^{-4} whereas most foods have a low thermal diffusivity, typical values for fresh meat being $1.80 \times 10^{-7}\ m^2\ s^{-1}$ and for frozen meat 1.2×10^{-6}. It is noteworthy that a frozen food has a much higher value than a fresh food. Thermal diffusivity is encountered in more detail in Chapter 4. More detailed compilations are provided by Mohsenin (1980), Rahman (1995), and Holdsworth (1997).

1.5.3 Convection—Heat Film Coefficient

Convection is inherently faster than conduction as a heat transfer mechanism. In convection, heat is distributed from the hotter to the colder parts by the bulk movement of the molecules. In natural convection, the driving force is due to density differences, whereas in forced convection, energy is supplied, for example, by agitating cans or by pumping.

The effectiveness of heat transfer by convection is given by the heat film coefficient; again this will depend upon a number of factors. The resistance to heat transfer is considered to take place in a film of the fluid in contact with the solid. All the temperature gradient occurs across this film. The heat film coefficient (h) describes the rate of heat transfer from the bulk of the fluid (θ_b) to the solid surface (θ_s) in terms of the surface area (A) and temperature driving force. The SI units are W m^{-2} K^{-1}.

$$Q = h A (\theta_b - \theta_s) \tag{1.6}$$

The heat film coefficient depends mainly upon the type of fluid and the degree of turbulence. In general, gases are poor heat transfer fluids, liquids have intermediate values, and boiling liquids and condensing vapors have very high values. Some typical values are given in Table 1–10.

As heat transfer fluids, steam is more effective than hot water, which in turn is more effective than hot air. The presence of air in steam will be detrimental for two reasons: it will reduce the heat film coefficient and it will lower the temperature at a fixed pressure. Both situations may lead to a reduced rate of heat transfer and the possibility of underprocessing. One might suspect the presence of air if steam pressures and temperatures do not coincide (see steam Table 1.4). This has been discussed in more detail by Holdsworth (1997). For example, inclusion of 10% air in steam, by volume, will reduce the temperature from 100° to 97.2°C at atmospheric pressure and from 126.1° to 122.8°C at 1.38 bar (g). Some typical values for h (W m^{-2} K^{-1}) are: steam, 12,000; steam with 3% air, 3,500; and steam with 6% air, 1,200. Thus inclusion of about 6% air by volume results in an approximate 10-fold reduction in the heat film coefficient.

There are many empirical relationships for predicting heat film coefficient values for different flow situations (see Section 1.5.6).

1.5.4 Multiple Resistances, Overall Heat Transfer Coefficient

In most heat transfer processes, there will be more than one resistance to heat transfer. In a simple heat exchanger, where heating is indirect, there are three resistances, two due to convection and one to conduction. These can be combined in an overall heat transfer coefficient (U) (OHTC).

For thin-walled tubes, U can be calculated as follows:

$$1/U = 1/h_1 + 1/h_2 + L/k \tag{1.7}$$

Each term (e.g., $1/h_1$) in this expression is termed a resistance. In principle, the rate of heat transfer can be increased by reducing any of the resistances. However, this approach also

Table 1–10 Order of Magnitude of Surface Heat Film Coefficients for Different Systems

Situation	*h (W/m² K)*
Free convection	
Gases	6–23
Liquids	114–681
Boiling water	1,136–23,000
Forced Convection	
Gases	11–114
Viscous fluids	57–568
Water	568–11,360
Condensing vapors	1,136–113,600

Source: Reprinted with permission from S. Rahman, *Food Properties Handbook*, p. 399, © 1995, CRC Press.

helps to identify the limiting resistance, which is the largest of the resistance terms. This is one that controls the overall transfer of heat and would be the one to focus upon in order to gain the largest increase in heat transfer efficiency. Where fouling occurs, the additional resistance associated with the fouled deposit may well become the limiting resistance and control the overall process (see Section 8.2.3).

The OHTC also provides a simple expression for the rate of heat transfer in terms of the temperature in the bulk of the hot fluid and cold fluid, which eliminates surface temperatures that are difficult to measure. The following equation is the basic design equation that is used to estimate the surface area of a heat exchanger (A) in terms of the duty (Q′), the OHTC, and the log mean temperature driving force ($\Delta\theta_m$).

$$Q' = U\,A\,\Delta\theta_m \qquad (1.8)$$

Note that the use of countercurrent flow, whereby the two fluids flow in opposite directions is usually employed, as it gives a larger log mean temperature difference and allows a closer approach temperature.

The duty Q′ (or rate of heat transfer, J s^{-1}) is calculated from

$$Q' = m\,c\,\Delta\theta \qquad (1.9)$$

where m = mass flow rate (kg s^{-1}), c = specific heat (J kg^{-1} K^{-1}), and $\Delta\theta$ = temperature change (K).

1.5.5 Direct Heating/Biot Number

A special case is direct contact heating, where only two resistances are involved. The rate of heat transfer can be expressed as the ratio of h L/k, where L is a characteristic dimension (i.e., half thickness for a slab or radius for a sphere). This ratio is known as the Biot number (Bi).

If the Biot number is below 0.2, convection is the limiting resistance, and most of the temperature gradient takes place over the boundary layer. If the Biot number is above 0.2, the limiting resistance is conduction and most of the temperature gradient takes place between the surface and center of the particle.

One direct application of this is in the heat treatment of suspended particles or particulate systems. When these are heated, energy is transferred into the bulk of the fluid, from there to surface of the solid, and then to the internals of the solid. In consequence the liquid heats much more quickly than the solid, and uneven heating occurs. In order to ensure that the solid is adequately sterilized, the liquid phase may be overprocessed. These problems increase as the Biot number decreases.

It is also not quite so straightforward to predict the relative velocity between the solid and the liquid phase, which introduces extra uncertainty into the estimate of the heat film coefficient. In continuous processing, the residence time distribution for the particles may not be the same as that for the liquid phase.

To summarize, it is much more complicated to heat treat particulate systems, and some interesting methods have been developed to overcome the problems (see Section 4.15).

1.5.6 Dimensionless Groups

In fluid flow and heat transfer problems, use is made of a technique known as dimensionless analysis. Some important dimensionless groups, which result from this work are as follows:

The Reynolds number (Re) is defined as $vD\rho/\mu$. Provided that consistent units (either SI or Imperial) are used, it can be seen that this group is dimensionless; it tells us whether the flow is streamline or turbulent. More detail is provided in Section 3.23. This will influence heat transfer rates and residence time distributions.

Two other important dimensionless groups are the Nusselt number (Nu) and the Prandtl number (Pr). The Nusselt number is defined as the ratio of heat transfer by $h\,D/k$ and relates the heat film coefficient to the thermal conductivity and the tube dimensions. The Nusselt number is used mainly for estimating the heat film coefficient. The Prandtl number is defined as $c\,\mu/k$; it takes into account the factors affecting the heat transfer into a fluid. One widely used correlation that is used for estimating the heat film coefficient for a fluid passing along a tube is as follows:

$$hD/k = 0.023\,(v\,D\,\rho/\mu)^{0.8}\,(c\,\mu/k)^{0.4}$$

It can be alternatively written as

$$Nu = 0.023\,Re^{0.8}\,Pr^{0.4}$$

It is a general correlation that can be used for any fluid, provided its physical properties, flow rates, and tube dimensions are known. A large number of other common flow situations are encountered, for example, the flow of fluids through tubes and between parallel plates and heat transfer from fluids to suspended particles. Many of these are also based on dimensionless analysis. A thorough review of these correlations has been given by Rahman (1995).

Other dimensionless groups that will be encountered later in the text are the Biot number, Fourier number, and Froude number (Chapters 3 and 4).

1.6 WATER ACTIVITY

Water activity (a_w) is a measure of the availability of water for microbial, chemical, and enzymatic reactions. It is usually defined as the p/p_s, where p = water vapor pressure exerted by the food and p_s is the saturated water vapor pressure. One method to measure water activity involves equilibrating a sample of the food in a sealed container with a relatively small free volume and measuring the equilibrium relative humidity (RH).

The water activity is given by

$$a_w = RH / 100 \tag{1.10}$$

Water activity may influence thermal processing in three main ways:

1. It will affect the heat resistance of microorganisms and hence the extent of inactivation during processing.
2. It will affect the growth rate of any organisms surviving the heat treatment (of more importance for short–shelf life products).
3. It will affect the rate of both chemical reactions during processing and subsequent storage.

Most of foods that pass through a heat exchanger will have a relatively high water activity (i.e., greater than 0.95). Some values are given in Table 1–11. Water activity is reduced by the presence of low-molecular-weight solutes. Humectants will lower the water activity when added to food systems.

1.7 ELECTRICAL PROPERTIES

The electrical properties of foods are of interest in continuous heating, as they will influence direct electrical heating (Ohmic heating) processes and microwave heating processes. Both of these can be used for single-phase materials, but their main benefits arise from the opportunity of generating energy directly within the solid phase in a particulate system and thus accelerating the heating of the solid phase.

1.7.1 Electrical Conductance (Resistance)

Electrical resistance and conductance provide a measure of the ability of a material to transport an electric current; resistance is usually preferred for solids and conductance for liquids. The specific conductance (K) is the inverse of resistivity, with units of S m^{-1}. *Note:* S is Siemen, which is the same as reciprocal ohm (ohm^{-1}). Specific resistance is measured by resistance techniques, the cell usually being calibrated with a liquid of known specific conductance. Some values for different foods are given in Table 1–12.

Table 1–11 Some Typical Water Activity Values for Foods

a_w	*Food*
0.98–1.00	Fresh vegetables, fruit, meat, fish, poultry, milk, cottage cheese
0.93–0.96	Cured meats, most cheese varieties
0.86–0.93	Salami, some dry cheeses
0.8–0.87	Flour, cakes, rice, beans, cereals, sweetened condensed milk
0.72–0.88	Intermediate-moisture foods, jams, old salami
0.6–0.66	Dried fruits
0.6	Dehydrated foods

Source: Reprinted with permission from M.J. Lewis, *Physical Properties of Foods and Food Processing Systems*, p. 249, © 1990, Woodhead Publishing Ltd.

Electrical conductivity increases as temperature increases, according to the following equation:

$$K_\theta = K_{25}(1 + M[\theta - 25]) \tag{1.11}$$

where K_θ = electrical conductivity at θ °C (Sm^{-1}), M = proportionality constant, usually about 0.02 C^{-1}, and θ = temperature (°C). Applications for Ohmic heating are discussed in Section 4.23. Conductivity measurement is used to benefit in other applications: for checking water purity in deionization and for monitoring the presence of detergents in cleaning processes.

Table 1–12 Electrical Conductivity of Selected Food Materials at 19°C

Material	*Electrical Conductivity (S/m)*
Potato	0.037
Carrot	0.041
Pea	0.17
Beef	0.42
5.5% starch solution	
With 0.2% salt	0.34
With 0.55% salt	1.3
With 2% salt	4.3

Source: Reprinted with permission from H.J. Kim et al., Validation of Ohmic Heating for Quantity Enhancement of Food Products, *Food Technology*, May 1966, pp. 253–261, © 1966, Institute of Food Technologists.

1.7.2 Dielectric Constant and Loss Factor

The dielectric properties of foods have recently received more attention with the emergence of microwave and dielectric heating. Microwave systems are available for continuous thermal processing. Permitted frequencies are 915 and 2,450 MHz.

The two properties of interest are the dielectric constant (ε') and the dielectric loss factor (ε''). The dielectric constant is a measure of the amount of energy that can be stored when the material is subjected to an alternating electric field. It is the ratio of the capacitance of the material being studied to that of a vacuum or air under the same conditions. In an AC circuit containing a capacitor, the current leads the voltage by 90°. When a dielectric is introduced, this angle may be reduced. The loss angle (θ) is a measure of this reduction and is usually recorded as the loss tangent ($\tan\theta$). Energy dissipation within the dielectric increases as the loss tangent increases. A second property, known as the dielectric loss factor (ε''), is a measure of the energy dissipated within the sample. The relationship between these properties is:

$$\varepsilon'' = \varepsilon' \tan\theta \tag{1.12}$$

Values for the dielectric constant and the dielectric loss factors for a wide variety of foods are reported by Mohsenin (1984) and Mudgett (1982). Both these properties are affected by the temperature and moisture content of the sample, as well as the frequency of the electric field. The level of salts will also have a pronounced effect. Dielectric properties of materials are measured over a wide range of frequencies, using a variety of instrumental methods. Some values are given in Figure 1–6.

During microwave and dielectric heating, the power dissipated (P_o) within the sample is given by

$$P_o = 55.61 \times 10^{-14} f E^2 \varepsilon'' \tag{1.13}$$

where P_o = power dissipated (W cm^{-3}), f = frequency (Hz), and E = electric field strength (V cm^{-1}).

Materials with a high dielectric loss factor absorb microwave energy well and are sometimes termed "lossy" materials. It should be noted that the rate of heating will also depend upon the specific heat capacity. In general, foods with high moisture contents will also have higher specific heats. Special fiber optic probes have been developed to measure temperatures during microwaving processes.

1.8 DIFFUSION PROPERTIES

Although mass transfer operations do not play a major role in heat processing, they assume significance during the blanching of foods and during the leaching of nutrients from solid to liquid during storage. There may be significant loss of the major nutrients during solids blanching as well as a transfer of solids from foods stored in brines or syrups; note that these nutrients will not be utilized if the liquid phase is not consumed.

There may also be some leaching of plasticizers or metals from packaging materials to food components. For example, there was recently some concern (largely unfounded) about the amount of phthalate plasticizer in some milk powders.

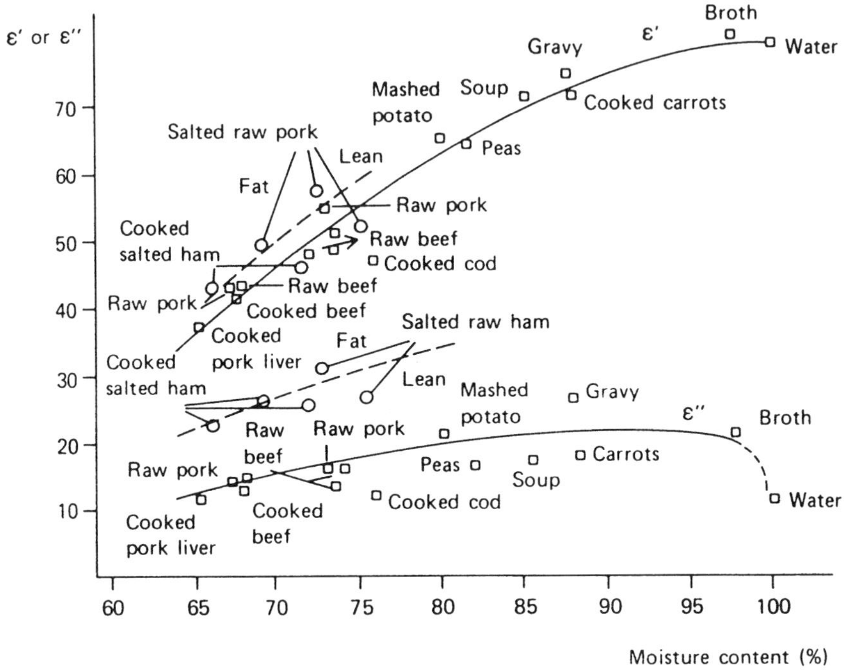

Figure 1–6 Dielectric properties of food at 20°C: the relationship between dielectric properties and moisture content. *Source:* Reprinted with permission from Bengtsson and Risman, Dielectric Properties of Food at 20°C: The Relationship between Dielectric Properties and Moisture Content, *Journal of Microwave Power*, Vol. 6, No. 7, p. 107, © 1971, International Microwave Power Institute.

During processing, there may be oxygen transfer from entrained air (as bubbles) into the bulk solution, followed by utilization of this dissolved oxygen for microbial and chemical reactions. During storage, there may also be further gas transmission (especially oxygen) through the packaging material.

Some loss of volatiles will also occur during flash cooling and vacuum cooling operations.

1.9 SURFACE PROPERTIES

Milk can be used as a typical example of a complex colloidal system. Casein micelles are stabilized by colloidal calcium phosphate and the dispersed fat phase by the fat globular membrane. The surface area to volume ratios of the dispersed phase components are very large (see Walstra & Jenness, 1984), so it is important to understand the forces acting at these interfaces. Processing operations such as homogenization will result in a further massive increase in exposed surface area. Also, changing the pH or addition of ions will affect surface charge and may affect the stability of the milk.

All fluids are subject to intense agitation during processing, during which air may be incorporated into the fluid. The forces acting at air/liquid interfaces are also important.

Surface tension is concerned with the forces acting within the fluid. Molecules at the surface of a fluid will be subject to an imbalance of molecular forces and will be attracted into the bulk of a fluid. Consequently the surface is said to be under a state of tension. Surface tension can be regarded in two ways; either as the force per unit length acting on a given length of surface, or as the work done in increasing its surface area under isothermal conditions. The SI unit is Nm^{-1} or $J\ m^{-2}$. It is the surface tension forces that cause most finely dispersed liquids to form spherical droplets, which is the shape that has the minimum surface area to volume ratio.

The surface tension of a range of fluid is given in Table 1–13. Water has a high value, although it readily becomes contaminated with surface-active agents; milk and other materials have much lower values. In general, surface tension decreases slightly as temperature increases.

Surface-active agents will lower surface tension; they are used for emulsification and detergency. Foaming can be a problem with some substances, especially in pumping and filling operations.

Interfacial tension deals with the forces acting at an interface, usually between two immiscible liquids; it assumes importance in emulsions.

Other examples of surface reactions concern the attachment of food components to the walls of heat exchangers (fouling) and their subsequent removal (cleaning). Other biofilms may assume importance, for example, algae and other materials in cooling water.

1.10 OPTICAL PROPERTIES AND SENSORY CHARACTERISTICS

In general, the optical properties of foods have received less attention than thermal or rheological properties. Of immediate importance is the appearance of the food, for example, for milk the presence or absence of visible defects such as fat separation or coagulation and

Table 1–13 Surface Tension Values of Some Liquids at 20°C

Liquid	*Surface Tension ($mN\ m^{-1}$)*	*Liquid*	*Surface Tension ($mN\ m^{-1}$)*
Water*	72.75	Milk*	42.3–52.1
Ethyl alcohol*	22.75	Skim milk 0.04% fat)†	51.0
Methyl alcohol*	22.65	Whole milk (2.4% fat)†	46.7
Chloroform*	27.14	Cream (34% fat)†	44.8
Carbon tetrachloride*	26.95	Cottonseed oil‡	35.4
Glycerol*	63.4	Coconut oil‡	33.4
Mercury*	435.5	Olive oil‡	33.0
		Oleic acid‡	32.5

* From the data of Weast (1982).
† From the data of Jenness et al. (1974).
‡ From the data of Powrie and Tung (1976).

Source: Reprinted with permission from M.J. Lewis, *Physical Properties of Foods and Food Processing Systems*, p. 171, © 1990, Woodhead Publishing Ltd.

in particular its color. Color, like texture, is not strictly a true physical property, but a sensory or psychological characteristic. In this sense it should be evaluated by human assessors. This can be extremely laborious, so a number of instruments have been designed to measure both texture and color; those measuring color are based on the spectral signal that results from light transmitted through or reflected by the sample. Other optical properties include refractive index and light scattering.

Refractive index for a transparent material is defined as the ratio of the velocity of light in air to that in the material. It is normally measured using light at a constant wavelength (589.3 nm—sodium D line) to four decimal places; values are usually quoted at a constant temperature of 20°C. It is affected by both temperature and wavelength.

The refractive index of some materials are as follows: water, 1.3330; milk, 1.3380; and butterfat, 1.4620. The presence of dissolved or suspended solids (below 0.1μm) will increase refractive index above that of water. Larger particles such as fat globules, air bubbles, and sugar crystals will have no effect.

Contrary to this, light scattering is influenced (caused) by larger particles whose refractive indices are different from that of the surrounding medium. Such scattering is a random process, but it is possible to measure the intensity of the light at some angle to the incident beam, as well as the intensity of the light transmitted through the sample. The same principles apply to other forms for infrared and ultraviolet radiation. Both transmitted and scattered light (radiation) form the basis of a range of techniques for analysis of milk and dairy products. Laser scattering techniques form the basis for particle size measurements, fat globule distributions in emulsions, or particle size determinations for powders.

The sensory characteristics of appearance, color, flavor, and texture are all influenced by many of the physical properties described above. Changes in sensory characteristics resulting from heat treatment and storage are discussed in more detail in Sections 6.4.9, 9.2, and 9.5.

1.11 OTHER PHYSICAL VARIABLES

As mentioned earlier, safety and quality are two key issues in thermal processing. In terms of ensuring safety and quality, the emphasis has moved toward understanding and controlling the process. This includes the quality of all the raw materials used, control of all the processes that affect the rate of heat transfer into the product, the microbial inactivation processes, and the level of postprocessing contamination. Fundamental to this control are certain key physical measurements: for example, temperature, pressure, and flow, and—to a lesser extent—level. It is important that these can be measured, recorded, and controlled. In fact, temperature measurement is a critical control point in both pasteurization and sterilization.

Some other important measurements variables are levels of ingredients, especially those that influence pH and ionic strength. In general, there has been a great deal of interest in sensor development for on-line measurement of parameters that can be related to product quality and safety. A comprehensive review of such sensors has been provided by Kress-Rogers (1993). A further interesting aspect is the collection of physical property data at temperatures greater than 100°C.

One practical problem is fouling of heat exchangers. A simple definition is the deposition of material on the surface of the heat exchanger. This will lead to a decrease in the overall heat transfer coefficient and an increase in the pressure drop over the heat exchanger. This limits the processing time. The heat exchanger must then be cleaned and disinfected. Fouling and cleaning are discussed in more detail in Chapter 8.

1.12 COMPETING TECHNIQUES

Heat processing may be regarded as a traditional method for pasteurizing and sterilizing foods. It is now encountering competition from a range of other methods, including irradiation, ultrahigh pressures (up to 10 kbar), pulsed light techniques, high-voltage methods, membrane processes, and other combination processes. A recent review of nonthermal preservation methods is provided by Barbosa-Canovas, Pothakamury, Palou, and Swanson (1998). Despite this, the continuous processes of HTST pasteurization and UHT sterilization provide and will continue to provide convenient, cheap, and effective methods for removing pathogens and providing high-quality nutritious foods.

REFERENCES

ASHRAE Handbook fundamentals. (1993). *Thermal properties of foods*. Chap. 30, pp. 1–26.

ASHRAE Handbook. (1994). *Refrigeration, systems and applications*.

Barbosa-Canovas G.V., Pothakamury, U.R., Palou, E., & Swanson, B.G. (Eds.). (1998). *Nonthermal preservation of foods*. New York: Marcel Dekker.

Bengtsson N.E., & Risman, P.O. (1971). Dielectric properties of foods at 3 GHz, as determined by a cavity perturbation technique, II, Measurement of food materials. *Journal of Microwave Power* **6**, 2, 107–123.

Biliaderis, C.G. (1983). Differential scanning calorimetry in food research—A review. *Food Chem* **10**, 239–266.

Bird, R.B., Stewart, W.E., & Lightfoot, E.N. (1960). In *Transport phenomena*. New York: John Wiley & Sons.

Burton, H. (1988). *UHT processing of milk and milk products*. London: Elsevier Applied Science Publishers.

Cowell, N.D. (1994). *An investigation of early methods of food preservation by heat*. PhD thesis, University of Reading, Reading, UK.

Cowell N.D. (1995). Who invented the tin can? A new candidate. *Food Technology* **12**, 61–64.

Dairy facts and figures. (1996). London: National Dairy Council.

EC *Dairy facts and figures*. (1994). Residiary Milk Marketing Board, Thames Ditton, UK.

Grandison, A.S., & Lewis M.J. (1996). *Separation processes in the food and biotechnology industries*. Cambridge, UK: Woodhead Publishers.

Hayes, G.D. (1987). *Food engineering data handbook*. Harlow, UK: Longman Scientific and Technical.

Holdsworth, S.D. (1997). *Thermal processing of packaged foods*. Blackie Academic and Technical.

James, A. (1996). The UK market for aseptically packaged foods and drinks. In *Training course notes, principles of UHT processing*. Leatherhead, UK: Leatherhead Food Research Association.

Jenness, R., Shipe, W.F. Jr., & Sherbon, J.W. (1974). Physical properties of milk. In B.H. Webb, A.H. Johnson, & J.A. Alford (Eds.), *Fundamentals of dairy chemistry*. Westport, CT: AVI Publishers.

Karel, M. (1975). Dehydration of foods. In O.R. Fennema (Ed.), *Principles of food science, Part 2, Physical principles of food preservation*. New York: Marcel Dekker.

Kessler, H.G. (1981). *Food engineering and dairy technology*. Freising, Germany: Verlag A Kessler.

Kim, H-J, Choi, Y-M, Yang, T.C.S., Taub, I.A., Tempest, P., Skudder, P., Tucker, G., & Parrott, D.L. (1996). Validation of Ohmic heating for quality enhancement of food products. *Food Technology* **5**, 253–261.

Kress-Rogers, E. (Ed.). (1993). *Instrumentation and sensors for the food industry.* Oxford, UK: Butterworth-Heinemann.

Lewis M.J. (1990). *Physical properties of foods and food processing systems.* Cambridge, UK: Woodhead Publishers.

Lewis, M.J. (1993). Physical properties of dairy products. In R.K. Robinson (Ed.), *Modern dairy technology* Vol. 2 (pp. 331–380). Elsevier Applied Science.

Milson, A., & Kirk, D. (1980). *Principles of design and operation of catering equipment.* Chichester, West Sussex, UK: Ellis Horwood.

Mohsenin, N.N. (1980). *Thermal properties of foods and agricultural materials.* London: Gordon and Breach.

Mohsenin, N.N. (1984). *Electromagnetic radiation properties of foods and agricultural materials.* New York: Gordon and Breach.

Mossel, D.A.A. (1975). Water and micro-organisms in foods—a synthesis. In R.B. Duckworth (Ed.), *Water relations of foods.* London: Academic Press.

Mudget, R.E. (1982). Electrical properties of foods in microwave processing. *Food Technology* **36**, 109–115.

McCance & Widdowson, (1991). *Composition of foods tables* (5th ed.). London: Royal Society of Chemistry, Ministry of Agriculture Fisheries and Food.

Powrie, W.D., & Tung, M.A. (1976). In O.R. Fennema (Ed.), *Principles of food science, Part I, Food chemistry.* New York: Marcel Dekker.

Rahman, S. (1995). *Food properties handbook.* Boca Raton, FL: CRC Press.

Riedel, L. (1955). Calorimetric investigations of the melting of fats and oils. *Fette. Seifen Anstrichmittel* **57**, (10), 771.

Sweat, V.E. (1995). Thermal properties of foods. In M.A. Rao & S.S.H. Rizvi (Eds.), *Engineering properties of foods.* New York: Marcel Dekker.

Toledo, R.T. (1980). *Fundamentals of food process engineering.* Westport, CT: AVI Publishers.

Walstra, P., & Jenness, R. (1984). *Dairy chemistry and physics.* New York: John Wiley.

Weast, R.C. (Ed.). (1982). Handbook of physics and chemistry (63rd ed.). Cleveland, OH: CRC Press.

Wilbey, R.A. (1993). Pasteurization of foods: Principles of pasteurization: In *Encyclopedia of food science, food technology and nutrition* (pp. 3437–3441), Academic Press.

SUGGESTED READING

Holmes, Z.A., & Woodburn, M. (1981). Heat transfer and temperature of foods during preparation. *CRC Critical Reviews in Food Science and Nutrition* **14**, 231.

Jowitt, R., Escher, F., Hallstrom, B., Meffert, H.F. Th., Spiess, W., & Vos, G. (1983). *Physical properties of foods.* London: Applied Science.

Lamb, J. (1976). Influence of water on physical properties of foods. *Chemistry and Industry* **24**, 1046.

Polley, S.L., Snyder, O.P., & Kotnour, P. (1980). A compilation of thermal properties of foods. *Food Technology* **11**, 76.

Rha, C.K. (1975). *Theory, determination and control of physical properties of food materials.* Dordrecht, Netherlands: D. Reidel.

Tschubik, I.A., & Maslow, A.M. (1973). *Warmephysikalische, konstanten von lebensmitteln und halbfabrikaten.* Leipzig, Germany: Fachbuchverlag.

CHAPTER 2

Kinetics for Microorganism Death and Changes in Biochemical Components

The rate at which microorganisms die at high temperatures is an obviously important study central to continuous-flow thermal processing. The different microorganisms themselves have inherently different resistances to high temperatures, in that vegetative cells and yeasts are generally the most susceptible while endospores are much more resistant, with viruses between these two extremes. The medium that surrounds the microorganisms also has an extremely large influence, especially its pH, water activity (a_w), and the concentration and type of most food components, especially simple carbohydrates, fats, or particular chemical ions. The importance of pH is often more associated with the death of vegetative cells or the ability of spores to germinate and grow, rather than for microbial spores. The type of acid and its pK value, however, can have a large effect on the survival of vegetative microorganisms (Brown & Booth, 1991). A low a_w value, or the presence of high levels of fat in a food, can have an extremely large effect on microorganism death, especially in butter or chocolate, where the resistance of vegetative bacteria can be 100 to 1,000 times greater than in aqueous environments (Hersom & Hulland, 1980).

The type of foodstuff to be heat treated will often have associated microorganisms with a high thermal resistance which it is important to inactivate to ensure sterility of that product (e.g., in milk, *Bacillus stearothermophilus*, *B. subtilis*, or *B. cereus*; for fruit products, *Byssochlamys spp*; for soy protein infant formulae, *Desulphotomaculum nigrificans*; and for vegetables, *B. stearothermophilus* and *Cl. thermosaccharolyticum*). Additionally, both specific intra- and extracellular enzymes must often be denatured while nutritional components must be maintained where possible to ensure the nutritional value of the product. All of these factors require a knowledge of the rate of thermal death or biochemical degradation as a function of time and temperature.

2.1 DEATH OF MICROORGANISMS AT CONSTANT LETHAL TEMPERATURE

The death of microorganisms at elevated temperatures is generally accepted to be a first-order reaction; that is, at a constant temperature, the rate of death of the organisms is directly proportional to the concentration present at that particular time. The result of first-order kinetics means that there is a defined time during which the number of microorganisms falls to one-tenth of the number at the start of that time interval, irrespective of the actual number. This can be described by following a particular number of microorganisms, held at a constant lethal temperature so that the number present at one time is falling. After one time interval, the number of microorganisms will have fallen to one-tenth of the original number and after a second time interval, the number of microorganisms falls to one-tenth of this new value, which is one-hundredth of the original value, and so on. The number of microorganisms remaining at any time is shown in Figure 2–1.

This time interval taken to decrease the number of microorganisms by a factor of 10 (or to reduce their number by 90%) is known as the decimal reduction time (D) for those microorganisms, and is constant at a constant temperature for a particular strain of microorganisms.

If we take an initial number of microorganisms, N_0, heated instantly to a lethal temperature, eventually as the number of microorganisms drops, there will be one microorganism left. After the next decimal reduction time interval, theoretically there will be one-tenth of a microorganism remaining. This obviously cannot actually happen but can be expressed as the chance of survival of that one remaining microorganism is 1 in 10 (i.e., on average, for every 10 times this happens, the microorganism will survive intact once). Similarly, in two decimal reduction time intervals, the chance of that single microorganism's surviving is 1 in 100, and so on.

This leads to the direct consequence that the number of microorganisms cannot ever be guaranteed to be zero, and there is always a finite chance that a microorganism may survive

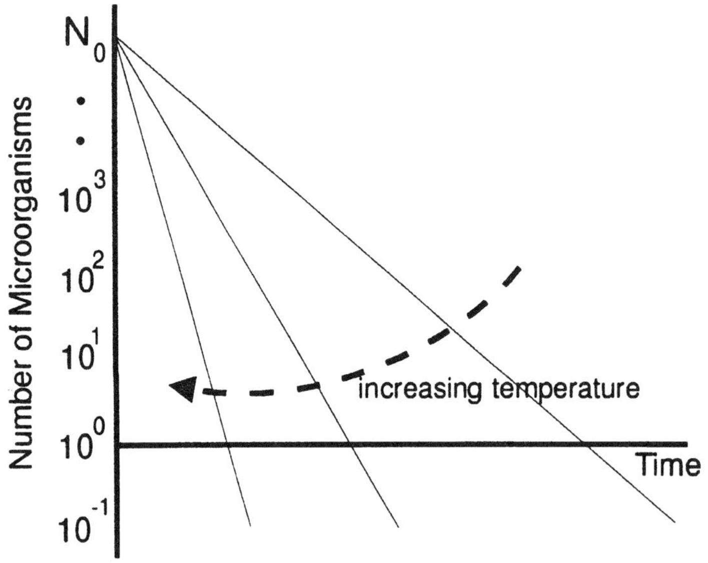

Figure 2–1 The number of microorganisms surviving at lethal temperatures at any time

and grow to cause spoilage of the container. The aim of the food technologist is therefore to ensure that the chance of a microorganism's surviving a given process in a given food product is reduced to an acceptably small value, ensuring specifically that the consumer is unlikely to come into contact with food poisoning microorganisms, or more generally, an unsterile product. To establish the reduction in the number of microorganisms required to achieve this due consideration must be given to the initial number of microorganisms in the food and the number of units of the food being produced. There is a convention that, in low-acid foods, the acceptable level of survival of a spore of *Clostridium botulinum* is 10^{-12}; that is, 1 spore in 10^{12} initially (1 in 1 million million or 1 in 1,000,000,000,000) will survive the thermal process. *Cl. botulinum* is important as it is the most heat-resistant spore-forming bacterium that can cause food poisoning and, if reduced to this acceptably low level, will ensure that the low-acid food will not cause food poisoning in consumers at any measurable level. This survival rate of 10^{-12}, or 12 log cycles reduction, in *Cl. botulinum* is known as the 12D concept and was proposed by Stumbo (1965). In the United Kingdom, the 12D concept has been modified to one in which the chance of survival of *Cl. botulinum* is 1 spore in 10^{12} containers rather than 10^{12} initial spores. This still translates as an $F_0 3$ process.

A 12D thermal treatment for *Cl. botulinum*, however, may not have reduced nonpathogenic spoilage bacteria to a level that will ensure commercially sterile products, as the latter often have a larger D value (i.e., are more thermally resistant). Higher levels of heat treatment are often necessary to ensure a commercially sterile product. In addition, the thermal resistance of thermophilic spores is generally higher than that for mesophilic spores, but their optimum incubation temperature is also higher, usually greater than 35°C. Where ambient conditions would allow thermophilic spores to grow, the acceptable survival level is often taken as between 10^{-5} and 10^{-7} of the most resistant spores (5D to 7D reduction), assuming that this results in a minimum of 12D for *Cl. botulinum.*

The thermal death of microorganisms can be expressed mathematically in terms of concentration of microorganisms (C) by the equation:

$$-\frac{dC}{dt} = k.C \tag{2.1}$$

where k is the reaction rate constant and t is time.

At different times t_1 and t_2, the respective concentration of organisms C_1 and C_2 is given by integrating Equation 2.1:

$$\ln\frac{C_1}{C_2} = k(t_2 - t_1) \tag{2.2}$$

It is often more useful, instead of considering the concentration of microorganisms, to consider the actual number of microorganisms N present in a container or batch of product. Equation 2.2 can then be written as:

$$\ln\frac{N_1}{N_2} = -k(t_2 - t_1) \tag{2.3}$$

or:

$$\log_{10}\left(\frac{N_2}{N_1}\right) = k'(t_2 - t_1) \text{ where } k' = \frac{k}{2.303} \tag{2.4}$$

where N_2/N_1 is the proportion of surviving microorganisms or spores, that is, the proportion of the initial number of microorganisms that will survive after the time interval $(t_2 - t_1)$, and $\log_{10}(N_2/N_1)$ is the number of log cycles reduction in microorganism numbers in the time interval (i.e., a log cycle reduction of 1 means a survival rate of 1 in 10 microorganisms, and 2 means 1 in 100, etc.).

The term $1/k'$ is replaced by D, the decimal reduction time, which is defined as the time taken for the number (or concentration) of microorganisms to fall to one-tenth of the original value. Substituting D for $1/k'$ in Equation 2.4 gives

$$\frac{N_t}{N_0} = 10^{-t/D} \tag{2.5}$$

where N_0 is the initial number of spores at time = 0 and N_t is the number of spores at time t. This is the key equation linking the number of microorganisms or spores with time at a constant lethal temperature. If experimental data were plotted on a graph of $\log_{10}(N_2/N_1)$ against time, the result should be a straight line of slope 1/D, as shown in Figure 2–1. This often does not occur, however, for a variety of reasons, which are covered in Section 2.7. Methods for obtaining experimental data are given in Section 2.5.

2.2 EFFECT OF TEMPERATURE ON DECIMAL REDUCTION TIME

As the temperature of the foodstuff is increased, the rate at which the microorganisms in it die increases and so the decimal reduction time value decreases. However, the exact nature of the relationship between the decimal reduction time and temperature has been the subject of much experimental work and theoretical debate. There are two major models used to quantify the relationship, the traditional Canners' model (constant-*z*) and the Arrhenius model, which is theoretically more acceptable. These two are covered in detail but it must be noted that several other models have been proposed.

2.2.1 Canners' (Constant-*z* Value) Model

Early work by Bigelow (1921) and Bigelow and Esty (1920) showed a linear relationship between the logarithm of the decimal reduction time for spores and their temperature. For many years food technologists, especially those in the canning industry, have used this model whereby a certain temperature rise, defined as the *z*-value, will change the decimal reduction time of the microorganisms by a factor of 10, and that this *z*-value is constant for all temperatures used.

Mathematically, this constant *z*-value theory is expressed by:

$$\frac{D_1}{D_2} = 10^{(\theta_2 - \theta_1)/z} \tag{2.6}$$

where D_1 and D_2 are decimal reduction times at temperatures θ_1 and θ_2, respectively.

This kinetic model assumes that the *z*-value is a constant value for a given spore population. It is usual to define a temperature as a reference temperature (e.g., 250°F or 121°C for F_0 value, 85°C for Pasteurization Unit, etc.) depending on the temperature range used for the thermal process, and the thermal death of any microorganism can be given by a D value at the reference temperature and its *z*-value. The equation is obviously only valid for temperatures greater than those at which thermal death starts to occur.

In a similar fashion, but less commonly used, one can define a Q_{10} value as the ratio of decimal reduction time values at a temperature interval of 10°C, that is:

$$Q_{10} = \frac{D_\theta}{D_{(\theta+10)}} \tag{2.7}$$

It can be shown that Q_{10} is related to the *z*-value by:

$$Q_{10} = 10^{10/z} \text{ or } z = \frac{10}{\log(Q_{10})} \tag{2.8}$$

For a single strain of microorganisms, a graph of $\log_{10}(D)$ against temperature (θ) will give a straight line (Figure 2–2) and the *z*-value can be determined from the slope of that line (m) where:

$$z = -\frac{1}{m} \tag{2.9}$$

2.2.2 Arrhenius Equation Model

The Arrhenius equation has been used extensively to describe the effect of temperature on chemical reaction kinetics and can be used for thermal death. The Arrhenius kinetic equation is

$$k = A.e^{-(E_a/R\theta_k)} \tag{2.10}$$

or

$$\ln(k) = \ln(A) - \frac{E}{R\theta_k} \tag{2.11}$$

where A is a constant, E_a is the activation energy for the reaction, R is the gas constant, and θ_k is the absolute temperature. It can be shown that

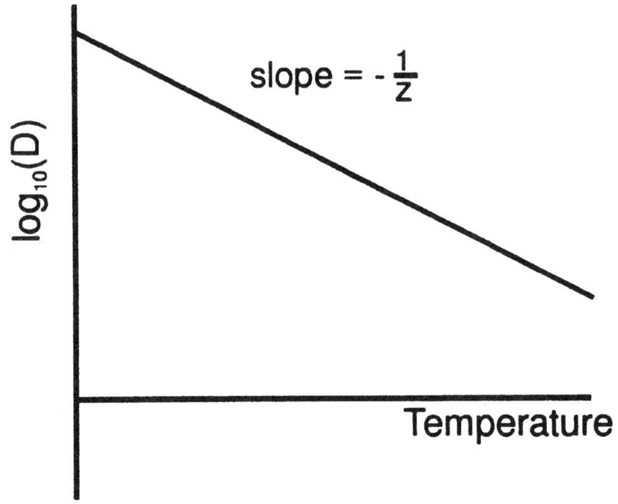

Figure 2–2 Graph of $\log_{10}(D)$ against temperature for Canners' (constant-*z*) model for single strain of microorganisms.

$$D = \frac{2.303}{k} \tag{2.12}$$

This kinetic model assumes that the activation energy for the reaction is constant and therefore a graph of ln(k) or ln(D) against $(1/\theta_k)$ would give a straight line.

2.2.3 Comparison between Models

The constant-*z* and Arrhenius kinetic models are both used to evaluate thermal death kinetics of microorganisms, but theoretically they are mutually exclusive. The constant-*z* model gives the ratio of decimal reduction times at two temperatures θ_1 and θ_2 by Equation 2.6 as:

$$\frac{D_1}{D_2} = 10^{(\theta_2 - \theta_1)/z} \tag{2.13}$$

whereas the Arrhenius model gives

$$\frac{D_1}{D_2} = 10^{\frac{E_a(\theta_2 - \theta_1)}{R(\theta_1 \theta_2)}} \tag{2.14}$$

The inaccuracy involved by assuming one model in preference to the other can easily be calculated from these two equations. As an example, if we take the decimal reduction value

Table 2–1 Prediction of D Values Using Constant-*z* and Arrhenius Models

	111°C	*121°C*	*131°C*	*141°C*	*151°C*
Constant-*z*	1,000	100	10	1.0	0.1
Arrhenius	1,000	100	11.21	1.396	0.192

of a hypothetical spore population at 111°C as 1,000 s and at 121°C as 100 s with the *z*-value to be 10°C, the following table of D values can be calculated (Table 2–1).

It can be seen that by extrapolating thermal death kinetics by only 20°C to 30°C, a substantial inaccuracy between the two models is evident. This error may appear as significant, but the thermal death kinetics of microorganisms cannot usually be determined over a great temperature range due to factors that are outlined later, and the inaccuracy in the measured data (especially concentration of microorganisms) is usually much greater than the inaccuracy due to choice of model.

There is some theoretical basis for preferring the Arrhenius model to the constant-*z* model in that the death of microorganisms is thought to be linked to denaturation of one or more biochemical components within the spore or vegetative cell and therefore would behave more like a chemical reaction for which it is known that the Arrhenius model is more accurate. Several workers (e.g., Jonsson et al., 1977; Perkin et al., 1977) have attempted to fit data to both models, generally without finding one model more accurate than another. Most workers in this area tend to provide data in both forms (i.e., D and *z*-values as well as activation energy), a range of which is given in Table 2–2. Other thermal death theories have been proposed, particularly the vitalistic theory, which is covered in Section 2.8.

There is a significant danger in predicting thermal death kinetic data at high temperatures by extrapolating from data collected at lower temperatures. Wherever possible, data must be measured around the temperature range of interest; then the use of either kinetic model fitted to the data would not lead to significant errors. Should extrapolation be necessary, however, then the Arrhenius model is generally preferred.

Table 2–2 Typical Ranges of D, z, and Activation Energy (E_a) Values for Spore-Forming Bacteria of Importance

	Medium/pH	*E_a kJ/mol*	*D_{121} value (s)*	*z-value (°C)*
B. stearothermophilus	water/buffer pH7	256–513	120–380	7.3–12.3
Cl. thermosaccharolyticum	phosphate buffer pH7	211–476	50–192	6.9–14.7
Cl. sporogenes	meats	340–390	9–90	8.9–11.1
B. subtilis	phosphate buffer pH7	220–374	22–30	8.3–14.1
Cl. botulinum	phosphate buffer pH7	310–376	1.1–8	7.8–10.8
	various food purées	260–370	1.9–160	7.4–11.1

2.3 OVERALL EFFECT OF A CHANGING TEMPERATURE

The relationships derived above make it easy to calculate the proportion of surviving spores, or biochemical change, for a process with a constant temperature. Practical thermal processes rarely operate at a single lethal temperature throughout, however, and the heating and cooling times to and from the required process temperature can be very long. For some in-container sterilization processes, the final process temperature may never actually be attained in the product. The effect of a varying temperature can be determined by equating this time-temperature curve for the whole thermal process to a single holding time at a reference process temperature that would give the same proportion of surviving microorganisms or biochemical change. Equation 2.6 can be rearranged to calculate this holding time at a different temperature that would give an equal thermal treatment, or equal proportion of surviving spores:

$$t_2 = t_1 \cdot 10^{\frac{(\theta_1 - \theta_2)}{z}} \tag{2.15}$$

where t_2 is the time at temperature θ_2 which will give an equivalent proportion of surviving microorganisms or biochemical change to a process at temperature θ_1 for a time of t_1 for a given z-value.

This can also be expressed as the rate of thermal death of the microorganisms at one temperature (θ) compared to a reference temperature (θ_{ref}), termed the Lethality (L) at that temperature.

$$L = 10^{\frac{(\theta - \theta_{ref})}{z}} \tag{2.16}$$

Using these equations, the overall proportion of surviving microorganisms or biochemical change for the whole process with varying temperature can be related back to a single holding time at a steady reference temperature by using one of two approaches:

1. Divide the process into a series of time-temperature combinations and then relate each of these combinations to the equivalent time at the reference temperature using Equation 2.15 or 2.16. These equivalent times can then be summed to give the total equivalent time at the reference temperature.
2. Or calculate the integral $\int_0^t 10^{(\theta - \theta_{ref})/z} \cdot dt$ for the whole process either graphically or by using a numerical integration computer program.

The first method assumes either that the time-temperature curve is divided into a series of temperatures (usually at 1°C intervals) and the holding time at that temperature or a series of temperatures at a set time interval, similar to the output from a data logger. Either of these will result in errors due to dividing the curve this way (especially at high heating rates), but the error will be reduced as the temperature or time intervals used are reduced.

2.4 THERMAL TREATMENT CRITERIA

Several criteria are used to determine the level of heat treatment that would be required for a particular foodstuff (i.e., the times and temperatures required to give an "adequate" thermal treatment for the food). The criteria can also be used to compare different processes that may have different holding temperatures, holding times, or different heating rates.

In the case of commercial sterilization, the most commonly used criterion is the F value, introduced by Ball (1923). The F value can be considered to be the holding time at an arbitrary reference temperature (assuming instantaneous heating and cooling) to which the whole process is equivalent, that is, it will give the same proportion of surviving spores or biochemical to the whole thermal process. The F value is the function:

$$F = \int_0^{\infty} L \cdot dt = \int_0^{\infty} 10^{(\theta - \theta_{ref})/z} \cdot dt \tag{2.17}$$

When quoting F values, the reference temperature and z-value used should be noted in the form $F^{Z}_{\theta\ ref}$. It should be noted that different mathematical values of F for the same process will be obtained for different θ_{ref} and z-values. In the case of *Cl. botulinum* spores, where the z-value is 10°C, and when the reference temperature is 121.1°C (250°F), the F value is designated F_0. Other designations may occasionally be used, for example, for the center of a container only (F_c) or for the bulk average of the container contents (F_B).

There are many other thermal process criteria that are currently used, depending on the thermal death kinetics of the microorganisms and the operating temperature of the process. In addition to the criteria based on microorganism death, there are also criteria based on biochemical degradation or even on cooking of the foodstuff. Most are in the form of Equation 2.17, using different *z*- and θ_{ref} values. The reference temperature is a very important aspect and, to give an accurate value, should be close to the range of temperatures normally used in that process, to reduce errors from extrapolating over too high a temperature range (see Section 2.2.3). The *z*-value used in the criterion must be that of the spores, biochemical, or other reaction of interest. Typical criteria are given in Table 2–3.

2.4.1 Establishing the Value of Thermal Process Criterion Required

Table 2–3 Reference Temperatures and *z*-Values for Different Thermal Treatment Criteria

Value	θ_{ref} *(°C)*	*z-Value (°C)*	*Reference*
Pasteurization unit (PU)	60, 65	10	Section 5.3
p* (for pasteurization)	72	8	Section 5.3
Cooking value (C_0)	100	33.1	Section 2.10
B* value for ultrahigh-temperature (UHT) sterilization of milk	135	10.5	Horak (1980)—Section 6.4
C* value for thiamin loss in UHT milk	135	31.5	Horak (1980)—Section 6.4

Using the methods described above, it is easy to calculate the value of each of these thermal treatment criteria from the temperature-time history that any foodstuff has undergone in the thermal process. What remains to be done is to define the value of the thermal process criterion that we actually required from a process that would give a product that has been "adequately" heat treated, for example, to define the F_0 value that a particular food product must have received to be sure it has been commercially sterilized.

For low-acid foods where *Cl. botulinum* must be considered for public safety, the 12D process in Section 2.1 relates to a minimum heat treatment (the botulinum "cook") of an F_0 value of 3 minutes. This level of heat treatment has been proved over decades of use to give an acceptably low risk to public health and no food so treated (and not recontaminated after thermal processing) has ever been found to have caused an outbreak of food poisoning. Brown (1991) states that the 12D concept and $F_0 3$ process are not necessarily the same, and the 12D process is beyond the level it is presently possible to measure.

For most foodstuffs, heat treatment greater than this minimum botulinum "cook" must be achieved to prevent spoilage of the food by more resistant, but nonpathogenic, microorganisms. The level of heat treatment required here may be calculated if the initial microbial contamination level, the thermal death kinetics of the individual strains of microorganisms, and the final microbial level required to ensure acceptable pack failure rate are all known, plus a small degree of overprocessing to give a safety factor. A secondary estimation is based on a desired reduction of the most resistant spore or microorganism. For example, for low-acid foods, a survival level of between 10^{-5} and 10^{-7} of the most thermally resistant spores present in the product (5D to 7D reduction) is commonly used (see Section 2.1).

However, evaluating the required level of heat treatment commonly requires many trials using a range of heat treatments with extended or accelerated storage trials and enumeration of contaminated packs. Methods similar to that adopted by Horak (1980) for evaluating the UHT conditions for milk may be used for other products but would be time consuming and expensive (see Section 6.4). Initial trials may be based on values that are used for similar products as a first estimation. For commercial sterility, for example, Table 6–2 gives some F_0 values that have been used for in-container processes for different foodstuffs. Although these could be used as a first estimate of conditions required for UHT processing, great care should be taken because of the uncertainty of the basis for extrapolation to higher temperatures, as outlined in Section 2.2.

The thermal process criterion that should be used to determine the thermal processing conditions is required must be appropriate to the microorganisms most likely to survive the process and to the process operating temperature. For example, for *Cl. botulinum* in an in-container process, the F_0 value is most appropriate, but if other spores are more likely to survive the process (through greater heat resistance or higher levels in the raw food) then an F-value based on the *z*-value of those spores should be used. For a continuous-flow thermal process with a higher operating temperature, a different reference temperature should be adopted.

Once the required heat treatment has been determined for a particular foodstuff and the holding temperature is decided, the holding time that would be required for a continuous-flow thermal process can be calculated. This topic is further covered in Chapters 3 and 6.

2.5 METHODS FOR DETERMINATION OF THERMAL DEATH KINETICS

The thermal death kinetics may be measured for single specific strains of microorganisms, usually those of highest heat resistance present in a particular food, or occasionally for the whole natural spore population associated with that food. It is important to note that whole foodstuffs are most likely to be heterogeneous and the spore environment may vary locally within the food in terms of pH, a_w, redox potential, and concentration of protein, carbohydrate, and fat. As the thermal death kinetics for any strain of spore may depend on its environment, the mixture of heat resistances in the food may give non-linear kinetics (see Section 2.7). To overcome this problem, it is common to measure the thermal death kinetics in a well-defined buffer (usually phosphate buffer) which is useful in ascertaining whether the kinetics are linear but makes it difficult to apply the information obtained to the heat treatment of actual foods.

The broad aims of any method used are to heat the spores to a uniform known temperature, preferably immediately or with the minimum heating-up time, hold at that temperature for a known measured time, and then cool rapidly before enumeration of surviving spores, usually by a colony count. Ideally, there must be no differences in temperature within the apparatus and no differences in holding time at the high temperature; it is relatively easy to achieve one or the other of these conditions but not both. There are only a few methods used for measurement of thermal death kinetics. The earliest, and still used, method is to seal a suspension of the microbial population in an ampule, tube, or capillary tube and to immerse it in a hot water or oil bath at a steady temperature for a measured time, followed by immediate cooling in cold or iced water (Bigelow & Esty, 1920). The ampule can then be washed, the outside sterilized by a chemical sterilant and washed in sterile water the ampule is then broken and the contents plated out in a suitable microbial growth medium. After incubation, the colonies formed can be counted.

An alternative method was used by Stumbo (1965), who developed an apparatus called a thermoresistometer, which consisted of a large, pressurized steam chamber at the required temperature into which small volumes of microorganism suspension on shallow metal trays were introduced for a range of holding times. A thermoresistometer has also been used by Pflug and Esselen (1953) and Gaze and Brown (1988); the latter was used to measure the thermal death kinetics of *Cl. botulinum* spores up to UHT temperatures. As a method for determining thermal death kinetics, however, the thermoresistometer has major disadvantages of being large and expensive to build, as commercial systems are not available.

Other methods involve small-scale direct-heating UHT plants where the spore suspension is injected into a steam atmosphere, held in the heated zone for a length of time, then passed into a vacuum chamber where it is flash cooled. Heating and cooling are achieved in an extremely short time [less than 0.1 second (Burton et al., 1977)], but there is usually a distribution of holding times in the liquid as it flows through the equipment. Such methods have been used by Burton, Perkin, Davies, and Underwood (1977) and Heppell (1985). Measurement of the residence time distributions of different liquid foodstuffs in the heated zone is not easy, however, and calculation of the thermal death kinetic data from experimental results requires care. Daudin and Cerf (1977) used a modification of the equipment to allow rapid injection of a small volume of spore suspension into the steam chamber, holding the sample in the chamber for the required time, then rapid ejection into the collection apparatus, eliminating the residence time distribution present in Burton et al. (1977).

2.6 PROBLEMS WITH THERMAL DEATH MEASUREMENTS

Many workers have measured thermal death kinetics for a wide range of microorganisms and determined decimal reduction times at fixed temperatures from the slope of the straight line fitted to a graph of the logarithm of the number (or concentration) of microorganisms versus time at that temperature.

In practice, there are many factors affecting the accuracy of the data obtained, and care must be taken in experimental detail, and indeed in interpreting results of other workers. One factor of major importance is in the enumeration of surviving spores. Spores are dormant, apparently inert bodies, designed to survive adverse conditions of temperature, nutrient availability, or presence of deleterious chemicals. The only method that will differentiate between spores that are viable and those that are dead is to force them to form vegetative cells and grow when plated onto some nutrient agar medium under optimum growth conditions and therefore give colonies that can be counted (colony count). The conditions for the colony count require that all the viable spores germinate and grow simultaneously to form colonies at the same time. A delay in germination of even a few hours will result in delayed formation of the microorganism colony, and it may not be visible when counting is performed.

There are three distinct stages that a spore undergoes successively in forming a vegetative cell: activation, germination, and outgrowth (Keynan & Evenchick, 1969). Unactivated spores will not germinate spontaneously until activation occurs. Activation is usually a reversible process and the activated spore retains nearly all the spore properties, whereas germination is irreversible and involves loss of all spore properties, especially heat resistance. During outgrowth, the embryo vegetative cell emerges from the spore coat and forms the full vegetative organism.

Activation can be brought about in different ways, mainly by the following:

- sublethal heat treatment (Curran & Evans, 1945)
- exposure to chemicals, particularly calcium dipicolinate, low pH (1 to 1.5), thiol compounds, and strong oxidizing agents
- aging of the spore population, by storing spores at a low temperature for a period of time

2.6.1 Heat Activation

When using sublethal heat to activate spores, the optimum temperature and exposure time for a maximum colony count depend on the species and strain of organism and may also vary from batch to batch. Other factors are known to be the composition of the medium on which the spores were grown, the medium in which activation occurs, the age of the spore suspension, and the temperature used for sporulation (Gould & Hurst, 1969).

Although a heat treatment will activate the spores, it also has two other effects that may decrease the number of colonies formed on plating out:

1. heat killing of spores, which may be of importance with *B. stearothermophilus* spores activated at 115°C (Gould & Hurst, 1969)
2. heat-induced dormancy, where short exposures to 80° to 100°C were found to have actually increased the dormancy of *B. stearothermophilus* spores, for which heat activation at 115°C was optimal (Finlay & Fields, 1962).

The effect of the latter can be determined, since, if other activation methods are employed or heat levels are increased, heat-induced dormant spores will reactivate. Heat-killed spores, however, will never produce a colony and cannot be differentiated from unactivated spores unless a different activation method to heat is employed.

However, generally all optimum conditions for heat activation are lower than the temperatures used in UHT processing. Spores that have passed through such a process do not require further heat activation (Keynan & Evenchick, 1969).

2.6.2 Activation by Calcium Dipicolinate

The use of a chelate of calcium dipicolinate (Ca-DPA) to germinate bacterial spores was first demonstrated by Riemann and Ordal (1961) and can be achieved by addition of solutions of calcium chloride and dipicolinic acid (2,6-pyridine dicarboxylic acid) to the plating medium in a ratio of 1:1 or higher at levels around 40 mM. This treatment causes rapid and complete germination of viable spores of a range of species without heat activation. This effect was examined for *B. stearothermophilus* spores by Kirk and Hambleton (1982), showing that activation occurred and was independent of the age and degree of dormancy of the spore suspension and also of the degree of lethal heat treatment received at 121°C.

2.6.3 Aging and Activation

It has been long known (Powell, 1950) that spores that have been recently prepared will not germinate without being activated first, but after storage at low temperatures will act as though they have been heat activated. This aging phenomenon has been demonstrated in a variety of *Bacillus* spores. Aging is thought to be the same in principle as heat activation, using low temperatures and very long holding times in line with Arrhenius kinetics. The major difference between the two, however, is that dormancy is lost permanently with aging, whereas with heat activation dormancy loss is only temporary.

Spore populations used for heat inactivation measurements need to be aged for a period of time before reliable data can be obtained; for example, for *Bacillus stearothermophilus* spores, 6 months is required (F.L. Davies, personal communication, 1980).

2.7 NONLINEAR THERMAL DEATH KINETIC DATA

Thermal death kinetic data are often collected that do not give a straight line on a graph of log(colony count) versus time and several practical factors are possible that could account for these discrepancies. They have been well reviewed by Cerf (1977). The possible patterns are given in Figure 2–3.

Curves that are concave downward (or have "shoulders"), such as those in Figure 2–3(a) and 2–3(b), have several possible explanations:

- Clumping of microorganisms: Spores and vegetative cells may be present in clumps, where each clump would only give a single colony on plating out. The actual number of microorganisms present is greater than that given by a colony count. After heat treatment, the death of several organisms in the clump will still result in a single colony, and hence little difference in colony count is observed until high levels of organism death are reached. The presence of clumping can easily be seen by a simple microscopic examination and may be minimized by gentle mixing or homogenization.
- Heat activation: As covered in detail above, spores usually require a heat shock to break dormancy so that they all simultaneously germinate on plating out to give colonies. As the severity of thermal treatment increases, spores will progressively become heat activated and germinate more readily on plating out, which causes the apparent numbers to increase. However, simultaneously, an increasing number of spores may be killed, causing a decrease in numbers. Depending on the rate at which these two factors occur, short heating times may give a shoulder, or even an increase in surviving spores until the optimum heat shock conditions have been well exceeded. Some guide to the concentration of spores that may be expected in a non–heat-treated sample may be obtained from the use of a spore-counting chamber under a microscope, with allowance for the proportion of phase-dark spores, which are likely to be dead. Another method may be to use calcium dipicolinate in the plating medium (Gould & Hurst, 1969), which has the effect of breaking spore dormancy and causing simultaneous germination, giving the maximum count expected. The optimum heat shock conditions for a given spore population can be determined with this maximum count as a goal.

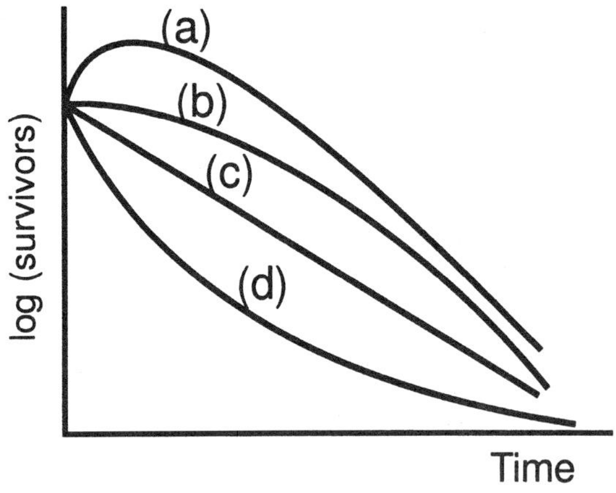

Figure 2–3 Nonlinear thermal death curves (a) activation "shoulder," (b) concave downward, (c) linear thermal death, (d) concave upward.

- Heat transfer into ampule: At high temperatures and short immersion times, the heating-up time for an ampule can be a large proportion of the total immersion time, resulting in a shorter exposure time at the bath temperature and substantially greater survival of microorganisms than expected. With increasing immersion times, the constant heating time becomes a smaller proportion of immersion time and the survival of microorganisms approaches that expected, resulting in a "shoulder." This was demonstrated by Davies et al. (1977) and Perkin et al. (1977), who investigated the thermal death kinetics of *Bacillus stearothermophilus* at temperatures of 120° to 160°C using 1-mm diameter thin-walled capillary tubes, preheated to 100°C before transfer to the oil bath, in order to maximize the rate of heating. Even under these conditions, the rate of heating of the capillary tube required a compensation of about 2 seconds to the immersion time for a bath temperature of 145°C. The rate of heating of a solid in a liquid is covered in detail in Chapter 4.

Curves that are upwardly concave (they have a "tail") are represented by Figure 2–3(d). The causes of this are usually attributed to the use of a genetically heterogeneous spore population of mixed heat resistance. The spores with low heat resistance will die relatively quickly, causing a rapid decrease in numbers at the shorter heat treatment times, but those with a higher heat resistance will only die relatively slowly, forming the tail.

One other explanation of the presence of a "tail" is whether there is a proportion of spores in any given spore population with greater heat resistance than the others, in a way similar to the explanation of heterogeneous populations above. It is thought that the highly resistant spores may simply have an innately higher heat resistance or may acquire greater heat resistance during heating (Figure 2–4). There has been some attempt to subcultivate these "super-resistant" organisms and measure the heat resistance of this population, which has not generally been found to be significantly improved.

Another reason for the presence of a "tail" arises from contamination of the spore population by a chemical sterilant. It is possible that this would occur, as chemical sterilization of the ampule surface is necessary following heat treatment and before breaking the ampule and performing a colony count to prevent contamination of the surviving spores. The colony count requires an optimum number of colonies on each plate (approximately 25 to 250) and therefore high numbers of surviving spores will be diluted more than low numbers. Any accompanying chemical sterilant will therefore also be diluted much more for large survivor numbers and affect the resulting colony count much less, if at all, than if there were very few surviving spores that were not diluted, or diluted only one decimal dilution, before performing the colony count. It is essential that, after any chemical sterilization of the ampule surface, the ampule is washed several times in sterile water, a procedure that adds considerably to the time taken for the experimental method.

2.8 LOG-LOGISTIC THEORY OF MICROBIAL DEATH

The theory of microbial death presented in Section 2.2 is called the mechanistic theory of microbial death and has gained wide acceptance. The basis of the theory is that the microor-

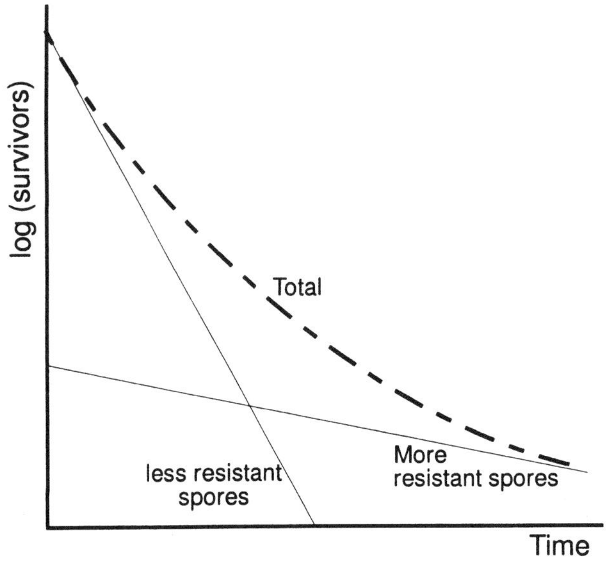

Figure 2–4 Upward concave heat resistance curve caused by different heat resistances in spore population

ganisms all have the same heat resistance but that death is due to the result of "collisions" of quanta of heat on some critical, heat-sensitive part of the cell on which death (or its life) depends. The first-order reaction kinetics come about in a manner similar to those for chemical reactions leading to the Arrhenius theory, in that as the number of cells decreases and therefore the chance of the collisions of heat quanta on the critical target decreases, a decreasing rate of cell inactivation occurs. There is a substantial body of work in which nonlinear death kinetics have been obtained, but those investigators supporting the mechanistic theory are of the opinion that the deviations are explained by physical and microbial factors outlined in Section 2.7.

More recently, an alternative theory has been proposed, called the vitalistic theory, in which the heat resistance of the microorganisms is not assumed to be the same for each individual, but rather the individuals have a range of resistances. Withell (1942) proposed a log-normal distribution of heat resistance in any population that fitted the thermal death of a range of microorganisms. This model, the log-logistic model, has recently been used by other workers to describe the thermal death of other vegetative microorganisms (Cole et al., 1993; Little et al., 1994; Ellison et al., 1994). Anderson et al. (1996) applied the theory to the death of *Cl. botulinum* 213B spores and concluded that the thermal death data fitted the log-logistic theory better than the mechanistic theory, contrasting with the results from Gaze and Brown (1988), using a Stumbo thermoresistometer with the same strain of *Cl. botulinum*, who found a good agreement with the mechanistic theory. Another conclusion of the work is that, with the log-logistic theory, an $F_0 3$ process will only give a 7D, rather than 12D, reduction in *Cl. botulinum* spores, but the proven safety of the $F_0 3$ process still fixes this as a standard process for public safety.

The theory currently remains controversial, and further work is most definitely required before the theory can be accepted or rejected with more certainty.

2.9 CHANGES IN BIOCHEMICAL COMPONENTS AT HIGH TEMPERATURES

In addition to the thermal death of microorganisms, there are other changes that occur in foods subjected to high temperatures, some of which are desirable, such as the degradation of enzymes, but others that are undesirable, such as the formation of brown components, loss of texture and other quality factors, and the degradation of vitamins and other nutritional components. It is desirable that we know the kinetics of these changes so that any given thermal process can be evaluated for food quality as well as microbial safety.

The major benefit of any aseptic process is in the improved product quality over in-container sterilization, especially as the process overall is inherently more expensive, is difficult to operate, and has a greater possibility of sterility failure. Great attention must be paid to this improvement in quality if the aseptic process is not to give a product barely indistinguishable from a canned product but at greater cost.

The difference between final aseptically processed products must be considered. Aseptic processes can be divided into two types, each of which should be considered differently:

1. preservation of raw products such as vegetables, fruits, fruit and vegetable juices, milk, etc., where the aim is a sterile product with minimal biochemical changes
2. manufacture of a convenience product to be reheated by the consumer (retail or catering)

It must be remembered that the vast majority of the second type of aseptic product will be reheated before use by the consumer, and this should be taken into account when designing the thermal process. One aim would be, arguably, to provide an undercooked but commercially sterile product with high nutrient availability (especially micronutrients) but no enzyme activity. The contribution of the aseptic process to cooking of the product must be taken into account but should certainly aim to be significantly less than that for in-container processing. Preparation of soups, stews, and sauces should involve minimal heating or cooking before the actual aseptic processing stage.

2.10 KINETICS OF BIOCHEMICAL CHANGES

There is a substantial body of work in which the biochemical changes during processing have been measured, particularly micronutrient losses, and the overall percentage loss of nutrient in a food after it has been subjected to an overall process reported. There is, however, a great shortage of actual kinetic data for these reactions, especially at elevated temperatures, required to be able to predict nutrient loss given a defined process temperature profile or residence time distribution for a particular food and nutrient.

Again, as with microorganism inactivation, it is also possible to use the constant-z or Arrhenius kinetics models to express the rate of reaction of chemical components and the same arguments outlined above as to the validity of each model apply. Although it is easier to measure the rate of chemical reaction over a much wider temperature range than for microorganism inactivation and perhaps to differentiate between the two models, chemical reaction rates change much less with temperature, and the same inaccuracies eventually result. The *z*-value for chemical change is generally in the order of 30°C as opposed to 10°C for microbial inactivation (i.e., they are much less sensitive to changes in temperature).

2.11 GENERAL COOKING CHANGES

The cooking of foodstuffs, especially solid foods such as potatoes, vegetables, and fruit, will occur during thermal processing and is characterized by a variety of reactions such as a reduction in firmness and increase in palatability and digestibility. Overcooking of a continuous-flow thermal processed foodstuff, however, is definitely to be avoided to maintain the quality image of the product as superior to an in-container processed one.

Overall, cooking may be characterized by first-order kinetics similar to those for thermal death or biochemical degradation. A cooking value, C, has been defined by Mansfield (1962) in a way similar to the definition of lethality, in which rate of cooking at one temperature is related to an equivalent time at a suitable reference temperature (usually 100°C) and can be used to determine the overall equivalent cooking time at the reference temperature for a changing process temperature:

$$C = 10^{\frac{(\theta - \theta_{ref})}{z_c}} \tag{2.18}$$

where z_c is the *z*-value for cooking characteristics. When the reference temperature is taken as 100°C and a generalized z_c value of 33.1°C is used, it is designated the C_0 value:

$$C_0 = 10^{\frac{(\theta - 100)}{33{\cdot}1}} \tag{2.19}$$

The C_0 value therefore defines the equivalent heating time at 100°C, that is, it relates to the time that the foodstuff would have been boiled in water. The z_c value of 33.1°C appears to be high, in the light of further work. Ohlsson (1980) showed that the z_c value ranged from 13° to 24°C, averaging at 23°C for meat and vegetables. The C value required for optimum cooking has been determined for several vegetables, for example, for different potato species, a C value of 8 to 12 minutes (Dagerskog & Osterstrom, 1978).

Obviously, with such a wide variety of factors implicated in cooking, not only in the sensory assessment but also the variability in raw material quality (e.g., variety, weather conditions during growing, age, storage conditions) and the different reactions occurring, there is likely to be some variation in kinetic data, especially as vegetables cooked at elevated temperatures have different sensory characteristics from those cooked at 100°C. Holdsworth (1992) has calculated the relationship between sterilization and cooking value and shows that, as the process temperature is raised, it is possible to obtain a sterile but undercooked product. This may be of value for products that are to be further heated by the consumer before consumption, so that the final product is at its optimum texture when actually consumed. The effect of any blanching must also be taken into account.

2.11.1 Enzyme Inactivation

The presence of enzymes such as lipase, protease, chlorophyllase, oxidases, and peroxidases, whether naturally occurring in the foodstuff or as a result of microbial action, can cause off-flavors; formation of brown compounds; breakdown of fats, proteins, and colors; and many other storage problems. Their denaturation is therefore obviously desirable, and even low levels of their survival can result in a very much shorter product shelf life than expected since storage is at ambient temperatures (see Chapter 9). However, a large number of naturally occurring enzymes are denatured at relatively low temperatures, and it is likely that a blanching operation or other preliminary heat treatment would be used to achieve this. Some microbial enzymes, though, are extremely thermally resistant and could cause considerable problems, for example, *Pseudomonas* protease with a D value of 1.5 minutes at 149°C (Adams, Barach, & Speck, 1975). To reduce these enzymes to acceptable levels would require a much more severe process than necessary for bacterial inactivation. Fortunately, some enzymes have unusual inactivation kinetics in that, above 100°C, their degradation follows Arrhenius kinetics but below this temperature, inactivation is abnormally faster (Barach, Adams, & Speck, 1976). The basis of this effect is not known, but some autolysis effect has been suggested (Barach, Adams, & Speck, 1978). It is easy to ensure that foodstuffs in which survival of these enzymes is likely to be a problem are subjected to a low-temperature treatment before the aseptic process to inactivate them if they have not already been so. The survival of enzymes, even at low levels, is likely to finally limit the shelf life of the foodstuff, for example, age gelation in milk (see Chapter 9).

2.11.2 Vitamin Degradation

The loss of vitamins from foodstuffs during a thermal treatment is not a simple matter, as degradation can be due to a variety of agents in addition to the thermal treatment, particularly oxygen and light, or losses arise from other factors, such as leaching of water-soluble vitamins. Degradation usually depends on pH and may be catalyzed by the presence of chemicals, metals, other vitamins, enzymes, etc. In addition, vitamins are often not only a single chemical but a group of compounds that may interconvert, each of which has vitamin activity and may degrade at a different rate.

One good example of this is vitamin C. Both L-ascorbic acid and its oxidized form, dehydro-L-ascorbic acid, possess vitamin C activity (but not D-ascorbic acid) but have different heat stabilities. Dehydro-L-ascorbic acid may be further degraded irreversibly by oxidation to 2,3-diketogulonic acid, which does not have vitamin C activity, and anaerobic degradation of ascorbic acid may also take place. Losses of the vitamin are therefore related to the conversion of L-ascorbic acid to dehydro-L-ascorbic acid by oxygen in the product before or during the thermal process, as well as by the severity of the process itself. In addition, degradation of L-ascorbic acid is catalyzed by traces of heavy metal ions, particularly Cu^{2+}, Fe^{2+}, and Zn^{2+}, which may be obtained from contact with any bronze, brass, or steel in the process equipment but not stainless steel, and can also be lost by oxidation in the package from headspace gases and leakage through the packaging material. To complicate matters even further, L-ascorbic acid degradation can also be catalyzed by ascorbic acid oxidase, phenolase, cytochrome oxidase, and peroxidase enzymes, all of which may be denatured by a thermal process and hence, after the same storage period, give a product with a higher ascorbic acid level than one that has not been heat treated.

There is a substantial body of work on the loss of vitamins in thermal processes but the majority of work comprises measuring the vitamin level before and after processing and expressing a percentage loss. There is surprisingly little actual kinetic data on the vitamin degradation, due perhaps to the difficulties covered above.

The thermal stability of vitamins is summarized in Karmas and Harris (1988) and Ryley and Kajda (1994). The heat-labile vitamins are the fat-soluble vitamins A (with oxygen present) D, E, and β-carotene (provitamin A), and the water-soluble vitamins are B_1 (thiamin), B_2 (riboflavin) (in acid environment), nicotinic acid, pantothenic acid, and biotin C. Vitamin B_{12} and folic acid are also heat labile but their destruction involves a complex series of reactions with each other. Vitamin B_6 is generally little affected by heat, but storage after heat treatment can cause high losses.

Of the heat-labile vitamins, thiamin appears to have the most stable denaturation kinetics and has been most studied. Inactivation of thiamin has been generally found to follow first-order kinetics, although second-order kinetics have been reported (Horak & Kessler, 1981). Degradation of vitamin C and folic acid are also reported as first-order reactions (Ulgen & Ozilgen, 1991; Barrett & Lund, 1989).

2.11.3 Browning Reactions

There are several browning reactions that occur in foods:

- enzymatic browning, caused by polyphenyl oxidase enzyme (PPO) and oxygen. The reaction can be prevented by thermal denaturation of the PPO enzyme, by exclusion of oxygen, or by the presence of sulfite.
- nonenzymic browning, involving either the Maillard reaction or caramelization of sugars

2.11.4 Maillard Reaction

The Maillard reaction is a complex series of reactions between reducing sugars and proteins, which form molecules of the type 1-amino-1-deoxy-2-ketose via an Amadori rearrangement. This type of compound may then develop one of three ways:

1. A strong dehydration gives rise to furfural, and dehydrofurfural derivatives, especially 5′-hydroxymethylfurfural (HMF) and deoxy-lactulosyl-lysine.

2. The molecules split into carbonyl compounds such as diacetyl, pyruvaldehyde, and acetol.
3. Moderate dehydration produces reductones and hydroreductones, which may form characteristic odor compounds such as aldehydes via further reaction with amino acids.

The Amadori compounds are colorless and, after taking any of the three routes above, will polymerize and lead to the formation of brown or black insoluble compounds called melanoidins. Although this reaction can be desirable in generating flavor compounds associated with cooking, the nutritional value of the product can be reduced by protein damage and a loss of amino acids, especially lysine and arginine.

There has been much research work on the kinetics of formation of intermediate compounds formed during the reactions, particularly 5′-hydroxymethylfurfural (HMF), which may be determined relatively easily using the method of Keeney and Bassette (1959). Depending on the procedure used, either free HMF or free plus potential HMF derived from other browning intermediates may be measured: the latter is sometimes called "total" HMF. HMF has often been used as a measure of the severity of heat treatment received by a product, especially milk in an in-container or UHT process (Fink & Kessler, 1986; Burton, 1983). pH is important and the rate of browning increases with increasing pH (Priestley, 1979) and also increasing oxygen level.

2.11.5 Lactulose

The formation of lactulose is important in dairy-based products only and has recently been used in the European Union (EU) to define the heat treatment level received by milk, distinguishing between pasteurized, UHT-sterilized, and in-container sterilized milk (EU Milk Hygiene Directive, 1992).

Lactulose is an epimer of lactose formed during heating by the Bruyn-van Ekenstein transformation, catalyzed either by the amino acid groups of milk proteins or by inorganic salts (Adachi & Patton, 1961; Andrews, 1986). Although lactulose is known to have a laxative effect at about 2.5 g/kg, most commercially produced milks fortunately have much lower levels than this, the maximum being around 2 g/kg for in-container sterilized milks with most other milks well below this value. Andrews (1984) showed that pasteurized, UHT-sterilized and in-container sterilized milks could be distinguished from each other on the basis of lactulose level and even direct and indirect UHT-sterilized milks could also be separated. Andrews (1985) showed a high correlation between lactulose level and integrated time-temperature for direct or indirect UHT-sterilized and in-container sterilized milks using activation energy of 152 ± 20 kJ/mol ($z = 21°C$). Andrews and Prasad (1987) measured lactulose formation in milk in sealed capillary tubes at 70° to 130°C and found the activation energy was 128 ± 6 kJ/mol ($z = 24.8°C$). Following this, the EU and International Dairy Federation (IDF) have proposed a definition of UHT milk as containing <600 mg/kg lactulose and >50 mg/kg undenatured β-lactoglobulin.

2.11.6 Color Compounds

The color of foods greatly influences its acceptability by the consumer as it is the first stimulus received, usually well before any other sensory experience. The loss of pigments by heat denaturation has also been the subject of many studies, the major compounds being chlorophyll, anthocyanins, flavanoids, betalains, etc. As with most other chemical factors, the degradation of these compounds is complex, with many reactions causing loss of, or change in, color in addition to thermal degradation, particularly enzymic- or photo-oxidation. Of the major compounds, on heating;

- chlorophyll may convert to pheophytin, a green-brown-olive color, by substitution of the magnesium ion at the center of its tetrapyrrole nucleus by a hydrogen ion but may also lose the phytol tail due to chlorophyllase activity to give pheophorbide, a brown-olive color, both reactions being strongly dependent on the pH of the medium. Pheophorbide may then be oxidized to give chlorins and purpurines, which are brown. The conversion of chlorophyll to pheophytin can be used as an indicator of the heat process received.
- Anthocyanins degrade due to thermal treatment and are strongly pH dependent, being more stable at low pH and unstable in the presence of oxygen.
- Carotenoids are generally more thermostable than the other color compounds due to their distribution in foods, but thermal treatment may induce changes that result in a loss of carotenoids on storage.

The thermal degradation kinetics of nutritional and other components have been studied in foodstuffs, but in light of the above, must be interpreted cautiously, as the subject is complex. Further information on chemical and nutritional changes in specific products during pasteurization and sterilization is given in Chapters 5 and 6. For a deeper treatment, the reader is directed toward more specific publications, such as Karmas and Harris (1988), Fennema (1996), and Priestley (1979).

REFERENCES

Adachi, S., & Patton, S. (1961). Presence and significance of lactulose in milk products. A review. *Journal of Dairy Science* **44**, 1375–1393.

Adams, D.M., Barach, J.T., & Speck, M.L. (1975). Heat resistant proteases produced in milk by psychrotrophic bacteria of dairy origin. *Journal of Dairy Science* **58**, 828–834.

Anderson, W.A., McClure, P.J., Baird Parker, A.C., & Cole, M.B. (1996). The application of a log-logistic model to describe the thermal inactivation of *Cl. botulinum* 213B at temperatures below 121.1°C. *Journal of Applied Bacteriology* **80**, 283–290.

Andrews, G.R. (1984). Distinguishing pasteurized, UHT and sterilized milks by their lactulose content. *Journal of the Society of Dairy Technology* **37**, 92–95.

Andrews, G.R. (1985). Determining the energy of activation for the formation of lactulose in heated milks. *Journal of Dairy Research* **52**, 275–280.

Andrews, G.R. (1986). Formation and occurrence of lactulose in heated milk. *Journal of Dairy Research* **53**, 665–680.

Andrews, G.R., & Prasad, S.K. (1987). Effect of the protein, citrate and phosphate content of milk on formation of lactulose during heat treatment. *Journal of Dairy Research* **54**, 207–218.

Ball, C.O. (1923). Thermal process time for canned foods. *National Research Council* **7**, 37.

Barach, J.T., Adams, D.M., & Speck, M.L. (1976). Low temperature inactivation in milk of heat-resistant proteases from psychrotrophic bacteria. *Journal of Dairy Science* **59**, 391–395.

Barach, J.T., Adams, D.M., & Speck, M.L. (1978). Mechanism of low temperature inactivation of a heat-resistant bacterial protease in milk. *Journal of Dairy Science* **61**, 523–528.

Barrett, D.M., & Lund, D.B. (1989). Effect of oxygen on thermal-degradation of 5-methyl-5,6,7,8-tetrahydrofolic acid. *Journal of Food Science* **54**, 146–149.

Bigelow, W.D. (1921). The logarithmic nature of thermal death time curves. *Journal of Infectious Diseases* **27**, 528–536.

Bigelow, W.D., & Esty, J.R. (1920). The thermal death point in relation to time of typical thermophilic organisms. *Journal of Infectious Diseases* **27**, 602–617.

Brown, K.L. (1991). Principles of heat preservation. In J.A.G. Rees & J. Bettison (Eds.), *Processing and packaging of heat preserved foods*. Glasgow, Scotland: Blackie.

Brown, M.H., & Booth, I.R. (1991). In N.W. Russell & G.W. Gould (Eds.), *Food preservatives*. London: Blackie.

Burton, H. (1983). Bacteriological, chemical, biochemical and physical changes that occur in milk at temperatures of 100–150°C. *Bulletin of the International Dairy Federation*, Document 157, pp. 3–16. Brussels, Belgium: International Dairy Federation.

Burton, H., Perkin, A.G., Davies, F.L., & Underwood, H.M. (1977). Thermal death kinetics of *Bacillus stearothermophilus* spores at ultra high temperatures. III. Relationship between data from capillary tube experiments and from UHT sterilizers. *Journal of Food Technology* **12**, 149–161.

Cerf, O. (1977). Tailing of survival curves of bacterial spores. *Journal of Applied Bacteriology* **42**, 1–19.

Circular (1972). Circular FSH 4/72 (England and Wales). Ministry of Agriculture, Fisheries and Food. London: HMSO.

Cole, M.B., Davies, K.W., Munro, G., Holyoak, C.D., & Kilsby, D.C. (1993). A vitalistic model to describe the thermal inactivation of *Listeria monocytogenes*. *Journal of Industrial Microbiology* **12**, 232–239.

Curran, H.R., & Evans, F.R. (1945). Heat activation inducing germinating in the spores of thermotolerant and thermophilic aerobic bacteria. *Journal of Bacteriology* **49**, 335–346.

Dagerskog, M., & Osterstrom, L. (1978). Boiling of potatoes. An introductory study of time-temperature relations for texture changes. *Slk Service Series* No. 584.

Daudin, J.D., & Cerf, O. (1977). The effect of thermal shock on the thermal death of spores. *Lebensmittel Wissenschaft und Technologie* **10**, 203.

Davies, F.L., Underwood, H.M., Perkin, A., & Burton, H. (1977). Thermal death kinetics of Bacillus stearothermophilus spores at ultra high temperatures. I. Laboratory determination of temperature coefficients. *Journal of Food Technology* **12**, 115–129.

Ellison, A., Anderson, W., Cole, M.B., & Stewart, G.S.A.B. (1994). Modeling the thermal inactivation of *Salmonella typhimurium* using bioluminescence data. *International Journal of Food Microbiology* **23**, 467–477.

EU Milk Hygiene Directive. (1992). Council Directive 92/46/EEC, 16 June 1992.

Fennema, O.R. (1996). *Food chemistry* (3rd ed.). New York: Marcel Dekker.

Fink, R., & Kessler, H.G. (1986). HMF values in heat treated and stored milk. *Milchwissenschaft* **41**, 638–641.

Finley, N., & Fields, M.L. (1962). Heat activation and heat-induced dormancy of *Bacillus stearothermophilus* spores. *Applied Microbiology* **10**, 231–236.

Gaze J.E., & Brown, K.L. (1988). The heat resistance of spores of *Clostridium botulinum* 213B over the temperature range 120 to 140°C. *International Journal of Food Science and Technology* **23**, 373–378.

Gould, G.W., & Hurst, A. (Eds). (1969). *The bacterial spore*. London: Academic Press.

Heppell, N.J. (1985). Comparison of the residence time distributions of water and milk in an experimental UHT sterilizer. *Journal of Food Engineering* **4**, 71–84.

Hersom, A.C., & Hulland, E.D. (1980). *Canned foods: Thermal processing & microbiology*. Edinburgh, Scotland: Churchill Livingstone.

Holdsworth, S.D. (1992). *Aseptic processing and packaging of food products*. London: Elsevier Applied Science.

Horak, P. (1980). *Uber die Reaktionskinetik der Sporenabtotung und chemischer Veranderungen bei der thermischen Haltbarmachung von Milch.* Thesis, Technical University, Munich, Germany.

Horak, P., & Kessler, H.G. (1981). Thermal destruction of thiamine—a second order reaction. *Zeitschrift.fur Lebensmittel Untersuchung und Forschung* **173**, 1.

Jonsson, U., Snygg, B.G., Harnulv, B.G., & Zachrisson, T. (1977). Testing two models for the temperature dependence of the heat inactivation of *B. stearothermophilus* spores. *Journal of Food Science* **42**, 1251–1252, 1263.

Karmas, E., & Harris, R.S. (1988). *Nutritional evaluation of food processing.* (3rd ed.). New York: Van Nostrand Reinhold.

Keeney, M., & Bassette, R. (1959). Detection of intermediate compounds in the early stages of browning reaction in milk products. *Journal of Dairy Science* **42**, 945–960.

Keynan, A., & Evenchick, Z. (1969). Heat activation. In G.W. Gould & A. Hurst (Eds.), *The bacterial spore* (pp. 359–396). London: Academic Press.

Kirk, B., & Hambleton, R. (1982). Optimized dipicolinate activation for viable counting of *B. stearothermophilus* spores. *Journal of Applied Bacteriology* **53**, 147–154.

Little, C.L., Adams, M.R., Anderson, W.A., & Cole, M.B. (1994). Application of a log-logistic model to describe the survival of *Yersinia enterocolitica* at suboptimal pH and temperature. *International Journal of Food Microbiology* **22**, 63–71.

Mansfield, T. (1962). High-temperature, short-time sterilization. *Proceedings of the 1st International Congress on Food Science and Technology*. London, 4, 311–316.

Ohlsson, T. (1980). Temperature dependence of sensory quality changes during thermal processing. *Journal of Food Science* **45**, 836–839, 847.

Perkin, A.G., Burton, H., Underwood, H.M., & Davies, F.L. (1977). Thermal death kinetics of *Bacillus stearothermophilus* spores at ultra high temperatures. II. Effect of heating period on experimental results. *Journal of Food Technology* **12**, 131–148.

Pflug, I.J., & Esselen, W.B. (1953). *Food Technology* **7**, 237.

Powell, J.F. (1950). *Journal of General Microbiology* **4**, 330.

Priestley, R.J. (1979). *Effects of heating on foodstuffs*. London: Applied Science Publishers.

Riemann, H., & Ordal, Z.J. (1961). Germination of bacterial spores with calcium dipicolinic acid. *Science, N.Y.* **133**, 1703–1704.

Ryley, J., & Kajda, P. (1994). Vitamins in thermal processing. *Food Chemistry* **49**, 119–129.

Stumbo, C.R. (1965). *Thermobacteriology in food processing*. New York: Academic Press.

Ulgen, N., & Ozilgen, M. (1991). Kinetic compensation relations for ascorbic-acid degradation and pectinesterase inactivation during orange juice pasteurization. *Journal of the Science of Food and Agriculture* **57**, 93–100.

Withell, E.R. (1942). The significance of the variation on shape of time survivor curves. *Journal of Hygiene* **42**, 124–183.

CHAPTER 3

Continuous-Flow Thermal Processing Plant

The purpose of a continuous-flow thermal process plant must be, as we have shown, to heat the product to the required process temperature, hold it at that temperature for the required time, and then cool it to a suitable temperature for filling, without recontamination. Since the chemical changes in the product are minimized by using the highest possible temperature for the shortest time needed to give the required sterilizing effect, it follows that the best overall product quality will be obtained if the rate of heating to the final temperature is as rapid as possible, and the cooling after the holding period also as rapid as possible. There are many factors in the design of the process plant that affect the quality of the product, which is covered in this and succeeding chapters, and these factors are obviously important in the selection of equipment and process parameters for such processes.

However, as well as product quality, there are other important considerations for any commercial, practical process that must be considered. The first of these are the economics of the process, which can be split into two major parts:

1. the capital cost of the process, that is, the initial cost of purchasing and installing the process equipment, including service equipment such as a steam boiler, vacuum pump or air compressor, plumbing, instrumentation, and any building costs
2. the operating costs, which are the costs of running the process in terms of consumption of raw materials and packaging (including waste materials), steam, water, electricity, effluent etc., equipment preheat treatment, process cleaning, labor, and maintenance

The overall cost of processing the product must be determined very early on, even during the process and product development stage, before a decision is made as to whether to proceed with the purchase and installation of the whole process. The cost of producing a unit of product (e.g., a 1-liter pack of pasteurized milk) is made up of the capital and operating costs and can be calculated when the two have been estimated. The operating costs are directly related to each unit, with an allowance for process cleaning, sterilization, and so forth. The capital cost can be converted to the unit cost by assuming that the initial capital cost depreciates to the scrap value of the plant over the operating life of the process equipment, usually between 7 and 15 years. The cost per unit product is calculated from the annual depreciated cost divided by the annual output of the plant. From the final cost per unit, an approximate cost for the retail product can be estimated and therefore whether the product is likely to be financially viable. The process economics are often determined for different production rates and, together with market research, are a useful tool in determining the production rate required for the process.

Process economics are considerably more complex than this, though, and are covered in more detail in other publications (IChemE, 1988).

A major consideration in the operating costs of a process is the ability of the process to recover heat put into it. The heat energy required to heat the product to the process temperature required is likely to be large, especially for large throughputs and a high temperature rise, but heat recovery, or heat regeneration, in which the heated product is used to heat the cold incoming liquid will result in reducing the total heat energy required for the process by between 50% and 90%. In addition, as the heated product is also being cooled by the incoming product, the cooling water requirement is reduced by the same amount. Heat regeneration can therefore severely reduce operating costs, but it increases the capital cost of the process slightly as larger heat exchangers are required, and can affect the quality of the product, as is demonstrated later in this chapter.

There are many different types of thermal processing plants, all of which aim to give a satisfactory product, but which may differ in other characteristics so that one type of plant is more appropriate for a given process and application than another. The principles of operation and construction of the main types of thermal process equipment are summarized in Figure 3–1. The heating medium in most equipment is steam or hot water, produced from the combustion of solid fuel, oil, or gas. In a few cases, the steam or hot water may be obtained from an electrically heated boiler, but this is an exceptional situation because of the relatively

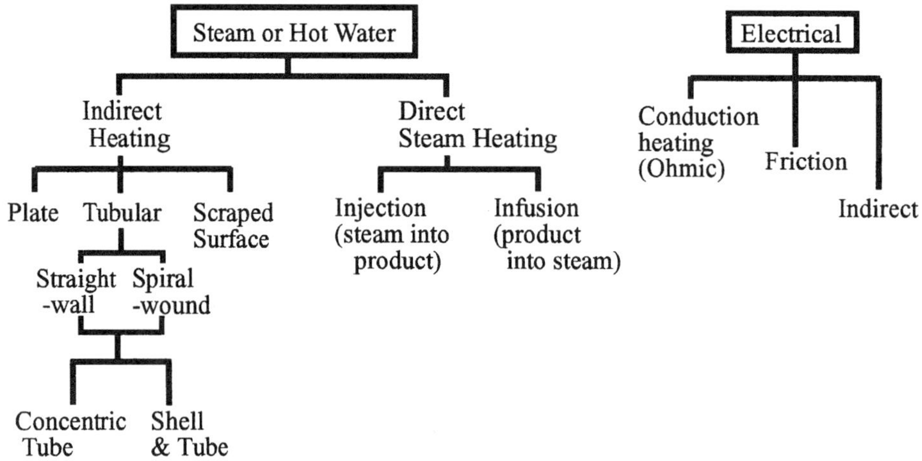

Figure 3–1 Types of continuous-flow thermal processing plant and heating energy source. *Source:* Adapted with permission from H. Burton, *Ultra-High-Temperature Processing of Milk and Milk Products*, p. 78, © 1988.

high cost of electricity used merely for heating. The overall efficiency of electricity, generated thermally itself from a fossil fuel–fired power station and with distribution system losses, is in the region of 30%, and compares badly with the efficiency of a boiler using these fuels directly on a factory site of perhaps 70% to 80%. An electrically heated boiler is only likely to be used where cheap electricity from hydroelectric sources is available or where other fuels cannot be obtained easily. In a few types of sterilizers, electricity is used in a more direct way without the generation of steam or hot water as an intermediate. Again, special circumstances will be necessary to justify the use of these types (see Chapter 4).

3.1 PROCESS EQUIPMENT FOR CONTINUOUS-FLOW THERMAL PROCESSES

3.1.1 Process and Equipment—General Requirements

The most important aspects of the equipment in a continuous-flow thermal process are to ensure that the foodstuff attains both a minimum specified process temperature and also maintains that temperature for a specified time. The equipment itself must also be of a standard of construction that allows it to be presterilized, that is, heat treated to a time-temperature combination greater than the foodstuff is to receive just before operation with the foodstuff, to prevent recontamination of the product during cooling and storage before aseptic packaging. All continuous-flow thermal processes ensure that the minimum holding time is achieved by designing the holding tube to be of the correct length for the designed volume throughput of foodstuff and then controlling the flowrate through the process to ensure that at no time does it exceed the designed value. The minimum temperature is ensured by a temperature sensor at the end of the holding tube, which can operate a flow diversion valve and alarm, recycling underprocessed product back to the feed tank.

3.1.2 Pumping and Flow Control

The selection of a suitable pump for the continuous-flow thermal process depends on the characteristics of the foodstuff to be processed. For low-viscosity liquids, with a viscosity less than about 600 mPa.s, a centrifugal pump is sufficient and can easily be selected for the required flowrate and pressure from characteristic curves available from manufacturers. For higher-viscosity liquids, the selection of pumps is covered in Section 4.10. It is especially important that a centrifugal pump be used if there is a homogenizer in the process, as the use of two positive displacement (PD) pumps in series is to be avoided. The flowrate of the process in this case is set by the speed of rotation of the homogenizer.

If using a centrifugal pump as the sole pump in the process, it is essential that some other flow control mechanism be used, as the flowrate delivered by centrifugal pumps is very dependent on the back-pressure. If, for example, a deposit forms in the heat exchangers, the back-pressure will increase and the flowrate from the pump will fall, increasing the residence time and giving a poor-quality product. Several types of flow control are possible, depending on the pressure involved, and can be a conventional sensor and control loop operating a back-pressure valve, or a flow control device as used in pasteurizers. The latter, shown in

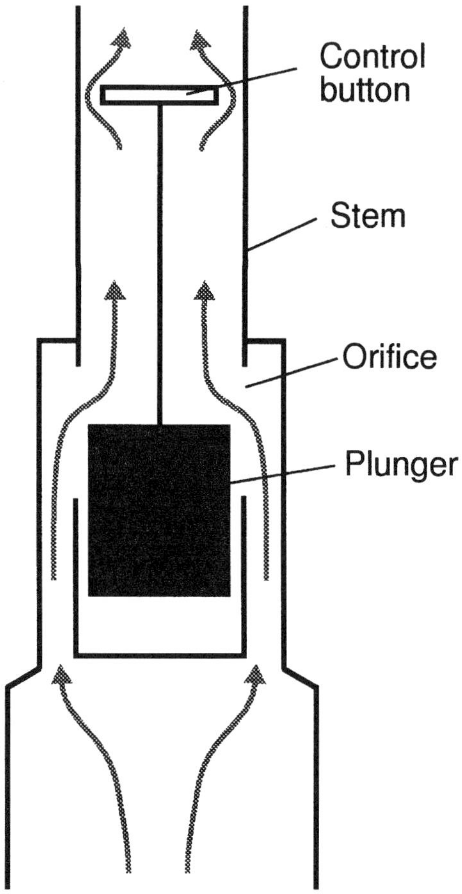

Figure 3–2 Flow control device used in pasteurizing plant

Figure 3–2, depends on a horizontal control plate of smaller diameter than the vertical tube in which it is situated, connected to a valve that closes the inlet ports as it rises. Liquid enters the tube via the inlet ports and, on rising up the tube, raises the control plate, which restricts the flow into the tube. The flow eventually balances out to a constant value and, on disturbance of the flow downstream, will rebalance to the same value.

3.1.3 Homogenization

Certain products require homogenization, particularly milk, creams, and other dairy-based products in order to stabilize fat, improve mouthfeel, or other effects. A high-pressure type of homogenizer is almost exclusively used over the other types available and must be incorporated into the process. The high-pressure homogenizer is basically a piston pump consisting of three or five pistons that pressurize the foodstuff typically to 20 MPa (3,000 psi) and force the liquid through a narrow slit, in the form of a homogenizer valve, at velocities up to 250 m/s. The narrow slit is only just larger than the fat globules in the food product, about 10 μm, which are broken down by a mixture of shear, cavitation, and microturbulence.

A diagram of a typical three-piston homogenizer is given in Figure 3–3 (a) and (b). The liquid feed enters through an inlet valve into the chamber formed as the piston moves backward, then is forced through the outlet valve on the forward stroke of the piston into the high-pressure region. Both inlet and outlet valves only allow passage of liquid one way and are kept closed when the pressure on the downstream side is greater than the feed side. High-pressure homogenizers cannot be used for foodstuffs containing solid particles or fibers, as these may become trapped between the valve and its seat, allowing leakage from the high-pressure side to the feed zone and preventing a pressure from being developed. The three pistons are operated at 120° phase shift to even out pressure pulsations on the piston forward stroke. Five-piston homogenizers are also available and give an even more constant pressure.

The homogenizer valve (Figure 3–3(b)) is pressed with an adjustable force onto a valve seat and when the pressure in the liquid generated by the three-piston pump exceeds this force, the homogenizer valve lifts, forming the slit through which the liquid passes and homogenization occurs. It often happens, however, that reagglomeration of the fat globules

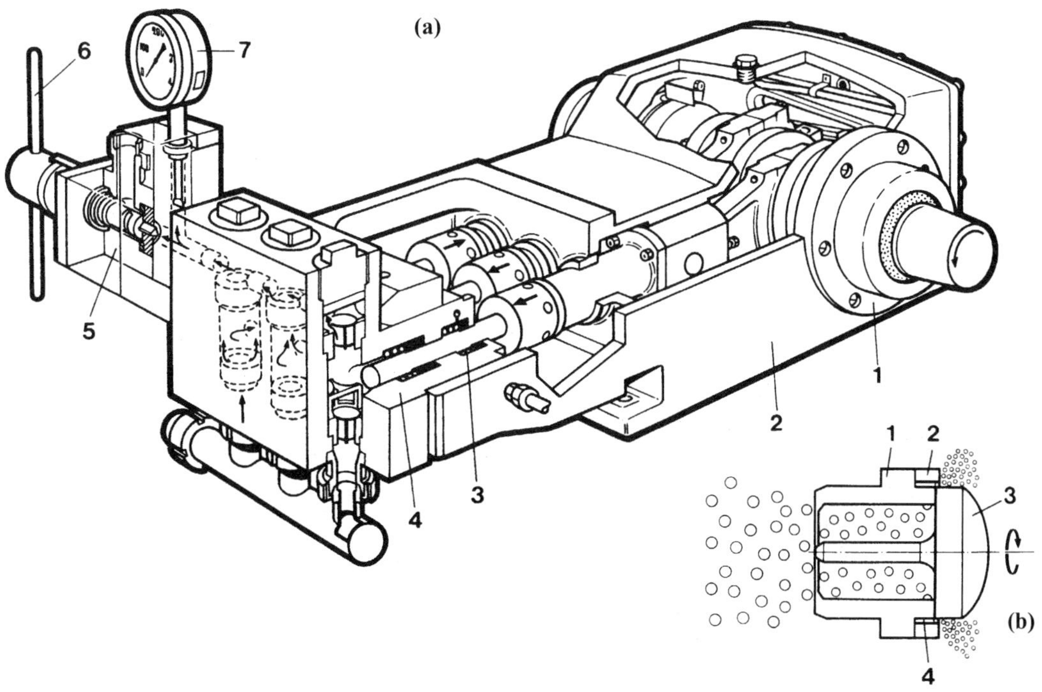

Figure 3–3 Cutaway view of cylinder block. (a): (1) Crankshaft. (2) High-pressure cylinder block. (3) Piston rings. (4) Pistons. (5) Homogenizer head. (6) Pressure-setting lever. (7) High-pressure gauge. (b): (1) Outer ring. (2) Homogenizer ring. (3) Core. (4) Gap. *Source:* Reprinted with permission from *Tetra Pak Dairy Handbook*, p. 142, © 1993, Tetra Pak, Lund, Sweden.

occurs rapidly after homogenization and a second homogenizer valve would be required, situated immediately after the first homogenizer valve and operating at a much lower pressure of about 2 MPa. The second valve breaks down clusters of fat globules formed immediately after the first homogenizer valve and disperses them, stabilizing the globules and decreasing the viscosity. Clusters will form more frequently with increasing first-stage homogenization pressure, temperature, and fat content.

High-pressure homogenization is only effective when both phases in the foodstuff are liquid, which, for dairy fat, is at a temperature greater than 45°C. The homogenizer must be situated in the process at a point where the product is at a suitable temperature, but consideration must be given as to whether homogenization is better before or after the sterilization heat treatment. Some homogenized products are stable to further heat treatment and the homogenizer can be situated at the feed end of the process, but others give a more stable product if homogenized after sterilization and so the homogenizer must be somewhere in the sterile part of the process (see Section 6.3). To prevent contamination of the product for the latter, the homogenizer must be of an aseptic design, with sterile barriers fitted where contamination is likely, especially at the rear of the main cylinders where there are oscillating piston shafts. Aseptic homogenizers have a significantly higher capital cost and require much more care in operation than nonaseptic versions. High-pressure homogenization is covered in much more detail by Phipps (1985).

3.2 HEAT EXCHANGE EQUIPMENT

Those thermal processing plants in which steam or hot water is the heating medium can be further divided into two major groups according to the heat-exchange system that is applied. These are termed "direct" and "indirect" heating systems. In indirect heating systems, the product being processed is kept separate from the heating medium, which is either hot water or steam, by a heat-conducting barrier usually made of stainless steel so that there is no physical contact between the product and the heating medium. Conversely, for direct heating systems, the heating medium is steam and the product is mixed directly with the steam so that it condenses and its latent heat of vaporization is transferred to heat the product very rapidly. At the same time, the product is diluted by the condensed steam, which can be compensated for either in the product makeup or it can be brought back to its original concentration using flash cooling. The consequences of this are discussed later in this chapter.

With both direct and indirect heating systems, heat recovery is possible but it is always by indirect means, so that even in a direct heating system a large proportion of the heat exchange is indirect and only the last stages of heating to the sterilization temperature are by a direct method.

3.3 PROCESS EQUIPMENT/PREPROCESSING HEAT TREATMENT

An important part of the process is to prevent recontamination of the product after it has passed through the holding tube by microorganisms on the surface of the process equipment. To prevent this, all surfaces that will come in contact with the heat-treated product must have been heated to a temperature/time combination greater than the normal processing conditions before the product is introduced into the process. In addition, careful, hygienic design of the equipment is necessary to prevent sources of contamination, especially reduction or elimination of cracks and crevices and areas of no flow, such as pockets, square bends, and so forth. This subject is covered in more detail in Chapter 8.

Provision must be made to allow all the process equipment after the holding tube to attain an appropriate temperature, such as 95°C for pasteurization or 135° to 145°C for ultrahigh-temperature (UHT) processes, and to be maintained for 15 to 30 minutes, then cooled before contact with processed product. Pressurized hot water is normally used for preprocessing heat treatment, but pressurized steam may be used for individual items, such as aseptic tanks, and for sterile barriers. The water is heated up to the required temperature with the usual heat exchangers, but it must pass through the cooling and chilling sections, and any heat recovery sections, at this elevated temperature; hence a valve arrangement and often a sterilization postcooler are required. Details of the arrangements for plant preprocessing heat treatment are given in the individual plant descriptions.

INDIRECT HEATING SYSTEMS

For any heating or cooling application, the rate of heat exchange required (the heat exchanger "duty") can easily be calculated from the mass flowrate of the foodstuff (m), its inlet and outlet temperatures (θ_{in} and θ_{out}, respectively), and its specific heat, (c_p) between these two temperatures.

$$\text{Heat exchanger duty } (Q) = m.c_p.(\theta_{out} - \theta_{in}) \tag{3.1}$$

The heat is provided by a heat transfer medium, which is usually pressurized, saturated steam, or water or, for cooling duties, a liquid refrigerant or brine or, for heat recovery sections, the heated or cold foodstuff itself. As heat losses from the system are normally insignificant, the heat gained by the product is equal to the heat lost by the cooling medium.

The rate at which heat is transferred from the heating medium to the product through the walls of the heat exchanger is defined by the general heat transfer equation:

$$Q = U.A.\Delta\theta_{LM} \tag{3.2}$$

where A is the heat transfer area of the heat exchanger heat transfer medium, U is the overall heat transfer coefficient, and $\Delta\theta_{LM}$ is the log mean temperature difference between the product and the heating medium:

$$\Delta\theta_{LM} = \frac{\Delta\theta_1 - \Delta\theta_2}{\ln(\Delta\theta_1 / \Delta\theta_2)} \tag{3.3}$$

$\Delta\theta_1$ and $\Delta\theta_2$ are the temperature differences between product and heating medium at each end of the heat exchanger.

The heat exchanger system selected not only must provide the rate of heat transfer required but must do so at the lowest capital cost, which, for any one type of heat exchanger, is related to the heat transfer area. As can be seen from Equation 3.2, the heat transfer area for any given duty can be minimized by maximizing both temperature difference and heat transfer coefficient. For most food products, however, fouling of the heat exchanger surface is a problem, and the log mean temperature difference $\Delta\theta_{LM}$ must be kept low. Therefore, the overall heat transfer coefficient U is the major factor over which the equipment designed has control and which must be maximized to reduce the capital cost. This is achieved predominantly by designing for a sufficiently high fluid velocity in the heat exchanger flow channel to minimize the boundary layer thickness, using a small-channel cross-sectional area. If the cross-sectional area is too small, however, the fluid will require an excessive pressure drop to maintain this velocity, possibly exceeding the pressure limitations of the equipment, especially if the product viscosity is high. The heat exchanger is therefore arranged in a series of parallel flow channels, called passes, where the number of channels is a minimum, decided by a compromise between maintaining a high fluid velocity and a sufficiently low pressure drop.

A heat exchanger system with a high heat transfer coefficient will maintain the maximum heating rate possible in the product, subject to the temperature difference requirement. It is shown later in this chapter that the faster the rate of heating, the higher the product quality; it is therefore important to ensure that the heat transfer coefficient is at the highest value possible.

Selection of the most suitable type of heat exchanger depends on the properties of the foodstuff, especially its viscosity, presence of solid particles, tendency to deposit or burn-on, and whether crystallization may occur in the product, as in margarine production. For products containing solid particles, a rule of thumb is that the size of the smallest part of the flow channel must be about three times the size of the largest solid particle, to prevent blockage.

The types of heat exchanger available can be subdivided according to the form taken by the heat transfer surface; plate, tubular, or scraped-surface heat exchangers can be used.

3.4 PLATE HEAT EXCHANGERS

Plate heat exchangers have been used for many years and are well established in continuous-flow thermal processes such as the pasteurization and UHT processing of milk, cream, ice cream mix, fruit juices, and so forth. The advantages of plate heat exchangers for these purposes are well established and these, with recent developments, have been reviewed by Müller-Steinhagen (1997). Plates can be designed to give high levels of induced turbulence in the flowing streams of product and heating or cooling medium through suitable profiling of the plates with corrugations: this, with the small channel size between the plates, gives high heat transfer coefficients, so that plate heat exchangers are compact for any required heating duty. Several separate heat exchanger sections performing different duties (e.g., heating, cooling, or regeneration) can be incorporated into a single frame assembly with suitable interconnections.

The construction of the plates usually requires plate-plate contact on parts of the corrugations to maintain rigidity of the plates and a uniform plate spacing. If the heat exchanger is used for heat treatment of liquids containing cell debris and fibers, such as fruit juices, the fibers will accumulate on these contact points and can lead to blockage of the flow channel. Special plate designs are available for these products where there are no contact points, but they require a slightly thicker metal plate to maintain rigidity.

One of the original advantages of the plate heat exchanger for the processing of liquid foods was that the stack of plates could be separated to give easy access to the heat exchange surfaces for manual cleaning, or to allow the effectiveness of circulation cleaning systems to be assessed routinely by visual examination. However, with improvements in cleaning-in-place (CIP) systems, plate heat exchangers are now not opened for routine cleaning, but are only opened infrequently for inspection. This is true even with UHT heat exchangers where fouling from the heated product and its effective removal during cleaning may be a problem.

3.5 PASTEURIZATION PROCESS

The simplest type of indirect thermal processing plant using plate heat exchangers is shown in Figure 3–4. Systems of this kind were among the earliest to be used for commercial pasteurization processes and are still widely used for low-viscosity liquids because of their characteristics such as simplicity and low cost.

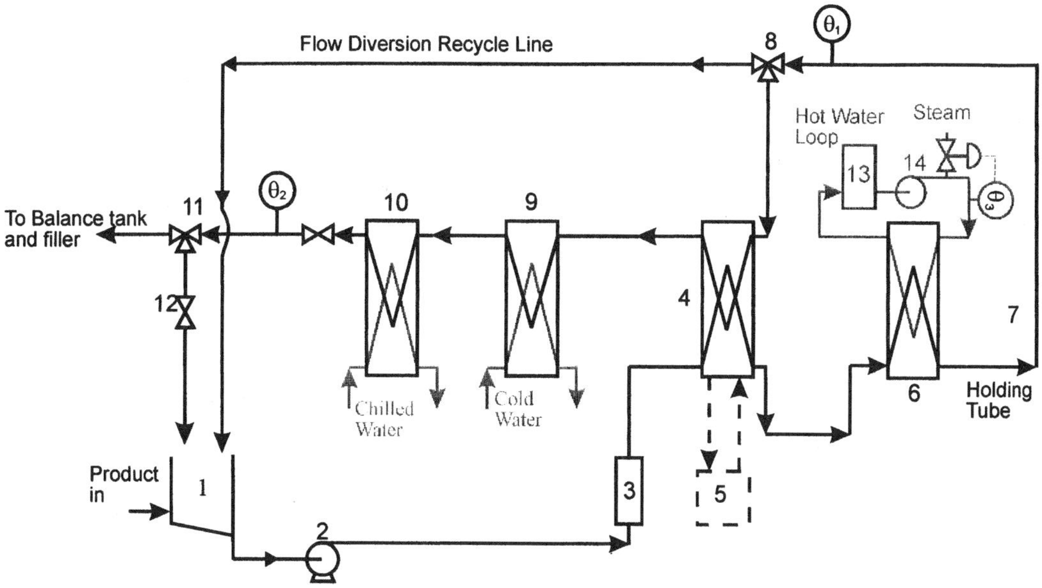

Figure 3–4 Pasteurization plant using plate heat exchangers

The untreated product is fed to a level-controlled balance tank (1) from which it is pumped by a centrifugal pump (2). The product then passes through a flow control device (3) to the first plate heat exchange section (4), where it is heated regeneratively by the outgoing product and, via filters if required, to the final heat exchange section (6) in which it is heated by recirculating hot water in countercurrent flow, or sometimes steam, to give the required process temperature. The temperature sensor (θ_1) and associated recorder, and the flow diversion valve (8), are situated at the end of the holding tube (7) and will recycle under-temperature product back to the balance tank. In normal operation, the product then passes to the regeneration section again (6), where it loses heat to the incoming product then passes to one or two cooling sections where it is cooled by cold (mains) water (9) and recirculating chilled water (10), if a low outlet temperature is required. The hot water in the final heat exchange section (6) heats the product in countercurrent flow and returns to the hot water balance tank (13), through the hot water feed pump (14), and is reheated to the required temperature (θ_3) by injection of steam before it enters the final heat exchange section again.

For products where homogenization is required, a homogenizer may be situated at point (5) at an intermediate point in the plate heat exchanger where the product temperature is suitable. In this case, the high-pressure homogenizer will act as a positive displacement pump and will control the process flowrate accurately, and the flow control device cannot be used and must be omitted.

The preprocessing heat treatment required for pasteurization plants is for temperatures of about 95°C for approximately 30 minutes and fortunately can be easily achieved at atmospheric pressure, but there is the complicating factor of the heat recovery section. To preheat-treat the equipment, water from the balance tank is heated using the final heating section (6), but as it passes from the holding tube through the heat recovery section again, its temperature falls to a few degrees of the incoming water temperature, and therefore as it passes through the two cooling sections (with cooling and chilled water turned off), its temperature is too low. This problem is overcome by diverting the water through valve (11) back to the holding tank and allowing the temperature in the holding tank build up to within a few degrees of that required for the preheat-treatment. The minimum temperature to which the process equipment has been exposed is shown by temperature (θ_2); once this has shown the minimum required temperature for the required time, the equipment can be cooled down by turning on the cooling and chilled water slowly and allowing the water temperature in the feed tank to return to ambient. Preprocessing heat treatment of the linkage between valve (11) and the filling equipment is covered in Chapter 7.

The level of water in the feed tank can now be allowed to fall and product can be introduced into the feed tank when it is nearly empty, but care must be taken as if its viscosity, thermal properties, and even its temperature are substantially different from the water, the temperature in the holding tube may fall below that required, and the flow diversion valve will operate. With experience, this fall in temperature can be predicted and allowed for.

3.6 STERILIZATION PROCESSES FOR LOW-VISCOSITY LIQUID PRODUCTS

The continuous-flow sterilization of low-viscosity foodstuffs is generally relatively well established and is not, in principle, too different from the pasteurization of such products outlined previously. Nearly all food products that can be classified as low viscosity can be sterilized using equipment similar to that used for pasteurization (also using plate heat exchangers), but the higher temperature-time heat treatments used mean that sterilization processes have more stringent requirements. The differences between the processing equipment required for pasteurization and for sterilization are now detailed, but the technology of the sterilization of low-viscosity products is covered in more detail in Chapter 6.

3.6.1 Heat Exchanger Specification

The principal difference between plate heat exchanger equipment used for pasteurization and that used for a sterilization process is that the latter must be able to withstand higher processing temperatures (up to 150°C, as against 65° to 85°C) and the associated higher internal pressures. The product in a UHT plant must not boil at the highest temperature reached, which will normally be the required sterilization temperature. Boiling with the evolution of vapor causes several problems. Vapor bubbles will displace liquid in the flow system, and thus reduce flow times: this will reduce the sterilization performance of the plant. In an indirect heating plant, boiling begins with the development of small vapor bubbles on the indirect heating surface where the temperature is highest. In a foodstuff prone to fouling, such as milk-based products, these bubbles encourage the formation of a thick layer of precipitated milk solids, which restricts product flow and interferes with heat transfer (see Chapter 8).

However, in addition, many foodstuffs may have a high content of dissolved air. Air dissolved in water displays reverse solubility, that is, as the temperature rises the air comes out of solution, until on boiling it is all removed. In an indirect heat exchanger, even before the equivalent boiling point is reached, small bubbles of dissolved air separate on the heating surfaces. The high induced turbulence in plate heat exchangers encourages this separation by causing local zones of high product velocity where low pressure and cavitation may occur. These bubbles of air have the same adverse effects, particularly on surface fouling, as do bubbles of steam at the boiling point. It is therefore necessary to apply a back-pressure over and above the equivalent vapor pressure in order to prevent the separation of air. Experience suggests that a pressure of at least 1 bar over and above the pressure determined by the sterilization temperature is needed (Burton, 1958).

Internal pressures of the order of 4 bar (60 lb/in^2) will therefore be required in the highest temperature sections of plate-type heat exchangers where the internal temperatures may be above 140°C. Because of the hydrodynamic pressure drop resulting from the product's being pumped through the heat exchanger, significantly higher pressures may be reached in other parts, particularly if the product is of higher viscosity. The highest internal pressure in the heat exchanger may reach 6 to 8 bar (90 to 120 lb/in^2).

To withstand these more severe conditions of temperature and pressure, the gasket materials used with plates for sterilizer duties must be more sophisticated, and more expensive, than those used for lower-temperature processes (Shore, 1970). New types of gasket material having improved physical properties at high temperature are introduced from time to time to give better performance. Medium nitrile rubber is suitable for temperatures up to about 135°C. For higher temperatures, up to 155° to 160°C, resin-cured butyl rubber or ethylene propylene diene methylene (EPDM) is suitable. Modern plates almost exclusively have gaskets mechanically held onto the plate by a variety of clip-in or clip-on designs. Replacement is very much more easy and rapid than for the old design plates, where the gaskets were glued in place and needed careful temperature curing before use.

High internal pressures, and varying high pressures within the heat exchanger, may cause flexing and distortion of the stainless steel plates themselves. While different manufacturers use different designs of plates and patterns of corrugations to encourage turbulence and improve heat transfer rates, current practice is to use designs with multiple contact points between adjacent plates, to give mutual mechanical support and increase the rigidity of the whole plate pack. For thermal treatment of liquid products containing fibers and cell debris, specially designed plates are available that do not have contact points on which the fibers can accumulate and cause eventual blockage of the flow channel.

3.7 STERILIZATION PROCESSES FOR LOW-VISCOSITY LIQUIDS

There are three differences in the heat exchanger arrangements for the pasteurization process given in Section 3.5 and a sterilization process, as follows:

1. To prevent boiling, a back-pressure must exist that is equal to the vapor pressure of the product at its maximum temperature. Since the product usually has a high water content, this is approximately equal to the vapor pressure of water at the sterilization temperature. At a temperature of 135°C, a back-pressure of about 2 bar (30 lb/in^2) is needed to prevent boiling. At a temperature of 150°C the pressure will have to be about 3.75 bar (55 lb/in^2). As discussed earlier, an excess pressure of 1 bar is desirable to prevent dissolution of air. Back-pressure is maintained by use of a back-pressure valve situated at the end of the regeneration section, when the temperature of the product is well below 100°C.
2. There is optionally a cooling section after the holding tube and before the regeneration section, which increases the rate of cooling and improves product quality, but reduces the regeneration efficiency.
3. Arrangements for preprocessing heat treatment of the process equipment are more complex to allow the higher temperatures required to be attained. Provision must at least be made for a poststerilization cooler and its back-pressure valve; the exact procedure for this is covered in Section 3.3.

A diagram of a typical sterilization process suitable only for low-viscosity liquids is given in Figure 3–5. The untreated product is fed to a small level-controlled balance tank (1) from which it is pumped by a suitable pump (2). The product then passes through the first heat exchanger (3), where it is heated regeneratively by the outgoing product. After this section, if required, the product may be homogenized (4). The next section is the final heat exchanger, where the required product temperature is achieved by steam heating or recirculating pressurized hot water. After the holding section, the product is partly cooled in a water cooling section (6) and then finally cooled to the storage and filling temperature by regeneration in section (3) so that the product leaves the heat exchanger at a tempera-

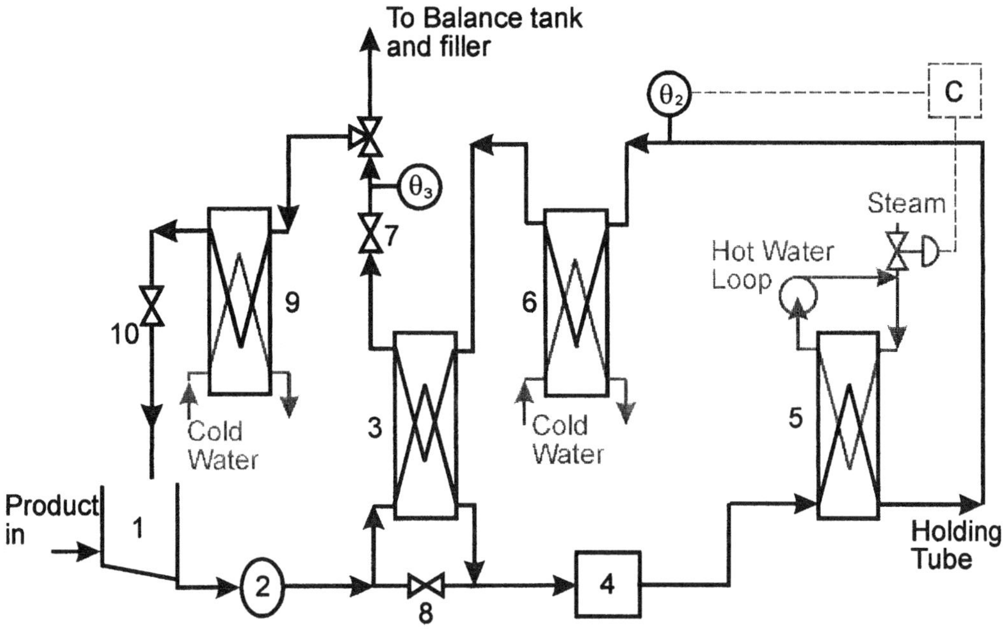

Figure 3–5 Typical sterilization process suitable for low-viscosity liquids. *Source:* Adapted with permission from H. Burton, *Ultra-High-Temperature Processing of Milk and Milk Products*, p. 82, © 1988.

ture a little above that of the untreated product. In a system of this kind, the degree of regeneration is relatively small, typically 60% to 65%.

As the product leaves the heat exchanger, it passes through a restriction (7), which may be either a preset valve or a suitably proportioned orifice plate. The pressure drop caused by the product flow through this valve provides the back-pressure, which, as already discussed, is needed to prevent boiling and air separation within the high temperature sections of the sterilizing system.

The holding time of the product at the required process temperature must be closely controlled and is determined by the flowrate of the product and the dimensions of the holding tube. The flowrate through the plant must be kept constant and is best achieved by a homogenizer, if used. If not, the feed pump (2) must be a positive displacement type; a constant flowrate is often achieved by a flow measurement system that controls the speed of rotation of the pump by a conventional control loop. The processing temperature is determined by a closed-loop temperature controller (C), which senses the temperature at the holding tube (θ_2) and controls the injection of steam into the hot water circulation system to maintain θ_2 constant. Variation of homogenizing temperature (θ_2), if required, is effected by varying the cooling water flow through the cooling section (6), to control the temperature reached by the incoming product at the end of the regeneration section (3).

The control system for the sterilization temperature of θ_2 can also be used to prevent underprocessed product passing to the filling system and so avoid the risk of keeping quality failures. In some early UHT systems, if the temperature was too low, a flow diversion valve similar to that used in pasteurization plants was actuated, and the underprocessed product was returned to the balance tank (1) for reprocessing. However, it is now realized that such recirculation has a bad effect on the flavor of the product and it is now more usual to divert any such underprocessed product away from the system entirely for use in some other way.

Preprocessing heat treatment of the equipment with hot, pressurized water must be at a much higher temperature than that for pasteurization, usually not less than 130°C, and hence a pressure of about 4 bar. However, the recirculation is through the balance tank, which is unpressurized, so the recirculating water must be reduced to a temperature less than 100°C after the finished product outlet line and before the balance tank. This is done in the sterilizing cooler (9), which reduces the sterilizing water temperature to about 80°C. A back-pressure valve (10) after the cooler retains the back-pressure in the whole of the sterilizing circuit. The circulating water can be brought to the plant sterilization temperature more quickly if the regeneration is made ineffective. This can be done by bypassing the heating side of the regenerator during sterilization by a valve (8). Any such bypass valve must be on the heating side: if it were on the cooling side of the regenerator, some part of the downstream side of the plant that carries sterile product would not be effectively sterilized.

It is not unknown for pinholes to develop in heat exchanger plates through corrosion, and if a pinhole should occur in the regenerator section and the pressure of the product on the inlet side were to be higher than that on the outlet side, the processed product might be contaminated by the untreated product. With pasteurization plants, some countries require that an additional circulating pump be used to ensure that the pressure of the finished product is higher than that of the untreated product, so that contamination of the finished product by untreated product cannot occur. In the plant shown in Figure 3–5, a homogenizer (or an additional PD feed pump) is fitted in position (4) and therefore feed pump (2) must be of the centrifugal type. The restrictor (7) can then be used to ensure that the back-pressure is higher than the delivery pressure of the feed pump (2), but not the homogenizer/PD feed pump (4), so that any flow through undetected pinholes is from the sterile to the nonsterile side and contamination is avoided.

A more advanced type of indirectly heated plant in which plate heat exchange sections may be used is shown in Figure 3–6. In this design, all heating and cooling of the product is performed through a quite separate circulating water circuit.

The untreated product is pumped from the level-controlled balance tank (1) by a centrifugal pump or, for some products that may be damaged physically by shear, by a positive displacement pump (2), to the first plate heat exchange section (3). At the end of this heating section, there may be a holding tube (4) to provide a few seconds' holding at an elevated temperature before further heating, for those foodstuffs where this would reduce fouling of the heating surfaces by the product at higher temperatures later in the plant (see Chapter 8). A second heating section (5) brings the product to the final sterilizing temperature, where it is held for the required holding time in the holding tube (6). The first and second regenerative cooling sections (7 and 9), together with a final cold water cooling section (10), bring the product to a temperature around ambient, before it passes through a restrictor (11) to provide back-pressure, as described previously, to the aseptic tank and packaging system.

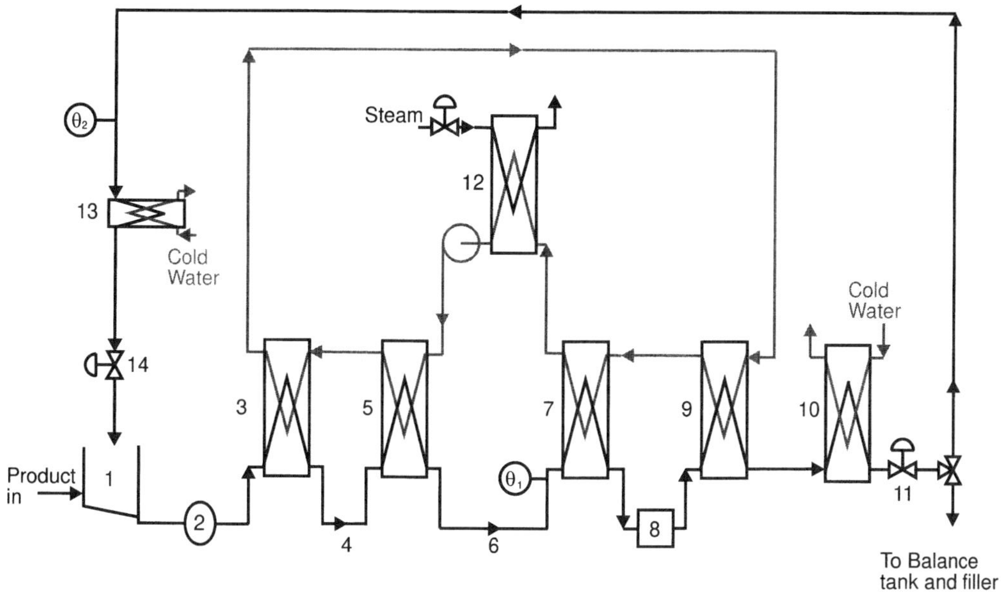

Figure 3–6 Advanced plate-type indirect heating plant. *Source:* Adapted with permission from H. Burton, *Ultra-High-Temperature Processing of Milk and Milk Products*, p. 86, © 1988.

For products that require homogenization, the homogenizer (8) may be situated before heating to the final processing temperature, as shown in Figure 3–5, or, where homogenization after heat treatment would give greater product stability, may be situated between cooling sections (7) and (9) as shown in Figure 3–6. This is a particular advantage in products such as cream, but, as the homogenizer operates on sterile product, it is complex and needs to be of an aseptic design; it also requires careful operation to avoid recontamination of the product.

In the heat exchanger sections (3), (5), (7), and (9), heat is transferred between the product and water circulating in countercurrent flow in a closed circuit. The water in this circuit is heated to a little above the required sterilization temperature by steam under pressure in a heat exchanger section (12). The water heats the product in section (5) to the sterilization temperature, and then passes to section (3) to provide preheating. The circulating water has now been cooled to a low temperature while heating the product, so it can be used in sections (9) and (7) to cool the outgoing product while being at the same time reheated. It therefore acts as an intermediate for regenerative heat transfer in the system, and the amount of heat to be supplied by steam in (12) is relatively small. The level of regeneration in such a plant can be above 90%, and the operating energy costs are correspondingly low.

In the system as shown in Figure 3–6, the heating of the recirculating water by steam is indirect through a heat exchanger section. However, an alternative is to heat the water by the direct injection of steam. In both of these variants, the sterilization temperature is controlled by sensing the product temperature at the holding tube (see Figure 3–5), and controlling the supply of steam that is heating, indirectly or directly, the circulating water. Other details are common to the simpler and more complex plants. For example, recirculation of water through the product lines during plant sterilization requires a sterilization cooler (13) and back-pressure valve (14) immediately before the return to the balance tank.

Contamination of sterile product through pinholes in the plates can be avoided in a simpler way than with the system of Figure 3–5, since untreated and treated product are never on opposite sides of the same plate. Transfer of contamination through the water circuit is prevented in two ways: the product pressure can be kept higher than the water pressure, and the water is resterilized at each recirculation.

The two main advantages claimed for this type of plant as compared with the simpler type shown in Figure 3–5 are, first, the higher level of heat regeneration that can be obtained and, second, the lower temperature differentials that can be maintained between the heating medium and the product being heated. It is claimed that the temperature differential need not be more than 3°C at any part of the plant. Such a low differential reduces the amount of fouling on the heat exchange surfaces and will allow longer operating times before the heat exchanger needs to be shut down to be cleaned. Operating times with these fouling products may be up to twice those for the simpler plate-type indirect systems. It has also been claimed that the low temperature differentials lead to a better-quality product because the amount of overheating is less. This claim must be considered unproven; in fact, the flavor of the product may be rather poorer because the product spends a longer overall time at elevated temperature. These effects are considered in more detail in Section 3.21 in a discussion of plant performance.

3.8 TUBULAR HEAT EXCHANGERS

There are many applications where plate heat exchangers are not suitable and a tubular heat exchanger system is to be preferred. The relatively large flow channel, compared to plate heat exchangers, means that the equipment can handle higher-viscosity liquids or liquids containing high levels of pulp or fibers or even particulate solids, but the heat transfer area is lower in relation to the volume of liquid in the equipment and therefore results in a slower heating rate.

When tubular heat exchangers are used in the food industry, as opposed to the chemical process industries, they are mainly of two types, concentric tubes or shell-and-tube heat exchanger. In addition, tubes may be either straight-walled or have a corrugated wall in a spiral-wound construction, as shown in Figure 3–7. The use of corrugated tubes instead of straight-wall tubes is becoming more common; the corrugations increase turbulence in the product and therefore increase heat transfer. For viscous products, the corrugations may actually induce turbulence where flow would normally be streamline, with an even greater effect on heat transfer rate. Increases of up to 30% in heat transfer coefficient are claimed, with the same reduction in area required for a given duty.

3.8.1 Concentric Tube Heat Exchangers

Concentric tube heat exchangers may be constructed of either straight or corrugated tubing and may consist of double, triple, or more concentric tubes with associated entry and exit

(a)

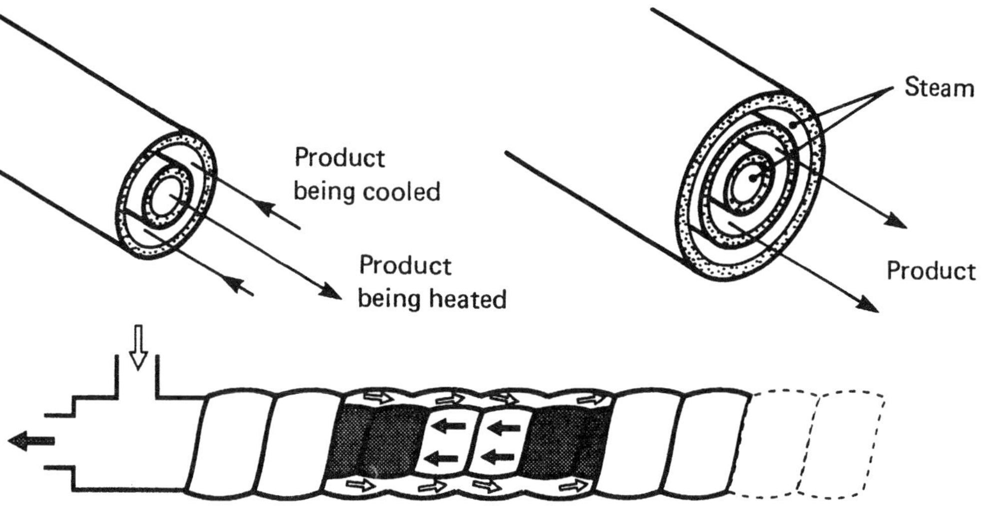

(b)

Figure 3–7 Types of concentric tube heat exchangers (a) plain-wall; (b) corrugated, spiral-wound. *Source:* Part a reprinted with permission from H. Burton, *Ultra-High-Temperature Processing of Milk and Milk Products*, p. 89, © 1988.

ports for the product and heating/cooling medium. Where there are three or more concentric tubes, the heating/cooling medium and the product flow in alternate annuli to give the largest heat transfer area between the two, with heat transfer into the product from both walls of its annulus. Double-tube systems are suitable for products with large particulate solids or high pulp or fiber contents, due to the large product channel, or are often used for the regeneration section.

One design of corrugated tube heat exchanger allows the individual corrugated tubes of the product channel to be rotated, allowing the corrugations to be in phase or out of phase. If out of phase, the convoluted channel shape permits greater turbulence in the product and therefore better heat transfer, but will result in a larger pressure drop and may cause blockage if the product contains fibers or small particulate solids. For the latter products, the corrugations would be in phase.

Concentric tube heat exchangers are formed by assembling two or three stainless steel tube lengths one inside another with a spacer in each inter-tube space to maintain them concentric. The tubes may be sealed with O-rings or other packing at each end, which will allow easy disassembly for cleaning and inspection, or may be of a fully welded construction to allow high pressures.

In one of the most common designs of straight tube, concentric tube heat exchangers, the concentric tubes are formed in a long length and then wound into a large-diameter helical coil, for assembly into an outer cylindrical housing for hygienic and mechanical protection. A triple-tube system is used for the final stage of heating to the sterilization temperature and in the final cooling sections, especially where cooling rates are restricted by high product viscosity, so that increased transfer area is used to compensate for reduced heat transfer coefficients. A double-tube system is used for the regeneration stage.

A typical flow diagram for an indirect continuous-flow thermal processing plant using concentric tube heat exchanger sections is shown in Figure 3–8. The product is pumped from the balance tank (1) by a centrifugal pump (2), through the steam-heated sterilizing heater, which, during product sterilization, is inactive. The product is first heated in the regenerative heater (4) by the outgoing product. If required, connections may be made to a homogenizer (5) at a suitable temperature point within the regenerator. However, an advantage of tubular heat exchangers is that the tubes can be strong enough to withstand full homogenization pressure (200 MPa), and it is therefore possible to install the reciprocating pump section of the homogenizer before the holding tube and the homogenizing valve somewhere further downstream. The reciprocating pump need not be of an aseptic design as it is sited in the nonsterile part of the process, but the homogenization valve can be positioned in the optimum position for homogenization, either pre- or poststerilization. As the major contamination in homogenizers is through the pump pistons and valve seats, not the homogenization valve, this arrangement eliminates potential contamination from this source. In addition, in a plant of the type shown, two homogenizing valves may be fitted, one at (5) in association

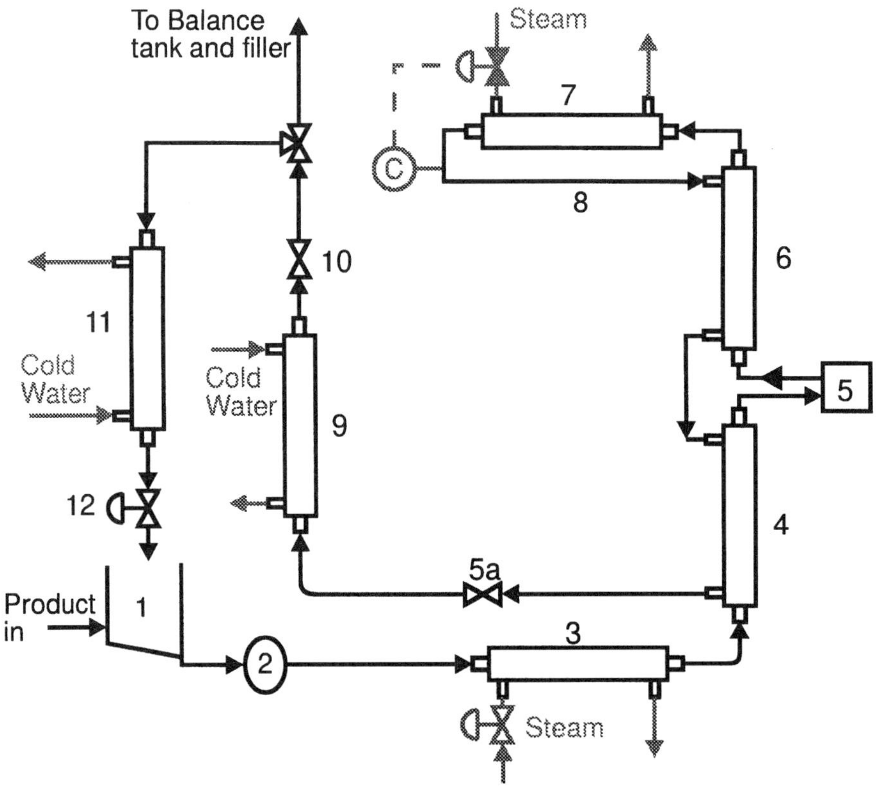

Figure 3–8 Typical concentric tube indirect heating continuous-flow thermal processing plant. *Source:* Adapted with permission from H. Burton, *Ultra-High-Temperature Processing of Milk and Milk Products*, p. 90, © 1988.

with the homogenizer pump and one after the sterile section where the product is at a suitable temperature during cooling, for example, at (5a). The product can therefore be homogenized before sterilization, after sterilization, or both.

After the first homogenizing point at (5), heating continues in the regenerator (6) and then in the steam-heated section (7) to the final sterilizing temperature. After the holding tube (8), the product returns to the regenerator for cooling. Final cooling to the outlet temperature is in a water-cooled section (9). Back-pressure is maintained by an orifice plate or back-pressure valve at (10). The required sterilization temperature in the holding tube is maintained by a controller (C), which senses the temperature in the holding tube and varies the steam supply to the final steam heating section (7) as necessary.

As in the plants we have considered previously, during plant sterilization by circulating hot water, the circulating water is cooled to below 100°C in the sterilization cooler (11) and passes through a restrictor (12) before the water returns to the unpressurized balance tank. In this system, during plant sterilization only, additional heat is supplied through the sterilizing heater (3): this allows the correct sterilizing temperature in the circulating water to be reached quickly and avoids cooling below the sterilization temperature in the regenerator. The mechanical strength of stainless steel tubes, and their resistance to corrosion, normally ensure that the risk of contamination of treated product by untreated product through pinholes is negligible.

3.8.2 Shell-and-Tube Tubular Heat Exchangers

In the second general type of tubular heat exchanger, several alternative forms of the shell-and-tube system may be used. In one type, straight lengths of small-diameter stainless steel tubes are assembled in multiples within an outer tube; five to seven of the smaller tubes having an internal diameter typically of 10 to 15 mm may be used in a single outer tube. The smaller tubes are connected by a manifold at each end of the large outer tube, and carry the product in parallel flow paths. The heating or cooling medium passes in countercurrent flow in the space round them (Figure 3–9(a)). Single straight units of tube assemblies can be connected in series with 180° bends at each end to give any required amount of heat exchange surface.

Similarly, this type of heat exchanger is available with corrugated tubes (Figure 3–9(d)); again, a decrease in heat transfer area of up to 30% over straight-walled tubes is claimed.

The flow system for a continuous-flow thermal processing plant using tubular heat exchange sections of this type is shown in Figure 3–10. This plant layout is very similar to that given for plate heat exchange sections in Figure 3–6, but using the steam-injection alternative as the method for heating the circulating water in the auxiliary circuit. A special feature is the use of a single drive (A) for the PD feed pump (or homogenizer) and for the pump that circulates the water in the auxiliary circuit. Change in the capacity of the plant by changing the feed pump speed to give a higher product flowrate is therefore associated directly with a corresponding change in the flowrate of the heating/cooling medium; the ratio of the flow velocities on the heating and cooling sides of the tubes is therefore constant. This is claimed to be an advantage in controlling the effectiveness of the heat treatment.

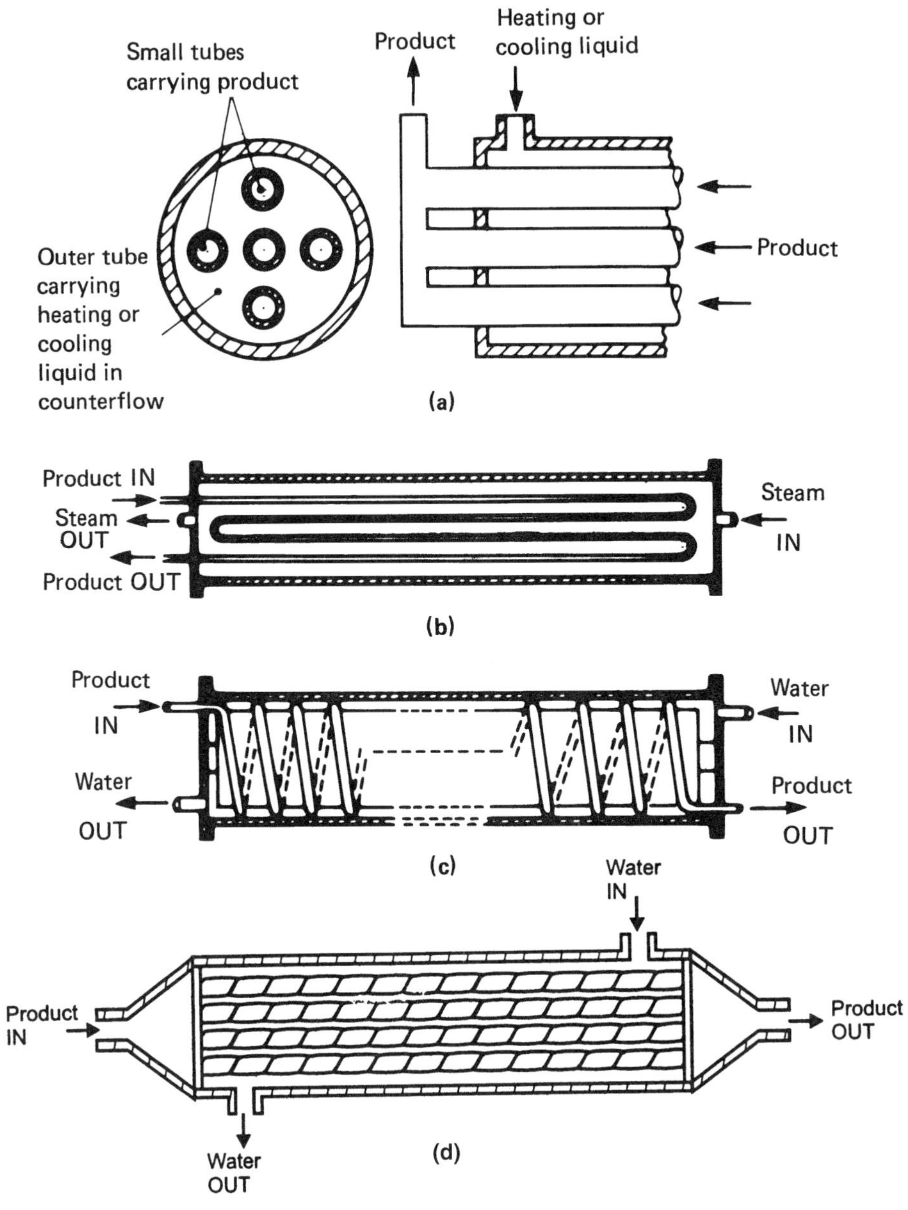

Figure 3–9 Types of shell-and-tube heat exchangers used in continuous-flow thermal processing plant. (a), Parallel-flow multiple plain-wall tubes. (b), "Trombone"-type single tube. (c), Spiral configuration single tube. (d), Parallel-flow multiple corrugated-wall tubes. *Source:* Parts a, b, and c reprinted with permission from H. Burton, *Ultra-High-Temperature Processing of Milk and Milk Products*, p. 92, © 1988.

Product is pumped from the balance tank (1) through the first preheating tubular heat exchange section (2) to give a homogenizing temperature of about 75°C at the homogenizer (A), if required. It then passes through a second preheating section (3); to a holding section (4) at a temperature of 90° to 95°C. A final heating stage (5) brings the product to the final

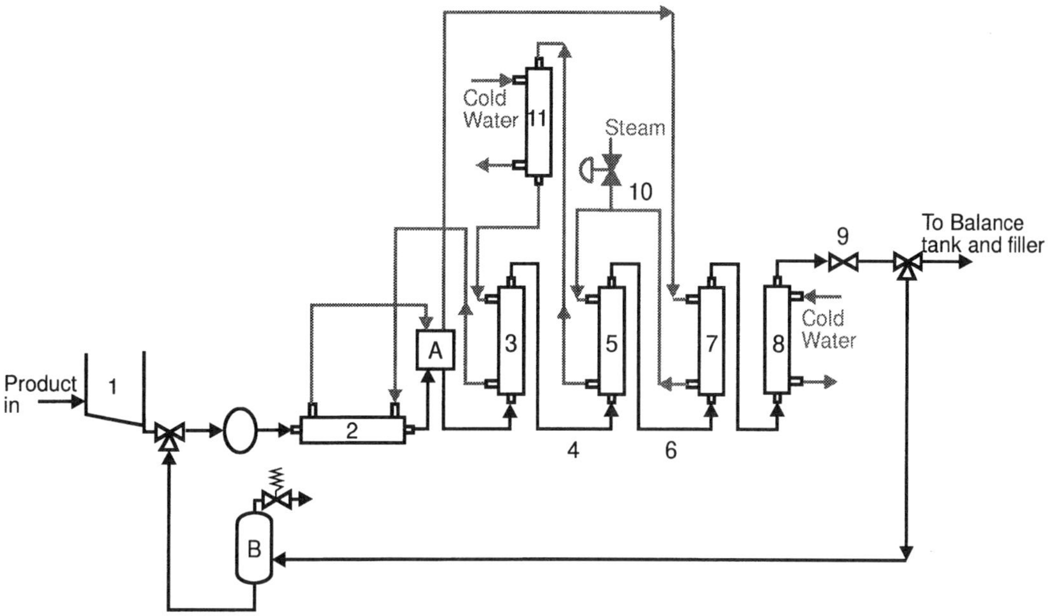

Figure 3–10 Continuous-flow thermal processing plant using shell-and-tube heat exchangers throughout. *Source:* Adapted with permission from H. Burton, *Ultra-High-Temperature Processing of Milk and Milk Products*, p. 93, © 1988.

sterilization temperature before it passes into the holding tube (6). Regenerative cooling is in section (7), and final water cooling in section (8). A restrictor (9) provides the necessary back-pressure in the sterilizer. Regenerative heat exchange is through the closed water circuit, which heats the product in heat exchanger sections (2), (3), and (5), and cools it in section (7). The water circulation rate is governed by the duplex pump (A), which is associated with the homogenizer. The final sterilization temperature is controlled by steam-injection heating (10) of the circulating water, and the temperature at the preholding section (4) is determined by means of a cooler (11). A regeneration level of 90% or more can be obtained with this system, as with the similar systems using plate heat exchange.

In the circuit shown, the plant cleaning and sterilizing return line does not pass through a sterilizing cooler to the balance tank as in the systems previously described. Instead, a pressurized circuit is completed through three-way valves, incorporating an expansion vessel (B) to allow for expansion of the circulating liquid as the temperature rises. With this type of tubular plant having multiple tube assemblies that are separate but connected in series, change of flow rate need not increase or decrease the flow time, thus increasing or decreasing the severity of the heat treatment. The necessary tube assemblies can be added or removed to keep the overall processing time the same.

In another system, tubular heat exchange sections are used in the high-temperature sterilizing parts of the plant and for the subsequent cooling of the sterile product where freedom from contamination is essential. Different types of heat exchanger are used, according to whether they are used for heating or cooling, but both are forms of shell-and-tube construction. The heat exchangers used for the final stages of heating consist of lengths of stainless steel tube of suitable diameter folded into "hairpin" or "trombone" forms, and contained in a cylinder pressurized with steam (Figure 3–9(b)). The exchangers used for cooling of the sterile product have spiral-wound tubes in an annular space at the outside of a cylinder (Figure 3–9(c)). The cooling medium, normally water, passes in countercurrent flow along the annular space.

A typical continuous flow thermal processing system composed of these units is shown in Figure 3–11. The product is pumped from a level-controlled balance tank (1) by a centrifugal pump (2) through a plate-type heat exchanger (3), where it is preheated to about 120°C by water flowing in a closed auxiliary circuit providing regeneration. The product is then pumped with a second centrifugal pump (4) to a high-pressure piston pump (5), which supplies both the hydrodynamic pressure for the rest of the system and the homogenization pressure. A tubular heater (6) of the type shown in Figure 3–9(b) heats the product to the final sterilizing temperature by means of steam under pressure. After the holding tube (7), the product passes to a series of heat exchangers of the type shown in Figure 3–9 (c), connected in series, where the product is cooled by circulating water (8). The homogenization of the product takes place after sterilization, making use of the ability with tubular plants to separate the homogenizing valve from a nonaseptic high-pressure pump as described earlier. The valve (9) is placed between two of the cooling units, where the product temperature is about 70°C. The product leaves the final cooling unit for filling at about 25°C.

With the homogenizing valve in this position, the back-pressure in the highest-temperature sections of the plant is the full homogenizing pressure, so no separate back-pressure valve is needed. The circulating water transfers heat from the tubular coolers (8) to the plate-

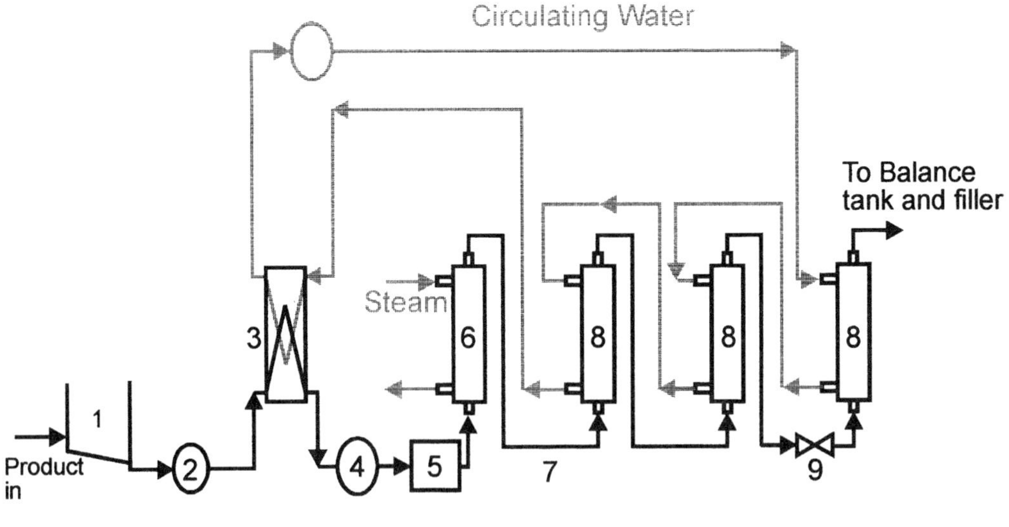

Figure 3–11 Indirect UHT plant using tubular heat exchangers for steam heating and cooling of product but plate heat exchanger for heat recovery. *Source:* Adapted with permission from H. Burton, *Ultra-High-Temperature Processing of Milk and Milk Products*, p. 95, © 1988.

type preheater (3), to give up to 90% heat regeneration. Since the auxiliary circuit does not have to heat the product to the full sterilization temperature, no additional heating of the circulating water is needed. In fact, if the degree of regeneration is to be limited in order to increase the rates of heating and cooling of the product as it passes through the system, heat may be extracted from the recirculating water with an additional tubular cooler. A water cooler will then be needed for the product after the regenerative coolers.

3.9 SCRAPED-SURFACE HEAT EXCHANGERS

The scraped-surface heat exchanger (SSHE), sometimes called a swept-surface heat exchanger, consists of a cylinder in which is the foodstuff and an outer jacket through which the heating or cooling medium flows. In the center of the cylinder is a driven shaft (the mutator) supporting scraper blades. The product is pumped through the space between the cylinder and shaft, contacts the heated (or cooled) cylinder surface, and is continuously scraped off and mixed into the bulk of the product by the scraper blades (Figure 3–12). The heat exchanger has several disadvantages and should only be used when the foodstuff to be heated is extremely viscous or sticky, contains large solid particulates, or is one in which crystallization may occur. The maximum viscosity that can be handled is very high, and is generally accepted as the maximum viscosity that can be physically pumped. The mutator shaft can be of different sizes; a large diameter gives a small annular space and is best for viscous liquids and liquids with small solid particulates, while a small diameter shaft can accommodate large solid particulates in the liquid, usually up to a maximum of about 25 mm. The largest-size shaft possible should be selected with regard to the product, since the larger the shaft, the smaller the product holdup and the more rapid the heating rate.

Disadvantages of the SSHE are in the higher capital and operating costs of the units. They are physically more complicated to construct, with moving parts and seals in the product area, giving greater wear, and hence require a greater degree of maintenance. Seals may need to be of aseptic design, especially if used for cooling duties downstream from the holding tube when recontamination is a danger. Energy utilization is generally poor, as heat recovery is not normally possible, and would be quite poor anyway as high differential temperatures between product and heating or cooling medium are generally necessary. Energy is also required to drive the mutator and will heat the foodstuff to an extent due to viscous dissipation. For cooling duties, this adds to the heat load required for the heat exchanger and can reduce the apparent heat transfer coefficient (Page, Grandison, & Lewis, 1997).

When processing particulate products, there is a possibility of physical damage to the solids from the rotating mutator, especially from the supports for the blades, and from sudden acceleration from the slow-moving inlet port through a right angle into the scraped-surface section. This can be minimized by slowing the mutator and increasing the number of blades, reducing the number and profile of the blade supports, and by careful design of the inlet ports so that the product enters and leaves at a tangent to, and with slow acceleration into, the scraped-surface zone. The speed of the mutator must normally be optimized to get the best compromise between product damage and a fast rate of heating.

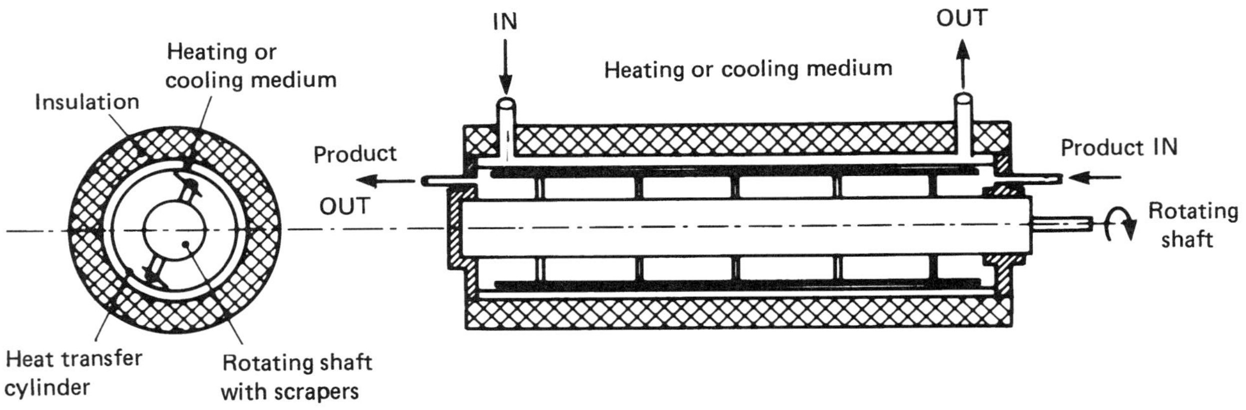

Figure 3–12 Scraped-surface heat exchanger. *Source:* Reprinted with permission from H. Burton, *Ultra-High-Temperature Processing of Milk and Milk Products*, p. 97, © 1988.

The materials used for the heater barrel and the scraper blades must be compatible with the product to be heated and with each other. For example, the cylinder materials can be chromium-plated nickel, stainless steel, or a suitable alloy. The scraper blades must not cause wear when in contact with the cylinder; for example, stainless steel blades cannot be used with a stainless steel cylinder.

As mentioned above, scraped-surface heat exchangers cannot be used for regeneration and therefore the flow arrangements for a continuous-flow thermal processing system using them are of the simplest, as shown in Figure 3–13. The product is pumped from a suitable balance or mixing tank (1) by an appropriate pump, almost exclusively a PD type, to one or more heat exchangers (3), in which the heating medium in the annular chamber will be steam or pressurized hot water to give rapid heating. After the holding tube (4), the product passes through scraped-surface heat exchangers used for cooling by means of chilled water, glycol, or even liquid refrigerant as part of a refrigeration cycle (5). As the available temperature difference for cooling is considerably less than that available for heating, more heat exchanger units normally will be used for cooling to provide more heat exchange surface. Finally, the product passes through a back-pressure device (6), which is unlikely to be a simple orifice or valve but is either another PD pump operated at a speed that will maintain pressure, or a sterile tank pressurized with compressed sterile air to the required operating pressure to prevent boiling. These arrangements are covered in detail in Section 4.10.

3.10 COMPARISON OF INDIRECT HEAT EXCHANGERS

The selection of a suitable indirect heat exchanger for a given foodstuff is very much dependent on the properties of the foodstuff and the presence of solid particles. It is shown in Section 3.22 that the rate of heating is an important factor in the quality of sterile foods from continuous-flow thermal processing systems. Factors that are important in maximizing the rate of heating are the ratio of heat transfer area to volume of foodstuff held in the equipment and the degree of turbulence in the liquid, which affects the heat transfer coefficient described above. The broad differences between the generic types of heat exchangers available are described below, but the differences in performance arising from different time-temperature profiles within the same type of heat exchangers are discussed later in this chapter.

Where possible, plate heat exchangers would be the first choice on the grounds of product quality, energy efficiency, and maintenance costs. They provide large heat transfer areas within a physically small plant size and also have a small volume holdup due to the small gap between the plates through which the liquid flows. Furthermore, there is a high level of induced turbulence in the foodstuff promoted by the shape of this flow channel (a function of the plate profile), which gives high heat transfer coefficients. Plates therefore are, in general, the cheapest way of obtaining a given amount of heating or cooling performance. However, plates need gaskets that limit the temperature of operation and the pressure that can be retained in the heat exchanger system.

Tubular heat exchangers are generally more suited to medium- and higher-viscosity liquids, or liquids with small solid particles, up to about 10 to 25 mm size. The larger cross-sectional area of the tubes allows a smaller pressure drop for medium- and high-viscosity

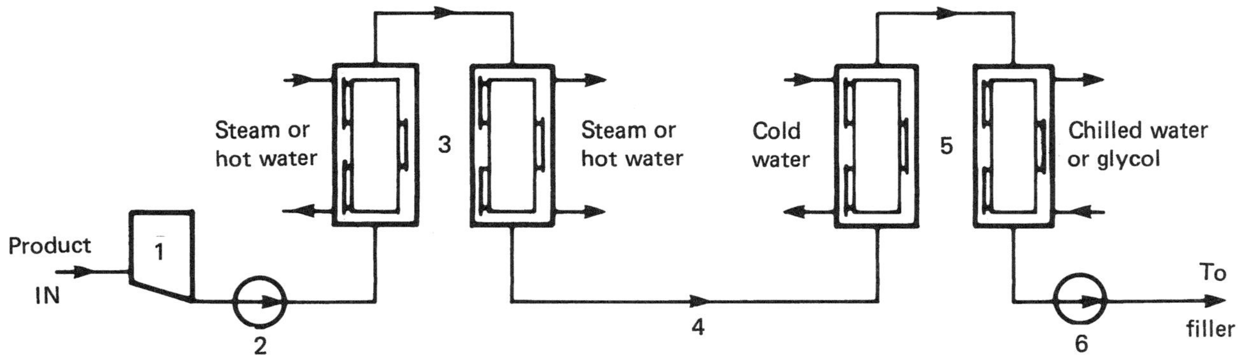

Figure 3–13 Continuous-flow thermal processing system using scraped-surface heat exchangers. *Source:* Reprinted with permission from H. Burton, *Ultra-High-Temperature Processing of Milk and Milk Products*, p. 98, © 1988.

liquids as compared to plates, and allows the passage of small solid particulates without the possibility of their blocking the channel. However, because of this, the volume holdup is generally larger for a given heat exchange area than for plates and results in lower rates of heating.

Tubular heat exchangers with plain walls rely solely on natural turbulence from high Reynolds numbers to obtain the highest heat transfer coefficient possible. The use of corrugated tubes can improve the heat transfer coefficient by up to 30% by inducing turbulence at the cylinder wall and is a definite improvement over plain-wall types. The tube walls are also normally thicker than heat exchanger plates and are much more robust, making them able to withstand much higher internal pressures. The lack of gaskets (or substantial reduction in size of the gasket, situated at the end of individual tube lengths) also allows high internal product pressures up to about 200 to 300 bar (3,000 to 4,500 psi gauge [psig]), which means that they will withstand full homogenization pressure and the homogenizing valve can be physically separate from the high-pressure driving pump. It is therefore possible to place the homogenizing valve at a point in the circuit after the product is sterilized, without the cost and maintenance problems of an aseptic high-pressure pump.

Not all products need homogenization after indirect sterilization; it is doubtful whether there is any advantage in this for normal products, for example. Cream, on the other hand, is often considered to benefit from homogenization after sterilization, to avoid reassociation of small fat globules as a result of high-temperature processing after homogenization. If flexibility in the placing of the homogenization valve is likely to be a commercial advantage, a suitable tubular heat exchanger may be considered preferable to a plate heat exchanger with an aseptic homogenizer.

Plate heat exchangers are sensitive to fouling by the deposition of solids from the product on hot plate surfaces. Because of the small spacing between plates, a fouling layer rapidly obstructs the flow of product through the heat exchanger. To maintain a constant product flowrate, the driving pressure must increase, and after some time the pressure may reach the limit of internal pressure set by the plate gaskets. Some action must then be taken, normally a break in processing to clean the system.

Tubular heat exchangers are equally liable to fouling, and perhaps are even more sensitive than plate exchangers because the temperature differential between product and heating medium may be higher. However, the effect of fouling on heat exchanger operation is less with tubes. First, the geometry of a tube system, whether it is formed from concentric tubes or whether it is one of the shell-and-tube types, makes it less sensitive to a layer of fouling material: there is more room for a layer to form without causing an undue restriction of the product flow passage. Second, when the fouling layer begins to restrict the flow and the pressure drop through the heat exchanger correspondingly increases, the internal pressure can rise much higher before problems arise. In fact, with tubular heat exchangers the limiting factor with fouling is inability to reach the required processing temperature because of reduction in overall heat transfer coefficient, rather than excessive internal pressures. If fouling has reached this level, problems may be encountered with plant cleaning. In an emergency, a plate-type heat exchanger can always be opened and cleaned by hand; this option is not always available with tubular systems.

The danger of contamination of product through corrosion pinholes in heat exchange surfaces has already been mentioned. The risk of pinholes is higher in plates than in tubes, because the metal is thinner and crevice corrosion may occur at the interplate contact points. Corrosion at this level may be unlikely with the thicker metal of tubular heat exchangers, but whereas it is relatively easy to examine individual plates and relatively cheap to replace them, long tubes are difficult to examine and must be replaced in their entirety. In situations where any pinhole leakage would cause a public health hazard, there are special plates designed to eliminate this. They consist of a pair of nonwelded plates with a gap between them, into which leakage will pass out to the atmosphere.

Scraped-surface heat exchangers would be used only where necessary, for handling high-viscosity liquids; liquids with large particulates; and sticky, temperature-sensitive, or crystallizing foodstuffs. They may, however, give more rapid heating or cooling than tubular heat exchangers under some circumstances with medium-viscosity products, and this may sometimes give a better-quality product. In particular, there may be a problem when cooling some products with a medium-viscosity at elevated temperatures but whose viscosity increases dramatically as its temperature is reduced, especially starch-based products. The product nearest the tube wall approaches the temperature of the cooling medium and its viscosity may then be so large that it becomes static and the remaining warm product channels down the tube center. Heat transfer is then very much reduced as the cold static product acts almost as a form of fouling. The use of scraped-surface heat exchangers obviously would be preferred for these products.

DIRECT HEATING SYSTEMS

With direct heating systems, the product is heated to sterilization temperature by mixing it with steam. Some of the steam condenses, giving up its latent heat of vaporization to the product and giving a much more rapid rate of heating than is available with any indirect system. The steam may be injected into the product, known as steam injection or steam-into-product, or the product may be pumped into a steam chamber as a curtain or spray, known as infusion or product-into-steam. These two methods give different characteristics and are considered separately.

This method of heating can cause considerable dilution of the product. For example, a 60°C temperature rise will involve about 11% addition of water as condensed steam. The design of the holding section for a desired residence time must therefore take into account that the flow rate within it will be greater than the inlet product flow rate.

Cooling of the product may be by any indirect heat exchanger that is suitable to the product characteristics as outlined above, and therefore the added water will remain in the product and allowance for the dilution must be made in the product formulation. The steam quality is important, and contamination of the food product by steam pipe corrosion or boiler feed water additives must be eliminated.

For those food products where the added water is not acceptable, or where very rapid cooling is required, "flash cooling" may be used. In flash cooling, the product is expanded through a restrictor into an expansion cooling vessel, usually a cyclone, where the pressure is at a level below atmospheric. Water in the product immediately boils, giving water vapor, which is separated from the remaining liquid, removed from the vessel, and condensed. The product goes through very rapid cooling due to loss of the latent heat of vaporization. In addition, dissolved gases and any volatile or flavor components from the product may also be removed. The amount of water removed, and the temperature the product is cooled to, may be controlled by the level of vacuum in the cooling vessel. It is obvious that, as the restriction device is either a valve or an orifice plate, products that are shear sensitive, or that contain discrete particulate solids, cannot be cooled using this method.

The thermal efficiency of adirect heating system may be increased by indirect regenerative heating before the product is mixed with steam; this may be achieved by heat transfer from the outgoing product after the expansion cooling stage, by the recovery of heat from the water vapor liberated in the expansion vessel, or by using the condenser cooling water on exit from the condenser, when it has been heated. For reasons that are explained later, the product temperature before mixing with steam must be closely controlled, and to do this there is a stage of indirect preheating when the outlet temperature can be closely controlled before the direct heating stage. It is also normal with direct heating to homogenize after sterilization, when homogenization is required. The overall process is shown in simplified form in Figure 3–14.

A typical temperature-time curve for a direct heating continuous-flow thermal process is similar to that in Figure 3–15: indirect heating is often used to raise the temperature to 80° to 85°C, followed by direct heating to give the final sterilizing temperature. After the designated holding time at the sterilization temperature, expansion cooling follows where the vacuum in the expansion vessel is normally controlled at a level corresponding to a boiling temperature a little above that of the product before mixing with steam. This vacuum level will remove the same amount of water added as condensed steam. This is explained later when the overall water balance during direct heating is considered.

The differences between the two direct heating methods, injection and infusion, are now covered in detail.

3.11 THE STEAM INJECTION SYSTEM

The steam injector, or steam-into-product system, is the heart of this system. There are many different injector designs, developed and used by different manufacturers, but they all aim to meet the same set of requirements with the least cost and minimum complication. The principal requirement is the rapid condensation of steam, to give rapid heating, and to prevent the passage of bubbles of uncondensed steam into the holding tube, which would reduce the effective holding time at the sterilization temperature by displacing liquid and increasing the effective volume throughput of product in the holding tube. Rapid condensation is achieved by introducing the steam into the liquid in the form of small bubbles or in the form of a thin sheet that has easy access to the liquid at both sides. It is also encouraged by having an adequate back-pressure in the liquid, above that needed to prevent boiling, at the steam injector.

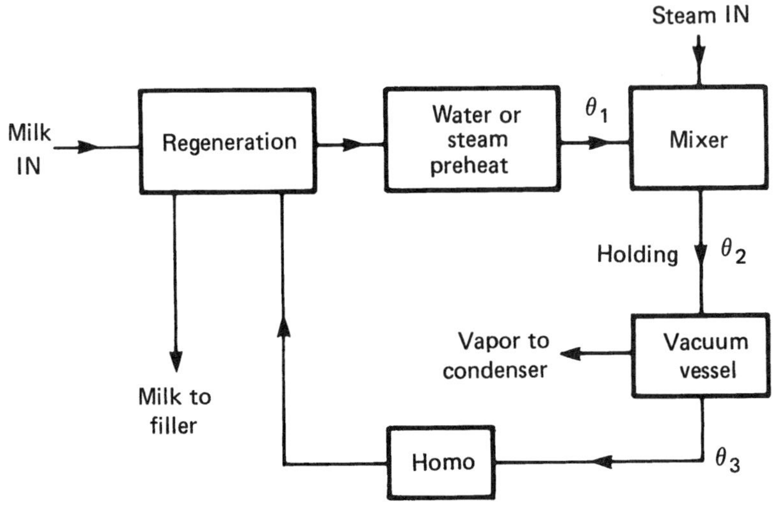

Figure 3–14 Diagrammatic representation of a direct heating system. *Source:* Reprinted with permission from H. Burton, *Ultra-High-Temperature Processing of Milk and Milk Products*, p. 103, © 1988.

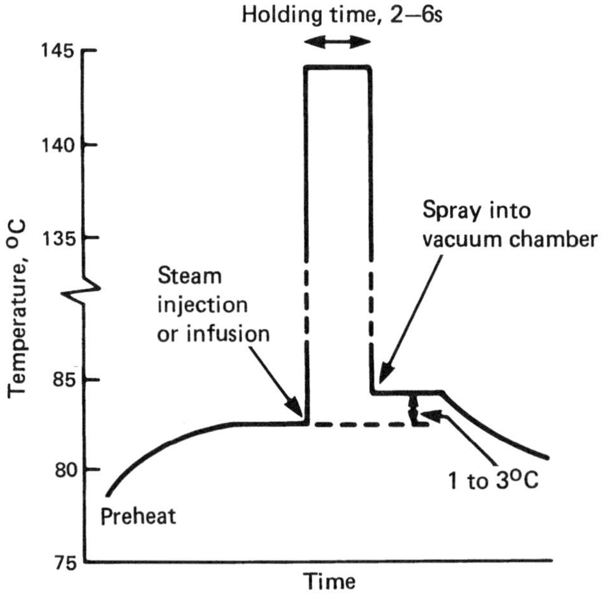

Figure 3–15 Time-temperature relationships during direct heating and expansion cooling. *Source:* Reprinted with permission from H. Burton, *Ultra-High-Temperature Processing of Milk and Milk Products*, p. 103, © 1988.

The rapid collapse of steam bubbles during condensation causes sharp local changes in pressure in the liquid, which can make the injectors noisy. This phenomenon is well known in other forms of steam injection (e.g., thermal recompressors in thermal evaporators), and some form of sound absorption may be necessary.

The second requirement is for the lowest possible pressure difference between the liquid and the steam. It has been said that higher steam pressures lead to higher product temperatures where the product first contacts the steam, with a consequent danger of excess damage to the product through overheating. However, Hallstrom (1981) doubts whether this happens to any significant extent. He suggests that as condensation is much more rapid than diffusion, a film of condensate is formed that separates the product from the injected steam and prevents overheating; the condensate is then mixed into the product by turbulence. In addition, the steam must be at a higher pressure than the product in the injector, up to about 1.5 bar extra, in order to give the flowrate required and good mixing characteristics, and will therefore be at a higher temperature. The steam and product must be thermally separated as far as possible until they reach the mixing zone, for products where fouling may occur. If parts of the injector carrying product are at high temperature through heat transfer from the steam, surface fouling will take place as in indirect heating systems, interfering with liquid flow within the injector and causing its performance to deteriorate. Injectors are therefore designed to minimize indirect heat transfer between steam and product.

Three typical steam injector designs are shown in Figure 3–16. In the first, Figure 3–16 (a), the product tube takes the form of a venturi, or the expanding part of a venturi. The steam is supplied through four tubes that surround the product tube. A series of small orifices along the lines of contact of the steam tubes with the product tube allow the steam to enter the product in small bubbles. Some of the orifices are drilled radially, and some may be drilled at an angle to encourage swirl and turbulence in the product to increase the rate of condensation. Although the steam chambers lie along the length of the product tube, transfer of heat from steam to product indirectly within the injector is minimized by the form of construction, with only a small heat transfer area along the lines of contact between the steam chambers and product tube.

The second design, Figure 3–16(b) appears to owe something to the design of steam atomizers for oil-fired boilers. The steam is injected at a sharp angle across the flow of the product, which is itself in the form of a thin, inwardly directed cone. This leads to rapid mixing and condensation. In the third design, Figure 3–16(c), the product passes through a venturi and the steam is injected in the expansion section of the venturi, where the product pressure is rising. The steam is injected in a thin annulus around the product. A second venturi section causes further fall and rise of pressure within the injector and encourages condensation and diffusion. In the last two designs, the steam supply is at right angles to the product flow so that indirect heat transfer is minimized.

A typical continuous-flow thermal process with steam injection and flash cooling is shown in Figure 3–17. The product is pumped from a level-controlled balance tank (1) through a heating section (2) where it is preheated by outgoing product. It is then further preheated to a constant temperature in the approximate range 75° to 85°C in a heat exchanger (3) by the condensation of steam under vacuum or by hot water circulation. The temperature

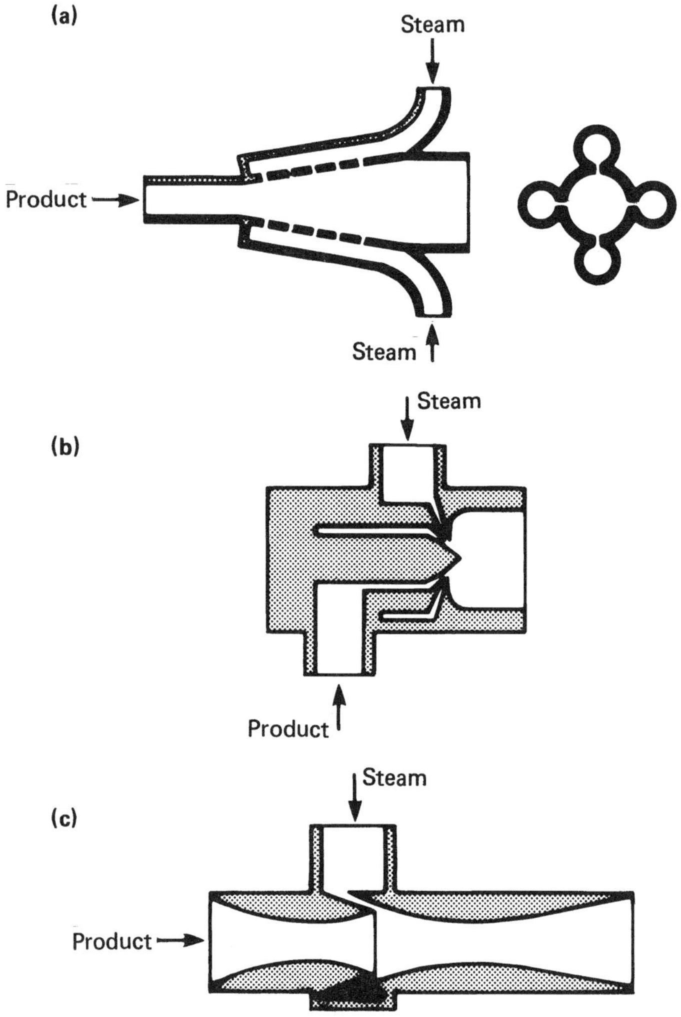

Figure 3–16 Typical steam injector designs. (a), Venturi-shaped product tube. (b), Steam injected at a sharp angle across flow of product. (c), Steam injected into expansion section of the product venturi tube. *Source:* Reprinted with permission from H. Burton, *Ultra-High-Temperature Processing of Milk and Milk Products*, p. 107, © 1988.

of the product at the outlet of this preheater is controlled by controlling the heating vapor supply or the temperature of the hot water. A high-pressure pump then supplies the product to the steam injector (4) and the holding tube, where the pressure must be high enough to prevent boiling of the product and the separation of dissolved air, as described in connection with indirect sterilizers, and also give satisfactory steam condensation within the injector. Too low a pressure will make it difficult to maintain processing temperature and holding time (Swartzel & Jones, 1980). A back-pressure of 1 bar as for an indirect plant should be adequate. Typical sterilization conditions are 144°C for a holding time of 5 seconds.

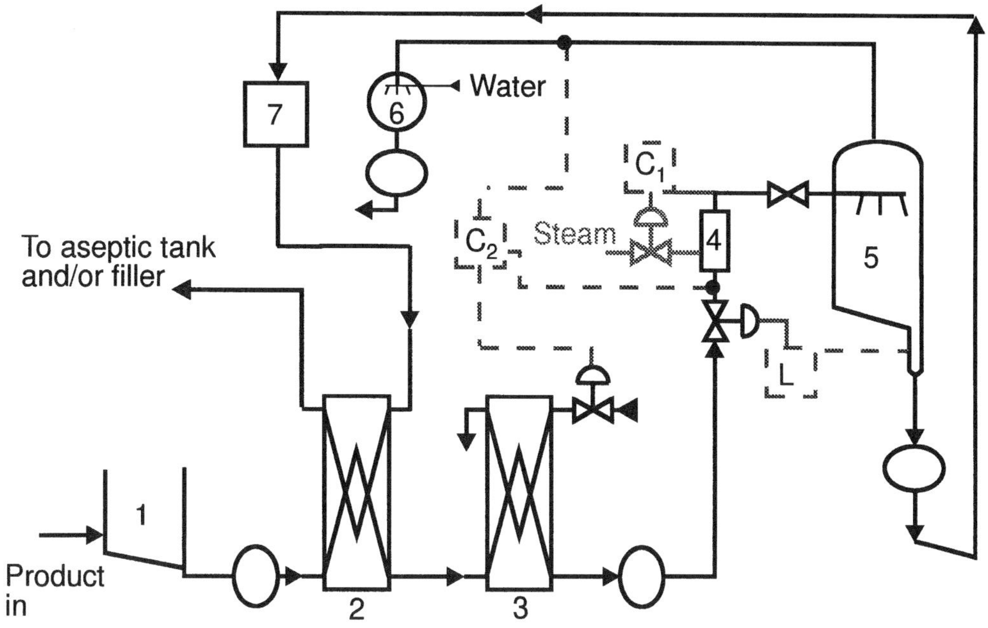

Figure 3–17 A typical continuous-flow thermal process with steam injection and expansion cooling. *Source:* Adapted with permission from H. Burton, *Ultra-High-Temperature Processing of Milk and Milk Products*, p. 108, © 1988.

The expansion cooling vessel (5) is kept at a suitable vacuum by condensation of the released water vapor in a water-jet or surface condenser (6) with a backup pump. The vacuum corresponds to a boiling temperature some 1° to 2°C higher than that of the product at the outlet of the final preheater (3); this temperature difference is needed to ensure that the correct amount of water vapor is removed to compensate for that added as condensed steam during heating, as is described later in this chapter. A restrictor valve at the end of the holding tube retains the high pressure in the injector and holding tube, and as the product passes through the restrictor into the low pressure in the expansion vessel, it boils and is rapidly cooled. The cooled product collects at the base of the expansion vessel, from which it is pumped using an aseptic pump, via a homogenizer (7) if required.

It has been shown that the turbulence involved in the direct heating of some homogenized products may cause breakdown of the homogenization. It has also been found with some dairy-based products that continuous-flow thermal processing by steam injection encourages the formation of casein aggregates, which can give a "chalky" or "astringent" mouthfeel to the product (Hostettler & Imhof, 1963). This is minimized if the aggregates are broken down by homogenization after it has been sterilized, and therefore an aseptic homogenizer is required. It is, of course, not possible in these systems to separate the high-pressure homogenizer pump from the homogenizing valve and so avoid the use of an aseptic homogenizer, as with indirect systems.

The product next passes through the regenerator (2) and leaves the plant to go to aseptic fillers or aseptic balance tanks. For plant cleaning and sterilization, a closed circuit is used in which detergents, rinse water, and sterilizing water are returned to the balance tank through a sterilizing water cooler and restrictor valve, as shown for an indirect plant in Figures 3–5, 3–6, and 3–8. During plant sterilization, the sterilizing water is heated to the required temperature by steam injection, but vacuum is not applied to the expansion vessel, so that the full sterilizing temperature is maintained in all the parts of the plant downstream from the injector. The expansion vessel is designed so that all parts of it are sterilized by the circulating water. The ways of preheating the product before steam injection, using heat exchangers (2) and (3) in the process plant shown in Figure 3–17, may vary in the designs of different manufacturers. For example, instead of a separate vacuum steam supply to preheater (3), a part of the vapor from the expansion vessel (5) may be used as the heating medium, diverted from the vapor line from the vessel to the condenser (6). Alternatively, where a water-cooled plate condenser is used, the cooling water can be used for preheating purposes. These are all different ways of making the maximum use of heat recovery. However, whatever the preheating system used, it must be possible to control accurately the temperature at the outlet of the preheaters and before steam injection.

The matching of flows in the different sections of a direct heating plant is important. The additional volume flow arising from condensed steam is compensated by the removal of vapor during expansion cooling. However, the presence of the expansion cooling vessel means that flowrates can be different before mixing with steam and after its removal, even after proper compensation, the difference being taken up by a change in the volume of product held in the vessel. The defined holding time for the process is normally determined by the flowrate given by the high-pressure homogenizer pump. However, in most direct heating continuous-flow thermal processing systems, this is after the expansion vessel, while the holding tube is before it. Some means is therefore needed to ensure that the flowrate before the expansion vessel, or more correctly the flowrate before its increase by the addition of condensed steam, is the same as that after.

In the type of plant shown in Figure 3–17, a level detector in the base of the expansion vessel is used to control a valve at the outlet of the pump supplying product to the injector, through controller (L). If the product flowrate into the injector and holding tube increases above that set by the homogenizer speed, the liquid level in the base of the vessel rises and the valve closes to reduce the flowrate to match the constant amount being extracted. Alternatively, the system can operate by varying the extraction rate from the expansion vessel by controlling the speed of the homogenizer drive motor.

In general, two independent temperature control systems are used in a direct heating continuous-flow thermal processing plant. One of these is a conventional temperature controller (C_1) used to control the product sterilization temperature: it senses the temperature in the holding tube after the steam injector and controls the supply of high-pressure steam to the injector. The second control system is used to match the amount of water removed during cooling in the expansion cooler to the amount added as steam during heating. With some products, it may be unimportant if some dilution or concentration takes place during the process. A fixed amount of dilution with water, or of concentration, may be acceptable and can be compensated for in product formulation, although variations throughout an operating run means that quality control is important. However, with some products, particularly milk, there may be legal penalties for dilution in some countries. Uncontrolled concentration will be commercially unacceptable to processors, as a loss of volume or weight will mean a loss of money.

It is shown later that the balance between water addition as steam and water extraction as vapors from the expansion vessel is determined by the difference between the temperature of the product before mixing with steam and the temperature of expansion cooling (1° to 3°C as shown in Figure 3–15). For any direct heating plant, there is a temperature difference at which the water content will remain unchanged overall. When this difference has been determined, the second control system is used to keep it constant during processing.

A differential temperature controller is used (C_2 in Figure 3–17) that senses the temperature of the product before mixing with steam and the temperature of the vapor leaving the expansion cooling vessel. The vacuum in the expansion vessel is normally held constant by a suitable valve, so that the corresponding temperature is constant. The differential temperature can therefore be controlled through C_2 by controlling the supply of heat to the preheating stage (3).

3.12 THE INFUSION SYSTEM

Infusion, or product-into-steam, systems are similar to injection systems in all aspects except the method of mixing product with steam. The designs of infusers used by different manufacturers differ. In every case there is a steam pressure vessel with a conical base into which the heated product falls and from which it passes to the holding tube. The size and proportions of the vessels differ, as do the methods adopted for distribution of the product into the steam, but the heating time as the product passes down through the steam atmosphere is not likely to be more than about half a second and therefore insignificant in the sterilization effect.

Two alternative distribution methods are shown in Figure 3–18. In the distributor shown in Figure 3–18(a), the product is supplied to a hemispherical bowl with a loose circular disc closing the top. The product flows from the space between the bowl and the loose disc to form a thin umbrella with access to the heating steam from both sides. For special applications, products containing particles can be heated by this method, as the loose disc can lift to allow particles to pass out of the bowl with liquid. More than one distributor may be fitted in a steam chamber. In the distributor shown in Figure 3–18(b), product flows into a series of parallel and horizontal distribution tubes. These have thin slits along the bottom, from which the product falls in thin, streamline free-falling films. All the distribution tubes can be used, or some can be closed off to give the infuser a variable heating capacity.

The infusion-heating and expansion-cooling stages of a typical infusion-type continuous flow thermal sterilizer are shown in Figure 3–19. The product passes from a balance tank through indirect preheating stages that have the same form, or combination of forms, as those used with steam injection systems. The purpose of these is to give some regenerative heating to conserve energy and to give accurate control of temperature before the infuser.

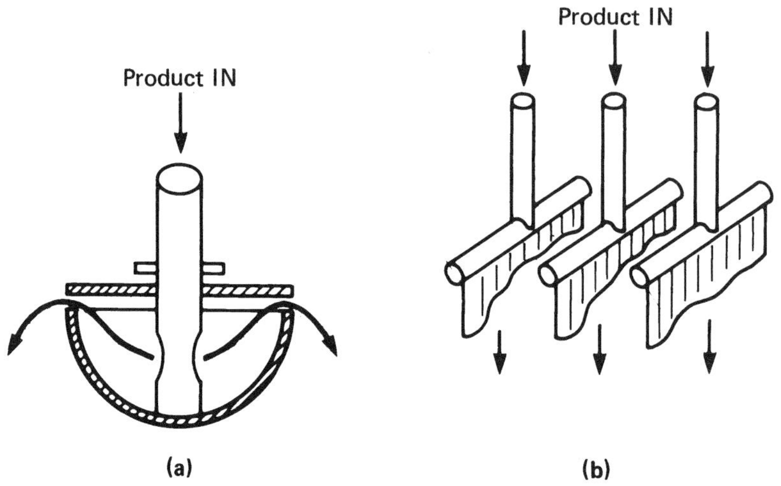

Figure 3–18 Product distribution systems for infusers. (a) Design to form umbrella-shaped curtain. (b) Design to form free-falling films. *Source:* Reprinted with permission from H. Burton, *Ultra-High-Temperature Processing of Milk and Milk Products*, p. 111, © 1988.

The product is then pumped (1) to the distributor in the infusion vessel (2), which is supplied with steam under the control of controller C_1 to give the required sterilization temperature in the holding tube. The heated product is forced by steam pressure through the holding tube and a restrictor (4) into the expansion vessel (5). The rate of product flow may be controlled by a PD pump (3), which may in turn be linked to a level controller (L) determining the level of the product pool in the base of the infuser; in this way the required holding time is held more precisely. The expansion vessel is maintained at the necessary vacuum for cooling by the condenser (6) and backup pump for the removal of noncondensable gases. The cooled product is removed from the vessel by an aseptic pump (7) for supply to an aseptic homogenizer (8), if required. To maintain a controlled pool of product in the expansion vessel, the level controller (L_2) in this case alters the speed of the homogenizer, or PD pump, as compared to the control system described above for the steam injection system. The later stages of cooling, regenerative or by cold water, may take any of the forms already described in relation to steam injection systems. It might appear that the infusion heating process is simple. Little work seems to have been done on what happens within an infusion vessel, but the situation is more complicated than might be imagined.

Because of the shape of an infuser, the volume of any pool of product in its base adds considerably to the total holding time as determined by the dimensions of the holding tube and the flowrate through it, and very much broadens the distribution of holding times as compared with those obtained with a holding tube. This volume therefore needs to be minimized by careful level control. In Figure 3–19 the level controller L serves this purpose. In the absence of a flow-control pump, the rate of product flow from the infuser into the expansion cooler depends on the pressure difference between the infuser and the expansion vessel.

The pressure relationships within the infuser are complex (Perkin, 1985). The saturated steam pressure within the vessel in theory need only be a little above that corresponding to

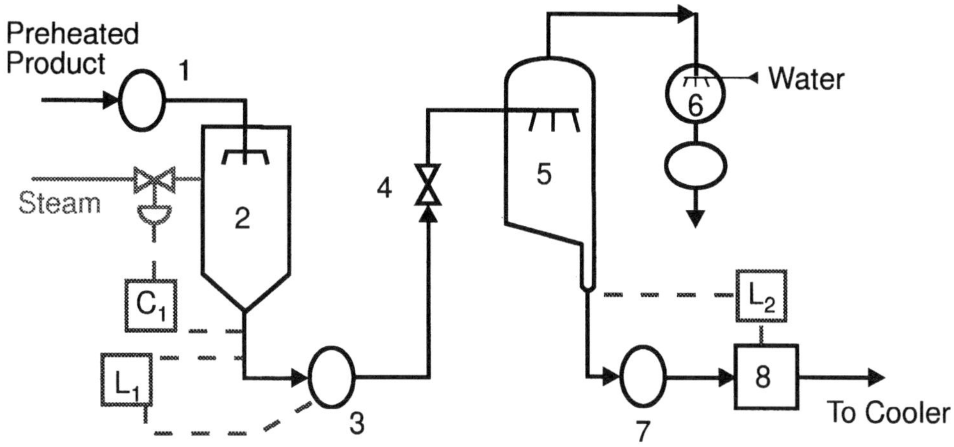

Figure 3–19 Infusion-heating and expansion-cooling stages of a typical infusion-type continuous-flow thermal process.*Source:* Reprinted with permission from H. Burton, *Ultra-High-Temperature Processing of Milk and Milk Products*, p. 112, © 1988.

---

the required product sterilization temperature, but in practice for operational stability this pressure should be at least 0.5 bar above that equivalent to the sterilization temperature.

Commercial infusion systems usually do not deaerate the product before processing, so dissolved gases enter the infuser with the product and often also in the steam supply. The solubility of the gases is lower in the product at sterilization temperature leaving the vessel than it is at the entering temperature, so that gases will leave the product and accumulate in the vessel to give a steam-air atmosphere. The effect of this on infuser performance will depend on the level control system used.

If a level controller varies the outflow to give a constant level, the gases remain in the atmosphere and accumulate within a constant volume. The partial pressure of the steam in the infuser must remain constant in order to give the necessary sterilizing temperature, so the total pressure will rise. The steam inlet pressure must therefore rise to allow steam to enter the vessel, which will give an equivalent temperature at the inlet above the required sterilizing temperature and a potential risk of overheating. Some of the air can be removed by venting the infuser atmosphere, but it cannot be removed selectively from a steam-air mixture, so the total pressure must always rise.

If the outflow from the infuser is not controlled by a pump, the product level will fall until a mixture of gases and steam can pass through the holding tube to the expansion vessel, maintaining a constant low level of product in the infuser, which depends on the outlet geometry of the vessel, but allowing an outflow of bubbles of the infuser atmosphere through the holding tube. In this case, there must be some uncertainty as to the holding time and hence a danger of underprocessing.

One manufacturer has improved the heat recovery of an infusion system operating on milk by siting a deaeration/flash cooling vessel before the infuser, instead of after it as normal (APV, 1996). An outline of the process is shown in Figure 3–20. The flash cooling vessel (1) operates with an inlet of 95°C and with sufficient vacuum to flash cool the milk to 75°C, therefore concentrating it to an extent, heating by regeneration (3) to 120° to 130°C, followed by infusion heating (4) to 140° to 150°C. The product is cooled by regeneration (5) in a tubular heat exchanger with the stream leaving the flash evaporator and again by regeneration (6) in another tubular heat exchanger with the incoming raw milk, both regeneration sections using a recirculating water loop as in Figure 3–6. A homogenizer (2) may be sited immediately after the flash cooling vessel, if required. Regeneration efficiencies of 75% are claimed for holding conditions of 150°C for 2 seconds, together with an operating time of over 20 hours before cleaning.

3.13 COMPARISON OF INJECTION WITH INFUSION SYSTEMS

Any comparison of these systems must be based on the characteristics of injection and infusion heaters. Since little is known about exactly what happens within them, many of the comments must be based on theory.

The heating process is gentler with infusers than with injectors, since the steam condensation does not involve the condensation of small bubbles and associated cavitation. The mechanical effects on the product, and the noise level, will therefore be less with infusers. Mechanically, infusers must always be much larger than the corresponding injectors and must be pressure vessels able to withstand the full operating pressure. The product distribution systems in an infuser can be considered mechanically similar and equivalent to a steam injec-

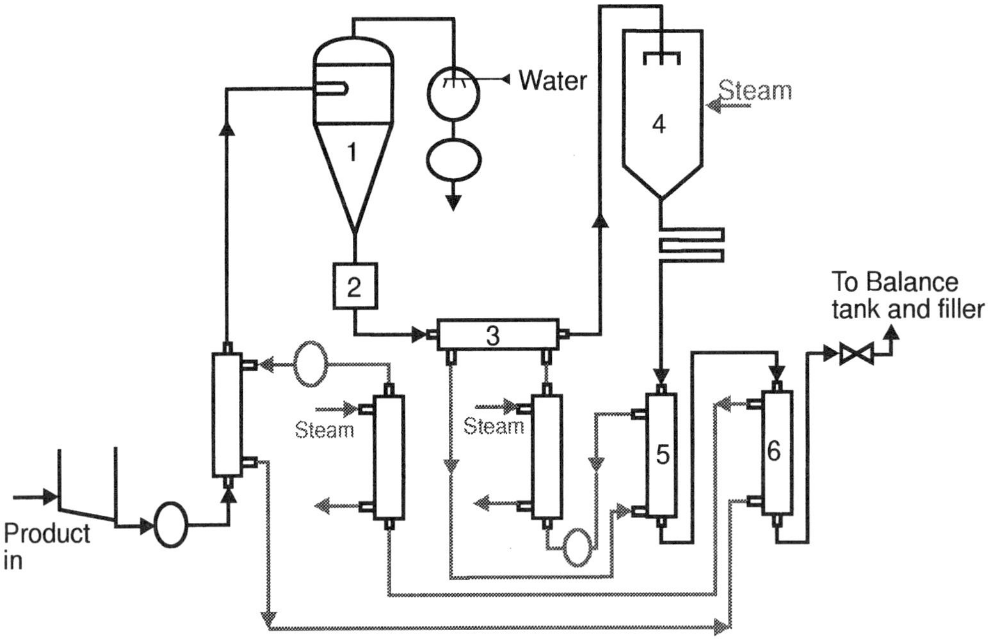

Figure 3–20 Improved heat recovery of an infusion system operating on milk with a deaeration/flash cooling vessel placed before the infuser instead of after it, as is the usual practice

tor. Therefore, from a constructional and cost point of view, an infuser involves a pressure vessel in addition to an injection device.

Both injectors and infusers give very rapid, but not necessarily instantaneous, heating. The rate of condensation in an injector determines the rate of heating, and this depends on the design of the injector and the back-pressure within it; for one commercial design of injector, full temperature was reached 0.9 second after steam injection (Burton et al., 1977). Comparable heating times for an infuser are difficult to measure; about 0.1 second has been estimated for a small infuser (Burton et al., 1977), and about 0.5 second has been given for a commercial plant. Both heating systems therefore appear to give similar heating times, which are short in comparison with holding times of the order of a few seconds.

It has been claimed that injectors overheat some products because the incoming steam is at a high pressure and therefore at a higher temperature than the product. We have seen, however, that similar arguments can be presented in relation to infusers because of the rise in pressure in the chamber from the release of dissolved gases.

The holding time is more predictable when an injector is used, although the effect of the increased flowrate through the holder must be considered as with all direct systems. Incomplete condensation could reduce the holding time by effectively reducing the volume of the holder by the volume of uncondensed vapor.

When an infuser is used, there seem to be several additional factors that can influence the holding time. The first of these is the effect of any pool of product held in the base of the vessel. The second is the possible effect of the passage of some of the vessel atmosphere through the holding tube to stabilize the atmosphere pressure: this will have the same effect as incomplete condensation after steam injection.

There is no evidence on the effect of infusion heating on the homogenization of milk, cream, or similar products, or on the development of the "chalky" or "astringent" texture in milk. It is generally assumed that the same effects occur with both types of direct heating systems, and that both need aseptic homogenization after the sterilization stage of the process.

3.14 CONTROL OF PRODUCT CONCENTRATION OR DILUTION DURING DIRECT HEATING PROCESSES

The amount of steam condensed during the heating stage of direct continuous-flow thermal processing is very significant in relation to the amount of product. If we assume a temperature before mixing with steam of 85°C, and a sterilization temperature of 145°C, the thermal energy to be provided by the steam is approximately $60 \times 4 = 240$ kJ/kg product, if the product is largely composed of water. The energy released by the condensation of steam at temperatures above 100°C is approximately 2.2 MJ/kg steam. The amount of steam to be condensed per kilogram of product is therefore about 0.11 kg or 11% of the product. If all this added volume is to be removed in the expansion cooling vessel, a small error in the balance will have a considerable practical and commercial effect on the total product volume leaving the plant, although the change in product composition may be impossible to detect with the present accuracy of analytical methods.

Hallstrom (1966) showed that, for accurate compensation, the temperature of the product or vapor leaving the flash vessel should be slightly higher than that of the product before mixing with steam. A detailed analysis of the problem made by Perkin and Burton (1970) showed the practical application of the theory to a steam injection system operating to process product.

By considering the heat balances between the total heat input to the injector, the heat content of product in the holding tube, and the total heat output from the expansion vessel, it can be shown that:

$$Z = \frac{i_v C_m \theta_1 - C_m C_{ml} \theta_1 \theta_2 - i_v C_{ml} \theta_2 + i_s C_{ml} \theta_2 + W C_{ml} \theta_2 + C_{ml} C_{m2} \theta_2 \theta_3 - i_s C_{m2} \theta_3 - i_s W}{i_v C_{ml} \theta_2 - C_{ml} C_{m2} \theta_2 \theta_3 + i_s C_{m2} \theta_3 - i_s i_v} \tag{3.4}$$

where

Z = proportionate increase in the apparent output of liquid, equivalent to proportionate dilution of product (+ for dilution, – for concentration) (kg/kg product)
θ_1 = temperature of liquid entering the injector (°C)
θ_2 = temperature of liquid leaving the injector (°C)
θ_3 = temperature of liquid leaving the expansion vessel (°C)
C_m = specific heat of product (J/kg°C)
C_{ml} = specific heat of diluted product leaving injector (J/kg °C)
C_{m2} = specific heat of product, in general of changed concentration, leaving the expansion vessel (J/kg°C)
i_s = heat content of steam injected (J/kg)
i_v = heat content of vapor leaving expansion vessel (J/kg)
W = heat losses from expansion vessel (J/kg product)

Putting Z = 0, an expression can be obtained from Equation 3.4 that gives the differential temperature at which there will be no concentration or dilution of the product as a result of the treatment:

$$\theta_3 - \theta_1 = \frac{1}{C_m}\left\{\frac{(W + C_m \theta_3 - C_{ml} \theta_2)(i_s - i_v)}{(C_{ml} \theta_2 - i_v)} - W\right\} \tag{3.5}$$

In these relationships, the heat loss associated with the expansion vessel is taken into account through the term W. It is assumed that the corresponding loss from the injector and its associated pipework can be neglected because the physical size of the injector is small compared with that of the expansion vessel. With an infusion system, the physical sizes of infusion and expansion vessels will be similar, and it will not be possible to neglect the additional heat loss. A similar analysis of an infusion system will therefore be more complex.

The heat loss term W appears in both equations. It is not a calculable quantity, and will in practice vary from sterilizer to sterilizer, and perhaps for the same sterilizer in different surroundings or different climatic conditions. The temperature difference to give no change in composition ($\theta_3 - \theta_1$) therefore cannot be found by calculation from Equation 3.5, but only by experiment. The best way to do this is to fill the sterilizer circuit with water, and to complete a closed circuit from the sterilizer outlet to the inlet with a section of transparent tube (a suitable temperature- and pressure-resistant plastic or glass) in which the water level can be seen. The plant can then be operated with the normal standard operating conditions and temperatures, for example, at the sterilizing temperature θ_2 and at the expansion temperature θ_3, and with the preheat temperature θ_1 set at any convenient value. When stable operating conditions have been reached, θ_1 can be varied in small increments until a setting is found at which the water level seen in the transparent tube remains constant over several minutes. The total circulating water content is then constant, and the value ($\theta_3 - \theta_1$) is that which will give no concentration or dilution of product. This value should be recorded. The differential temperature controller should be set and locked so that the setting is maintained under normal operation, and regular checks of the controlled differential temperature should be made and recorded.

Perkin and Burton (1970) checked the reliability of Equations 3.4 and 3.5 by experiment. The differential temperature for a small direct heating sterilizer to give no change in concentration of milk was found as described above. The value was found for this sterilizer to be 2.1°C, to give a calculated heat loss, W from Equation 3.5 of 4.7 kJ/kg milk. From Equation 3.4, using this value of W and the best obtainable values for other physical data, relationships between differential temperature and concentration or dilution were calculated for water and milk. These are shown in Figure 3–21(a). These relationships were checked experimentally by circulating milk or water as described above, and determining the decrease or increase in circulating volume for different differential temperatures. The experimental results are shown in Figure 3–21(b).

The correspondence between calculation and experiment is very close. The temperatures at the point of intersection with the axes must necessarily be the same, since this was used to determine the heat loss. The slopes are, however, very similar. The calculated and experimental values for water are identical: those for milk are not significantly different in view of the scatter of the experimental results. The theoretical analysis therefore represents practical plant behavior very closely.

Figure 3–21 shows that control of the differential temperature is a very precise way of controlling composition. For example, a deviation of differential temperature from the set point by 0.6°C gives rise to a concentration or dilution of only 0.1%; that is, the content of any constituent of the product would increase or decrease by a factor of 1.001, which could not be detected by the most accurate methods of chemical analysis.

Any change in other operating conditions of the plant will also lead to concentration or dilution. If the sterilizer is calibrated with the differential temperature to give proper balance at a sterilizing temperature of 144°C, a fall in the sterilizing temperature will cause a dilution of about 0.12% per °C, and a rise in sterilizing temperature will give a similar concentration. A change in the steam pressure needed to operate the injector also causes concentration changes. If the steam pressure at the injector rises by 1 bar from the pressure when the correct differential temperature was determined, perhaps because of a similar increase in pressure in

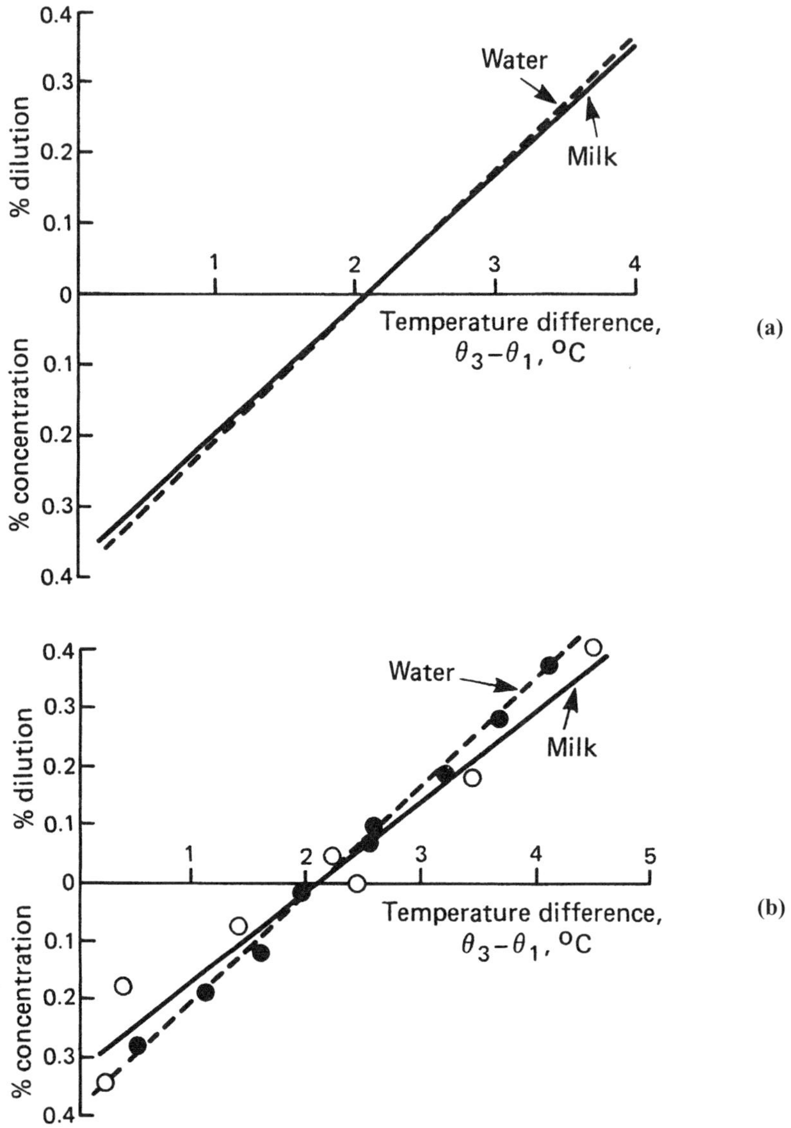

Figure 3–21 Calculated (a) and experimental (b) relationships between differential temperature and concentration/dilution in direct heating systems. *Source:* Reprinted with permission from H. Burton, *Ultra-High-Temperature Processing of Milk and Milk Products*, p. 118, © 1988.

the holding tube, the product will be concentrated by about 0.02%. Whether any of these changes are of practical importance will have to be determined by plant management, and will depend on the product's being sterilized and the framework of legislation within which the process is being carried out.

3.15 THE SUPPLY OF CULINARY STEAM FOR DIRECT HEATING PROCESSES

As the steam is to be condensed into the product, any additional compounds that it might be carrying will be transferred to the product, and while the added water will be removed in the expansion cooling vessel, the additional compounds will not be removed. It is therefore important that the steam should be carrying no compounds that might be toxic or undesirable to the consumer, or that might in some way spoil the product during storage. Steam that is free from these compounds is often called "culinary" steam.

Steam used for industrial purposes is normally generated in a boiler from feedwater to which water treatment compounds have been added. These compounds may be added for several reasons, for example, to prevent fouling of boiler surfaces by hardness compounds in the feedwater, to prevent foam generation in the boiler that will cause excessive carry-over of water with the steam, to remove oxygen from the feedwater to prevent corrosion within the boiler, or to supply compounds that are carried with the steam to inhibit corrosion in steam pipes and condensate return lines.

In practice, any boiler feedwater treatment compound is likely to be carried over with the steam. Compounds intended to inhibit corrosion in steam and condensate lines are designed to be carried with the steam. Other compounds enter the steam by accident, since steam can never be generated entirely free from suspended water droplets, which will themselves carry the treatment compound. To avoid the risk of undesirable or even dangerous compounds (e.g., compounds that might influence flavor or that are potentially carcinogenic) being introduced into a foodstuff from steam, some countries require that culinary steam should be produced only from water of drinking quality, specify a restricted list of boiler feedwater treatment compounds that may be used, and recommend ways of cleaning the steam to reduce to the minimum those that are allowed, before the steam is mixed with the foodstuff. Exhibit 3–1 gives a list of suitable compounds based on United Kingdom regulations for milk (UK Regulations, 1989), which have been strongly influenced by United States requirements for culinary steam. Current UK dairy legislation (UK Regulations, 1995) requires that steam shall be obtained from potable water and not leave deposits of foreign matter in the milk or affect it adversely. Legislation for other food products does not appear to be specifically concerned with direct heating but will come under normal legislation regarding addition of harmful compounds (e.g., the UK Food Safety Act, 1990).

Although these compounds are all considered to be safe, it is desirable to limit as far as possible their transfer to the foodstuff being heated. Since they are nonvolatile but soluble, they can only reach the product by being carried in the steam in entrained water. The way to prevent their reaching the product is therefore to make the steam as dry as possible.

Where the foodstuff is subject to compositional standards and the condensed steam is to be removed later (e.g., milk and creams), it is especially important to ensure that entrained water is removed before mixing with the product and that the steam is not superheated. If either of these occur, then compensation will not be accurate and legal problems will result.

Recommendations for steam cleaning have been made in the United Kingdom (UK Regulations Circular, 1972), based on the measured removal of a soluble tracer added to the boiler

Exhibit 3–1 Acceptable Boiler Feedwater Treatment Compounds

Potassium alginate
Sodium alginate
Potassium carbonate
Sodium carbonate
Sodium hydroxide
Sodium dihydrogen orthophosphate
Disodium hydrogen orthophosphate
Trisodium orthophosphate
Pentasodium triphosphate
Sodium polyphosphates
Tetrasodium diphosphate
Sodium silicate
Sodium metasilicate
Sodium sulfate
Magnesium sulfate
Neutral or alkaline sodium sulfate
Unmodified starch
Sodium aluminate
Polyoxyethylene glycol (minimum molecular weight 1000)

Source: Reprinted with permission from H. Burton, *Ultra-High-Temperature Processing of Milk and Milk Products*, p. 121, © 1988.

feedwater (Figure 3–22). After the valve controlling the steam supplied to the heater, there should be a length of uninsulated steam pipe to de-superheat the steam. A high-quality centrifugal separator is then recommended, discharging condensate through a suitable steam trap and an open tundish, so that proper operation of the trap can be monitored. After the centrifugal separator, a mechanical filter is recommended having a small-pore (approximately 3 μm) filter element made of polytetrafluoroethylene (PTFE). This filter has the dual purpose of removing any solid particles (e.g., of scale, from the steam) and removing further suspended water particles through the hydrophobic nature of the filter element. Water rejected by the filter is again discharged through a steam trap and an open tundish.

After the filter, the steam line should be made of stainless steel to prevent further contamination, and a nonreturn valve should be fitted to prevent product passing back into the steam line under reverse pressure. It is also desirable to fit between the filter and the nonreturn valve a pressure gauge and indicating thermometer to act as a check on plant and controller performance, and to permit a check on the dryness or supersaturation of the steam. A sampling point should also be fitted so that steam samples can be taken for checks of chemical quality if required.

The importance of having an effective and functioning steam trap at the end of the steam line must be emphasized. If the trap is not discharging, then the steam will almost certainly

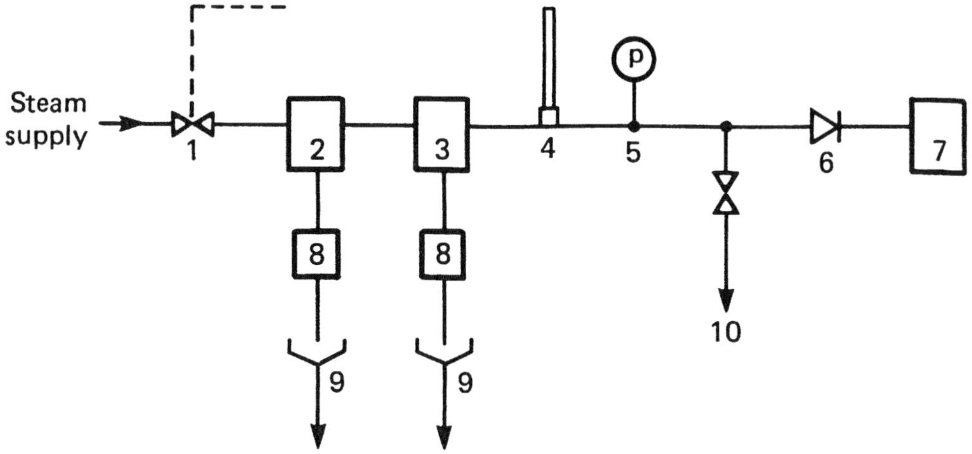

Figure 3–22 Recommended steam cleaning system for direct heating. (1) Automatic steam valve controlled by milk temperature. (2) Centrifugal separator (vortex pipeline separator). (3) Filter (PTFE elements). (4) Thermometer. (5) Pressure gauge. (6) Nonreturn valve. (7) Steam injector or mixer. (8) Condensate traps. (9) Tundishes. (10) Steam sampling point. *Source:* Reprinted with permission from H. Burton, *Ultra-High-Temperature Processing of Milk and Milk Products*, © 1988.

be very wet, and control of product concentration or dilution by differential temperature control will be ineffective.

Because of the limitations on feedwater treatment, it may not always be possible to take the steam for a direct heating processing system from the main factory boiler; satisfactory operation of the rest of the steam supply system may demand treatment compounds that are not permissible for the production of culinary steam. It may be necessary to install a separate small boiler to supply steam exclusively to the sterilizer. This may be a primary boiler, or it may be a secondary boiler heated by steam from the primary factory supply. In the latter case, the primary steam supply will have to be at a pressure 1 to 2 bar above that which is required by the supply and control system of the direct heating sterilizer.

COMPARISON OF INDIRECT AND DIRECT HEATING SYSTEMS

The most obvious difference between these two systems is the high rates of heating and cooling that are obtainable with direct systems. This means that the ideal of high temperatures held for short times, which is the basis of the continuous-flow thermal process, is more easily obtainable with direct than with indirect systems.

The product quality aspects of different processes are discussed in more detail in Chapter 6. Briefly, there are quality benefits with direct heating, but these are less than are generally believed. The advantages of the direct system lie more in other directions, and are counteracted by disadvantages, mainly economic.

The main practical advantages of the direct heating system are its ability to process more-viscous products, particularly those that cannot be satisfactorily handled in plate-type indirect plant, and comparative freedom from fouling. For example, a direct plant may be able to operate producing UHT milk without cleaning for twice the length of time of a plate-type indirect plant; a tubular indirect plant has intermediate characteristics.

For products where fouling is significant, it is not true that direct heating plant is unaffected, although fouling of heat exchange surfaces is insignificant. With injection types of sterilizers there is likely to be a buildup of deposit in the restrictor at the outlet of the holding tube. This causes an increase in the pressure in the holding tube, and requires an increase in the injected steam pressure to maintain the product temperature. With long processing runs, the steam pressure available from the supply may become limiting. With infusion systems, fouling is likely to occur on the walls and base of the infusion vessel; this will have little effect on plant operation, but may cause a cleaning problem.

A technical disadvantage of direct systems is, for those products where it is necessary, the need for an aseptic homogenizer after the sterilization stage. Apart from the higher capital cost of an aseptic homogenizer, careful maintenance is needed, particularly in the replacement of piston seals, if bacteriological contamination is to be avoided.

However, the factor that influences the choice between a direct and an indirect plant in most cases is that of comparative costs, both capital cost and running cost. A direct heating plant is relatively complex. Much of it is nonstandard and specially fabricated, unlike heat exchanger plates and tubes. It requires additional equipment such as expansion vessels and condensers. More pumps and control systems are needed when compared to the equivalent indirect system, as can be seen from the typical flow diagrams given above. Some parts of the system after product sterilization are at pressures below that of the surrounding atmosphere and, to avoid the ingress of bacterial contamination from the atmosphere, it may be necessary to incorporate steam seals on vessels and pumps. In all, a direct plant may cost about twice as much as an equivalent plate-type indirect plant, and almost twice as much as a tubular indirect plant. The running costs of a direct plant are also relatively high. The costs of heating energy are high because the possibilities for the regenerative use of energy are restricted. For example, if we consider a sterilizer in which the product is heated from 5° to 85°C by indirect means, then from 85° to 145°C by mixing with steam, and then cooled to 87°C by expansion cooling, no regeneration to a temperature higher than that of the expansion vessel is possible. That means that even in the best situation, only 80° of the total 140°C temperature rise can be by regeneration. The maximum level of heat regeneration is therefore 57%; in practice, regenerations between 50% and 54% may be obtained. Higher levels of regeneration can be obtained by increasing the preheat and expansion vessel temperatures: an increase in the preheat temperature of 10° to 95°C will give a theoretical level of 64%, but the additional indirect heating will nullify some of the advantages of direct processing. An indirect plant can have regeneration levels of 90% or more, and correspondingly lower thermal energy costs. However, this is at the expense of extended product residence times in the processing plant, with consequences for overall product quality that are considered later.

Similarly, the water consumption for cooling in a high-regeneration indirect plant during operation is negligible, because the cooling requirements are met by the incoming cold product. Even with an indirect plant with lower regeneration levels, the cooling water requirements are relatively modest. However, direct plants require large amounts of water, particularly for the operation of the condenser, and the demand may be of the order of 1,500 L of water per 1,000 L of product, or more. Water is increasingly an important cost factor, and the costs of water for a direct plant may be similar to the costs of heating energy. A cooling tower could be used with recirculated water, but this introduces additional capital and pumping costs.

Direct heating systems need more pumps than indirect systems. Only a single product pump may be needed with an indirect system, together with a homogenizer. In general, a direct system will need three product pumps of different types, as well as a homogenizer if the product requires it, and a vacuum pump for the removal of noncondensable gases from the vapor condenser. The electric power requirements for pumping and homogenization may therefore be 25% to 50% higher with a direct plant.

When total operating costs are considered, those of a direct heating plant may be twice as high as those of an indirect plant of equal throughput. These higher costs may be to some extent offset by the ability of direct plants to operate for a longer time than indirect plants, for example, with some products that are unstable and liable to cause heavy fouling of heat exchange surfaces. However, considerably shorter operating runs would be needed before the difference in operating costs became absorbed by increased cleaning costs.

In general, the increased complexity and costs of a direct system will be justified only in special circumstances, where product characteristics and required product quality make its use desirable. In recent years, increasing costs of energy and the increased costs of water, particularly in industrialized countries, have led to the installation of indirect sterilization plants of different types rather than direct plants. It is possible that energy and water costs may fall once more in real terms, and that new quality factors will become important. The direct heating processes may then become more widely used once more.

PROCESSES INVOLVING THE DIRECT USE OF ELECTRICITY

The following processes use electricity to heat the foodstuff directly, and can hence be used where a steam supply is not available or not feasible. There are other processes that use electricity where it has a sterilization effect other than heat generation, such as high-voltage electric pulse, but they are not considered here.

3.16 FRICTION HEATING

A sterilizer using friction for the heating of milk and milk products gives very rapid heating and has been used experimentally for UHT processing. The sterilizer consists of a stainless steel disc, 300- or 400-mm diameter, which rotates at a speed of 4,000 to 5,000 rpm within a stainless steel housing, the spacing between the disc and the housing being 0.3 mm (Alais, Humbert, & Chuon, 1978). Liquid enters at the axis and passes radially through the spaces between the disc and the housing, to leave at the periphery. The frictional shear forces in the liquid cause it to be rapidly heated (e.g., by 70°C in about 0.5 second). At the same time, the liquid pressure increases by some 14 bar.

Typically, the liquid is preheated to 70°C in a tubular regenerative heat exchanger and is heated to 140°C in the sterilizer. The heated liquid, without holding, passes to the regenerator and then to a final water cooler.

The heating power is constant, determined by the power of the drive motor, so that the temperature rise in the sterilizer and the liquid flow rate are interrelated. The temperature at the outlet of the sterilizer is therefore maintained constant during operation by a control system that operates a flow restrictor valve after the regenerator.

The sterilizing effect of this system depends in a complex way on the final temperature because no holding time is applied, so that the time of passage of liquid through the friction unit is a controlling factor, and this is itself a function of the final temperature. The unit also provides some homogenizing effect.

The friction system of heating does not seem to have been used commercially, and it does not seem to have significant advantages over more conventional methods of processing.

3.17 USE OF MICROWAVES

Microwave radiation has often been used for heating solid products and packaged liquids but has rarely been applied to continuous-flow thermal processes. The only effect of microwave radiation on microorganisms is due to the temperature developed in the foodstuff, and the main economic problem with its use is in the efficiency of conversion of electrical energy (itself produced inefficiently from fossil fuels), first to microwave energy and then to heat in the product. Cooling must obviously be achieved by conventional heat exchangers. In the laboratory, the use of a microwave-transparent tube in a microwave cavity has been used to heat milk continuously for pasteurization. Continuous cooking of meat and fish emulsions on a larger scale has also been achieved, in a similar way.

3.18 OHMIC HEATING

Ohmic heating is a technique in which electricity is passed directly through the food, which heats up solely due to its electrical resistance (i.e., heat is generated directly within the food). The food is usually in continuous flow through the Ohmic heater. This technique, developed at the Electricity Research Council, Capenhurst, UK, and developed commercially by APV International, is more commonly applied to liquid foodstuffs that contain particulates, as it has its greatest technical advantages there. It can, however, also be applied to liquids only and may have technical advantages over other heat exchangers if the liquid is very viscous, thermally sensitive, or prone to fouling due to the lack of heated surfaces in the equipment. In common with the other methods using electricity for heating, the additional operating cost means that this technique is rarely used, but Ohmic heating will be more efficient in conversion of electricity to heat (and therefore cheaper) than microwave heating or friction heating. Ohmic heating is covered in more detail in Section 4.2.

EVALUATION OF PROCESS PLANT PERFORMANCE

The primary performance required of a continuous-flow thermal process is in terms of a reduction in microorganism concentration, whether it is the production of an ambient, shelf-stable product with as low a pack failure rate as feasible, or simply to achieve a legal process defined in terms of a required temperature held for a minimum time, to inactivate pathogenic vegetative cells for public safety. While this specification cannot be compromised, there are other, secondary performances that require the product to be of the best quality (nutritional or organoleptic) and to be produced as economically as possible, in terms of energy consumption, low capital cost, and long operating times with minimum fouling. These secondary requirements, in particular, are complex and conflicting and have led to a wide range of continuous-flow thermal process plants being available with different configurations and different combinations of process temperature and holding time. It is necessary initially to evaluate the microbiological performance of these plants and would, in addition, be very useful to be able to evaluate and to compare their quality performances to give a basis for selection.

For the majority of lower temperature continuous-flow thermal processes, such as pasteurization, it is easy to evaluate the microbiological effectiveness of the plant, as all that is required is to ensure that the specified process temperature and holding time are achieved by the foodstuff. For sterilization operations, it is possible to evaluate the microbial performance in a purely practical way by processing a single product with its natural microorganism population and evaluating the number of pack failures. There are many problems with this approach, as the large batch size required to give a measurable number of failures requires a significant physical effort in evaluation, it is not easy to determine what safety factor (what degree of overprocessing) is present, and the fact that natural microbial flora of raw ingredients are rarely constant and may even be cyclic in nature means that credence could not be placed on a single evaluation. This approach is extremely costly and very time consuming.

Alternatively, assessment of the sterilization and quality performance of a process can best be accomplished in two ways:

1. theoretically, from an analysis of the time-temperature profile and residence time distributions of the process with the kinetics of microorganism death and of some indicator of quality (e.g., degradation of some biochemical or formation of reaction products)
2. practically, by direct measurement using high concentrations of microorganisms, marker microorganisms with high heat resistance, and biochemical components naturally in the product or added marker chemicals

3.19 ACCEPTABLE PACK FAILURE RATES

As shown in Chapter 2, it is a consequence of the kinetics of microbial sterilization that it is impossible to ensure that there will be zero microorganisms in a product after any thermal treatment and that there must be a failure rate due to survival of microorganisms. The level of this failure rate that is acceptable must therefore be established; it must maintain public health and be acceptable from a consumer point of view, but must not be so low that the product quality is compromised for no real consumer benefit.

A useful comparison would be to compare pack failure rates of commercially produced in-container products. Few published data are available but failures are thought to be at worst in the region of 1 container in 1,000 for the most difficult products (canned peas). Hersom (1985a, 1985b) suggested a failure rate of 1 in 5,000 as acceptable, and Wiles (1985) suggested 1 in 10,000. The concept of failure rates expressed as 1 failure per number of containers may not be considered a useful one, however, when the container size varies from 10 mL to 1,100 L!

The failure rate that would be acceptable will be dependent on several other factors, for example, (1) the customer's frequency of use of the product, (2) the production rate proposed, (3) the drop in quality of the product when subjected to more severe processing, or even (4) the value of the product.

In the instance of 10-mL pots of milk supplied in many catering establishments, the customer may use several of these per day and maybe as many as 1,000 per year; a poor failure rate would result in the customer's being exposed to at least an irritating or off-putting situation, if not a potentially more dangerous food poisoning incident, every year. As the production rate is high, a large number of complaints would also be received by the producer, even considering the poor return rate of contaminated samples by the general public. A lower failure rate is obviously desirable for this product than, for example, for a 1-L aseptic soup product.

In the case of milk, good-quality raw milk in Europe contains no more than 10/mL or 10^4/L of resistant mesophilic spores. Assuming 1-L packs, for a failure rate of 1 pack in 5,000 the proportion of surviving spores required by the process would be

$$\text{Proportion of surviving spores} = \left(\frac{1}{10^4}\right)\cdot\left(\frac{1}{5000}\right) = 2\times10^{-8} \tag{3.6}$$

The subject of establishing the level of thermal treatment that is required by a given food product is covered in more detail in Chapters 2, 5, and 6. In general, the pH of the foodstuff is the broadest indicator, as well as storage temperature, shelf life expected, and the natural microbial flora of the foodstuff. For low-acid products where *Cl. botulinum* is a problem, the minimum thermal treatment required for public health is an F_0 value of 3 minutes. Foods containing nonpathogenic microbial spores that are more resistant than *Cl. botulinum* but that will cause spoilage of the product require higher levels of thermal treatment, usually 5D to 7D of the most resistant spores present in the product and that will grow in it. A first guess for the level of heat treatment required for a given product can be estimated if the F_0 value usually used for an in-container process is known. Some values are given in Table 6–2.

The extension of the F_0 value data given above from its reference temperature of 121°C into the UHT range of temperatures must be done with caution, as discussed in Chapter 2.

It can be shown that an F_0 value of 5 minutes is equivalent to a residence time of 3 seconds at 141°C using constant-*z* kinetics, but using Arrhenius kinetics a residence time of 4.2 seconds would be required (see Chapter 2).

3.20 PREDICTION OF PROCESS PERFORMANCE USING MEASURED PLANT DATA

The ideal process would heat the product instantaneously from inlet temperature to the process temperature required, hold each element of it for the same time, then cool it instantaneously to the temperature required for packaging. Needless to say, no process equipment approaches this ideal, although direct heating and flash cooling will give temperature change close to instantaneous, but there will be a spread of residence times in the holding section. The evaluation of sterilizing performance is considered first from the process temperature profile, then the effect of residence time distribution is considered.

3.21 PREDICTION OF PROCESS PERFORMANCE FROM TIME-TEMPERATURE PROFILE

The sterilizing performance and biochemical change in a given product can be predicted from the thermal death data of the microorganisms or biochemical components and their time-temperature history as they pass through the plant. The method used is similar to that used for calculating the F_0 value for an in-container process from temperature readings at the slowest heating point in the container as a function of time (see Section 3.22).

The time-temperature profile of the process plant must first be determined; temperatures must be measured at as many different points as possible in the process, and the average residence time of the product between them determined from the volume of that part of the process divided by the volume throughput passing through it (using Equation 3.14). In this way, a graph of the temperature of the product against time can be constructed and the proportion of surviving microorganisms, or survival of biochemical components, calculated overall. The shape of the temperature profile depends on the type of heat exchanger used; direct heating or flash cooling will give a virtually instantaneous temperature change but indirect heating heat exchangers heat or cool in a finite time. The rate of temperature change is dependent on the value of the heat transfer coefficient and temperature difference between the product and heating or cooling medium.

As a first estimation, the performance of the thermal process can be calculated using the average residence time of product in each section but, because some elements of fluid will have a shorter residence time than this, more microorganisms will survive the process, leading to a potential public health problem. There is a good argument instead for using the minimum residence times for each section of plant (which may be predicted from the average residence time), which will lead to an underestimation of the sterilizing performance, but will at least err on the side of caution. The effect of residence time distribution on the sterilizing efficiency of a process is covered in detail later in Section 3.30.

3.21.1 Determination of Time-Temperature Profile for a Process

It is relatively easy to measure the temperatures of the product at the junctions between the individual items of process equipment (i.e., at the inlets and outlets of heat exchangers and the holding tube), but temperatures in between these points are much more difficult to determine. In plate heat exchangers, thermocouple wires may be inserted between plates and held in the product flow by a rubber spacer of the correct size near the thermocouple tip; the rubber gasket will form an acceptable seal to prevent leakage. For tubular heat exchangers, thermocouples may be mounted in the tubes at known points and the wires fed out through a suitable seal, while with scraped-surface heat exchangers, voltages from thermocouples mounted on the mutator may be connected to the thermocouple readout via slip rings. With the latter, inaccuracy from electrical resistance in the slip rings can be minimized by first amplifying the signal in a circuit attached to the mutator shaft (Maingonnat & Corrieu, 1983).

There are many difficulties with this approach, including the effect of fouling, which will change the temperature profile with time, and the temperature profile through heat exchangers is often predicted rather than measured.

The temperature profile through an indirect heat exchanger can be predicted using Equations 3.1 and 3.2. The heat exchanger duty (Q) is the rate that heat is gained by the product but this must also be the rate at which the heating medium loses it. In the regeneration section of the process, with product heating product in countercurrent flow, the mass flowrates and specific heats are approximately the same, so the temperature rise in the incoming product is the same as the temperature fall in the outgoing product, and the temperature difference is constant throughout the heat exchanger. In this case, the temperature profile for both liquids is linear.

However, this situation is unusual; normally the heating medium is water and has both a different mass flowrate and different specific heat to the product, and its temperature change will be different from that for the product. The rate of heat exchange is proportional to the temperature difference at any point and will therefore be different at different points in the heat exchanger, affecting the temperature profile of both liquids. This effect has been investigated by Kiesner and Reuter (1984), who calculated the profiles given in Figure 3–23.

The effect of fouling, however, is to reduce the rate of heat transfer, and the product temperature profile becomes more linear as fouling increases. In most cases where there is liquid-liquid heat exchange in countercurrent flow, a linear temperature change is a reasonable assumption for the purposes of determining process plant performance. In the case of steam or vaporizing refrigerant sections, it can usually be assumed that the temperature of the medium is constant throughout the heat exchanger, giving an exponential temperature profile in the product (Figure 3–24). The temperature profile will need to be predicted more accurately, especially since, in this section, the product temperature approaches the maximum process temperature and will contribute significantly to the integrated sterilization or biochemical change value being calculated.

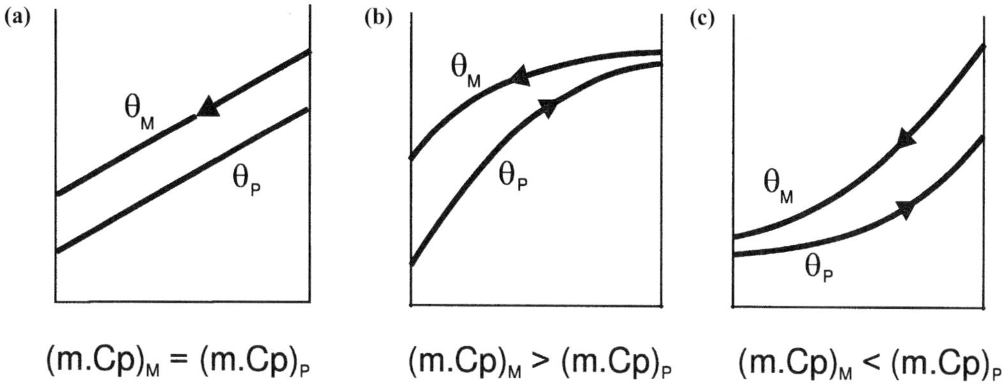

Figure 3–23 Temperature profiles for product and pressurized hot water heating medium in indirect heat exchanges. (a), For equal values of (mass × specific heat). (b), Where (mass × specific heat) of heating medium greater than for product. (c), Where (mass × specific heat) of heating medium less than for product. *Source:* Reprinted with permission from C. Kiesner and H. Reuter, *Kieler Milchwirtscaftliche Forschungsberichte*, Vol. 36, No. 2, pp. 67–79, © 1984, Verlag Thomas Mann.

3.22 CALCULATION OF PROCESS PERFORMANCE FROM TIME-TEMPERATURE DATA

Having established the time-temperature profile, any of the sterilization or biochemical change criteria defined in Chapter 2 can be determined (e.g., F_0, B*, or C* values) by integrating the appropriate function with respect to time, either using a numerical algorithm (such as Simpson's rule) or graphical integration, for example:

$$F_0 = \int_0^t 10^{(\theta-121)/10}\,dt$$

or (3.7)

$$B^* = \left(\frac{1}{10.1}\right)\cdot\int_0^t 10^{(\theta-135)/10.5}\,dt$$

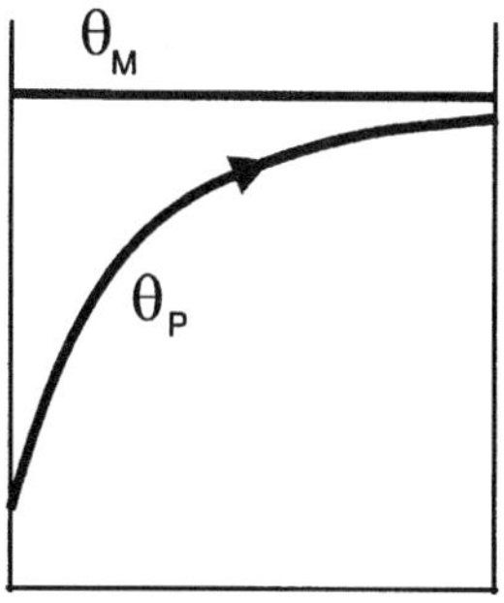

Figure 3–24 Temperature profile for product and steam heating medium in indirect heat exchange

The time-temperature profiles for a number of milk UHT plants have been collected in Germany by Reuter (1982) and Kessler and Horak (1981a, 1981b) and examples are given in Figure 3–25. Although these data are presented for milk, a similar analysis could be made for any food product using any thermal process criterion or chemical damage to give a similar analysis of the performance of a process.

Residence times through the different processes vary considerably, from approximately 90 seconds to 380 seconds, and the different sections such as preheating, final heating, holding, and cooling are easily visible. The primary difference between the curves is attributable to the use of direct heating or indirect heating heat exchangers as described above; direct heating and flash cooling give a virtually instantaneous temperature change similar to curves (1) and (3) and indirect heat exchangers heat or cool in a finite time, giving a sloping heating or cooling curve similar to curves (7), (9), and (12). Some of these processes will be designed to have a greater heat recovery, or regeneration, which requires a larger heat exchange area due to the lower temperature differences involved and therefore a greater volume holdup in the plant, resulting in a profile similar to curve (9). The product temperatures and times in the holding tube are given in Table 3–1.

The authors have also calculated the B* and C* values for these curves, which are given in Table 3–1. In addition, the B* and C* values for the holding tube only, without the remainder of the process, are presented in this table. It must be stressed that the values for time are

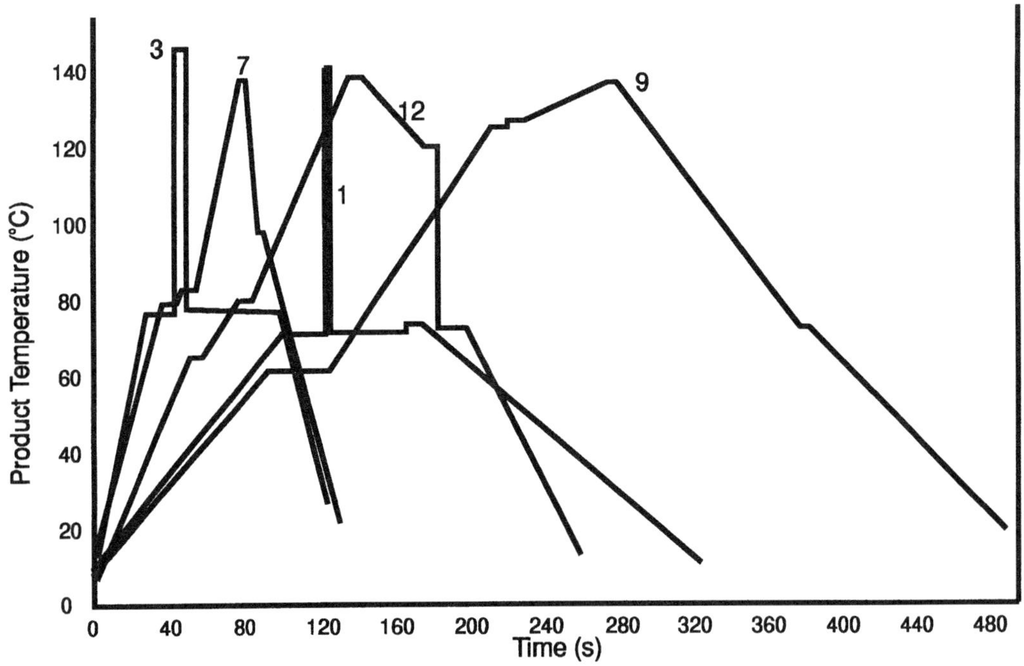

Figure 3–25 Time-temperature profiles for a number of milk UHT plants. Numbers refer to plants in Table 3–1. *Source:* Adapted with permission from H. Burton, *Ultra-High-Temperature Processing of Milk and Milk Products*, p. 174, © 1988.

average residence times and assume no distribution of residence time; therefore, the values calculated will be an overestimation of the values obtained in practice.

3.22.1 Microbiological Criteria

An examination of the calculated B* and C* values given in Table 3–1 is very useful as a comparison between the different processes. Very few processes operate at a B* value close to 1.0, proposed by Horak (1980) as the minimum for milk (see Chapter 2), and many are much higher, up to B* values of 9.4. The latter certainly are giving a heat treatment far in excess of that required and, for some, the operating temperature (or holding time) could be reduced without any practical risk of nonsterility. An examination of the calculated B* values for the holding tube alone demonstrates that for direct heating plants, nearly all the sterilization occurs in the holding tube, while for indirect heating plants, a significant proportion of the sterilization effect occurs in the final heating and the cooling/regeneration sections. It is interesting to note that the holding tube in plant 9 could be removed entirely and the process would still give a sterilization effect higher than necessary! The B* values would be lower in practice than those given in Table 3–1, as the effect of residence time distribution in each section has not been included and would result in a reduction of approximately 80% of the calculated value.

Table 3–1 Process Conditions and Calculated B* and C* Criteria for Different Commercial UHT Processes for Milk

		Nominal Holding Conditions		*Overall Process*		*Holding Tube Only*		
Plant Number	*Plant Type*	*Temperature °C*	*Time (s)*	*B* Value*	*C* Value*	*B* Value*	*C* Value*	*References**
1	D	142	2.4	1.1	0.17	1.10	0.13	1,3,4
2	D	140.5	5.6	1.9	0.31	1.85	0.27	1,3,4
3	D	147	6.7	9.4	0.57	9.22	0.53	1,3,4
4	D	140	4	1.2	0.19	1.19	0.19	2,3
5	I,p	139.5	2.2	1.9	0.68	0.58	0.10	1,3,4
6	I,t	138.5	2.6	1.6	0.54	0.55	0.11	1,3,4
7	I,p	139	3.4	1.7	0.6	0.81	0.15	1,3,4
8	I,p	138.5	5.2	2	0.87	1.11	0.22	1,3,4
9	I	138	6	6.7	3.03	1.15	0.25	1,3
10	I	138	1.7	1.5	0.67	0.32	0.07	1,3
11	I,p	137	4	1.9	0.83	0.61	0.15	2,3
12	I,p	139.5	6.5	9	2.23	1.73	0.30	1,3,4
13	I,t	138	1.1	1.2	0.67	0.21	0.04	5

**References:* 1, Reuter (1980); 2, Damerow (1980); 3, Kessler and Horak (1981a); 4, Reuter (1982); 5, Kessler and Horak (1981b). D, Direct heating; I, indirect heating; p, plate type; t, tubular type.

Source: Adapted with permission from H. Burton, *Ultra-High-Temperature Processing of Milk and Milk Products*, p. 175, © 1988.

3.22.2 Chemical Quality Criteria

Horak (1980) proposed a C* value below 1.0 as representing good-quality milk with thermal denaturation of thiamin below 3% (i.e., the lowest level he could measure). From Table 3–1, it can be seen that the C* values for the overall thermal process are generally lower for direct heating equipment than for indirect heating plants, representing the longer time the milk spends at temperatures lower than that at which the death of microorganisms becomes significant. Two of the processes (9, 12) have C* values significantly higher than the maximum 1.0 specified and have very long overall residence times within the process (and high regeneration efficiencies), which appear to be the reason for this value.

The C* values attributed to the holding tube alone are not very different for the two heating methods, and for indirect heating plants are only a small proportion of the value for the whole process, a result to be expected considering the time spent at temperatures where the rate of thermal denaturation is significant but the rate of microbial death is not.

3.22.3 Summary of This Work

Although relevant to milk and milk products only, this work demonstrates several basic principles that can be applied to all continuous-flow thermal processes. The major conclusion is that the more rapid the heating, the lower the biochemical damage incurred in the product, since a substantial contribution to biochemical damage (more than to sterilization) can be achieved in the heating up and cooling parts of the process, due to its lower dependence on temperature. Although this can be deduced from a knowledge of the sterilization and degradation kinetics (and indeed the basic principle of UHT processes) it is interesting to apply it to practical processes. The major difference is between direct and indirect heated plants, where the former has extremely rapid heating and cooling but has a relatively poor heat recovery and is hence more energy intensive with higher operating costs. For indirect heating plants, the factors that affect the heating rate are (from Equation 3.2) the heat exchange area, the overall heat transfer coefficient for the heat exchangers, and the temperature difference between the product and the heating medium. Of these factors, the temperature difference is usually limited to prevent fouling and therefore it is important to concentrate on improving the heat transfer coefficient in order to improve the rate of heating of the product. The relationship between the rate of heating and the heat transfer area is slightly more complex in that, for a given throughput, the larger the heat transfer area (i.e., the larger the heat exchanger), the larger the volume holdup of product in the heat exchangers, the longer the residence time in them (from Equation 3.16), and therefore the slower the rate of heating would be. If designing a process with a high rate of heat recovery (to minimize operating costs), a large heat transfer area would be required, as the temperature difference in the heat exchanger would be lower. The different temperature profiles shown in Figure 3–25 are generally due to the different heat recoveries of the processes. In addition, the different types of heat exchanger would give different rates of heating depending on the ratio of heat transfer surface to volume holdup; that is, plate heat exchangers will give more rapid heating than nonconcentric tubular heat exchangers but not necessarily for concentric tube types, all else being equal.

It can be concluded from an examination of these temperature profiles that processes with high heat recovery and hence low operating costs (but higher capital cost in terms of heat transfer area) will produce a product with greater biochemical damage. However, the design of a process requires that many more factors be taken into consideration, and the operating cost is an important one. What is not known is the exact effect of the extent of biochemical damage (e.g., C* values) on the final organoleptic quality of the product and whether the difference between these processes can actually be detected by the average consumer.

RESIDENCE TIME DISTRIBUTION

3.23 EFFECT OF RESIDENCE TIME DISTRIBUTION ON PLANT PERFORMANCE

As mentioned above, the ideal flow of product would be for each element of fluid to take the same time to pass through each section of the process. This ideal, known as plug flow, is almost never achieved in practice and there always is a range of residence times for the product, the range depending on the shape of the process section and the characteristics of the fluid.

The flow of any fluid through a pipe occurs in one of two ways:

1. By streamline flow (sometimes called laminar flow), in which the elements of fluid flow as concentric shells of liquid, with the fastest velocity down the center and slowest at the pipe surface (i.e., a parabolic flow profile). There is no mixing between the layers.
2. By turbulent flow, where there is substantial radial mixing across the pipe, flattening the velocity profile. There is a very thin layer of liquid in streamline flow next to the channel wall (the laminar sublayer) and between that and the turbulent flow, a thin buffer zone.

The point at which the flow regime changes from streamline to turbulent can be predicted using the Reynolds number (Re):

$$Re = \frac{D.U_m.\rho}{\mu} \tag{3.8}$$

where D is the diameter of the pipe, U_m is the average flow velocity, and ρ and μ are the density and viscosity of the fluid, respectively. The average flow velocity, U_m, is calculated from the volume flowrate passing through the pipe divided by the cross-sectional area of the pipe; that is, for a pipe of circular cross-section:

$$U_m = \frac{4.\text{Volumetric flowrate}}{\pi D^2} \tag{3.9}$$

For most practical situations where fluid flows through tubes of circular cross-section, streamline flow generally occurs at Re <2,100 and turbulent at Re >4,000, with a transitional stage between these values where it may be either. The Reynolds number at which the flow regime changes is called the critical Reynolds number, Re_{crit}. For foodstuffs with non-Newtonian viscosity, however, the critical Reynolds number is different. This is covered in Section 4.3.

It must be noted, however, that it is possible to induce turbulence at Reynolds numbers as low as 200 using a convoluted channel shape (as found in plate heat exchangers). In addition, using special techniques, it is possible to maintain streamline flow up to very high Reynolds numbers (about 30,000), although once the flow has turned turbulent, the Reynolds number must be reduced to below 2,100 for it to revert to streamline again.

The viscosity of the fluid is a very important property. The following generally applies to low-viscosity liquids similar to water, whose viscosity is Newtonian (i.e., their viscosity does not rely on the shear rate applied to the liquid). Liquids that have high viscosity and are non-Newtonian generally behave differently and have different residence time distributions and are covered in detail in Chapter 4.

3.24 VELOCITY PROFILES FOR LOW-VISCOSITY NEWTONIAN LIQUIDS

The flow of fluids through items of processing equipment results in many flow patterns, which gives a distribution of residence times for that fluid in the equipment. Generally, these residence time distributions must be measured, but may in some cases be calculated from the velocity distribution of the liquid across the pipe or other channel shape. These predictions, however, should be taken with care; inaccuracies are covered in Section 3.29.

In the case of streamline flow in a holding tube of circular cross-section, the velocity of the fluid (u_r) at any radius (r) in the tube measured from the center can be derived from a force balance on an element of fluid and integrating across the pipe (Coulson & Richardson, 1977). The resulting equation is

$$u_r = 2.U_m\left[1-\left(\frac{r}{R}\right)^2\right] \tag{3.10}$$

where R is the actual radius of the tube and U_m is defined by Equation 3.9. It can be seen from this equation that the flow is parabolic with the maximum velocity down the center of the tube ($r = 0$) at a value twice that of the mean velocity.

However, for turbulent flow, the velocity profile in a tube of circular cross-section is quite different and is dependent to some extent on the degree of turbulence present, and hence the degree of radial mixing, and the size of the laminar sublayer. At high Reynolds numbers, the velocity distribution in the liquid is given by the one-seventh power rule and

$$u_r = 1.22U_m\left[1-\frac{r}{R}\right]^{\frac{1}{7}} \tag{3.11}$$

except near the wall, in the laminar sublayer. The two flow regimens are compared in Figure 3–26.

A velocity ratio ξ can be defined, which is the ratio of maximum velocity to average velocity:

$$\xi = \frac{\text{maximum fluid velocity}}{\text{average fluid velocity}} \tag{3.12}$$

This velocity ratio, ξ, for turbulent flow is 1.22 as opposed to 2.0 for streamline flow.

3.25 RESIDENCE TIME DISTRIBUTION THEORY

The method of expressing residence time distributions was developed by Danckwerts (1953), who gave three functions representing the distribution. The most useful function for continuous-flow thermal processes is the exit age distribution, E(t), which represents the proportion of the fluid that leaves the relevant section in time t. This is measured by injecting a pulse of tracer at the inlet to the section of interest and detecting the concentration at the outlet, C(t), as a function of time (t). The graph of tracer concentration, C(t), versus time (t) is called the C curve and the area under the graph is the amount of tracer injected. The curve can now be normalized; that is, the area under it is equal to 1.0. This is achieved by dividing each C(t) value by the area under the original curve and the values obtained are now equal to the E(t) values mentioned above.

This can be expressed mathematically, as follows:

$$\text{Area under curve} = \int_0^\infty C(t)\,dt = \text{amount of tracer injected } (M_t) \tag{3.13}$$

and

$$\int_0^\infty \frac{C(t)}{M_t}\,dt = \int_0^\infty E(t)dt = 1.0 \tag{3.14}$$

There are two other functions that express residence time distribution, the F curve and the I curve. The F(t) curve is the fraction of liquid that has a residence time less than time (t) and can be obtained by making a sudden step change in the tracer concentration in the inlet

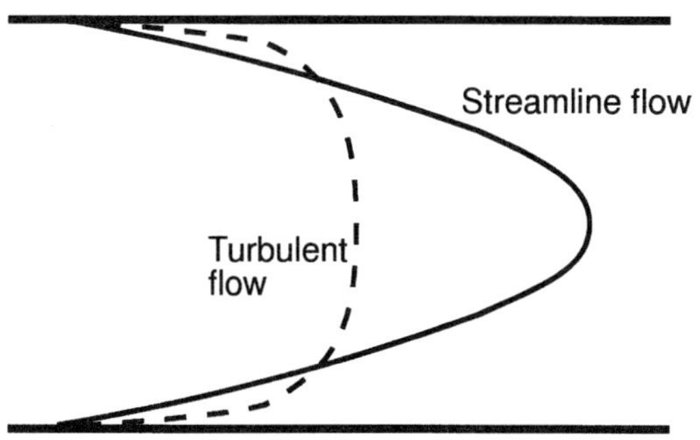

Figure 3–26 Comparison of velocity profiles of streamline flow and turbulent flow regimens in cylindrical pipe

stream. The concentration of tracer in the outlet stream, when normalized, is the F(t) curve. The I(t) curve (internal age) is the fraction of liquid that has a residence time greater than time (t) and is determined in a manner similar to the F(t) curve. The I(t) function is rarely used in food processing and will not be considered further here.

The different functions are related as follows:

$$F(t) = \int_0^t E(t)\,dt \qquad E(t) = \frac{dF(t)}{dt} \qquad I(t) = 1 - F(t) \tag{3.15}$$

The two functions E(t) and F(t) are compared in Figure 3–27 for a typical residence time distribution curve.

The mean residence time t_m of the liquid in the section under study is important. It can be calculated from the volume of the section (V) divided by the volumetric throughput of the product going through it (Q):

$$t_m = \frac{V}{Q} \tag{3.16}$$

The mean residence time can also be determined from the E(t) or F(t) curves. For the latter, t_m is given as the time when F(t) = 0.5, by definition. From E(t) curves, t_m is given by the first moment of the distribution:

$$t_m = \int_0^\infty t\,E(t)\,dt \tag{3.17}$$

The mean residence times calculated from the residence time distribution curve by Equation 3.17 and determined from Equation 3.16 should be the same value. If not, this demonstrates that there is an inaccuracy in the measurements taken, that the tracer is not inert, or that channeling is taking place, with areas of dead flow in the equipment (Levenspiel, 1972).

The second moment of the distribution is the variance, σ^2, which describes the measure of the spread of the curve about the mean value and is defined as

$$\sigma^2 = \int_0^\infty t^2\,E(t) - t_m^{\,2} \tag{3.18}$$

In addition to the mean and variance of the curves, the third and fourth moments of the distribution can also be calculated to describe the distribution accurately. These moments are the skewness and kurtosis of the distribution, respectively, which describe the degree of departure from a symmetrical curve and the type of peak of the distribution.

The residence time distribution can also be normalized with respect to time by dividing the time values by the mean residence time t_m to give dimensionless time (θ). The resulting distribution $E(\theta)$ allows comparison of distributions with different mean residence times.

3.25.1 Discrete Data

The tracer concentration may be determined in situ using a data logging system or by collecting samples at set time intervals for later analysis. The conversion of these data at

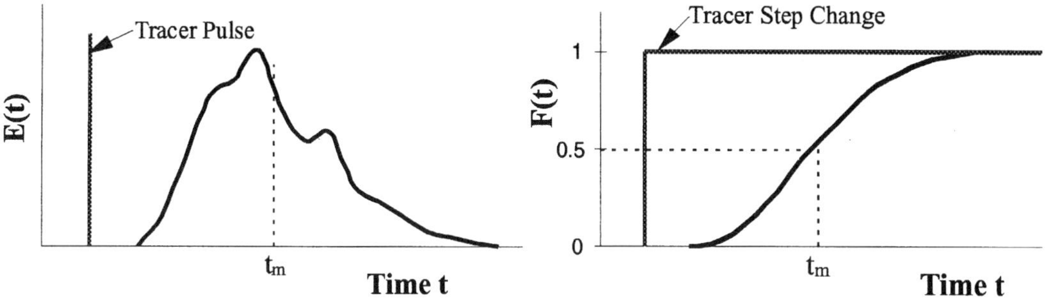

Figure 3–27 Comparison of E(t) and F(t) curves for an non–ideal flow system

discrete time intervals to E(t) values may be achieved using the following equations:

$$t_m = \frac{\sum_{i=0}^{\infty} t_i\, C_i(t)}{\sum_{i=0}^{\infty} C_i(t)} \tag{3.19}$$

and

$$\sigma^2 = \frac{\sum_{i=0}^{\infty} t_i^2 C_i(t)}{\sum_{i=0}^{\infty} C_i(t)} - t_m^2 \tag{3.20}$$

where C_i is the tracer concentration at time t_i.

3.26 MEASUREMENT OF RESIDENCE TIME DISTRIBUTION

The most common method of measuring residence time distribution (RTD) is to inject a tracer at the inlet to the flow system and detect its concentration at the outlet with time. On injection, the tracer must be as close as practical to an infinite concentration in zero time (a Dirac delta function); the detection may be achieved either by continuously monitoring the outlet by a suitable sensor and data recording system or by collecting samples at short time intervals for later analysis.

The tracer obviously must not affect the properties or flow behavior of the liquid. It must not adsorb onto surfaces, degrade, undergo reaction, or affect the behavior of the food product as it passes through the equipment. In addition, it must be capable of being injected in a small volume at high concentration (i.e., a good approximation to the Dirac function). Various tracers have been used for residence time distribution measurements. In order to construct an accurate E(t) curve, a continuous in-line detection system with fast data collection gives the most accurate results, especially when residence times are short. Sodium chloride is very commonly used and is detected by conductivity measurements, but to prevent polarization of ions around the conductivity probes, an alternating current bridge should be used. Sodium chloride cannot be used in foods with a naturally high conductivity, or in foods that contain added salt, as the sensitivity of the detector is reduced. Care must also be taken to ensure that the addition of sodium chloride does not affect the product chemically, or alter its viscosity, especially when injected into the foodstuff at high concentrations. It must also dissolve evenly in the foodstuff; this may cause a problem, for example, when fats or oils are present.

Dyes have also been used successfully and their concentrations detected in the product by a photosensitive semiconductor. The light emitter and detector can be arranged as two fiber optic probes, each mounted on either side of a pipeline, or as a single probe using a split fiber optic bundle in a Y-shape. One leg of the Y is connected to a light source and the other leg to a photosensitive semiconductor and logging system; the central leg is fitted into the pipe through which the product (and dye) flows. Light from half the fiber optic bundle is reflected back either from the interior pipe wall or from the product itself (for transparent and opaque foodstuffs, respectively) and passes down the detection leg to the detector and computer logging system (Heppell, 1985; Patel & Wilbey, 1990). It is important for sensitivity that the color of the dye is selected with the color sensitivity of the detector in mind. With the arrangement just described, absorption of light is measured and hence dye and detector must be sensitive to different colors (i.e., red dye and green-sensitive detector). Care must be taken when selecting a suitable dye for continuous-flow thermal processes that it is temperature stable in the food product; for example, methylene blue has been used in milk at UHT temperatures but will actually decolorize under the reducing conditions present at these temperatures in milk. Heppell (1985) screened many dyes before finding that Basic Fuchsin was stable under these conditions. The dye obviously must also be able to be detected; for example, if the foodstuff is a deep brown color, it will be difficult to detect any color dye. Fluorescent dyes may be of value in this situation, which can be stimulated by light at one frequency and emit light at a different one. Dyes are easy to inject in high concentrations without affecting the foodstuff, and can be detected at very low concentrations.

Other tracers that may be used are acids or bases, with change in pH measured, or other chemicals such as nickel chloride or sodium nitrite, which cannot be detected in situ, and samples must be taken for laboratory determination. Any system where samples must be taken will obviously be less accurate, depending on the number and size of samples, and at temperatures above 100°C pose severe problems in preventing "flashing" of the product. Radioactive tracers have also been used but pose serious safety problems in handling and plant decontamination after exposure and give few advantages over other tracers.

3.26.1 RTD for Any Input Signal

There are many difficulties encountered in ensuring a good injection pulse of tracer in high concentration in a very short time and a small deviation from the ideal Dirac function can induce errors in the final E(t) trace detected. An alternative method is to inject a tracer distribution of any shape into the flow system upstream of the section of interest and then to detect the tracer concentration at both the inlet and outlet of the section. The RTD of the section can be calculated by deconvolution of the inlet and outlet signal. This is achieved by taking Fourier transforms of each signal, dividing the outlet by the inlet signal and taking the inverse transform (Levenspiel, 1972). This technique is very useful either where a Dirac delta function cannot be obtained or where the residence time distribution of several sections is being measured simultaneously with one injection of tracer at the plant inlet, as shown in Figure 3–28.

If there is substantial "noise" on the signals, it may be necessary to smooth them first, either by using a weighted digital filter (Savitzky & Golay, 1965) or by filtering the signal in the frequency domain using Fourier transform methods. The subject of smoothing signals from a data logging system is covered in Teixeira and Shoemaker (1989).

3.27 MODELS FOR RTD DATA

There are two basic models to which the RTD data can be fitted: the tanks-in-series model and the dispersion model. In the former, the data are fitted to a model consisting of a number of continuously stirred tank reactors (CSTRs) in series, and the latter fits to a model that assumes that the liquid has a plug flow with a linear amount of diffusion or mixing. Although the data may be easily fitted to such models, interpretation of the model derived is not always significant, relevant, or useful! It may be possible, however, to identify non–ideal flow in the equipment, such as bypassing or "dead" areas, where no flow occurs.

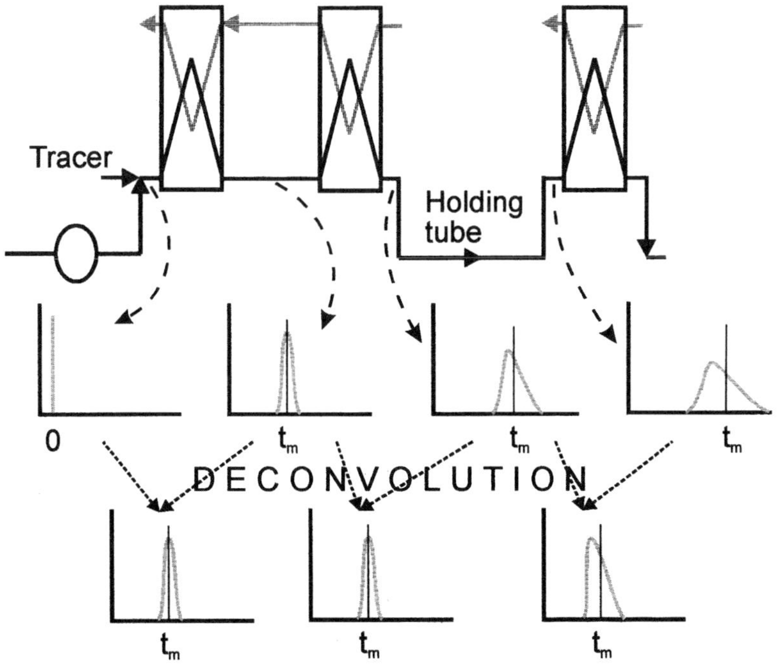

Figure 3–28 Calculation of residence time distribution for different process sections from single tracer input

3.27.1 Tanks-in-Series Model

The model described is a number of equal-sized CSTRs in series through which the liquid flows, with no lag time for transfer of liquid between each tank, as shown in Figure 3–29.

For the first tank, the tracer enters and, if perfectly mixed, is instantly distributed through the tank and appears instantly in the outlet then falls in concentration with time. The outlet from this tank appears immediately at the inlet to the second tank and so on, giving, for N equal-sized tanks

$$E(\theta) = \frac{N(N\,\theta)^{N-1}}{(N-1)!} e^{-N\theta} \tag{3.21}$$

The $E(\theta)$ curves for different numbers of CSTRs is given in Figure 3–30. It can be seen that as the number of tanks increases, the $E(\theta)$ curve approaches that of plug flow. The number of CSTRs can be determined for any $E(\theta)$ or $E(t)$ distribution from the variance of the distribution σ_θ^2 or σ_t^2 using

$$N = \frac{1}{\sigma_\theta^2} \quad \text{or} \quad N = \frac{t_m^2}{\sigma_t^2} \tag{3.22}$$

This model has been used by Sancho and Rao (1992) to model the residence time distribution in a holding tube; for water, the number of tanks was found to correspond to 14 to 20 for streamline flow and 61 to 103 for turbulent flow.

3.27.2 Dispersion Model

In this model, the flow is assumed to be plug flow through the equipment but, superimposed on this, there is a degree of diffusion or mixing in two directions, axial and radial, resulting in a dispersion of residence times (Figure 3–31). The mixing may be due to several factors, such as diffusion, the eddies present in turbulent flow, or mixing due to velocity gradients across the pipe. This model is called the Dispersion model or, more accurately, the dispersed plug flow model, where dispersion is expressed in terms of a diffusion coefficient, D, in the axial direction only. Radial diffusion is ignored as, in plug flow, the tracer concentration should be the same radially, but there is zero concentration on either side of the tracer axially, where diffusion will occur according to Fick's second law.

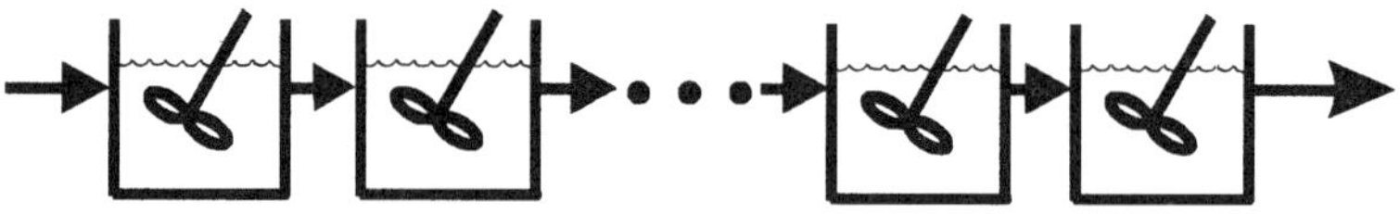

Figure 3–29 Model for equal size continuously stirred tank reactors

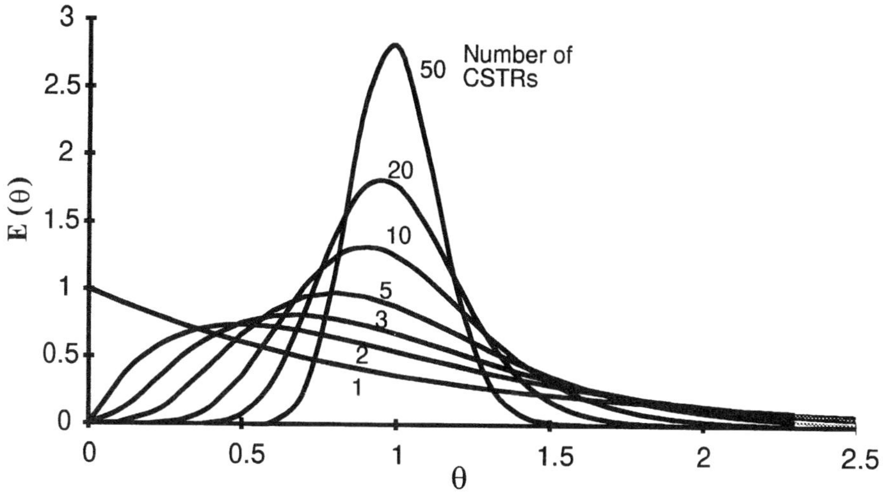

Figure 3–30 E(θ) curves for different numbers of CSTRs in tanks-in-series model

---

The diffusion coefficient is expressed in terms of a dimensionless group, (D/uL), where u is the fluid axial velocity and L is the length of the pipe. The inverse of this dimensionless group (uL/D) is the Bodenstein number (Bo) or the mass Peclet number (Pe). The degree of dispersion will vary from zero, when there will be plug flow and (D/uL) will be zero (or Bo = ∞), to a situation where the mixing is complete, where the value of (D/uL) will be infinite (Bo = 0), such as in the CSTR shown above. For small values of (D/uL), the tracer distribution is symmetrical about the mean residence time, and, as one assumption of the model is a constant (D/uL) value throughout the section, the spread of tracer increases down the flow section. For larger values of (D/uL), the tracer becomes increasingly skewed, as shown in Figure 3–32. The E(θ) distribution can be calculated from the (D/uL) value from the following (Levenspiel, 1972):

$$E(\theta) = \frac{1}{2\sqrt{\pi\theta(D/uL)}} \exp\left[-\frac{(1-\theta)^2}{4\theta(D/uL)}\right] \tag{3.23}$$

The method for calculating the (D/uL) value for a measured residence time distribution is given by Levenspiel (1972) and depends on the actual value of (D/uL) and the type of vessel for which the distribution has been measured.

For (D/uL) < 0.01 (approximately), the value of (D/uL) can be calculated from the σ^2 value of the E(θ) curve from:

$$\sigma_\theta^2 = 2\left(\frac{D}{uL}\right) \tag{3.24}$$

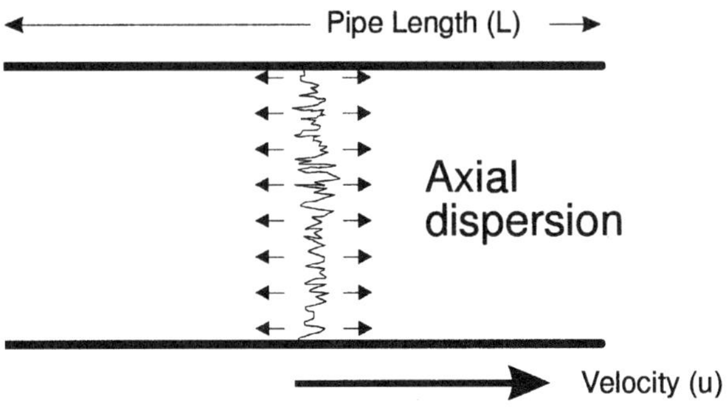

Figure 3–31 Axial dispersion (by diffusion from plug flow) used in Dispersion model

For larger values of (D/uL), the calculation of its value from the RTD curve is dependent on the type of section of interest, whether it is an open or closed vessel. An open vessel is defined as one into, and out of which, the fluid flows without disturbance, such as the mid-section of a pipe of constant diameter. For open vessels, the (D/uL) value can be calculated from:

$$\sigma_\theta^2 = 2\left(\frac{D}{uL}\right) + 8\left(\frac{D}{uL}\right)^2 \tag{3.25}$$

There are substantial problems in measuring RTD in this situation. Both the introduction of tracer and detection at the end of the section must be made without disruption of the flow and without creating a barrier across the section. Injection of the tracer must occur instantaneously, and both injection and detection must take into account the whole cross-section of the flow. One solution to the injection problem would be to use a tracer injection upstream of the section to be measured and to detect tracer concentration at the start and end of this section. As the variances are additive, the variance of the section can be found from the difference in the variances of the two signals:

$$(\sigma_\theta^2)_{section} = (\sigma_\theta^2)_{end} - (\sigma_\theta^2)_{start} \tag{3.26}$$

Alternatively, a closed vessel is one where there is a change in flow pattern at the start and end of the section of interest. This may be due to a change in diameter in the flow channel, a right-angle bend, mixing section, orifice plate, or other similar obstruction. For closed vessels, (D/uL) can be calculated from the following:

$$(\sigma_\theta^2) = 2\left(\frac{D}{uL}\right) - 2\left(\frac{D}{uL}\right)^2 \cdot \left(1 - e^{-\frac{uL}{D}}\right) \tag{3.27}$$

Sancho and Rao (1992) compared the number-of-tanks and dispersion models for flow in a holding tube. Neither model was found to be better than the other.

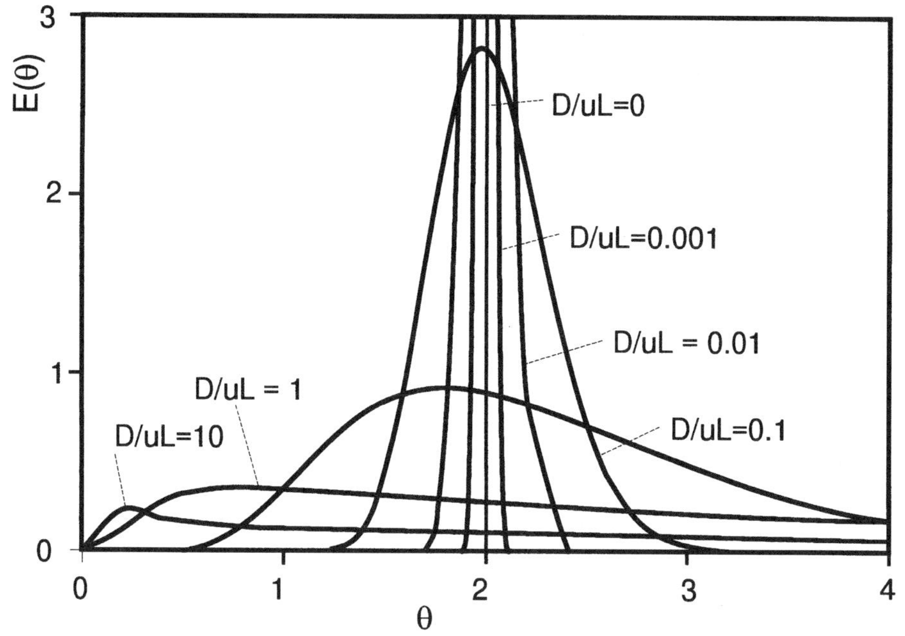

Figure 3–32 E(θ) curves for different values of D/uL in Dispersion model

3.28 FURTHER ANALYSIS OF PRACTICAL RTD CURVES

RTD curves can be further analyzed to detect faults in the flow patterns such as channeling or stagnant areas. This can be detected easily from a comparison of the mean residence time of the curve obtained with the mean residence time expected from Equation 3.16, where the volume of the fluid in the flow section can be calculated or determined experimentally. If the former is substantially shorter than the latter, then there is a stagnant area through which the fluid does not flow. If there are several peaks to the curve, this indicates the possibility of parallel flow channels or internal recirculation. The RTD data can be further analyzed by a variety of techniques to fit flow models that are combinations of CSTRs, plug flow, diffusion, and so forth, and are covered in more detail by Levenspiel (1972).

3.29 ERRORS IN PREDICTION OF RTD

Although all the above equations are very useful in designing process equipment, it must be stressed that they should be treated as approximate only, and RTDs should still be determined experimentally wherever possible using techniques outlined in Section 3.26. There are problems associated with developing the above equations, in particular that the assumption must be made that flow is fully developed. At the entry to a pipe, the velocity is the same across the whole section, and it takes some distance down the pipe, L, for the flow to become fully developed and achieve the velocity profile calculated from the equations. This distance, the entry length, L, can be calculated for streamline flow of a Newtonian liquid (Kay & Nedderman, 1985):

$$L = 0.0575.D.Re \tag{3.28}$$

For turbulent flow, profile development is much more complex, but a distance of 50 to 100 times the pipe diameter is assumed (Kay & Nedderman, 1985). For a 50-mm diameter pipe, the entry length for streamline and for turbulent flow will be in the order of 2½ to 5 meters, which is long enough for a process of about 7 tonnes of product per hour! If the holding tube is shorter than the entry length, the RTD will be more uniform than expected.

Another effect is present in turbulent liquids where the element of foodstuff that flows fastest in the center of the pipeline will not maintain its position in the center, but the radial mixing will move it into zones with different velocities. The minimum residence time will therefore not correspond to the fastest velocity but will be longer than expected from this. In other words, the holding tube efficiency, defined as the ratio of the minimum residence time to average residence time of the fluid, will be different from that predicted from the velocity ratio, defined in Equation 3.12.

In addition, no holding tube is straight and either contains bends or is coiled, and the presence of valves, instrumentation pockets, expansions, contractions, and other fittings make the theoretical prediction of RTD inaccurate, and hence practical measurements of the RTD should be carried out wherever possible. This was shown by Sancho and Rao (1992), who measured minimum and average residence times for water, Newtonian, and non-Newtonian liquids in a trombone-type holding tube with 18 bends and compared them to the values predicted from the velocity profiles using Equations 3.10 and 3.11. In all cases, the measured minimum residence times were longer than the predicted times, leading to an overprocessing of the liquid and hence poorer-quality, if safe, product.

3.30 EFFECT OF RTD ON STERILIZATION EFFICIENCY

The effect of a distribution of residence times of the product in the high-temperature sections of a continuous-flow thermal process can be quite severe. If we take a residence time distribution curve for a holding tube at constant temperature, as shown in Figure 3–33, the proportion of incoming microorganisms that would be expected to leave the tube in the short-time interval δt at time t, if no thermal death had taken place, would be expressed by the value of E(t). However, as these microorganisms have been in the high-temperature section for time t, the number of survivors in this time interval would be, from Equation 2.5,

$$N = E(t).10^{-t/D} \tag{3.29}$$

where D is the decimal reduction time for the microorganisms at the temperature of the holding tube. The total number of microorganisms surviving the process would be obtained by integrating Equation 3.29 to give:

$$N = \int_0^{\infty} E(t).10^{-t/D}\, dt \tag{3.30}$$

The initial number of microorganisms is given by:

$$N_0 = \int_0^{\infty} E(t)\, dt \tag{3.31}$$

which, by definition, has the value 1.0. The proportion of surviving spores (PSS) is therefore given by Equation 3.27 divided by Equation 3.28 and, as the latter has the value 1.0, gives

$$PSS = \int_0^{\infty} E(t).10^{-t/D}\, dt \tag{3.32}$$

This can be evaluated from the area under the graph of $E(t)\cdot 10^{-t/D}$ against time t and can easily be calculated from the E(t) values at a small time interval Δt by computer program, using:

$$\sum_0^{\infty} E(t).10^{-t/D}.\Delta t \tag{3.33}$$

or by more accurate algorithms for determining areas under curves, such as Simpson's Rule.

Consider the RTD curves given in Figure 3–34, measured by Heppell (1986) for milk in a small-scale UHT infuser. The mean residence time for the curve, calculated from Equation 3.16 or 3.18, is 3.27 seconds but the minimum holding time is 1.1 seconds, with significant amounts passing after 9 seconds. The distribution of surviving spores can be constructed using Equation 3.26 underneath the tracer curve. Surviving spores at the minimum residence time end of the curve have only been in the sterilizer for 1 second and their numbers are relatively high, whereas those at the end of the curve have been in the holding section for, say, 9 seconds and may be several log cycles lower and may not contribute significantly to the number of surviving spores. The survivor curve therefore becomes skewed toward the minimum residence time; the effect becomes more pronounced as the D value for the spores becomes smaller. The *B. stearothermophilus* spores used in this work had D values in milk of

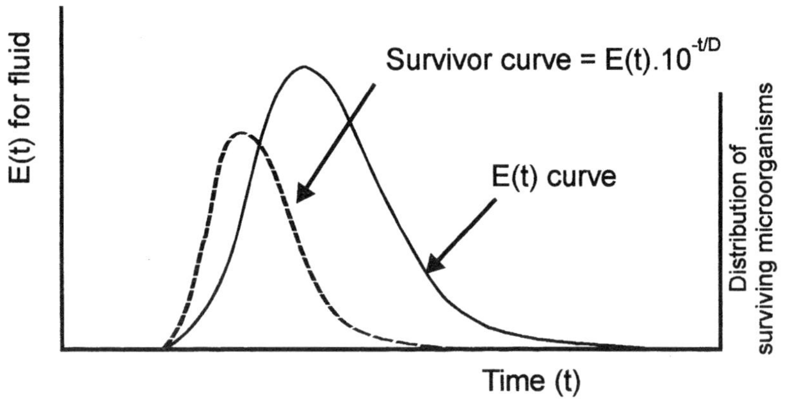

Figure 3–33 Typical RTD curves for product and for surviving microorganisms in a continuous-flow thermal process

5.5 seconds at 138°C and 0.6 seconds at 147°C. This effect can be clearly seen from Figure 3–34, where the higher the D value, the closer the distribution of surviving spores is to the minimum residence time. It is therefore the minimum residence time, and by implication, the spread of the RTD curve as typified by the σ^2 value for the curve, a large D/uL value, or a large number of CSTRs that is the important factor in determining the sterilization efficiency in a holding tube.

A bacteriologically effective mean holding time (t_b) can be defined as a single value of holding time that would give the proportion of surviving microorganisms obtained from Equation 3.29:

$$t_b = (-\log_{10}(PSS)) \,.\, D \tag{3.34}$$

This bacteriologically effective mean holding time is an average value of the RTD, which, it will be shown, is a function of the D value for the microorganisms, and therefore depends on the microorganisms, the temperature, and foodstuff. The value for this bacteriologically effective mean holding time will lie between the value of the mean residence time obtained from the E(t) curve and the minimum residence time for the product, depending on the relative D value for the microorganisms and the product residence time. Heppell (1986) calculated the mean residence times and bacteriologically effective mean holding times for the experimental UHT infuser plant used in this work, which are given in Table 3–2.

It must be noted that the mean of the survivor curve has no relevance for the proportion of surviving microorganisms, only the area under this curve.

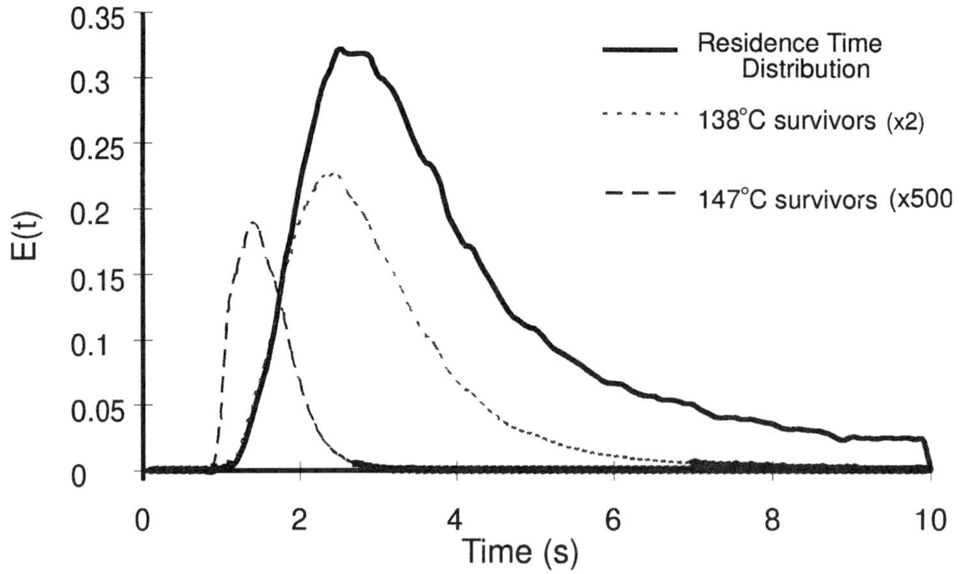

Figure 3–34 RTD curves for water and for surviving spores in a small-scale UHT infuser at 138°C and 147°C

3.30.1 Effect of RTD on Quality of Foods

When designing a holding section of a continuous-flow thermal process, the length is determined by the residence time distribution and decimal reduction time of the spores of most importance and must be long enough to give the required (small) proportion of surviving microorganisms. As can be seen from the above, the greater the spread of the residence time curve (σ^2 value), the longer the mean holding time must be to obtain the required proportion of surviving microorganisms. However, when considering the chemical changes associated with the thermal process, the longer mean residence time will result in a poorer-quality product (i.e., larger degradation of nutritional components or more browning). It is important for the quality of the product, therefore, to ensure that the σ^2 value for the residence time distribution curve is as small as practical.

There has been little study on techniques that may minimize the σ^2 value in a holding, or other, section and therefore give a flow profile closer to plug flow. The major improvement would be to ensure that flow is turbulent rather than streamline and therefore the diameter of the holding section must be selected to ensure the Reynolds number is well in excess of 2,100. It is unlikely that any significant reduction in σ^2 value could easily be obtained with turbulent flow, but for high-viscosity products in streamline flow, a more significant improvement may be expected. Techniques that induce turbulence, for example, by using a convoluted channel or by incorporating baffles, plates, or stirrer sections, to increase the number of effective CSTRs in the tube may have potential (see Section 4.5).

3.31 EVALUATION OF PLANT PERFORMANCE WITH TEMPERATURE DISTRIBUTION AND RTD

Many sections of a continuous-flow thermal process have an RTD and a simultaneous temperature profile in the fluid, such as in heat exchangers or even in unlagged holding tubes with a lower temperature around the periphery. The calculation of microbiological and chemical criteria in this situation is much more difficult as the RTD theory given in Section 3.30 is only valid for constant-temperature sections. Published work in this area is rare; Nauman (1977) proposed a thermal distribution time, $E(\gamma)$, which is the fraction of elements of liquid that has received a thermal load between γ and $\gamma + d\gamma$. The thermal load is defined as

Table 3–2 Comparison between Mean Residence, Minimum Residence, and Bacteriologically Effective Mean Holding Times (BEMHT) for Milk and Water in the Holding Section of an Experimental Infuser

	Process Temperature (°C)	*Spore D Value (s)*	*Mean Residence Time(s)*	*Minimum Residence Time(s)*	*BEMHT (s)*
Milk	138	5.5	3.29	1.1	3.29
	147	0.6	3.29	1.1	2.16
Water	138	5.0	3.27	1.5	3.27
	147	0.4	3.27	1.7	2.43

$$\gamma = \ln \frac{C_0}{C} = \sum k_0 \exp\left(\frac{-E_a}{R\theta}\right) \Delta t_i \qquad (3.35)$$

The RTD may actually be affected by the temperature profile across the channel in the heat exchangers, and data collected under constant temperature conditions must be used with care. For heating sections, the liquid next to the heat transfer surface is at a greater temperature than the rest and hence its viscosity will be lower. Unless radial mixing across the channel is large, this may well result in a greater velocity than normal in this area, and therefore a slower velocity near the center of the channel (i.e., the flow profile is flattened). Similarly, for cooling, the cooled liquid at the outside of the channel moves more slowly than normally expected, resulting in a faster "core" of hot liquid moving quickly down the center of the channel. It is likely, however, that for low-viscosity liquids, mixing across the channel is good, especially in plate heat exchangers where turbulence is induced, and also that the temperature difference between the liquid and heating medium would not be large anyway (both as a result of the regeneration and also to minimize deposition of solids on the heat exchanger surface) and therefore it is unlikely that this effect will be significant, if it is indeed measurable.

3.31.1 Direct Process Evaluation

The direct measurement of thermal treatment received from a process can be directly measured using single cultures of microorganisms or marker chemicals for which the denaturation can be readily characterized. Selection of an appropriate microorganism or chemical is obviously critical, especially with regard to the accuracy to which its concentration can be measured. The *z*-value of the denaturation is also vital, and must match that of interest, almost inevitably a microorganism denaturation, and hence must be in the region of 10°C. The use of marker microorganisms will first be considered, then the use of chemical markers, which, if possible to use, would be preferable. The problems considered here are also the problems associated with process evaluation for particulate liquids, covered in Section 4.21.

3.31.2 Marker Organisms

The difficulty in using marker microorganisms is the sheer number required, especially if the throughput of the process is high, as most systems for homogeneous foods are. The sterilizing effect of the process should, by definition, be high and hence the concentration of surviving microorganisms will be low but must be high enough to measure accurately. The concentration of microorganisms in the inlet will need to be high even if thermophilic organisms are used, probably in the region of 10^6 spores per milliliter. When combined with a high throughput, even as low as 1,000 L/h, one determination will require at least 10^{11} spores. There is also a problem with subsequent contamination of the process equipment, especially if thermophilic microorganisms are used.

Previous workers have used different microorganisms to measure the sterilizing effect of process equipment, introduced by Galesloot (1956), and Galesloot and Radema (1957) for tubular and plate systems that gave a log (proportion of surviving microorganisms) of less than –6 with *Bacillus subtilis* spores in milk (i.e., a 6D sterilizing effect). Franklin et al. (1958) tested a plate-type system with spores of *B. subtilis* in milk and found a sterilizing effect of 7D at 130.5°C rising to over 10D at 135°C. When the same plant was tested with *B. stearothermophilus*, of much higher heat resistance, the sterilizing effect was found to be 2D at 135°C and 4D at 138°C using colony count tests. Using dilution techniques to enumerate the surviving spores, however, gave an 8D sterilizing effect at 138°C, thus demonstrating the inhibiting effect of milk on spore germination and outgrowth.

Burton (1988) examined the results of work of Franklin, Underwood, Perkin, and Burton (1970) on the sterilizing effect in different plants and showed that a direct heating plant required a process temperature 4°C higher than an indirect heating one to give the same sterilizing effect, due to the extra heat treatment in the latter within the heating and cooling sections because of their slower rate of heating.

The temperature of the product in the holding tube (the process temperature) is important in that it is easier and faster to change this than to change the holding time for any given process, the latter requiring some variation in product flowrate or mechanical change in the holding tube length. Using the results from sterilizing effect measurements, Burton (1988) proposed a linear variation between sterilizing effect for the process and the holding tube temperature for any given process plant, and demonstrated that a small change in the latter has a large effect on the overall sterilizing effect. A change in process temperature of about 1°C was shown to give an increase in sterilizing effect from 8D to 10D for *B. stearothermophilus* with an indirect heating process, and therefore would be the best way of compensating for variations in product microbial quality. The chemical quality of the product would also be better than if the holding time were extended to give the same improvement.

Heppell (1990) proposed the use of spores entrapped in calcium alginate spheres (see Section 4.20) of less than 0.5 mm diameter to evaluate the sterilizing effect of a process plant. The method consists of introducing about 1 mL of entrapped spores into the inlet of the process and collecting them by in-line sieve just before the exit, followed by recovery and enumeration of the spores. The advantage of this method is the low total number of spores required, as only about 1 mL of spore spheres is required per measurement, irrespective of the process throughput. A high concentration of spores can be used (approximately 10^7/mL of spheres), and therefore high values of sterilizing effect can be measured and several determinations can be made.

The basis of the method is that the small size of sphere behaves both hydrodynamically and thermally the same as the foodstuff being processed. The thermal behavior of the sphere can be predicted using the mathematical model developed by Heppell (1990) for a 0.5-mm particle containing *B. stearothermophilus* spores with the minimum liquid-solid heat transfer coefficient corresponding to a Nusselt number of 2.0 (see Equation 4.35) exposed to the time-temperature profile of an experimental plate-type process. The model output is given in Figure 3–35 and compares the temperature at the center of the sphere and in the liquid, along with the proportion of surviving spores predicted for thc whole sphere and for the liquid. The small size of the sphere and the fact that its density is very close to that for water (specific gravity of 1.024) would mean that the difference expected between the velocity of the sphere

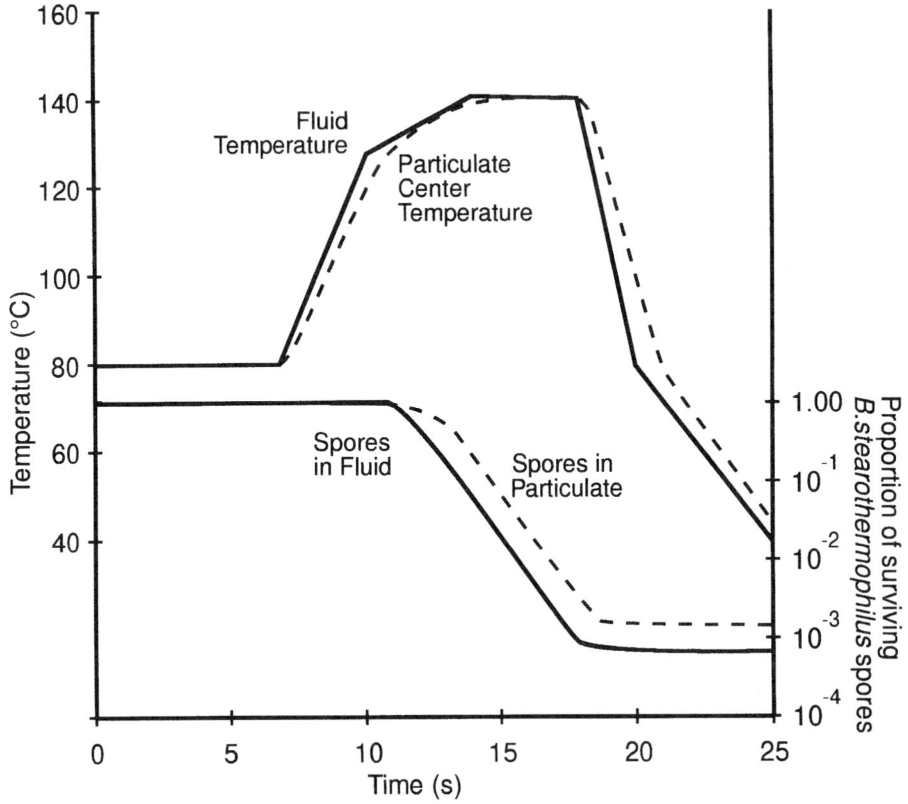

Figure 3–35 Prediction of temperature and proportion of surviving spores in 0.5 mm calcium alginate spheres compared to those for the product

---

and the liquid would be very small, especially in fully developed turbulent flow. The method obviously could not be used for a direct heating process where cooling is by direct expansion, due to clogging of the orifice plate or back-pressure valve.

3.31.3 Use of Chemical Markers

Chemical markers may be used to determine the overall sterilization efficiency of process equipment and have some advantages over the use of microorganisms, mainly in the accuracy to which their concentration may be determined. Selection of appropriate chemicals is difficult, however, as any reaction that occurs must be first order, with a *z*-value in the region of that for microorganisms (i.e., about 10°C), which is not typical of that for chemical reactions overall. Search for a suitable marker still continues, but work in this area is covered in more detail in Section 4.18.

REFERENCES

Alais, C., Humbert, G., Chuon, S.E. (1978). Ultra-flash sterilization of liquid foods by friction. *Annales de Technologie Agricole* **27** (4), 739–765.

Burton, H. (1958). *Journal of Dairy Research* **25**, 75.

Burton, H. (1988). *UHT processing of milk and milk products.* London: Elsevier Applied Science Publishers.

Burton, H., Perkin, A.G., Davies, F.L., & Underwood, H.M. (1977). Thermal death kinetics of *Bacillus stearothermophilus* spores at ultra high temperatures. III. Relationship between data from capillary tube experiments and from UHT sterilizers. *Journal of Food Technology* **12**, 149–161.

Coulson, J.M., & Richardson, J.F. (1977). *Chemical engineering.* Oxford, UK: Pergamon Press.

Damerow, G. (1980). Construction and operation of UHT installations. *Deutsche Milchwirtschaft* **37**, 1456–1463.

Danckwerts, P.V. (1953). Continuous flow system (distribution of residence times). *Chemical Engineering Science* **2**(1), 1–13.

Franklin, J.G., Underwood, H.M., Perkin, A.G., & Burton, H. (1970). *Journal of Dairy Research* **37**, 219.

Franklin, J.G., Underwood, H.M., Perkin, A.G., & Burton, H. (1970). Comparison of milks processed by direct and indirect methods of UHT sterilization. ii) sporicidal efficiency of an experimental plant for direct and indirect processing. *Journal of Dairy Research* **37**, 219.

Fredsted, L. (1996). In *Innovating for profit. Proceedings from the APV Technical Symposium at Anuga Foodtech*, November 2–12.

Galesloot, Th. E. (1956). *Netherlands Milk & Dairy Journal* **10**, 79.

Galesloot, Th. E., & Radema, L. (1957). *Rapporten Nederlandse Instituut voor Zuivelonderzoek*, No. 9.

Hallstrom, B. (1966). *Proceedings 17th International Dairy Congress*, Munich, Vol. B, p. 601.

Hallstrom, B. (1981). In *New monograph on UHT milk*, Document 130. Brussels: International Dairy Federation.

Heppell, N.J. (1985). Comparison of the residence time distributions of water and milk in an experimental UHT sterilizer. *Journal of Food Engineering* **4**, 71–84.

Heppell, N.J. (1986). Comparison between the measured and predicted sterilization performance of a laboratory-scale direct heated UHT plant. *Journal of Food Technology* **21**, 385–399.

Heppell, N.J. (1990). *Continuous sterilization processes.* PhD Thesis, Reading, UK: University of Reading.

Hersom, A.C. (1985a). In: *Symposium on aseptic processing and packaging foods.* Tylosand, Sweden: Lund Institute of Technology.

Hersom, A.C. (1985b). Aseptic processing and packaging of food. *Food Reviews International* **1**, 215–270.

Horak, P. (1980). *Uber die Reaktionskinetik der Sporenabtotung und chemischer Veranderungen bei der thermischen Haltbarmachung von Milch.* Thesis, Munich: Technical University.

Hostettler, H. & Imhof, K. (1963). *Milchwissenschaft* **18**, 2.

IChemE, 1988. *A Guide to the Economic Evaluation of Processes.*

Kay, J.M., & Nedderman, R.M. (1985). *Fluid mechanics and transfer processes.* Cambridge, UK: Cambridge University Press.

Kessler, H.G. & Horak, P. (1981a). Objective evaluation of UHT-milk-heating by standardization of bacteriological and chemical effects. *Milchwissenschaft* **36**(3), 129–133.

Kessler, H.G., & Horak, P. (1981b). *North European Dairy Journal* **47**, 252.

Kiesner, Ch., & Reuter, H. (1984). *Kieler Milchwirtschaftliche Forschungsberichte* **36**, 67.

Levenspiel, O. (1972). *Chemical Reaction Engineering.* Wiley International Edition. New York: Wiley & Sons Inc.

MAFF (1972). Recommendations for steam cleaning. Circular FSH 4/72 (England & Wales). London: HMSO.

Maingonnat, J.F., & Corrieu G. (1983). Etudes des performances thermiques d'un exchangeur chaleur à surface raclée. *Entropie* **No. 111**, 37–48.

Müller-Steinhagen, H. (1997). *Plate heat exchangers-past-present-future.* Engineering & Food at ICEF-7. Sheffield, UK: Sheffield Academic Press.

Naumann, E.B. (1977). The residence time distribution for laminar flow in helically coiled tubes. *Chemical Engineering Science* **32**, 287–293.

Page, T.V., Grandison, A.S., & Lewis, M.J. (1997). Engineering & Food at ICEF-7. Sheffield, UK: Sheffield Academic Press.

Patel, S.S., & Wilbey, R.A. (1990). Heat-exchanger performance—the use of food colourings for estimation of minimum residence time. *Journal of the Society of Dairy Technology* **43**(1), 25–26.

Perkin, A.G., & Burton, H. (1970). The control of the water content of milk during ultra-high temperature sterilization by a steam injection method. *Journal of the Society of Dairy Technology* **23**, 147–154.

Perkin, A.G. (1985). Review of UHT processing methods. *Journal of the Society of Dairy Technology* **38**, 69–73.

Phipps, L. (1985). *The high pressure dairy homogenizer*. Reading, UK: National Institute for Research in Dairying (Technical Bulletin No. 6).

Reuter, H. (1980). Evaluation of thermal efficiency of UHT installations for milk. *Milchwissenschaft* **35**, 536–540.

Reuter, H. (1982). UHT milk from the technological viewpoint. *Kieler Milchwirtschaftliche Forschungsberichte* **34**, 347–361.

Sancho, M.F., & Rao, M.A. (1992). Residence time distribution in a holding tube. *Journal of Food Engineering* **15**, 1–19.

Savitzky, A., & Golay, M.J.E. (1965). Smoothing and differentiation of data by simplified least squares procedures. *Analytical Chemistry* **38**, 1627–1639.

Shore, D.T. (1970). Engineering aspects of UHT processing. In *Ultra-high-temperature processing of dairy products* (p. 8). Huntingdon, UK: Society of Dairy Technology.

Swartzel, K.R., & Jones, V.A. (1980). Fail safe requirements for holding tube pressure in an UHT steam injection system. *Journal of Dairy Science* **63**, 1802–1808.

Teixeira, A.A., & Shoemaker, C.F. (1989). *Computerized food processing operations*. New York: Van Nostrand Reinhold.

UK Regulations. (1989). The Milk (Special Designation) Regulations 1989. Statutory Instruments No. 2383, London: HMSO.

Wiles, R. (1985). Aseptic filling into containers other than cartons, with special reference to sterile bulk packs. *Journal of the Society of Dairy Technology* **38**, 73–75.

CHAPTER 4

Continuous-Flow Thermal Processing of Viscous and Particulate Liquids

The aseptic processing of liquid food stuffs is reasonably well established, when those foods have a low viscosity and do not contain solid particulates of any substantial size, such as milk and fruit juices. However, when applying continuous-flow thermal processing technology to liquid foods with a high viscosity, there are some extra considerations that must be taken into account, mainly in establishing the correct residence time in the holding tube, but also in selection of the most appropriate heat exchangers. However, all elements of the food entering the holding tube can be assumed to be the same temperature, which is easily measured.

For foodstuffs that contain solid particulates, however, the process design of a continuous-flow thermal process becomes much more difficult. In order to transport the particulates through the system, the viscosity of the liquid is usually increased and hence we have the problems associated with high-viscosity liquids. On top of this, the liquid phase is heated first (in a conventional process) and heat must then transfer to the solid particulates as they are transported through the equipment, so that the point in the foodstuff with the lowest thermal treatment, the one around which the process must be designed, will be the center of one of the particulate solids. Two major areas of concern, and subject of much research effort, are the residence time distribution of both the liquid phase and of the solid particulates in the equipment and also the rate of heat transfer from the liquid to the particulates during continuous flow. In contrast to the processing of viscous liquids alone, the temperatures of all elements of the food at any point in the process are not the same and cannot easily be measured; therefore, the safety of the process cannot be established as easily as for purely liquid foodstuffs. This area is one of the most difficult areas in food processing today.

VISCOUS LIQUID PRODUCTS

Foods that fit into this category can be defined as liquids that have a viscosity substantially higher than water but do not contain solid particulates, consisting only of a homogeneous liquid phase. It can be shown, however, that liquid foods that contain suspended solid matter less than about 0.5mm size, such as fibers, cell debris, and comminuted solids, do not have the problems associated with discrete solid particulates outlined later in this chapter, and these foodstuffs can be treated as homogeneous liquids.

The higher viscosity of viscous liquid products is due to high levels of dissolved solids, such as evaporated milk, or the presence of comminuted solids, gelatinized starches, or thickeners as in cook-in sauces, gravy, homogenized vegetable soups, custard, passata, and so forth. The increased viscosity has implications in three areas of the process design and equipment selection.

1. There is usually a change in flow regime from turbulent to laminar flow in the holding tube, and this change, together with a change in the type of viscosity, will alter the residence time distribution.
2. The heat transfer to the foodstuff in indirect heat exchangers is generally lower.
3. The pressure drop through the equipment is greater, and heat exchanger and pump selection may be more difficult.

The first stage in designing a continuous-flow thermal process for viscous liquid products is to characterize the viscosity of the liquid and determine the Reynolds number (Re) in the holding tube at the required flowrate. If the liquid is Newtonian, the Reynolds number can be determined using Equation 3.8, or if non-Newtonian, Equation 4.6. If the Reynolds number is high enough for turbulent flow (see Section 4.7), then it is unlikely that any changes to a process used for low-viscosity liquids are required, with the exception of a greater pressure drop through the process equipment and a reduced heat transfer rate in the heat exchangers. If the flow is laminar, then the first priority is to calculate the length of holding tube required using the expected residence time distribution (Equation 4.12). As a second priority, the pressure drop and heat exchange rate should be investigated and appropriate heat exchange and pumping equipment selected.

4.1 EQUIPMENT SELECTION FOR VISCOUS LIQUID PRODUCTS

The continuous-flow thermal processing of viscous liquid foodstuffs is not significantly different from that used for low-viscosity liquids except in one or two respects. Plate heat exchangers may be used, but for thermal processing of liquids containing fibers and cellular debris, special plates without plate-to-plate contact points are required to avoid buildup of fibers at these points and blockage of the heat exchanger. In many instances, the higher viscosity of the liquid gives problems with increased pressure drop through a plate heat exchanger, and therefore the use of tubular or scraped-surface heat exchangers is preferred because of their higher pressure capability. The use of direct heating is a good alternative, assuming that the added water from condensed steam can be accommodated in the recipe or that direct expansion cooling to remove the water will not damage the product due to high shear rates in the expansion valve. The pumping of liquids with a viscosity greater than about 600 mPa.s requires the use of a positive displacement pump rather than a centrifugal type which, in addition, can be used to control the flowrate through the process.

4.2 CHARACTERIZATION OF VISCOSITY

The viscosity of a fluid has a very significant effect on the way it flows through continuous-flow thermal processing equipment, especially through the holding tube. As outlined above, the viscosity rise may cause flow in the holding tube to become laminar instead of turbulent. Viscous liquid products that are still in turbulent flow in the holding tube will not be affected by the viscosity increase and the liquid can be treated as low-viscosity foodstuffs as covered in Chapter 3. If the flow regimen does change to laminar, however, not only will the fluid velocity profile change to a parabolic one, as defined by Equation 3.3, but the exact shape of the velocity profile will depend on whether the liquid has a Newtonian or non-Newtonian type of viscosity. The viscosity of a viscous liquid product must therefore be characterized before a thermal process can be designed for it. Viscosity was introduced in Section 1.3.3, including its definition, effects of temperature, and distinctions between Newtonian and non-Newtonian behavior.

The viscosity of a liquid is defined in Chapter 1 as the ratio of the shear stress to the applied shear rate:

$$\text{Viscosity } (\mu) = \frac{\text{Shear stress } (\tau)}{\text{Shear rate } (\gamma)} \tag{4.1}$$

There are several models that have been used to characterize time-independent non-Newtonian liquids:

- Power law model

$$\tau = K\,.\,\gamma^{n} \quad \text{or} \quad \mu_a = K\,.\,\gamma^{(n-1)} \tag{4.2}$$

where μ_a is defined as the apparent viscosity, which is the viscosity at the shear rate g, K is defined as the consistency index, and n is the power law exponent. For pseudoplastic fluids (shear thinning), $0 < n < 1$, for dilatant (shear thickening) $n > 1$ and for Newtonian liquids, $n = 1$. The value of n shows the deviation of the viscosity from Newtonian behavior; the further the value is from 1, the more non-Newtonian the behavior. The power law model is also known as the Ostwald-de-Waele model.

- Bingham plastic

$$\tau = K\,.\,\gamma + \tau_0 \tag{4.3}$$

The Bingham plastic model assumes the existence of a yield stress, τ_0. Below the yield stress, the liquid behaves as an elastic solid and above the shear stress, behaves as a Newtonian liquid.

- Casson model

$$\sqrt{\tau} = K.\sqrt{\gamma} + \sqrt{\tau_0} \tag{4.4}$$

The Casson model has been particularly applied to molten chocolate and has been adopted as the standard viscosity model for this.

- Hershel-Bulkley model

$$\tau = K\,.\,\gamma^{n} + \tau_0 \tag{4.5}$$

Similar to the Bingham plastic model, the Hershel-Bulkley model assumes the existence of a yield stress τ_0 but in this case, above the yield stress, flow may be either pseudoplastic, dilatant, or Newtonian, depending on the value of exponent n. This model is now widely used as it can represent the power law model (if $\tau_0 = 0$) and Bingham plastic model (where $n = 1$).

- Ellis model

$$\mu = \frac{\mu_0}{\left(1 + (\tau / \tau_{1/2})^{\alpha - 1}\right)} \tag{4.6}$$

The Ellis model is used more in engineering design: μ_0 is the viscosity at zero shear, $t_{1/2}$ is the shear stress at which the apparent viscosity is half the value μ_0, and α is a factor related to the fluid's shear-thinning behavior.

Further information on these and other less common models, together with data for a large number of liquids, can be found Holdsworth (1993). Other useful references are Lewis (1990), Rao and Anantheswaran (1982), Fryer, Pyle, and Rielly (1997), and Charm (1978).

4.3 EFFECT OF NON-NEWTONIAN VISCOSITY ON CRITICAL REYNOLDS NUMBER

As identified in Chapter 3, it is desirable on terms of product quality, to ensure turbulent flow in the holding tube where possible. When a foodstuff has a non-Newtonian viscosity and hence it has no constant viscosity value, we require a new definition of Reynolds number (Re) for non-Newtonian liquids. For a power law fluid, defined by Equation 4.2 and characterized by K and n, the following can be used (Steffe, 1992):

$$Re = \frac{D^n u^{2-n} \rho}{8^{n-1} K} \left(\frac{4n}{1 + 3n} \right)^n \tag{4.7}$$

and this value of Re can be substituted for the normal Reynolds number in correlations for Newtonian liquids requiring a Reynolds number.

In addition, the Reynolds number at which laminar flow changes to turbulent flow, the critical Reynolds number (Re_{crit}) is a function of the power law exponent, n. For a Newtonian liquid, the accepted value of Re_{crit} is 2,100 but for power law fluids, relationships have been derived by Hanks (1963), and by Mishra and Tripathi (1973), respectively, as:

$$Re_{crit} = \frac{6464 (2 + n)^{(2+n)/(1+n)}}{(1 + 3n)^2}$$

and (4.8)

$$Re_{crit} = \frac{2100 (4n + 2)(5n + 3)}{3(1 + 3n)^2}$$

Measurements by Campos, Steffe, and Ofoli (1994) for carboxymethylcellulose solutions gave higher critical Reynolds numbers than predicted by these two equations, ranging from $Re_{crit} = 2{,}900$ for $n = 0.87$ to $Re_{crit} = 4{,}000$ for $n = 0.46$. Their data gave the correlation:

$$Re_{crit} = 3258(1.61 - n) \tag{4.9}$$

The increased critical Reynolds number resulting from the equations means that flow is even less likely to be turbulent, unless induced by some mechanical means. If turbulent flow does exist, its velocity profile is also affected by the exponent n, although to a small extent. Palmer and Jones (1976) give a maximum velocity of 1.136 of the average velocity for an n value of 0.5 compared to 1.22 for Newtonian liquids.

4.4 EFFECT OF NON-NEWTONIAN VISCOSITY ON LAMINAR FLOW VELOCITY PROFILE

For liquid products with a non-Newtonian viscosity in laminar flow, there is a substantial change in its residence time distribution (RTD) in the holding tube and also in heat exchangers over that for a Newtonian liquid. Instead of the parabolic velocity profile normally associated with laminar flow of the latter (Figure 3–26), the velocity profile of a non-Newtonian liquid is a different shape. The extent of the difference between the two depends on the model of non-Newtonian behavior to which the liquid can be fitted and is outlined below.

4.4.1 Modification of Velocity Profile for Power Law Fluids

There has been much work on calculating the velocity profile in a cylindrical tube for liquids fitting the power law and, from this, the RTD of the liquid can be predicted. Broadly, the velocity profile is derived by performing a force balance over an element of fluid in the pipeline and then integrating across the diameter of the pipeline (Fryer et al., 1997; Coulson & Richardson, 1977).

For a power law fluid in a tube of circular cross-section, the fluid velocity u_r at radius r (from the center) in a tube of overall radius R is given by:

$$\frac{u_r}{u_{ave}} = \frac{3n+1}{n+1}\left(1 - \frac{r^{((n+1)/n)}}{R}\right) \tag{4.10}$$

where n is the power law exponent and u_{ave} is the average velocity of the liquid based on the volume throughput and area of cross-section (see Equation 3.9). It can easily be shown that the velocity and residence time (t_{min}) of the fastest element of fluid at $r = 0$, in terms of the average velocity and average residence time (t_{ave}), respectively, are given by the following equations:

$$u_r = \left(\frac{3n+1}{n+1}\right) \cdot u_{ave}$$

or (4.11)

$$t_{min} = \left(\frac{n+1}{3n+1}\right) \cdot t_{ave}$$

The RTD in terms of E(θ) can be derived from Equation 4.10, where θ is defined in Section 3.25:

$$E(\theta) = \frac{1}{\theta^3}\left[\frac{2n}{3n+1}\right]\left[1 - \frac{n+1}{3n+1} \cdot \frac{1}{\theta}\right]^{(n-1)/(n+1)} \quad \text{for } (\theta) > \frac{(n+1)}{(3n+1)} \tag{4.12}$$

$$\text{and } E(\theta) = 0 \quad \text{for} \quad \theta = 0 \text{ to } \frac{(n+1)}{(3n+1)}.$$

For Newtonian liquids, where n = 1, this equation reduces to:

$$E(\theta) = \frac{1}{2\theta^3} \quad \text{for } (\theta) > 0.5 \tag{4.13}$$

The velocity profile of the fluid and its RTD can be plotted from Equations 4.10 and 4.12 and is given in Figure 4–1 and Figure 4–2 for "n" values covering the range typical for foods.

It can be seen from Figures 4–1 and 4–2 that, as the exponent n decreases and the viscosity becomes more non-Newtonian, the velocity profile flattens and the RTD becomes more uniform. This fortunately decreases the spread of RTDs and flow tends toward plug flow. This has been shown in Chapter 3 to improve overall product quality and nutrient retention for a given sterilization efficiency.

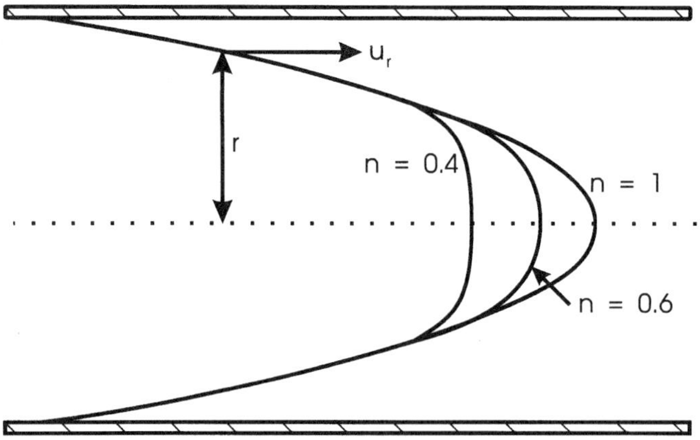

Figure 4–1 Velocity profiles in a holding tube for power law fluids with exponents n of 1.0, 0.6, and 0.4, calculated using Equation 4.10

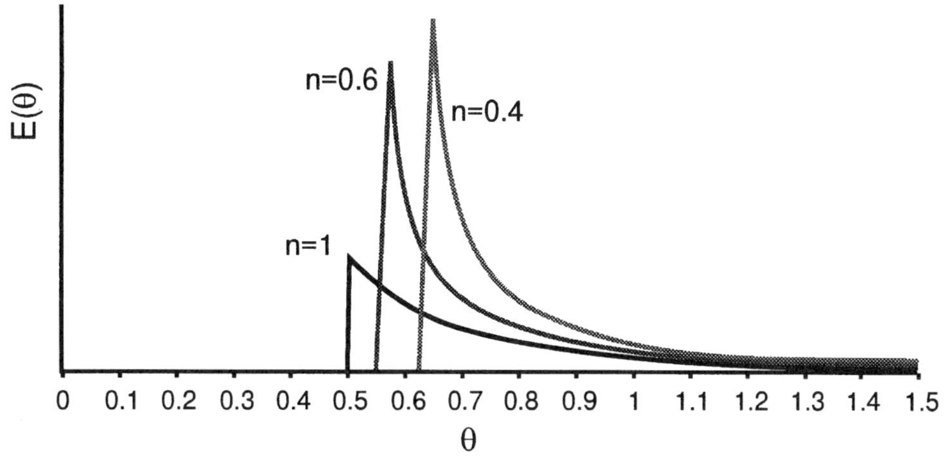

Figure 4–2 RTD curves in a holding tube for power law fluids with exponents n of 1.0, 0.6, and 0.4, calculated using Equation 4.12

4.4.2 Modification of Velocity Profile for Hershel-Bulkley Fluids

In a similar fashion, the flow behavior of liquids with a yield stress can be derived (Fryer et al., 1997). Below a certain radius, the yield stress required for flow is not exceeded and the fluid flows as a central "plug," with the remainder of the fluid at the periphery acting as a liquid with a parabolic velocity profile associated with the value of the exponent n. The radius below that yield stress (τ_0) is not exceeded (r_y) is given in Equation 4.14.

$$r_y = \frac{4\tau_0}{\Delta P / L} \tag{4.14}$$

where ΔP and L are the pressure drop and length of pipe, respectively. The velocity profile expected for Hershel-Bulkley liquids is shown in Figure 4–3.

4.5 RESIDENCE TIME DISTRIBUTION FOR NON-NEWTONIAN LIQUIDS IN DIFFERENT-SHAPED SECTIONS

The flow velocity profile, and hence RTD, can be developed for other-shaped flow sections with Newtonian or non-Newtonian fluids. One technique of note in this area is the use of computational fluid dynamics in which the basic equations of flow are solved by computer for defined channel shapes and the RTDs calculated. An area little explored is whether the most efficient shape for a holding tube is the conventional cylindrical tube or whether, for example, a tube with other cross-sectional shape (such as a square), flow between parallel plates or in an annulus or even the use of a spiral-wound tubular heat exchanger as a holding

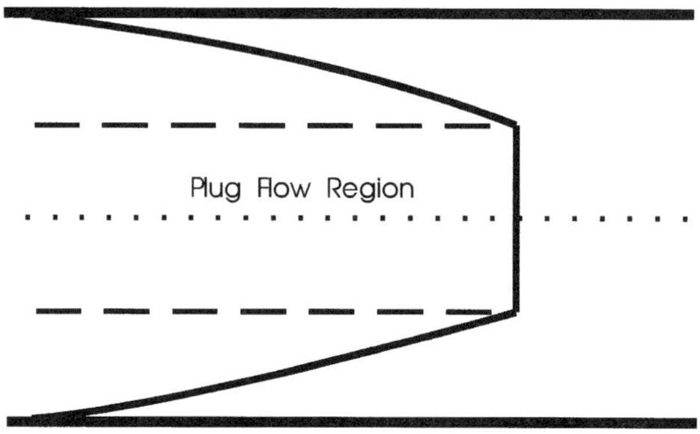

Figure 4–3 Velocity profile in a tube expected for Hershel-Bulkley fluids

tube would give a flatter velocity profile and improved product quality. In addition, other techniques to change the flow profile have been little investigated. Methods of inducing radial dispersion (as occurs in turbulent flow) should be of benefit, and the use of static mixers is known to help (Mutsakis, Streiff, & Schneider, 1986).

Another technique that may prove useful refers back to the modeling of RTD data as a number of stirred tanks (Section 3.27). If the number of continuously stirred tank reactors (CSTRs) to which the distribution is equivalent can be increased, the flow profile will approach that of plug flow. Zhang et al. (1990) investigated dividing a holding tube into small CSTRs using plates set across the pipe and/or adding driven stirrers. For water flowing in laminar flow, a decrease in RTD was found in some arrangements, especially for stirrers operating between 10 and 60 rpm, and further improvement if the stirrer was reciprocating rather than continuous.

4.6 RESIDENCE TIME DISTRIBUTION IN PROCESS EQUIPMENT

The only items of equipment in which RTD may be important, apart from the holding tube, are in the heat exchangers as the liquid foodstuff is heated to and cooled from the final process temperature. This is only really of interest when trying to model the sterilization efficiency for the whole thermal process, which is fraught with difficulty, as covered in Chapter 3. Heat exchangers appropriate for use with viscous liquids have a lower ratio of surface area: liquid volume as well as lower heat transfer coefficients and therefore generally have a slower rate of heating those for low-viscosity liquids, although this may be offset by the fact that heat recovery is usually not used. The sterilization effect generated within the heat exchangers is likely to be greater than for low-viscosity liquids, and it would be useful to determine this, in an attempt to reduce overprocessing of the foodstuff.

For tubular heat exchangers, the RTD for viscous Newtonian and non-Newtonian fluids passing through them is likely to be affected by the nonlinear temperature profile across the channel. For heating, the liquid next to the heat transfer surface will be at a greater temperature than the bulk of liquid, and hence its viscosity will be lower. As mixing across the pipe is much reduced, the localized lower viscosity will result in a greater velocity than normal near the wall. The velocity in the central part of the channel will therefore be reduced over that expected (i.e., the flow profile is flattened). Similarly, for cooling, the liquid at the outside of the channel moves more slowly than normally expected, resulting in a faster "core" of hot liquid moving quickly down the center of the channel.

This effect obviously will not be present in scraped-surface heat exchangers, as the mechanical stirring action generated by the mutator and blades will induce radial mixing and minimize local temperature effects, but will result in a complex temperature and flow pattern in the liquid.

4.7 PRACTICAL EVALUATION OF RESIDENCE TIME DISTRIBUTION

Although all the above equations are very useful in designing process equipment, it must be stressed that they should be treated as approximate only, and RTDs should still be determined experimentally wherever possible using techniques outlined in Section 3.25. There are problems associated with developing the above equations, in particular that their accuracy relies on the assumption that flow is fully developed. At the entry to a pipe, for example, the velocity is the same across the whole section and it takes some distance down the pipe, L, for the flow to become fully developed and achieve the velocity profile calculated from the equations. For turbulent flow, a distance of 50 to 100 times the pipe diameter is assumed, and for laminar flow of a Newtonian liquid, can be calculated from (Leniger & Beverloo, 1975):

$$L = 0.0575 \cdot D \cdot Re \tag{4.15}$$

which, for a Reynolds number of 2,000, is 120 times the pipe diameter. For a 50-mm diameter pipe, L, called the entry length, will be in the order of 5 m, which is somewhere about the length of the holding tube expected for the whole process! In addition, the presence of bends, valves, instrumentation pockets, expansions, and contractions makes the theoretical prediction of RTD inaccurate, and hence practical verification is very wise.

4.8 HEAT TRANSFER COEFFICIENTS FOR VISCOUS LIQUIDS IN HEAT EXCHANGERS

The heating and cooling of viscous liquids in continuous-flow thermal processing is usually achieved in practice by the use of either tubular heat exchangers or scraped-surface heat exchangers, although plate heat exchangers can be used. In the former, if straight-wall tubes are used, flow is usually laminar for viscous liquids, especially as flowrates for these liquids are low. For laminar flow, there is no mixing between the laminar layers across the pipe, and heat is transferred by conduction from the heated wall through the laminar layers to the center of the pipe, giving a temperature distribution as well as a velocity distribution across the tube. The heat transferred is therefore dependent on the diameter to length ratio (d/L) of the tube, as well as other factors. The classic correlation for the individual heat transfer coefficient on the tube side in this situation, for Newtonian fluids, is given by Sieder and Tate (1936) as:

$$Nu = 1.86\left(Re \cdot Pr \cdot \frac{d}{L}\right)^{1/3} \cdot \left(\frac{\mu}{\mu_w}\right)^{0.14} \tag{4.16}$$

where the physical properties used are those at the mean bulk temperature except for μ_w, which is the wall temperature. The correlation is valid for Re, Pr, and L/d >100, and is still valid at ± 20% if Re, Pr, and L/d >10. The dimensionless parameters Reynolds number (Re), Nusselt number (Nu), and Prandtl number (Pr) are defined in Chapter 1.

For turbulent flow, the following correlation is used :

$$Nu = 0.023\ Re^{0.8}\ Pr^{1/3}\left(\frac{\mu}{\mu_w}\right)^{0.14} \tag{4.17}$$

This is valid for Re > 10,000, 0.7 < Pr < 700, and L/d > 60.

The calculation of temperature at different radii in the tube for a step change in tube wall temperature is the classic Graetz-Nusselt problem for which solutions are available, an analytical solution by Jakob (1949) and a numerical one by Adams and Rogers (1973).

Alternatively, scraped-surface heat exchangers may be used for viscous liquids. Heat transfer coefficients for these heat exchangers are more complex than those for tubular heat exchangers, and are usually correlated as a function of both radial and axial Reynolds numbers, taking account of the speed of rotation of the shaft and scraper blades. There are many correlations that have been proposed, the main work being by Maingonnat and Corrieu (1983a, 1983b) and Härröd (1987, 1990a, 1990b, 1990c, 1990d, 1990e) in which the heat transfer coefficient is correlated with the radial and axial Reynolds numbers as well as the Prandtl number and physical dimensions of the heat exchanger. For further information on this subject, the reader is directed to these publications.

4.8.1 Effect of non-Newtonian Viscosity on Heat Transfer Coefficient

For heating or cooling of a non-Newtonian (power law) liquid in laminar flow in a tubular heat exchanger, where natural convection is negligible, the individual heat transfer coefficient is given by Metzner and Gluck (1960) from:

$$Nu = 1.75\left(\frac{3n+1}{4n}\right)^{1/3} Gz^{1/3}\left(\frac{K_b}{K_w}\right)^{0.14} \tag{4.18}$$

which is valid for Gz > 20 and n > 0.1. Gz is the Graetz number, which may be defined for a circular tube as:

$$Gz = \frac{G \cdot C_p}{k \cdot L} = \frac{\pi}{4} Re \cdot Pr \cdot \frac{d}{L} \tag{4.19}$$

where G is the mass flowrate, C_p is the specific heat, k is thermal conductivity, L is the length of the tube, and K and n are the power law exponents given in Equation 4.2.

The effect of non-Newtonian viscosity can be seen by substituting different values of n in Equation 4.17. The individual heat transfer coefficient for a fluid with an n value of 0.3 will be greater by a factor of nearly 60% over that for a Newtonian fluid (n = 1.0), all else being equal.

LIQUIDS CONTAINING SOLID PARTICULATES

The introduction of solid particulates into a liquid imposes considerable problems, making the design of the process very much more difficult to ensure a microbiologically safe product. The difficulties may be summarized in the following areas:

- The equipment used to process these foodstuffs obviously must have passages large enough to allow the particulate solids to pass through it without blocking the equipment or damaging the particulates in any way.
- Pumping a liquid containing particulate solids without damaging the solid phase is difficult, especially as a steady, pulse-free flow is required at pressures of up to 4 bar. Even more difficult is releasing the pressure after thermal treatment to atmospheric pressure for filling, where the solid will be more fragile and microbial recontamination is a major concern.
- Transport of the particulate solids through the process plant: To prevent the solid particulates from settling out wherever horizontal or upward flow occurs, the liquid viscosity is increased using suitable ingredients, such as starch, gums, pectin, or other biopolymers. Raising the viscosity will give the problems covered in the first part of this chapter, especially in terms of encouraging laminar flow and its associated RTD, but the presence of solid particulates will be an extra complicating factor. The RTD of the particulate solids themselves is especially critical, as the slowest heating point in the whole foodstuff is at the center of one of the particulates, which must have a sufficiently long holding time to receive an adequate heat treatment.
- Heating of the solid particulates: The temperature history at the center of the particulates, the slowest heating point, must be known to ensure that adequate heat treatment has been received by the foodstuff overall. In all conventional processes, the liquid is heated first and heat must then transfer from the liquid to the cooler solid through a boundary layer around the solid particulate, which can itself be a large resistance to heat transfer, then through the solid to its thermal center. The rate of heating at the center of a particulate can be predicted mathematically from a knowledge of the size, shape, and thermal properties of the solid and the rate of heat transfer across the boundary layer. Actual measurement of the temperature at the center of a particulate is much more difficult, however, when the solid is actually being transported by the liquid. Some form of

mobile temperature indicator is required that will not affect the way the solid particulate behaves. An extra additional problem is that, in the holding tube as the particulate solids heat up, conservation of heat means that the liquid temperature falls!

These problem areas are considered in depth in the remainder of this chapter as well as the problems of verifying the overall process. There are other techniques (i.e., the use of Ohmic heating or Rota-Hold) that may circumvent the problems altogether, and these are also covered in Sections 4.21 and 4.22.

4.9 PROCESSING EQUIPMENT FOR LIQUIDS CONTAINING PARTICULATE SOLIDS

As mentioned above, all equipment used for continuous-flow thermal processing of liquids containing particulate liquids must allow the particulate solids to pass through without blocking the channel or damaging the particulates in any way. There is a rule of thumb that says that, to prevent blockage, the internal passages of equipment in any direction must be three times the size of the largest particulates in the foodstuff. For the heat exchangers, this means the choice between tubular or scraped-surface units. The former would usually be of the spiral-wound type because of the improved heat transfer rate over straight-wall tubes, and the induced turbulence would improve liquid-particulate heat transfer. If using scraped-surface heat exchangers, a small-diameter mutator is required to allow passage of the particulates between it and the heat transfer surface, which reduces their rate of heat transfer, but it will still usually be faster than any tubular heat exchanger. The scraper blades, and especially their supporting pins on the mutator, can however cause physical damage to solid particulates, as can the method of introducing the solids into this region from a slow-moving inlet pipe and at right angles to it. Some manufacturers have considered these in their scraped-surface heat exchanger design for these foods and have reduced the number and also the profile of scraper blade supports, as well as profiling the ports to give a tangential entry and hence relatively gentle acceleration into and from the scraped-surface zone.

4.9.1 Separate Processing of Liquid and Solid Phases

Processes have been proposed in which the solid phase is thermally processed in a batch system and later mixed with a liquid phase processed in a more conventional continuous-flow thermal process, before the whole is aseptically packaged. One such process, the APV Jupiter process, is shown in Figure 4–4. It consists of a double-cone processing vessel in which the particulates are heated by pressurized steam injected into the vessel and by conduction through a steam jacket on the outside of the vessel while the vessel rotates at speeds between 2 and 20 rpm. When the required F_0 value has been reached, the conventionally processed liquid phase is introduced into the vessel and mixed with the solid phase, simultaneously cooling it. The product can then be discharged from the vessel, using sterile compressed air, to the aseptic filler.

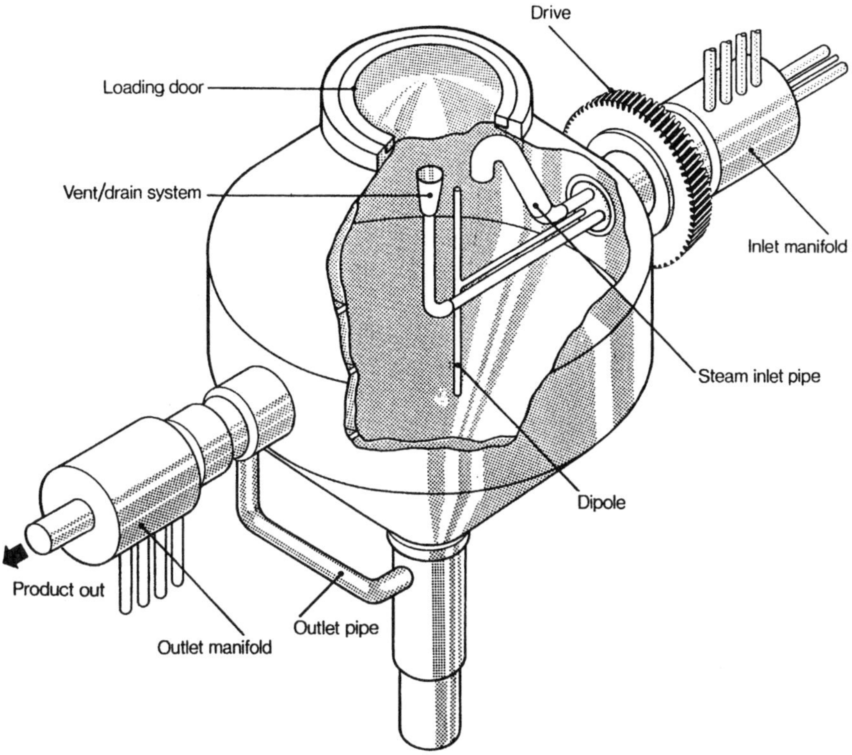

Figure 4–4 The APV Jupiter processing vessel. Courtesy of APV Company, Crawley, Sussex, United Kingdom.

The process has the major advantage that both the solid phase and liquid phase can each be thermally treated to the correct level independent of each other, without the overprocessing of the liquid phase normally associated with heating the foodstuff as a whole. However, this type of process has rarely been taken up by the food industry, possibly due to a high capital cost and complex operating system for a relatively small output, but the principle remains a sound one.

4.10 PUMPING OF LIQUIDS CONTAINING PARTICULATE SOLIDS

The pumping of liquids containing particulate solids is a very demanding aspect of continuous-flow thermal processing and should certainly not be underestimated. The severest specification for a full aseptic process would require a feed pump to raise the pressure of our foodstuff, commonly containing particulates up to 25mm size, to a pressure of about 4 bar at a steady, known, relatively low flowrate. The combination of the requirement for large chambers to accommodate the large particulate size and the relatively low flowrates used means that slow rotational speeds are normal, which, coupled with the high pressure required, means that leakage from the high-pressure zones to the low-pressure zones may well be a problem and must be minimized by careful design.

Pumps that are suitable feed pumps for liquids containing large particulate solids would be the positive displacement type, not the centrifugal type used for low-viscosity liquids, and without valves, as the particulates can be trapped between the valve and its seat and the system pressure will not be maintained. Of the piston pump types, the Marlen twin piston pump is excellent as a feed pump and can handle large particulate solids at low, constant flowrates and high pressures, due to its unique hydraulic drive mechanism, but it is expensive. Another piston-type pump has been developed by Metal Box Engineering specifically for aseptic processing of particulate liquids and can be used as either feed or back-pressure pump but currently is not available commercially.

Rotary pumps are probably the most commonly used type for aseptic processing of these foodstuffs, especially for smaller solid particulates of about 15mm size. In particular, the D-wing rotor (Waukesha Inc.) and progressive cavity (Mono range) types are used, the former often as a back-pressure pump as well as feed pump. Sine-type rotary pumps have also been used successfully as feed pumps. Of the other types of pump available, both peristaltic and diaphragm pumps are capable of handling particulate liquids, but the former have major problems with pulsation, temperature limitations, and failure of the tubing, and the latter have problems in that they contain valves.

In addition, provision must also be made to allow the foodstuff to return to a pressure just above atmospheric pressure after processing while maintaining the required pressure within the process. An orifice plate or control valve, as used for low-viscosity homogeneous liquids, obviously cannot be used when solids are present, and so pressure release must be achieved in another way. Two systems are commonly in use, shown in Figure 4–5.

1. The use of another pump (back-pressure pump) on the outlet to the process, operating at a slightly slower speed, which can be used to control pressure in the system. In this situation, the pump acts as a rotary valve and would require the same specification as the feed pump with the extra requirement of aseptic operation.
2. The use of a pair of sterile balance tanks. The product would pass to one of the tanks pressurized to the required process pressure using sterile air, until it was full, then flow would be diverted to the second tank, also pressurized to the process pressure. The pressure in the first tank would then be reduced by decreasing the overpressure of sterile air to the pressure required by the filling equipment and the contents discharged to the aseptic filler. When this first tank is empty and the second tank full, the flow would be diverted to the first tank, repressurized to the process pressure, while the second tank was discharged at the lower pressure, and the cycle repeated. This arrangement also satisfies the need for an aseptic balance tank to provide a buffer in case of packaging machine breakdown.

Skudder (1993) recommends the use of the former for flowrates over 2,000 kg/h and the latter for up to 750 kg/h. The problem with both systems is sterilization of the equipment

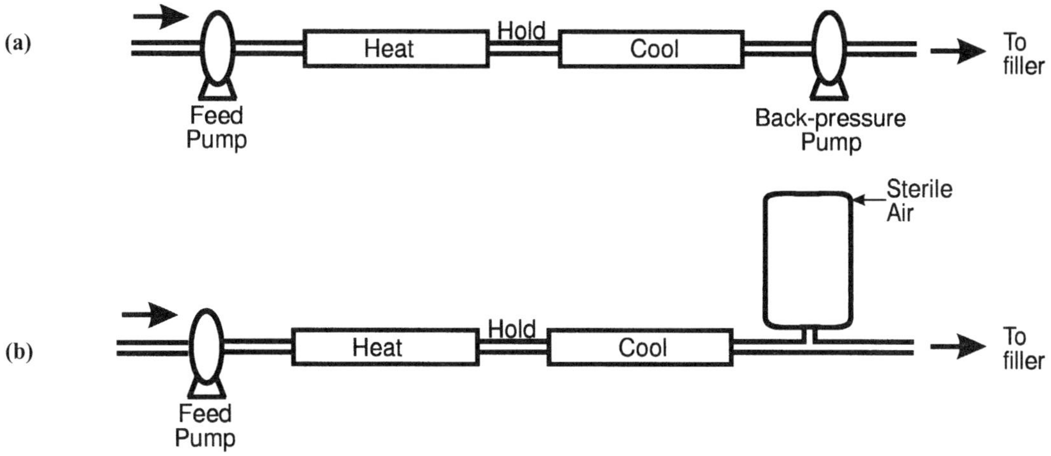

Figure 4–5 Back-pressure arrangements for liquids containing particulate solids. (a), Use of a back-pressure pump. (b), Use of a pair of pressurized sterile balance tanks.

before processing. The second arrangement, using sterile tanks, is obviously more difficult to sterilize as there are a greater surface area, valves, and sterile air filters to contend with. The sterilization of back-pressure pumps is more difficult than first envisaged, as any suitable pump must have small clearances between moving and static parts (as covered later), and heating to about 130°C may result in expansion of the metal, which will eliminate these clearances. If the pump is operated, seizure is very likely to occur and so it must be sterilized without moving, which may give cold spots and potential contamination unless care is taken.

Back-pressure pumps must maintain the required pressure in the process equipment and therefore need to have a very small liquid leakage rate through them, and hence small clearances and long liquid paths. The reason is that, as mentioned above, the pump is likely to be operating at a low rotational speed anyway, to accommodate the size of particulate. If there was a high leakage rate from the high process pressure at the inlet, through the pump to the low outlet pressure, the pump would need to operate very slowly (or even stop!) to maintain the required pressure in the process. Although the correct liquid flowrate would be leaving the system to maintain the required process pressure, the flowrate of solid particulates through the slowly rotating pump may not be adequate, and particulates would build up in the process equipment, eventually blocking it. Fortunately, the high liquid viscosity helps reduce leakage in this respect and, where possible, a pump with relatively long leakage path should be selected. Rotary pumps are most often used as back-pressure pumps, especially the D-wing rotor type, which has a lower leakage rate than the trilobe rotor shape. The Marlen twin-piston pump cannot be used as a back-pressure pump, and the progressive cavity (Mono range) types are rarely used.

Another difficulty may occur on start-up of the process, after presterilization of the equipment, if water is used before the foodstuff is introduced. The leakage rate through the back-pressure pump with such a low-viscosity liquid may be too high to maintain the process pressure required to prevent boiling, the pressure in the holding tube will fall, and sterility of the equipment may well be lost. This situation can be prevented by using an additional back-pressure valve after the back-pressure pump to control process pressure—this may well be required for presterilization of the equipment anyway—but must be of such a design to allow passage of particulate solids on full process operation when the back-pressure pump operates effectively.

Particle damage by pumps has been investigated by Buchwald (1988), who has identified combinations of particle concentration and pressure resulting in large or small amounts of damage. For a 5-bar pressure change, a small amount of damage is found at 20% particle concentration, whereas a large amount of damage is found at particle concentrations over 65%. No experimental details are given.

4.11 TRANSPORT OF PARTICULATE SOLIDS BY LIQUIDS

The way in which solid particles are transported by a flowing liquid through the different sections of process equipment is an essential study in the design of an aseptic process. There has been research work, both historically and recently, in this area but the application of this work to aseptic processing must be carefully interpreted in the light of the problems specific to aseptic processing of these types of foodstuffs. Some of the important factors are as follows:

- The liquid is usually in laminar flow, due to the relatively high liquid viscosity and large pipe required to accommodate the particles, and generally has a non-Newtonian viscosity. The liquid would normally have the velocity flow profile as outlined in Section 4.4, giving a wider distribution of residence times and associated decrease in product quality. Incorporation of particles into the liquid may well disturb this velocity profile.
- There is a particle-particle interaction that affects the particulate flow patterns and also disturbs that of the liquid. Some research work has used single particles in a liquid, the results of which are unlikely to be valid for high solids concentrations.
- There is often radial migration of particulates to a radius equidistant from the pipe wall and the axis, although migration to the wall and to the axis has also been reported. This has been attributed, among other things, to rotation of the particle.
- The configuration of the holding tube is not a single straight horizontal pipe but will also have bends to give a trombone-type arrangement. It may be slightly inclined upward (as required by the US Food and Drug Administration [FDA] regulations) and/or have vertical sections, depending on the plant installation. The bends will have a significant effect on the flow pattern of both the solid and liquid phase.
- There will always be a variation in size, shape, density, and thermal properties of the particulates, even if cut from a single piece of raw material, which itself will not be a regular shape or have homogeneous properties (e.g., layers of fat, air, etc.).
- Heterogeneity of the particles: Most foodstuffs in this category will be composed of several particulate ingredients, each potentially with different physical and thermal properties and size (e.g., soup may contain meat, carrots, potato, mushrooms, etc.). It is

not always the fastest particle through the holding tube (or necessarily the largest particulate) that will define the sterility of the product overall.
- The part of the foodstuff that has had the lowest heat treatment must be determined, that is, the slowest heating point of all of the particles in the foodstuff. This task requires determining the temperature history at the center of all particles in the foodstuff and ensuring that the lowest overall heat treatment is satisfactory for the process required (see Section 4.15).

4.12 PHYSICAL FACTORS AFFECTING THE TRANSPORT OF PARTICULATE SOLIDS

When a solid particulate is placed into a column of static liquid it will, when released, fall (or rise) through the liquid due to the difference in density between the solid and the liquid. The particle will accelerate after release until the drag force on it by the liquid equals the gravitational force on it, whence it reaches a steady velocity, the terminal velocity. This terminal velocity is directly proportional to this difference in density and is also dependent on the viscosity of the liquid as well as the size and shape of the solid body. Perry, Green, and Maloney (1984) give the terminal velocity (u_t) in the Stokes' law region for Newtonian liquids as:

$$u_t = \frac{g\, d_p^2(\rho_s - \rho_f)}{18\mu} \tag{4.20}$$

This equation will be slightly different depending on the value of the particle Reynolds number, defined later. The subject is covered in more detail in Perry et al. (1984) and Lareo, Fryer, and Barigou (1997).

One of the important aspects of the continuous-flow thermal processing of particulate liquids is to enable transport of the particulates through the items of process equipment and associated pipework without their settling out and accumulating in the process. To prevent this, the terminal velocity would need to be very low or zero, and therefore, from the above equation, the only variables that we can manipulate to obtain this are either the density (ρ_f) or the viscosity of the liquid (μ). In practice, the liquid viscosity is usually the factor that is used, as it may well be high anyway for the types of foodstuffs that will be thermally processed in continuous flow, such as soups, stews, ice cream mix, and so forth. It would theoretically be possible to alter the density of the liquid to match that of the particulates (the system is often termed neutrally buoyant), but this is more difficult and rarely done. The density of the liquid would need to be adjusted by a foodstuff, usually sugar or salt, which, at the levels required, would usually make food unpalatable; for example, from data published by Lewis (1990), it can be shown that a salt solution of about 9% (w/w) would be required to match the density of meat. A meat particulate would be prevented from settling much more readily by increasing the viscosity of the carrier fluid, using gravy or stock made with starch, which would normally be the case anyway.

The situations where the liquid column is vertical are few in aseptic processing, being restricted mainly to any vertical interconnecting pipework, vertical portions of the holding tube, vertical scraped-surface heat exchangers, or to Ohmic heating columns. The majority of situations experienced are where the liquid and its particulates are flowing horizontally or slightly upward, which is the situation that most research work has investigated. Analysis of the flow patterns in this situation (Zandi, 1971) has identified five types of flow that may occur, as follows:

1. Homogeneous flow, for light, small particulates, or where the velocity is high enough to keep particulates in an even concentration across the pipe, and can be treated as a pseudofluid behaving as a single component. Many foodstuffs containing fine comminuted solids come into this category
2. Heterogeneous flow, for larger, denser particulates where there is partial separation of the solid particulates and therefore different layers of solids across the pipe, but no actual deposition of solids on the pipe wall
3. Intermediate flow, where conditions exist for both homogeneous and heterogeneous flows simultaneously, especially where there is a wide range of particle sizes and densities
4. Saltation, where particulates form a bed at the bottom of the pipe and then proceed along the pipe by rolling, sliding, or discontinuous jumps. Spherical particulates tend to roll; long, thin, regular shapes tend to slide; and irregular solids tend to jump.
5. Capsule flow, where solids are packed into cylindrical capsules of slightly smaller diameter than the pipe and travel in series

These five regimes have been developed in the chemical engineering literature for hydraulic conveying of mineral slurries in water, with a large density difference and small particulates, so care must be taken in interpreting their use in a continuous-flow thermal process where liquid viscosity is higher and non-Newtonian and the densities of particulate and liquid are very close (i.e., the particulates are nearly neutrally buoyant). The relevance to aseptic processing of these five categories has been reviewed by Sastry and Zuritz (1987).

Another mechanism relevant to continuous-flow thermal processing is the radial migration of particulates, first described by Segré and Silberberg (1961). They investigated the flow of neutrally buoyant particulates in a viscous liquid and found that the particulates congregated in an annulus just over halfway between the axis and the wall of the pipe. This radial movement of particulates appears to be due to some radial force moving them away from the axis and also away from the walls. On the other hand, other workers instead found migration to the axis or to the wall, or neither, and many workers have attempted to explain this phenomenon. One major effect appears to be rotation of the particulate brought about by the velocity profile in the liquid, with one side of the particulate in a stream of liquid moving faster than that on its other side. Rotation of the particulate gives rise to the Magus force, which moves the particulate radially across the liquid flow. The Magus flow is very evident in a variety of sports, where a spinning ball gives a rising, falling, or curving trajectory, as any cricketer, golfer, or tennis player knows well. Rotating particulates in pipes tend to come to equilibrium at some intermediate point between axis and wall, and nonrotating particulates tend to come to equilibrium nearer the axis (Oliver, 1962). Those particulates that travel near the axis tend to start rotating if displaced away from it and would then migrate to the intermediate radius. All particulates traveling near the wall tend to migrate inward to this intermediate radius.

4.12.1 Critical Velocity

When designing the holding tube for a continuous-flow thermal processing system, it is important to select a diameter (and hence a fluid velocity) that will prevent particulate solids from settling out on the lower surface of the pipe (i.e., prevent the saltation described above). The minimum velocity of liquid that will prevent this is called the critical deposit velocity (u_d); at this velocity, the turbulence in the liquid will maintain the particulates in suspension. The critical deposit velocity can be determined using a dimensionless number, the modified particle Froude number (Fr′). This is defined as:

$$Fr' = \frac{u_m}{\sqrt{2 \cdot g \cdot D\left(\frac{\rho_s}{\rho_f} - 1\right)}} \tag{4.21}$$

where u_m is the mean flow velocity of liquid and solid mixture, g is the acceleration due to gravity (9.81 m/s^2), D is the pipe diameter, and ρ_s and ρ_f are the densities of solid and fluid, respectively.

Sedimentation of solids on the lower surface of the pipe occurs below an Fr′ number of 1.35 for particles over 2mm size (Perry et al., 1984) therefore, by substitution into Equation 4.21 and rearrangement,

$$u_d = 1.35\sqrt{2 \cdot g \cdot D\left(\frac{\rho_s}{\rho_f} - 1\right)} \tag{4.22}$$

The minimum Fr′ value for deposition in general is affected very little by the particle diameter, assuming it is over 0.5-mm size, and by the concentration of particles. These two factors therefore have very little effect on the critical deposit velocity (Bain & Bonnington, 1970).

The incipient, or start-up, velocity has been investigated by Grabowski and Ramaswamy (1995) for foodstuffs under aseptic processing conditions. The incipient velocity is the minimum average liquid velocity required to cause movement of solid particulates and below which they become stationary on the pipe wall, measured in their case by determining the velocity required to cause a single particulate on a pipe wall to move. For carrot, parsnip, or potato cubes and nylon spheres in water and starch solutions, they obtained the following correlation:

$$Re_0 = 0.0056\ Ar^{0.615}(d_c / D)^{-0.07}\Psi^{-8.5} \tag{4.23}$$

where Re_0 is the generalized incipient fluid Reynolds number, Ar is the fluid Archimedes number given for this work as $gD^3(\rho_s - \rho_L)\rho_L/\mu^2$, and ψ is a sphericity factor (*surface area of particulate*/surface area of sphere of equal diameter).

4.13 EFFECT OF DIFFERENT PARAMETERS ON THE RESIDENCE TIME DISTRIBUTION OF PARTICULATES

Although there is a substantial body of work on the transport of solid particulates by a liquid, little of it applies to the situation in continuous-flow thermal processing where the particulates are relatively large, nonspherical, and nearly neutrally buoyant in the liquid, and the liquid is viscous and usually non-Newtonian. Studies in this area are becoming more common and have concentrated on examining the effect of a range of parameters on the RTD of a single particulate or multiple particulates in cylindrical pipes and other pieces of process equipment. Particulates used have been either natural, mainly cubes of carrot or potato, or peas, or have been simulated food particles manufactured from polystyrene, calcium alginate, or nylon with density around 1,050 kg/m^3. Liquids used are usually water, sugar solution, or a non-Newtonian solution such as starch, carboxymethylcellulose (CMC), or other gums. The difficulties encountered in noninvasively determining the velocities and RTD of particulate solids in process sections has resulted in a wide variety of interesting and novel techniques, including visual observation and manual timing, use of photoelectric sensor, radioactive tracer, metal detector, laser methods, magnetic tracer, and Hall-effect sensor, or by electrical conductivity. Several workers have used a video camera and subsequent frame-by-frame analysis, often with mirrors surrounding the tube to enable the position of the particulate in the tube to be determined. Experimental work has been summarized in Table 4–1.

Several factors affect the velocity of a single particulate (u_p), and have been identified by using the technique of dimensionless analysis by Fryer et al. (1997) as diameter of pipe and particle (D and d_p), particle and fluid densities (ρ_s and ρ_f), fluid viscosity (μ), mean flow velocity (u_m), and the acceleration due to gravity (g). He (1995) also includes the shape of the particle, solid-liquid ratio, shape, surface, and configuration of the pipe and its orientation as additional factors. Although work with single particulates is of some fundamental value, particle-particle interactions affect the movement of particulates significantly, and research work needs to concentrate on multiple particulate systems.

Important aspects of the results from the experimental work are the mean and minimum residence times of the particulates, and the spread of the RTD curve (i.e., the σ^2 value for the distribution as defined in Section 3.25). It is most likely that the process will have to be designed around one of the particles with the minimum residence time in the holding tube as the one that will receive the least heat treatment. The spread of the residence times of the particulates, however, will dictate the quality of the product to a lesser or greater extent, as particulates with a longer residence time than required for sterility will be overprocessed (as occurs with the residence time distribution of liquids covered in Section 3.30).

Some studies have reported only the minimum residence time, expressed as a function of the average liquid velocity, not the maximum liquid velocity, as a so-called efficiency factor,

Table 4–1 Some Research Work on Residence Time Distribution of Particulate Solids in Process Equipment

Particulate Used	*Fluid and Flow Conditions*	*Reference*	*Comments*
Holding tube (horizontal)			
Calcium alginate spheres; 6 mm diameter; 4%, 8% concentration	Water	Taeymans et al. (1985)	F(t) curves for both phases measured
Rubber cubes, 6,10,13 mm; up to 6% concentration	1.5% CMC	Berry (1989)	E(t) distributions; small particles, bimodal distribution
Polystyrene spheres, 0.95 cm, 0.2%–0.8% concentration density same as liquid	0.2%, 0.5%, 0.8% CMC; 0.448–0.883 L/s	Dutta & Sastry (1990a, b)	Math model developed for normalized velocity; distribution log normal
Calcium alginate spheres, 5 mm	Water, turbulent flow	Hong et al. (1991)	RTD normal distribution
Potato cubes 1 cm; 10%–30% concentration	10% NaCl, turbulent flow	Palmieri et al. (1992)	Bimodal RTD found, due to different groups of particles, in center and near wall of pipe; increasing particle concentration reduces RT of slowest group of particles but not fastest
Polypropylene spheres, 6.35–19.05 mm; potato and carrot cubes, 7.5, 12.7 mm	Water, 3% starch	Ramaswamy et al. (1992)	Larger particles faster than small
Polystyrene, acrylic spheres, 6.35, 9.52 mm; 1%, 3.1% concentration	Water, 0.3% CMC	Baptista et al. (1994)	E(t) curves given; interactive effects of parameters on RTD evaluated
Polystyrene spheres, 7,12 mm; 4%, 7%, 10% concentration	CMC; 3 concentrations used (values not given)	Sandeep & Zuritz (1994,1995)	Correlations presented for mean, minimum, and standard deviation of RTs
Carrot cubes, 8, 15 mm; 3.25–11.85% concentration	4% starch	Tucker & Withers (1994)	Mean particle velocities smaller than mixture for 8-mm particles, converse for 15-mm
Meat cubes, 10, 15, 20 mm; carrot cubes, 6, 13 mm; 5% concentration	3%, 4%, 5%, 6% starch; 3 flowrates, 4 tube lengths	Abdelrahim et al. (1995)	Correlations presented for fastest particle
0.3, 0.4 of tube diameter, 1% and 3%	Various flowrates, viscosities	Baptista et al. (1995)	Presence of particles increased mean RT and decreased dispersion of liquid RTD
Carrot, parsnip, potato cubes, 7.6–12.7 mm; nylon spheres, 6.35–19.1 mm; single particles used	Water, 2%, 4% starch; tube diameters, 29.5, 41.3, 54.6 mm	Grabowski & Ramaswamy (1995)	Minimum velocity for transport of single particles measured; correlations given

continues

Table 4–1 continued

Particulate Used	*Fluid and Flow Conditions*	*Reference*	*Comments*
Meat cubes, 10, 20 mm; carrot cubes, 6, 13 mm	4%, 6% starch; 3 flowrates, 4 tube lengths	Abdelrahim et al. (1997)	E(t) curves presented; data fitted to generalized logistic model
Potato, carrot, turkey, peas (cube, sphere, cylinder); polystyrene, 10%, 20%, 30% concentration	Water, 0.5%, 1% CMC; 3 flowrates	Alhamdan & Sastry (1997)	Effect of parameters on RTD: RT of cylinders greater than cubes; no effect of SSHE rpm on RT in holding tube
Holding tube (vertical)			
ABS spheres, 6 mm and 0–20.6% solids	Water, Re 6,320—26,100; 2 upflow, 2 downflow legs	Fan & Wu (1996)	Particle distribution normal; RTD correlations presented
Carrot cubes, 6 and 10 mm; single and up to 10% solids	Water, 0.3%, 0.5%, 0.8% CMC	Lareo et al. (1997)	Nonuniform radial distribution found
Carrot cubes, 6–10 mm diameter; 25% solids	Water, 0.3%, 0.8% CMC, 4% starch	Lareo & Fryer (1998)	Single or double peak distributions found; mean slip velocities presented
Holding tube; curved section			
Agar gel cylinders; diameter and length, 20 mm; 20%, 30%, 40% solids	0.5% CMC; 47 mm inside diameter; bend radius, 89, 220, 280 mm	Salengke & Sastry (1996)	Correlations presented; mean RT increases as bend radius increases
Commercial SSHE-based system			
Carrot cubes, 6, 13 mm	3%, 5% starch at 15, 20 kg/min; 80°C and 100°C; 1.5, 17.5, 26.7 m length	Abdelrahim et al. (1993a, 1993b)	All factors investigated important; effect on fastest; mean particle time and variance given
Potato cubes, 12.7 mm, chicken/alginate cubes; 15% solids	6% starch; 135°–140°C holding temperature	Chandarana & Unverferth (1996)	Distributions for heat, hold, and cool sections given; data fitted gamma and log normal distributions

continues

Table 4–1 continued

Particulate Used	*Fluid and Flow Conditions*	*Reference*	*Comments*
SSHE (horizontal)			
Calcium alginate spheres, 6 mm diameter	Water	Taeymans et al. (1985)	Effect of viscosity and rpm on F(t) given
Polystyrene spheres, 10, 14.38, 22.17 mm diameter, single spheres	Sodium CMC solutions, 3 concentrations, 3 flow rates; shaft 70 to 190 rpm	Alcairo & Zuritz (1990)	Viscosity did not affect RTD; increased particle diameter, mutator speed, and flowrate decreased distribution
Potato, carrot, turkey, peas (cube, sphere, cylinder); polystyrene, 10%, 20%, 30% concentration	Water, 0.5%, 1% CMC; 3 flowrates; 30, 60, 90 rpm shaft	Alhamdan & Sastry (1998)	Effect of parameters on RT and RTD; fastest particle velocity 3.23 times average bulk velocity
SSHE (vertical)			
Potato cubes, 1, 1.5, 2 cm; 5%, 10%, 20%, 40% concentration	0.4%, 0.9%. 1.2% CMC; 3 flowrates; 60,110, 160 rpm	Lee & Singh (1993)	Distributions given; tailing due to particle-particle interaction
Meat cubes, 10, 15, 20 mm; carrot cubes, 6, 3 mm	4%, 5%, 6% starch; 4 flowrates	Ramaswamy et al. (1995)	All variables affected results; data fitted generalized logistic model

where

$$\text{Efficiency factor } (\xi) = \frac{\text{Residence time of fastest particle}}{\text{Average fluid residence time}} \tag{4.24}$$

Efficiency factors less than 1.0 mean that the fastest particle has a residence time less than the mean residence time for the liquid, which would itself have a minimum residence time, if Newtonian, of 0.5 times the average residence time. Expressing the RTD results in this way, although of immediate practical value in ensuring overall product sterility, does not give any indication of the effect of the parameters on the spread of the RTD of the particulates or give any insight into flow behavior. It is important to note that in no study has a particle velocity faster than the fastest liquid element been reported and is nearly always slower to a greater or lesser extent.

Other workers have fitted the RTD of the particulates to different probability models, predominantly:

- Normal distribution

$$F(t) = P(z \leq (t - \mu)/\sigma)$$

or

$$E(t) = \frac{1}{\sigma\sqrt{2\pi}} e^{\frac{-(t-t_m)^2}{2\sigma^2}}$$

- Log normal distribution

$$F(t) = P(z \leq (\ln(t) - \mu)/\sigma)$$

- Gamma distribution

$$F(t) = \int_0^\infty \frac{t^{(c-1)} e^{-t}}{\Gamma(c)} dt$$

where Γ is the gamma function.

This at least gives a measure of the distribution of particulate residence times in the system under study but it is difficult to relate this information to any fundamental particulate flow behavior.

In an attempt to obtain correlations that may be applicable to the continuous-flow thermal processing situation, and therefore useful for design of practical processes, researchers have often expressed particle velocity or RTD results using the particle Froude number, Fr_p, which differs from the modified Froude number (Fr') given above, and also differs from the general Froude number used in chemical engineering (Fr). The particle Froude number is defined as

$$Fr_p = \frac{u_m}{\sqrt{g \cdot d_p \cdot \left(\frac{\rho_s}{\rho_f} - 1\right)}} \tag{4.25}$$

where the main difference between this and the modified Froude number, Fr′, is the use of the particle diameter (d_p) instead of the tube diameter D. One such correlation is that proposed by Liu et al. (1993) for the average velocity of single particles (u_p):

$$\frac{u_p}{u_f} = 1.16 - \frac{0.7234}{Fr_p} \quad (4.26)$$

He (1995) argues for the use of the modified Froude number instead of the particle Froude number for developing general relationships, which allows the d_p/D ratio to be accommodated in the data, as the particle diameter itself is not important for sedimentation when it is over 0.5 mm. Using his experimental results, He (1995) concluded that for Fr′ values greater than 1.5, the standard deviations of the residence time distribution curves was small and did not decrease with further increase in Fr′.

Other correlations relate the mean and minimum residence times and the σ^2 of the residence time distribution to the modified particle Reynolds number, the volume fraction ϕ, and the power law index (n). One example is that developed by Sandeep and Zuritz (1995) for multiple particles in a non-Newtonian fluid in a circular holding tube:

$$\frac{t_{min}}{t_{ave}} = 1.35 Re_p'^{(0.3-0.31/n)} \phi^{0.076} n^{0.29} \quad (4.27)$$

where

$$Re_p' = \frac{\rho \, u_{fp}^{2-n} \, d_p^n}{K}$$

Correlations for the mean particle residence time and spread of the particulate residence time distribution are also given. Research work in which other correlations have been developed is detailed in Table 4–1 and covered in very much greater detail in Lareo et al. (1997).

4.13.1 Summary of Particulate Residence Time Distribution Research

The last decade or so has shown a buildup in research work in particulate RTD appropriate to aseptic processing, rather than hydraulic conveying of solids. The majority of this work takes an academic chemical engineering approach in an effort to understand the basic principles involved and provide information applicable to any situation. Some of the research work to date has been summarized in Table 4–1 and it is not intended to review all the details of this work in this book. Readers interested in this are directed to the excellent review by Lareo et al. (1997) on flow in pipelines, and to the review by Ramaswamy, Abdelrahim, Simpson, and Smilt (1995).

There is a body of work involved in measuring the velocity of particulates in pipelines, which, although doubtless of value in contributing to an understanding of the principles concerned, seems less applicable to the design of practical processes, in the same way that the prediction of the velocity profiles of pure liquids in tubes does not accurately predict the RTD of that liquid in a holding tube. Rather, work on RTDs themselves is required for process design and operation, especially within scraped-surface heat exchangers and tubes, both horizontal and vertical, and especially for multiple particulate systems. The latter is a more complicated system, as particle interactions affect the distribution, and very little work involves mixtures of particulates of different sizes, shapes, and densities. The complexity of the problem should not be underestimated and a substantial amount of research work is still required in this area.

Of importance are the minimum residence time of a defined particulate, on which the sterilization criterion will be based, and the spread of the distribution of the particulates, on which the organoleptic quality will depend. It must be noted that it is not necessarily the particulate with the shortest residence time overall that will define the sterility of the product, but a combination of residence time and heating rate for one particulate, which must be identified, especially within foodstuffs containing a mixture of particulates. Here, the work on heat transfer to the particulate in flow from the heated liquid is important, as the heating rate depends on the liquid-particulate heat transfer coefficient as well as heat conduction through the particulate to its center. This author has, in past work, found that the particle on which the sterility of the product depends is not the largest or the fastest, but one with a lower thermal conductivity that reduced the rate of heating of the center more than any of the other parameters.

Although the experimental work does not always agree, it is, however, useful to review the basic principles shown by work to date.

- The factors that affect the RTD of particulates in pipelines are size, shape, density, and concentration of particulates; the density, viscosity, mean flow velocity, and non-Newtonian behavior of the liquid; the diameter of the pipe; orientation of flow; and number and position of bends.
- Residence time distributions may be unimodal or bimodal (or more) in laminar flow as the flowrate increases. The multiple modes derive by a combination of the different modes for different groups of particulates moving in the pipe at different radii.
- As the mean velocity in the pipe increases, the mean particle residence time decreases, as does the spread of the RTD as the flow regime becomes more plug-flow. Faster velocities therefore minimize overprocessing of particulates but require a longer holding tube with its associated greater pressure drop.
- Larger particles have shorter mean residence times, but the difference reduces as the flow velocity increases. The effect of increasing particle size is to give a small decrease in the spread of the RTD curve.
- Denser particulates have a longer residence time and a wider RTD spread, unless the bulk flow velocity is high enough, where there is little effect.
- Increasing the concentration of particulates decreases the mean residence time and decreases the spread of the RTD. In addition, the mean residence time of the liquid phase is increased, which reduces the velocity of the liquid and therefore has implications for the liquid-particulate heat transfer coefficient (see Section 4.15 onward).
- An increase in fluid viscosity decreased the mean residence time of the particulates but increased the spread of the RTD curve.

No particulate solid has been found with a velocity that exceeds the fastest element of liquid through the holding tube; generally the fastest particulate found is substantially slower than this. Designing a process based on a "worst-case" assumption of the fastest particulate being the same as for the fastest liquid element is considered by many investigators to be too conservative and would lead to considerable overprocessing.

The current research obviously needs to be increased in volume and in scope, especially in application to multiple particulate systems with different sizes and shapes of solid together, and also the effect of the particulates on the hydrodynamics of the liquid phase. The effect of physical factors on the movement of particulates needs to be understood, not just measured, to allow the food processor to manipulate the system and the food characteristics at the process design stage, which will reduce the RTD of the solids to gain the optimum product quality (within organoleptic acceptability) as well as to ensure safe heat treatment of the food as a whole. Design correlations produced from research work are unlikely to be extremely accurate (like most correlations) and cannot be relied on absolutely without verification in actual commercial processing systems, using techniques suitable for such systems, for example, embedded magnets used by Chandarana and Unverferth (1996). It is important that techniques like this are also developed to allow easy, rapid, and accurate measurements with somewhere around 100 particulates at a time.

In addition, full verification of the complete thermal process would be required, probably using one of the intrinsic marker systems covered in Section 4.20.

As we have seen, the problems in measuring particle residence times, and devising correlations for its prediction, are considerable and, once developed, there may not be very much that could be manipulated to give an optimum RTD obtained for a given system. It is here that systems such as the Stork Rota-Hold (see Section 4.22) would be of potential benefit, as the residence time of the particulates can be set and controlled accurately. The development of a device that would take the whole foodstuff (liquid with its particulates), divide it into small quantities and hold each for a set, predefined time may well be of benefit and would circumvent all of the problems mentioned, and potentially may give better product quality than a standard holding tube.

It is tempting to consider the residence time of particulates and the liquid-particulate heat transfer problems as separate, but they are inextricably linked, and the mechanism of this linkage must be determined before our knowledge of aseptic processing of solids can be said to be complete.

4.14 EFFECT OF PARTICULATES ON HEAT TRANSFER COEFFICIENT IN HEAT EXCHANGERS

The presence of particulate solids in the liquid has a definite positive effect on thc heat transfer coefficient for flow inside tubular heat exchangers. There has been very little experimental work investigating the effect, except for that of Sannervik, Bolmstedt, and Tragardh (1996), who found that the heat transfer coefficient for the inside of spiral-wound tubular heat exchangers, which were heating and cooling a pseudoplastic 4% starch solution, increased dramatically as particles were introduced into the liquid. An increase in concentration of 5.6-mm diameter calcium alginate particles from 0% to 10% gave an increase in heat transfer coefficient of about 100%, and successive increases from 10% to 20% and 20% to 30% each gave increases of approximately 25% for the inner heat transfer coefficient in the heating section and 35% in the cooling section. It is thought by the authors that the movement of particles through the heat exchanger boundary area causes an exchange of fluid between the areas, increasing the heat transfer.

There is no other work on other types of heat exchanger but it would appear unlikely that an increase would be obtained for scraped-surface heat exchangers as the boundary layer area is already under agitation.

4.15 EFFECT OF SURFACE HEAT TRANSFER ON RATE OF HEATING OF PARTICULATE SOLID

In order to know the time-temperature history achieved by a solid particulate as it is transported through the aseptic process, it is important to understand the heat transfer mechanisms involved. The slowest heating point in the particulate (the thermal center) is the point at which microorganism and enzyme survival is most likely and around which the whole aseptic process must be designed. For regular-shaped bodies, the thermal center will be its geometric center but for irregular shapes, it will be elsewhere and must be determined by other techniques such as finite difference or finite element heat transfer methods.

The rate of heating or cooling of a solid body immersed in a fluid is controlled by two factors: (1) the rate of heat transfer from the fluid to the surface of the solid, controlled by the surface heat transfer coefficient associated with the boundary layer around the solid, and (2) heat transfer through the bulk of the solid, to its thermal center, controlled by the size, shape, and thermal properties of the solid.

Given any particulate food product, there is little that can be done to increase the rate of heating of the solid, as the shape, size, and physical and thermal properties will be fixed. However, the effect of the boundary layer can be substantial and, if large, will severely reduce the heat transfer from the liquid to particulate surface and therefore reduce the rate of rise of the temperature at the center of the particulate. Any mechanism that will reduce the size of the boundary layer will increase the rate of heat transfer to the particulate and therefore through it to the thermal center. Such mechanisms are the difference in velocity between the liquid and the particulate, rotation of the particulate, particulate-particulate interaction, a change in direction, or maybe even pulsation of the liquid flow. As mentioned earlier, when processing liquid foodstuffs containing particulate solids, the viscosity of the liquid is increased to help transport the solid particulates through the equipment and to reduce the difference in residence time distribution between the two phases. This has been shown to have a substantial effect on the rate of heating of the particulate, as the reduction in the liquid-solid velocity increases the boundary layer and therefore decreases the heat transfer to the particulate surface (Heppell, 1985).

The relative rates of the two mechanisms outlined above dictate the way in which the solid particulate heats up or cools down. Using classical chemical engineering, the ratio of the two mechanisms is expressed using the dimensionless Biot number (Bi):

$$\text{Biot number (Bi)} = \frac{xh}{k_s} \tag{4.28}$$

where x is the characteristic half-dimension of the body, h is the liquid-solid heat transfer coefficient, and k_s is the thermal conductivity of the solid. The characteristic half-dimension depends on the body; for an infinite plate heated on both sides, it is half the thickness; for a sphere and cylinder, the radius.

For situations where the Biot number is less than about 0.2, heat transfer from the liquid to the solid surface controls the overall rate. The internal resistance to heat flow can be considered negligible and the body can be assumed to be isothermal. In this situation, the temperature of the body can be calculated from

$$\frac{\theta_F - \theta_T}{\theta_F - \theta_I} = e^{-hAT/\rho c_p V} \tag{4.29}$$

where c_P is the specific heat, A the surface area, V the volume, ρ the density, θ_I the initial temperature of the body, θ_F the temperature of the fluid, and θ_T the temperature of the body at time T. This situation is most common for bodies with a high thermal conductivity, such as metals and for small particulates.

If the Biot number is greater than about 5,000, heat transfer throughout the body controls and the heat transfer to the surface can be considered negligible. In this case, the rate of heating at any point can be calculated by classical heat conduction (Carslaw & Jaeger, 1959). This situation is most common when solid bodies are heated by steam or for bodies with a low thermal conductivity.

However, for situations where the Biot number is between these two values, both heat transfer mechanisms are important and must be taken into account. In this situation, the rate of heating within a body can be derived from the Fourier equation. For any solid body, this is given as:

$$\frac{\partial}{\partial x}\left(k\frac{\partial\theta}{\partial x}\right)+\frac{\partial}{\partial y}\left(k\frac{\partial\theta}{\partial y}\right)+\frac{\partial}{\partial z}\left(k\frac{\partial\theta}{\partial z}\right)+Q_l=\rho c_p\left(\frac{\partial\theta}{\partial T}\right) \tag{4.30}$$

Equation 4.30 has been solved by many workers and solutions are available either in chart form or by analytical solution to the above equation. In order to simplify these solutions, several assumptions have been made: (1) the bodies are a simple shape (i.e., an infinite plate, infinitely long cylinder, and a sphere); (2) the thermal and physical properties of the body are uniform and constant; (3) the body has uniform temperature initially; and (4) the liquid temperature rises from this initial value to the final, constant value instantaneously.

4.15.1 Graphical Solutions

Graphical solutions to Equation 4.30 plot the dimensionless temperature against Fourier number for different values of the Biot number. The Fourier number (Fo) is defined as:

$$Fo = \frac{k_S T}{\rho c_p x^2} \tag{4.31}$$

where T is time.

The dimensionless temperature change θ_{dim} is

$$\theta_{dim} = \frac{\theta_F - \theta_T}{\theta_F - \theta_I} \tag{4.32}$$

where θ_T is the temperature at time T.

Charts for different-shaped bodies, and for different situations, are given by Gurney and Lurie (1923), Heisler (1947), and Schneider (1963).

4.15.2 Analytical Solutions

Analytical solutions to Equation 4.30 appear complex but can be easily solved using a computer program and therefore are easier and more accurate to use than graphical solutions.

For an infinite slab, heated from both sides, the center temperature can be obtained from:

$$\theta_{dim} = \sum_{n=1}^{\infty} \frac{2 \sin \mu_n}{\mu_n + \sin \mu_n \cos \mu_n} \cos\left(\mu_n \frac{x}{X}\right) \cdot \exp(-\mu_n^2 Fo) \tag{4.33}$$

where μ_n is the nth root of

$$\cot(\mu) = \mu / Bi \tag{4.33a}$$

Equation 4.30, in spherical coordinates, reduces to (Rohsenow & Choi, 1961):

$$\frac{\partial \theta}{\partial T} = \frac{\alpha}{r^2}\left[2r\frac{\partial \theta}{\partial r} + r^2 \frac{\partial^2 \theta}{\partial r^2}\right] \tag{4.34}$$

which has been solved for an infinitely long cylinder of radius R. The temperature at radius r is given by

$$\theta_{dim} = \sum_{n=1}^{\infty} \frac{2J_1(\mu_n)}{\mu_n [J_0^2(\mu_n) + J_1^2(\mu_n)]} \cdot J_0\left(\mu_n \frac{r}{R}\right) \exp(-\mu_n^2 Fo) \tag{4.35}$$

where μ_n is the nth root of

$$J_0(\mu) / J_1(\mu) = \frac{\mu}{Bi} \tag{4.35a}$$

J_0 and J_1 are the relevant Bessel functions.

The analytical solution for a sphere is also available. The temperature at radius r in a sphere of radius R is given by

$$\theta_{dim} = \sum_{n=1}^{\infty} \frac{2(\sin\mu_n - \mu_n \cos\mu_n)}{\mu_n - \sin\mu_n \cos\mu_n} \cdot \frac{\sin(\mu_n r / R)}{(\mu_n r / R)} \exp(-\mu_n^2 Fo) \tag{4.36}$$

where μ_n is the nth root of

$$\tan\mu = \frac{-\mu}{(Bi - 1)} \tag{4.36a}$$

Computer programs to calculate the temperature at different positions in the above bodies are given by Thorne (1989).

Solutions for shapes other than those above, such as a cube, finite bar, finite cylinder, and others, may be made by combining these solutions. This can be achieved by calculating the θ_{dim} for each dimension x, y, and z (or radius R), using the relevant sizes in that dimension, then multiplying the θ_{dim} values together. For example, for a finite cylinder of length L and radius R, θ_{dim} at a set Fourier number is calculated from θ_{dim} calculated for an infinite cylinder of radius R at that Fourier number multiplied by the θ_{dim} for an infinite plate of thickness L (using x = L/2) at the same Fourier number, that is, $(\theta_{dim})_{cylinder\ R} \cdot (\theta_{dim})_{plate\ L}$.

In a similar way, the following can be calculated:

- for an infinite bar of length size L and M,
  $(\theta_{dim}) = (\theta_{dim})_{plate\ L} \cdot (\theta_{dim})_{plate\ M}$
- for a rectangular parallelepiped of sides L M and N
  $(\theta_{dim}) = (\theta_{dim})_{plate\ L} \cdot (\theta_{dim})_{plate\ M} \cdot (\theta_{dim})_{plate\ N}$
- for a cube
  $(\theta_{dim}) = (\theta_{dim})^3_{plate\ L}$

4.15.3 Mathematical Models

The limitations imposed on the chart and analytical solutions are clearly restrictive when using them to estimate body temperatures in practical situations. The thermal and physical properties of the body are not often uniform, due to different layers in the foodstuff. In addition, they will change with temperature, especially when reactions involving latent heat occur such as starch gelatinization, protein denaturation, and melting of fats. It is also unlikely that the liquid temperature will rise instantaneously unless direct steam heating is used.

In order to simulate the whole thermal process, with different values of temperature and heat transfer coefficient in the different heating, holding, and cooling sections, as well as compensating for changes in thermal and physical properties, a mathematical model would have many advantages. Models have been developed by Heppell (1985) for a sphere (see Appendix A), Teixeira et al. (1969b) for a short cylinder, and McKenna and Tucker (1990).

4.16 HEAT TRANSFER COEFFICIENT BETWEEN LIQUIDS AND SOLID BODIES

The heat transfer coefficient between a fluid and a solid body immersed in it depends on a large number of factors, not only the physical properties of the fluid and solid but especially anything that affects the size of the boundary layer around the solid, such as natural turbulence or shear forces in the liquid, rotation of the solid, particulate-particulate interaction, and induced agitation, for example in heat exchangers or even pulsing of the flow from positive displacement pumps.

Dimensional analysis gives an arrangement of dimensionless groups as:

$$Nu = f(Re_p.Pr) \tag{4.37}$$

where Re_p is the particle Reynolds number.

Correlations are generally expressed as:

$$Nu = K.Re_p^n.Pr^m \tag{4.38}$$

where constants K, n, and m are determined experimentally.

In the special case of a sphere, it can be shown mathematically that the Nusselt number has a minimum value of 2.0 for steady-state transfer when the Reynolds number is zero (i.e., when the liquid is stationary). The above equation can therefore be arranged as:

$$Nu_p = 2.0 + K.Re_p^n.Pr^m \tag{4.39}$$

The classical relationship (Ranz & Marshall, 1952) gives $K = 0.6$, $n = 0.5$, and $m = 0.33$; therefore:

$$Nu_p = 2.0 + 0.6.Re_p^{0.5}.Pr^{0.33} \tag{4.40}$$

and is valid for $1 < Re < 450$ and $Pr < 250$.

These heat transfer correlations have been refined and modified other workers, for example, Rowe, Claxton, and Lewis (1965) and Whitaker (1972), but all are relevant to static bodies held in flowing liquids with well-developed turbulent flow and steady-state heat transfer. Correlations for regions with parameters relevant to continuous-flow thermal processing (i.e., laminar flow and non-Newtonian liquids) are much less common and unlikely to apply to the situation where the particulates are actually being transported by the liquid. Correlations for non-Newtonian flow usually use the generalized Reynolds number as given in Equation 4.7, and a generalized Prandtl number. For the former, the particle diameter d_p is used for D, and U_m is substituted by the velocity difference between the particulate and liquid. The generalized Prandtl number (Pr_G) for a power law fluid is given by Holdsworth (1992) as:

$$Pr_G = 2^{(n-3)} K \left(\frac{3n+1}{n} \right)^n c_p . \left(\frac{d_p}{u_m} \right)^{(1-n)} \left(\frac{1}{k_f} \right) \tag{4.41}$$

4.16.1 Minimum Value for Heat Transfer Coefficient

It is interesting to investigate this concept of minimum heat transfer further. It can be proved mathematically that in the case of a sphere alone, the minimum heat transfer rate is equivalent to a Nusselt number of 2.0 for steady-state only (Bird, Stewart, & Lightfoot, 1960). The situation in continuous-flow thermal processing, however, is one of unsteady-state heat transfer; steady-state heat transfer does not apply until the temperature difference between the solid particulate and the fluid surrounding it becomes a small value. This becomes important when attempting to predict the heat transfer to a particulate in aseptic processing. Obviously, before heating, the particulate and fluid are at approximately the same temperature but, during indirect heating to the process temperature, it is possible that a particulate will pass rapidly through the heat exchanger into the holding tube (as a worst case) and will be surrounded by liquid at the maximum process temperature. This is certain to happen for direct heating, where the temperature rise is extremely rapid. The heat transfer is in unsteady state, as there is a large temperature difference initially and the rate of temperature rise in the particulate will be faster than the steady-state solution. As the particulate-liquid temperature difference falls, the heat transfer rate will approach the minimum Nusselt value of 2.0. This will be the case for all shapes of particulate; the worst heat transfer situation will correspond to that where a body at one temperature is exposed to an infinite conductive environment at a different temperature and thermal properties. Only for a sphere does this situation reduce to an easy numerical solution.

Clift (1978) gives a solution for mass transfer to a sphere in a stagnant continuous phase subjected to a step change in concentration. By similarity of mass and heat transfer, the equation gives the instantaneous Nusselt number for a step change in fluid temperature as:

$$Nu = 2.0 \left(1 + \frac{1}{\sqrt{\pi Fo}} \right) \tag{4.42}$$

where Fo is the Fourier number as defined in Equation 4.31. Table 4–2 shows the Nusselt number predicted at different Fo values.

Other solutions would be available for other patterns of fluid temperature change, such as a “ramp” increase. Heppell (1990) has used this equation to predict the rate of heating in a 3.1-mm diameter spherical particle and found that it corresponded overall to a constant Nusselt number of 4.4 at large values of Fourier number. As shown in Figure 4–6, this gives a significant increase in particle heating rate over the normally accepted minimum Nusselt number of 2.0.

Table 4–2 Predicted Nusselt Number as a Function of Fourier Number

Fourier Number	*Nusselt Number*
0	∞
0.01	13.28
0.1	5.57
1.0	3.13
10.0	2.36
∞	2.0

4.17 EXPERIMENTAL METHODS FOR DETERMINATION OF LIQUID-PARTICULATE HEAT TRANSFER COEFFICIENT

In order to measure the heat transfer coefficient between the liquid and the solid particulate (the surface heat transfer coefficient), the temperatures of both particulate and liquid must be known as a function of time. There is obviously difficulty in attempting to measure the temperature in a body that is being transported by a fluid down a pipeline without disturbing the flow pattern of the particulate. Several approaches to the problem have been tried and can be classified broadly into static particles in a moving liquid, moving thermocouples, the use of intrinsic markers (e.g., microbial spores or thermolabile chemicals) as a time-temperature integrator (TTI), and thermochromic paints. A summary of the published work using these techniques is given in Table 4–3.

1. *Static particles in a moving liquid:* There have been several attempts to measure liquid-solid heat transfer by simulating flow down a pipe using stirred water baths or pumping liquids past objects held in the flow (Chang & Toledo, 1989; Chandarana, Gavin, & Wheaton, 1989; Clement et al., 1997). One of the more comprehensive pieces of work was by Zuritz, McCoy, and Sastry (1990), who measured the heat transfer coefficient to a mushroom-shaped aluminium casting immersed in a power law pseudoplastic liquid pumped past it in laminar flow. The resulting correlation was:

$$Nu = 2.0 + 28.37\, Re^{0.233}\, Fr^{0.143} (d_M / d_T)^{1.787} \tag{4.43}$$

where d_M and d_T are the diameters of particle and tube, respectively.

The accuracy of this type of correlation must be in doubt, as the static particle is not subject to the same conditions as it would be in a flowing liquid, where the particle-liquid velocity may be extremely low and movement to different radii, and even rotation of the particulate, may be present. However, the technique is relatively easy and will give an indication of the order of heat transfer coefficient likely to be present, but confirmation of the data will still be required in a continuous-flow thermal processing system.

2. *Moving thermocouples:* A refinement to this method has been achieved by Sastry, Heskitt, and Blaisdell (1989) and Sastry et al. (1990), who used a fine-wire thermocouple attached to a transducer particle with the thermocouple wire being withdrawn from the

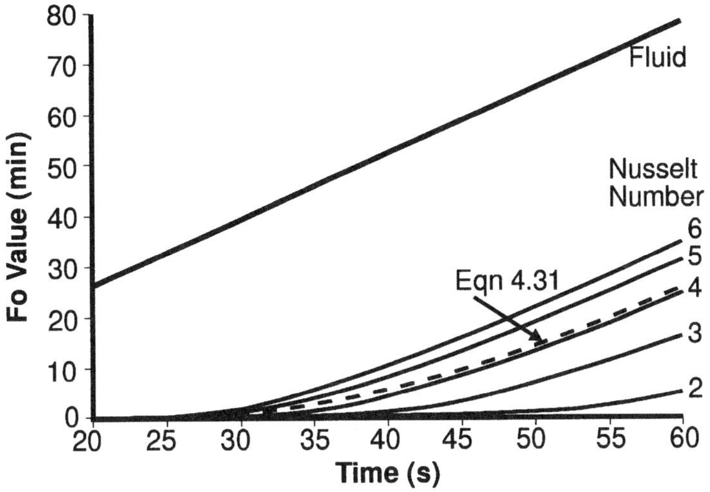

Figure 4–6 Effect of Equation 4.31 and constant Nusselt numbers on the Fo value at the center of a 3.1-mm diameter sphere

downstream end of the holding tube at the same speed as the particle would travel, having been determined beforehand. Although this is an improvement on the static particle apparatus, the presence of the thermocouple wires will modify the particle behavior compared to a free-moving particle.

3. *Markers:* Markers have been developed and used by several workers. Microbial spores with a high heat resistance are common, especially *Bacillus stearothermophilus* and *Clostridium sporogenes*, and work on biochemical markers is promising, especially enzyme-based ones. The immobilization technique of Dallyn, Falloon, and Bean (1977) for entrapping microorganisms into and releasing them from calcium alginate particles is extremely useful in this respect; the ability to recover the marker microorganism and determine the number of survivors accurately is vital. The technique, as used for measuring surface heat transfer coefficients (HTC), is to pass the marker immobilized in some particle through the system of interest, collect it, and evaluate the number of surviving spores (or concentration of chemical surviving). This is then compared to the results from mathematical calculation of surviving spores expected for different surface HTC values using a prediction of the temperature profile in the body from the physical properties of that material together with the thermal inactivation kinetics measured for the marker in the particle material. Details of this technique are given in Section 4.18. The technique can also be used to evaluate the sterilization efficiency of the thermal process overall.

4. *Thermochromic paints:* The use of thermochromic paints in food processing has been reviewed by Balasubramaniam and Sastry (1995) and developed for use in measuring the liquid-particulate HTC by these authors in a series of publications. The basis of the technique is the use of thermochromic paints, which change color depending on their temperature,

coated onto a test particulate and injected into a test system. The color of the paint is monitored as a function of time, usually by video camera, and the color translated into the relevant temperature. The time-temperature relationship can be used to evaluate the liquid-particle HTC, either from a mathematical model or by using Equation 4.29 if appropriate. As the color of the particle must be measured, the whole system must be transparent and its use is therefore limited to model systems. However, the technique is a powerful one. Mwangi, Rizvi, and Datta (1993) used a similar technique, but with a melting point indicator that changes color at a set temperature, immobilized in polymethylmethacrylate.

Another technique with potential, but presently under development, is magnetic thermometry, in which a magnetic metal ball is passed through a magnetic field; the magnetization, which is temperature dependent, is measured by a conducting coil (Ghiron & Litchfield, 1997).

At present, there are only two techniques used that can measure the temperature of a particle in a situation where behavior of the particulate is not affected by the measuring system—the use of markers and thermochromic paints (or melting point indicators). The latter (which can also be considered a marker) gives an instantaneous temperature, while TTI markers give the total time-temperature treatment received, without indicating the maximum temperature achieved.

4.18 USE OF TTI MARKERS

The accuracy of information measured using a marker depends on the accuracy with which the concentration of marker can be determined, the accuracy of the degradation kinetics in the carrier material used, the *z*-value (or activation energy) of the marker used, and the position of the marker in the carrier particle. The use of TTI markers to measure overall process sterilization efficiency is covered in Section 4.20 but, when used to determine the surface HTC, additional factors must be considered. These are the accuracy of the measurements of physical properties of the immobilization material itself, especially its thermal properties; the residence time of the particle in the section of interest; and the accuracy of the mathematical model.

For determination of surface HTC, the use of a marker that has a *z*-value similar to that of microorganisms (around 10°C) would be preferable to the higher *z*-values common for biochemical changes (around 30°C). These higher *z*-values would be less sensitive to temperature change and, unless its concentration could be determined very accurately, would lead to errors in predicted sterilization efficiency; therefore, the use of microorganisms is preferable to use of biochemicals, unless one with a lower *z*-value than usual were found. It is important to contrast this with the use of markers for overall process evaluation, where it is important that its *z*-value is close to 10°C.

There are problems with use of microorganism-based markers, however, that are related to the level of experimental work involved, variability of heat resistance from batch to batch, and problems in optimizing spore recovery and outgrowth outlined in Chapter 2. Unpublished results by Heppell show that dissolution of *Bacillus stearothermophilus* spores immobilized in calcium alginate (using the method of Dallyn et al. [1977]) can affect the number of spores recovered, especially at low numbers of survivors, but the addition of calcium dipicolinate to the recovery medium increased the survivor count by up to two log cycles. This was thought to be due to removal of calcium from the natural calcium dipicolinate in the spore by the sodium citrate used to convert the calcium alginate back to sodium alginate for release.

It is important that a particulate made from the carrier material behave in a manner similar to that of a real food particle, otherwise measurement of the surface HTC will not be representative of that in practical processes. The carrier material must be selected carefully, not only in terms of its density, but also its thermal conductivity and specific heat, to enable the marker to react sensitively to changes in surface HTC. Depending on the conditions present in the system where surface HTC is being measured, especially the Biot number, size of particulate used, and residence time, the survival of the marker may not be sensitive, and results with poor accuracy may be obtained. It may be possible to alter the position of the marker within the carrier particulate to improve accuracy; for example, by using the calcium alginate technique it is possible to distribute the marker organism evenly throughout the particle or, using a double casting system, only at its center or only near the surface. For conditions with a low surface HTC, short residence time, and poor thermal conductivity of the carrier particulate, a marker situated at or near the surface will give more accurate results than if distributed throughout the particulate or positioned at the center, where the temperature rise will be low and survival of the marker high. For long residence times and high heat transfer coefficients, a marker situated at the surface will undergo severe degradation and survival levels will be extremely low, especially if the D value of the marker is low. For each situation, therefore, a balance must be determined between marker degradation kinetics, residence time, particulate size, carrier thermal properties, and the likely surface HTC to obtain results with high accuracy.

A similar argument applies to the use of the thermochromic paint and the moving thermocouple techniques in that the temperature change to be measured should be rapid enough to give an accurate value for the surface heat transfer coefficient. If the temperature sensor is placed at the center of a large particulate of low thermal conductivity, the temperature would change only very slowly and would not be sensitive to changes in surface heat transfer coefficient. In this situation, placing the thermocouple near the surface and increasing the thermal conductivity of the body gives a much more sensitive system. Workers with these techniques achieve this using metal particles, such as brass or aluminium, with the temperature determined at the inside or outside surface. The overall density of the particulate is matched to that of foodstuffs by carefully machining the inside of the hollow particle until the correct weight is achieved. As metals have high conductivities, the Biot number of the particulate is low and therefore the surface HTC can be calculated from the temperature change much more easily and accurately using Equation 4.29 than for markers that require a mathematical model and accurate marker thermal kinetics.

Markers other than microbial cells or spores, such as enzymes or other biochemicals, generally have the advantage of more rapid assay methods and more immediate results, rather than the lengthy plating procedures and 2- to 3-day incubation required for a colony count. Evaluation of marker concentration is also likely to be more accurate, without the complicating spore germination factors described above and in Section 2.6, but is also likely to have higher z-value than spore markers, making the reaction less sensitive to temperature change. These two factors may, however, compensate for each other to some extent.

For further information, the reader is directed to Maesmans et al. (1994), who excellently reviewed the use of markers and further investigated the sensitivity of results from experimental work.

4.19 EXPERIMENTAL MEASUREMENT OF LIQUID-PARTICLE HEAT TRANSFER COEFFICIENT

There is a body of work in which the liquid-particle HTC has been measured where the particle is held static and correlations have been presented that relate HTC to the relative particle-fluid velocity. As mentioned above, the absence of free movement of the particle must cast some doubt on the relevance of the results from these experiments to the continuous-flow situation, especially for heat exchanger sections. It may be possible, however, to use correlations developed this way for a practical situation, using the liquid-particulate relative velocity (the slip velocity) in the Reynolds number. Balasubramaniam and Sastry (1996a) have compared this approach, using three conventional correlations, with measurements from the moving thermocouple and liquid crystal techniques (outlined below) and found significant differences in the predicted and measured values. Determining the relative velocity between liquid and particulate for a practical situation would not be easy, but may be related loosely to the sedimentation velocity (see Section 4.12). Especially in light of this work, correlations obtained in this way should be considered as a guide to the approximate values likely, but for process design, they should be used with great caution.

The most relevant approach to the problem is where free movement of the particle is possible, where markers or thermochromic methods (such as methods 3 and 4 of Section 4.17) are used. There is little work using this approach because of the difficulties involved, but it has the advantage of in situ measurements and the results incorporate the fluid flow patterns within the equipment.

The earliest attempt to measure liquid-particle heat transfer during aseptic processing was by Hunter (1972), who used *Bacillus anthracis* spores immobilized in polymethylmethacrylate (Perspex) spheres of 1/8-inch (3 mm) diameter passing through a direct heating, indirect cooling system operating on water. This work established an extremely useful technique, and was modified and refined by Heppell (1985), where spores of *B. stearothermophilus* were immobilized in calcium alginate spheres of 3.1-mm diameter. Measurements with water showed a decrease in mean surface HTC from 7,900 to 2,120 W/m^2 K (corresponding to Nusselt numbers of 37.6 to 9.2) with decrease in fluid Reynolds number from 51,800 to 5,250. Heppell (1990) also measured surface HTCs for a range of starch solutions and found surface HTCs of 1,916 W/m^2 K (Nusselt number of 8.32) for the 2% starch solution and values between 902 and 1,520 W/m^2 K (corresponding to Nusselt numbers between 3.9 and 6.6) for 4% to 10% starch. These particulates, at 3.1-mm diameter, are too small for continuous-flow thermal processing and would not give the same heat transfer coefficient as larger particulates of about 10 or 20 mm.

Few other workers have used microorganism markers due to the large amount of experimental work required and inherent inaccuracies in determining microbial survival. More productive methods are to use thermochromic paint-type systems or an immobilized enzyme marker. The former has been used extensively by Balasubramaniam and Sastry (1996a,

1996b, 1996c; 1995; 1994a, 1994b, 1994c) to measure the liquid-solid HTC for cubic particles in a holding tube and a scraped-surface heat exchanger. They used a thermochromic paint on the outside of a hollow aluminium cube (the weight of the cube was designed to give a particle relative density approximating to that for foodstuffs), which was videotaped as it passed through the apparatus and the hue value of the paint measured by an image analysis system. This was translated to the temperature of the body from a calibration curve previously determined and the liquid-particulate HTC calculated from Equation 4.29, as the conductivity of the aluminium ensured that the Biot number was below 0.1. Liquid-particulate HTCs were measured in tube flow under various conditions in a series of papers (Balasubramaniam & Sastry 1996a, 1996b; 1994a, 1994b) and were found to correspond to Nusselt numbers from 3.6 to 54.6, where the HTC increased with decreasing liquid viscosity, increasing fluid velocity, and decreasing particle/tube diameter ratio.

Heat transfer correlations developed from the experimental work in this field are given in Table 4–3, but must at present be treated with caution. There is still a great need for further work in this area.

4.19.1 Other Sections

There has been a substantial amount of work on HTC measurement in the holding tube section, and rightly so, since this is likely to have the poorest heat transfer rates in the whole process. However, the heating and cooling parts of the process consist of heat exchangers in which the liquid is in forced convection to give good heat transfer, due to channel shape or induced from rotating scraper blades. It is expected that the liquid-particle heat transfer in this situation would be higher than in a holding tube due to increased liquid turbulence, and therefore the particle could have been heated to a substantial degree before it entered the holding section. To base the thermal process on heating in the holding section alone, assuming that a particulate at its inlet temperature would suddenly appear at the start of the holding tube, and then base the length of holding tube on this particle would be to overprocess the food, possibly to a large degree. Studies on the heating and cooling sections are therefore an important part of the process, and HTC and the minimum residence time of the particulates in these sections require evaluating. It is important to consider the whole process, not just the holding section, unless the worst-case situation requires it. Only one study has measured particulate-liquid heat transfer coefficient for single cubic particles in a scraped-surface heat exchanger (Balasubramaniam & Sastry, 1994c), giving values corresponding to Nusselt numbers of 9.9 to 72.6. The HTC values increased with increasing rotational speed of the scraper blades as well as the liquid flowrate and decreased with increasing carrier viscosity and particle size.

4.19.2 Conclusion

Although the above work is useful and gives broad guidelines to process parameters that would improve the liquid-particulate HTC, verification of these results and further measurements in this area are obviously necessary before they can be used comprehensively in process design. Measurement of the liquid-particle heat transfer is especially required for the

Table 4–3 Some Research Work on Liquid-Particle Heat Transfer in Process Equipment

Particulate Used	*Fluid and Flow Conditions*	*Reference*	*Comments/Surface Heat Transfer Coefficients (HTC) ($W/m^2 {}^\circ K$)*
Static particulate			
Sweet potato cubes, 38 mm	Water, 35% sucrose; velocities 3.8–8.6 mm/s	Chang & Toledo (1989)	HTC zero velocity; 239 and 146 for water and sucrose, respectively; 303 at highest flowrate for water
Silicone cubes, 25.4 mm	Water and 5 starch concentrations; liquid-particle velocity zero	Chandarana et al. (1989)	HTC 8.1–35.9 with increasing starch concentration 51.1 for water
Silicone cubes, 25.4 mm	Water and 2%–3% starch; 4.4–11.6 mm/s fluid velocity	Chandarana et al. (1990)	HTC 55.63–89.5 (Nu 2.8–3.7) starch and 64.67–107.11 (Nu 2.5–4) water
Aluminium mushroom-shaped casting	1.3%–2.2% CMC; laminar flow, 62–287 g/s	Zuritz et al. (1990)	$Nu = 2.0 + 28.37Re^{0.233}Pr^{0.143}(d_M/d_T)^{1.787}$
Lead, polyacrylamide, calcium alginate spheres; 8- and 16-mm lead cubes; turnip sphere, 17 mm	10-L stirred flask of silicone fluid or starch (rotating fluid, thermocouple at different radii)	Åström & Bark (1994)	Nusselt number of 5–6 could be used to design system for safety of viscous liquids
Potato cube, 10 mm	Sodium chloride solution 10%; velocities 200, 230, 290 mm/s	Cacace et al. (1994)	Result validated by comparison with potato cubes inoculated with yeast cells
Aluminium sphere (9.5–30 mm), cube (10–32 mm), cylinder (12.7–32.5 mm diameter, 12.8–31.9 mm long)	Water, 0.8–1.5% CMC; Re_p 239–5,012 (water), 4.5–177 (CMC); open channel flow	Kelly et al. (1995)	Nu 14–76 for water, 17–53 for CMC
Teflon and potato cylinders (21–25.4 mm diameter, 24 and 25.5 mm long)	0%–1% CMC; $1–1.9 \times 10^{-4} m^3/s$ flowrates	Awuah et al. (1996)	HTC 100–700, Bi = 10–50

continues

Table 4–3 continued

Particulate Used	*Plant Section*	*Fluid and Flow Conditions*	*Reference*	*Comments/Surface Heat Transfer Coefficients (HTC) (W/m²°K)*
Moving thermocouple				
Hollow brass spheres (13.3, 16.8 mm), Aluminium spheres (22, 23.9 mm)	Holding tube	Water; 5 flowrates; Re_f 7,300–43,600	Sastry et al. (1989, 1990)	HTC 688–3,005; correlations presented
Melting point indicator				
Polymethylmethacrylate (8, 9.6, 12.7 mm diameter; 1%, 2.5%, 5% concentration)	Holding tube	Glycerine in water, 4 densities; 7 flowrates; Re_p 73.1–369.4	Mwangi et al. (1993)	58.3–1,301; $Nu = 0.1Re_p^{0.58}Pr^{0.33}$ for laminar flow, $Nu = 0.0336Re_p^{0.8}Pr^{0.32}$ for turbulent flow
Microbial marker				
Bacillus anthracis spores in polymethylmethacrylate spheres (3 mm diameter)	Holding tube	Water, Re_f 40,700, 42,900	Hunter (1972)	Nu 8.2–13.1
B. stearothermophilus spores in calcium alginate spheres (3 mm diameter)	Holding tube	Water, Re_f 51,800 and 5,250; starch 2%–10%; Re_f 1,170–4.9	Heppell (1985, 1990)	Water Nu 64 and 11.3 (high and low Re, respectively); starch Nu 8.4–4
Liquid crystal temperature sensor				
Aluminium hollow spheres (22.3, 16, 12.8 mm; single particle)	Holding tube	CMC (0.2%, 0.5%, 0.8%); 4 flowrates; Re_f 798–0.0058	Balasubramaniam & Sastry (1996a)	Compares HTCs from liquid crystal, moving thermocouple, and relative velocity methods; HTC 1,630–397 (Nu 64.15–8.25)
Polystyrene cubes (14.8, 11.7, 7.4 mm; single particle)	SSHE	Water, Re_f 2.5–53.7; CMC, 0.5%, 0.8%, 1%; 3 flowrates; 3 mutator speeds	Balasubramaniam & Sastry (1994c)	HTC 500–2,938 (Nu 9.9–72.6); decreased with increasing CMC concentration; increased with mutator speed

heating stage of the process, as outlined above, in scraped-surface heat exchangers and corrugated-tube heat exchangers. Together with measurements of the minimum residence time in these sections, these data could allow the whole thermal process to be modeled.

Consideration must also be given to the link between particulate size, its residence time, and its liquid-particle HTC. It is important to identify the slowest-heating particulate, around which the safety of the process will be defined, and any link between these factors will affect the decision as to which particulate is the "worst-case."

Studies on methods of actually maximizing the liquid-particulate HTC are required and, perhaps, techniques that would alter the particulate-liquid relative velocity, such as changes in direction by use of a series of U-shaped holding tubes (trombone sections), static or dynamic mixers, or pulsed flow. Other techniques, such as sonic or ultrasonic waves, have been known to improve heat transfer rates by reduction of the boundary layer around the particulate and may give an improvement in heat transfer without affecting the flow of the fluid. The only method at present that uses this type of approach is the Stork Rota-Hold, covered in Section 4.22, where the particles are retained on a series of slowly rotating blades and the liquid phase can flow over them at a higher velocity.

4.20 MATHEMATICAL MODELING OF THE THERMAL PROCESS

Using a mixture of theory covered in Section 4.19 and data presented in the above publications, it is possible to model the whole thermal process using a mathematical model, such as that used by Heppell (1985), to give an estimate of the order of sensitivity of the sterility of the foodstuff to parameters such as size and material of the particulate, particle RTD, process temperature profile, and liquid-particulate HTC.

As an example, the output from a mathematical model for a sphere written by Heppell (1985) (see Appendix A) is presented for a particulate that is a sphere of beef of 10-mm diameter and passes through a continuous-flow thermal process consisting of two tubular heat exchangers for heating, a holding tube, and a single tubular cooling heat exchanger. A straight-line temperature profile through the heat exchangers is assumed, which is easily achieved if pressurized water is used as the heating and cooling medium (see Section 3.21).

In the following examples, the time-temperature profile of the fluid (Figure 4–7) is based on that of a semicommercial process. There is a rapid initial temperature rise to around 115°C, a short transfer time, and a slower heating rate in the second heating section to a process temperature of 130°C. Holding time is for 60 seconds followed by cooling to around 65°C over 30 seconds. The Fo value received by the liquid with this temperature profile is 9.2 minutes. For the beef particle, a worst-case assumption has been made where it is assumed that there is no difference in velocity between it and the liquid (i.e., a zero "slip" velocity).

The temperature and Fo value for the center of the beef particulate has been calculated using the mathematical model of Heppell (1985) for the following conditions: (1) an infinite liquid-particulate HTC throughout the process (corresponding to a Bi > 5000; see Section 4.15); (2) a minimum liquid-particulate HTC for a sphere throughout the process (corresponding to Nu = 2.0); and (3) a mixture of the two closer to that likely to be found in practice, that is, an infinite HTC for the heating and cooling sections and the minimum HTC for holding sections.

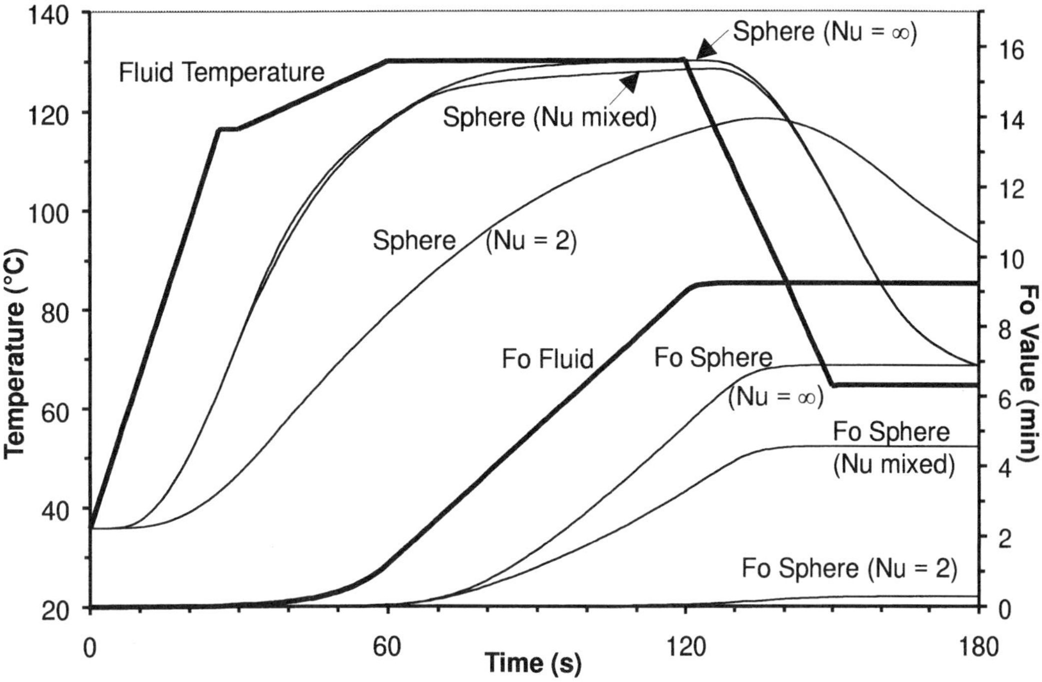

Figure 4–7 Prediction of the temperature and Fo value at the center of a 10-mm diameter beef sphere. (a), Infinite HTC throughout. (b), Minimum HTC throughout (from Nu = 2.0). (c), Infinite HTC in heating/cooling sections and minimum HTC in holding sections. Process temperature in holding tube 130°C.

The model output for the above conditions is shown in Figure 4–7. The difference between minimum and maximum HTC is a substantial one and may be extremely significant in terms of public safety. As can be seen, the final Fo values for these two conditions are 6.9 and 0.3 minutes respectively, where a value of 3 is the minimum required for public safety regarding *Cl. botulinum*. The third case, with the different HTC values, would be expected to be closer to the practical situation in that most heat exchangers use induced liquid turbulence to increase the rate of heating of the fluid, especially spiral-wound tubular and scraped-surface heat exchangers, and this turbulence would increase the liquid-particulate HTC (Balasubramaniam & Sastry, 1994c), although it is less likely to be close to infinity. The model output gives a temperature profile close to the infinite HTC case with a difference in temperature of less than 2°C between the two. This difference, however, results in a lower Fo value of 4.6 minutes.

The difference between these three cases illustrates what would be expected, namely, that conditions in the initial parts of the process are of great importance in determining the overall heat treatment received by the particulate. Evaluations of liquid-particulate HTCs in holding tubes are of obvious importance, but it is their values in passage through heat exchangers that may be more important to the overall process and are presently underresearched. In a practical situation, if the liquid-particulate HTC in the holding tube is poor and cannot be im-

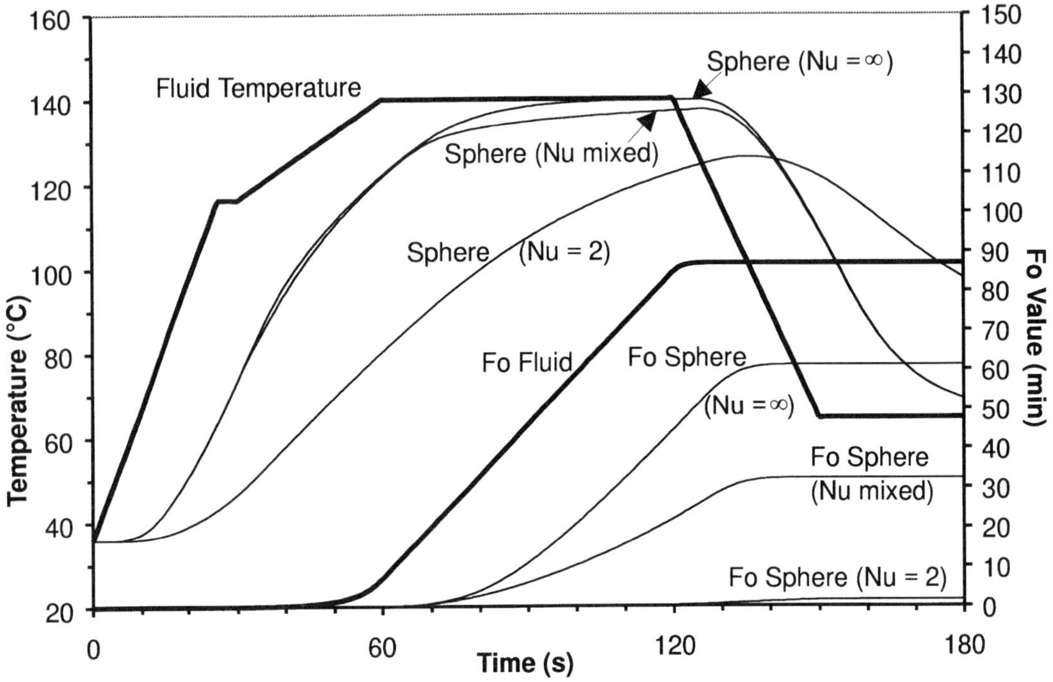

Figure 4–8 Prediction of the temperature and Fo value at the center of a 10-mm diameter beef sphere. (a), Infinite HTC throughout. (b), Minimum HTC throughout (from Nu = 2.0). (c), Infinite HTC in heating/cooling sections and minimum HTC in holding sections. All parameters are the same as for Figure 4–7 except that the process temperature in holding tube was raised to 140°C.

proved significantly, it is obviously important to concentrate on ensuring that the temperature of the particulate is as high as it can be before it passes into the tube.

A mathematical model such as that used here is useful in determining the effect of changing different process parameters to see which is likely to affect the overall thermal process received. The effect of increasing process temperature from 130° to 140°C, while keeping all other parameters constant, is shown in Figure 4–8. The effect on the case with the minimum HTC is minimal, the final Fo value at the particulate center is only slightly improved from 0.3 minutes to 1.6 minutes, whereas for the "mixed" HTC and infinite HTC cases the effect is somewhat more dramatic, increasing the Fo values from 4.6 to 32.3 minutes and from 6.9 to 61.2 minutes. The Fo value for the fluid is increased from 9.2 to 87 minutes.

This increase in Fo value of 7- to nearly 10-fold may appear dramatic, but once the Fo value required for commercial sterility has been reached, the effect of overprocessing on the product quality should be evaluated using a *z*-value appropriate to the biochemical and quality changes (e.g., 30°C rather than 10°C). The F value for the overall processes given in Figures 4–7 and 4–8 can be calculated to give the F^{30}_{121} value, as defined in Section 2.4. The F^{30}_{121} values for the fluid alone are 2.9 and 5.8 minutes for holding temperatures of 130°C and

140°C, respectively, and for the particulate center in the infinite heat transfer coefficient case are 2.5 and 4.9 minutes, respectively. Although any product produced using the higher process temperature would still be overprocessed, the effect of this overprocessing is not as severe as the increase in Fo value would suggest. Another complicating factor is that the F_{121}^{30} values given above are based on the center of the particulate, whereas overprocessing should ideally be assessed on the time-temperature history of the whole particulate (i.e., a bulk F_{121}^{30} value).

The effect of increasing the initial temperature of the foodstuff before entering the process can be seen in Figure 4–9, where the temperature has been raised from 36°C to 80°C, while maintaining the same outlet temperature from the first heating section of 116°C. The minimum HTC case shows a modest increase in Fo value at the center of the particulate from 0.3 to 0.94 minutes, while the infinite HTC case had an increase from 6.9 to 7.2 minutes, and the fluid had no significant increase. Increasing the input temperature gives, for this example, a small but significant increase in the Fo value of the beef particulate without increasing overprocessing of the liquid phase. This reinforces the importance of the heating rate at the start of the process, as discussed earlier, and may actually occur in the practical situation anyway, if the product is subjected to a degree of cooking during its preparation before continuous-flow thermal processing. Using a higher input temperature in this way can be seen as analogous to using a hot-fill process for a conventional in-container sterilization, with its similar benefits.

In a similar way, it can also be shown that reducing the rate of heating in the heating section can also minimize differences between the fluid and infinite and minimum HTC cases, but will reduce the organoleptic and nutritional quality of the product. This indicates a strategy for product development of continuous-flow thermally processed products, in which the rate of heating can be reduced and a lower process temperature used initially to prevent overprocessing and produce a product with a lower risk of underprocessing but lower organoleptic quality than optimum. When a knowledge of the product and process has been built up, and the safety (and failure rate) of the product assessed over a production time of months, for example, then the rate of heating and process temperature can be gradually increased over a long time period to give the improved quality expected, while simultaneously monitoring product safety.

Mathematical modeling in the process evaluation of continuous-flow thermal processes certainly has its place, but the model must be as flexible and representative of the practical situation as possible. The model ideally should be able continuously to change fluid temperature, liquid-particulate HTC, and particulate thermal properties and accommodate nonuniform initial temperature distribution and thermal properties to be able to give a reasonable accuracy. The model used for the simulations given above will allow all these conditions, but may require a small amount of extra programming. The model can also be made to simulate heating in scraped-surface heat exchangers with exponential temperature rise, rather than straight-line temperature rise, by incorporating an appropriate algorithm for the temperature rise or by inputting many more time-temperature points. All mathematical models, however, have limitations, and the above values for temperature and Fo calculated for the center of the particulate are only as accurate as the data used in the model. The heterogeneity of the prac-

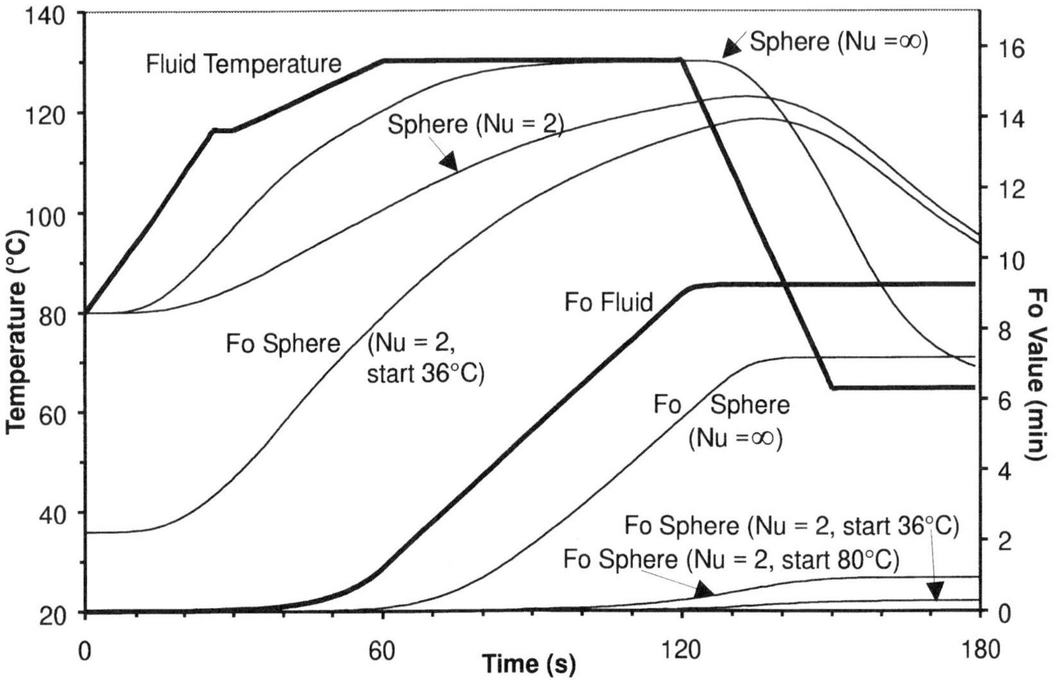

Figure 4–9 Prediction of the temperature and Fo value at the center of a 10-mm diameter beef sphere. (a), Infinite HTC throughout. (b), Minimum HTC throughout (from Nu = 2.0). All parameters are the same as for Figure 4–7 except that the inlet temperature to process was raised to 80°C.

tical system in terms of particulate size, physical and thermal properties of the foodstuff, RTD of fluid and particulates, and so forth, make the results of such models approximate at best and, although useful for analyzing the sensitivity of changing process parameters, it would be very unwise to rely on these without verification with extensive experimental data.

4.21 OVERALL PROCESS VALIDATION

The prediction of sterilization effect within a solid particulate as it passes through a given continuous-flow thermal process given above is becoming more of a reality, although more data are certainly required before reasonable confidence may be placed on the output of any mathematical model purporting to calculate it. Any such mathematical model would, however, be extremely effective both in the initial design of a continuous-flow thermal process and in helping to determine process parameters likely to give the best-quality product while maintaining the correct level of thermal treatment, but it would be foolish in the extreme to depend on prediction alone. Actual measurement of the thermal treatment received by the foodstuff, or at least demonstrating that it has received a certain minimum value, is essential before the product goes into the catering or retail chain and may indeed be mandatory by any licensing authority involved.

The ultimate test that the required heat treatment has been received by a foodstuff is by extensive storage trials of the product and destructive analysis of a large proportion, if not all, of the packages from many batches over a long time period. The expense and inconvenience of this approach are high, and it will not provide any data on whether the product is being excessively or minimally overprocessed (i.e., the extent of a safety factor in the process conditions).

Some other method of process evaluation is definitely required, and again the use of markers or TTIs is the most useful technique presently available, with a marker immobilized within the solid particulates. Markers may be intrinsic (already present in the foodstuff) or extrinsic (added to the food) and be a microorganism or spore; an enzyme; or a chemical or a physical marker whose degradation kinetics are first-order, are stable, and have been accurately measured. Intrinsic markers are less useful, as there may not be a measurable concentration surviving or their *z*-value is not that required (see below); therefore, extrinsic markers are most common. Methods of immobilizing the marker are important, and it must not diffuse out of the particulate; it is most sensitive when immobilized at the slowest heating point in foodstuff, rather than distributed evenly throughout it. For most heterogeneous foodstuffs passing through a continuous-flow thermal process, this will be the physical center of the particulate, except perhaps for the Ohmic process (see Section 4.23), and the marker should be immobilized there. For process validation, as opposed to measurement of liquid-solid HTC covered in Section 4.19, all physical and thermal properties of the particulate are very important. The TTI must simulate the food particulate in all respects if results are to be representative of real foodstuffs, that is, not only in size and density in order that the particulate moves through the process in a manner similar to the food particulate of interest, but also in thermal conductivity and specific heat to ensure that it heats at the same rate as the real particulate. The latter point is most important, as a small difference in thermal properties will result in a different heating rate and an inaccuracy in the estimation of sterilization efficiency, possibly leading to underprocessing and public health risk. If using immobilized vegetative microorganisms or spores, it is not necessary that they be representative of the natural microbial population or have the same thermal death rate (D value); rather, a higher initial count than usual or a larger decimal reduction time would be useful, as it would ensure that a countable number of surviving spores remain after passage through the thermal process.

In addition, it should be easy to prepare and immobilize the marker; it must be stable with a reasonable storage life, once prepared; and evaluation must be rapid and easy, as many evaluations will be necessary to ensure that a representative value of the minimum heat treatment for the foodstuff has been achieved. The validation will normally be repeated for several different time-temperature processes, making these properties even more desirable.

Whichever type of marker is used, it is vital, however, that it has the same *z*-value as the microorganisms it is to emulate. The marker will go through an unknown temperature-time process and will give a result in the form of an integrated, single, time-temperature combination (e.g., an overall thermal process equivalent to a certain time at 135°C or 121°C), and the lower the *z*-value of the marker, the larger the value of this time would be. It cannot be converted to another time-temperature combination for a different *z*-value without knowing the temperature-time history the marker has gone through, although if the process were only at a single lethal temperature (as in a direct heating process) it could.

The classes and types of TTI available, and their relative advantages and disadvantages, have been extensively reviewed by van Loey et al. (1996). The authors divide the markers into microbiological, enzymatic, chemical, and physical.

4.21.1 Microbiological Markers

Microorganisms have been the most commonly used marker for process validation in terms of sterilization effect, as they have the required *z*-value but also because it seems appropriate to measure sterilization with a microbial marker that is closest in form and behavior to it. These markers suffer from many disadvantages, however, as enumeration of initial and final microbial concentrations usually takes several days and requires trained personnel, but they also have the problems outlined in Chapter 2 and in Section 4.18, especially for germination of spores and the inherent low accuracy of the final count. It is especially important that the thermal death kinetics of the marker microorganism be measured in the immobilizing medium, as several workers have found that this affects the results, which is especially difficult at high temperatures and short D values. In addition, the method for release of microorganisms from the immobilization medium before enumeration must not affect their subsequent germination and growth (see Section 4.18).

Using the thermal death kinetics and the reduction in concentration of the marker, an equivalent holding time at an arbitrary reference temperature (with its associated D value) can be calculated from Equation 2.5. This holding time can be translated to an F_0 value by calculating the equivalent holding time in minutes at 121°C, using Equation 2.15 or more accurately using Arrhenius kinetics, thus establishing the heat treatment received and whether the foodstuff will be commercially sterile or not. Microorganisms such as spores of *Bacillus stearothermophilus*, *Clostridium sporogenes*, and *Bacillus coagulans* have been used with the alginate immobilization method of Dallyn et al. (1977) for process verification (Bean, Dallyn, and Ranjith, 1978, Brown et al., 1984). Of special consideration are the verification of the Ohmic process (Skudder, 1993) (see Section 4.22 for details) and validation of a conventional scraped-surface heat exchanger-based system for the US Food and Drug Administration (FDA) (Palaniappan & Sizer, 1997). The latter involved using 362 chicken-alginate cubes inoculated with *Cl. sporogenes* for each set of process conditions.

4.21.2 Enzyme Markers

The problems associated with using microorganisms has led to development of other biochemical and chemical-based methods. Markers based on enzyme degradation have looked promising, as assay methods are relatively rapid and highly accurate, and techniques for their immobilization onto solid substrates have been developed in the biochemical industry. Enzymes may also be immobilized using the alginate method of Dallyn et al. (1977) by reducing diffusion of the enzyme through the biopolymer network.

Several relatively thermostable enzymes have been isolated, which have high D values and hence can easily be used to validate full aseptic processes, as a significant postprocess enzyme concentration remains. The difficulty with their use, however, lies in obtaining a reaction with a *z*-value representative of microorganisms, around 10°C. The normal *z*-value for enzymes is around 30°C but manipulation of the enzyme environment can reduce this value by altering its pH, ion concentration, hydrophobicity, moisture content, or even the immobilization process itself. Several systems have been reported with low *z*-values, mainly α-amylases from *Bacillus subtilis* in tris buffer and with trehalose ($z = 6.2°$ to 12.8°C) (van Loey, 1996), from *Bacillus licheniformis* ($E_a = 302$ kJ/mol) (de Cordt et al., 1992), and immobilized peroxidase in dodecane (Weng et al., 1991), and for pasteurization, α-amylase from *Bacillus amyloliquefaciens* with a *z*-value of 7.6°C and $D_{80°C}$ value of 46.8 minutes (van Loey et al., 1996).

4.21.3 Chemical Markers

Chemical markers have greater potential than either enzymic or microbiological markers, but problems in identifying a reaction with a suitably low *z*-value are even more severe and, at present, no practical system is feasible. Care must be taken when identifying marker chemicals that no other reactions may occur in the food to affect the marker, such as may happen with the degradation of some thermolabile vitamins. However, the potential for a suitable chemical marker is great, and verification could be extremely rapid if, for example, the readout could be based on a color change system.

4.21.4 Conclusion

The use of time-temperature integrators is vital in the development of continuous-flow thermal processing systems and has the potential to give rapid evaluation of the heat treatment received by a particulate and also by the liquid phase independently. By coimmobilizing, or using several marker chemicals, each with different *z*-values, both microbiological and biochemical qualities of the product can be evaluated simultaneously and the process optimized relatively rapidly. A considerable number of determinations will have to be made, however, to ensure that the particle that receives the minimum heat treatment has actually been included in the assessment and the full range of heat treatments received within the foodstuff is known.

4.22 STORK ROTA-HOLD PROCESS

As outlined earlier in this chapter, one of the problems encountered by a conventional continuous-flow thermal process for liquid foods containing particulate solids is the long residence time required for the liquid phase, and its subsequent overprocessing, while the solid phase reaches the required heat treatment when using conventional heat exchangers.

One approach to this problem is to provide different residence times in a holding section for the two phases.

This approach has been taken by Stork, which has designed a device, the Rota-Hold, that can slow down solid particles while allowing the liquid phase to pass through normally. The first, the Rota-Hold, is shown in Figure 4–10 and consists of a series of fork blades spaced the required distance apart (5, 10, or 15 mm) mounted on a rotating shaft in a cylindrical vessel. There is also one series of static forkblades interleaved with the rotating blades. The foodstuff enters the cylindrical vessel by the inlet port, and any solids larger than the interblade gap are retained on the forkblades while smaller particles and the liquid phase pass through them. The residence time of the solid particles in the Rota-Hold section is dictated by the speed of revolution of the forkblades and, as they approach the outlet port, are removed from the rotating forkblades by the static interleaved blades and guided to the outlet. The interleaved blades also prevent short-circuiting of the device by large particulates.

The use of several Rota-Hold devices in the same process, each with different forkblade spacings, can be used to sterilize foodstuffs with a wide range of particulate sizes. A different residence time can be set for each particulate size range, reducing over-processing of the smaller-sized particulates and improving the organoleptic quality of the product.

The second device is the Stork Spiral-Hold consisting of a vertical tube down in which there is a mechanically-driven shaft fitted with a large number of wings arranged in a clockwise spiral configuration. These wings interleave with corresponding wings fitted to the inside of the tube in a static counter-clockwise spiral. The wing spacing decreases from, for example 15 mm at the inlet end to 5 mm at the outlet end. The rotating and static wings form a number of helically-rotating cages and slows the large particulates near the inlet and the smaller particulates further down the tube.

The advantages of these Selective Holding Sections are that the residence time of the particles can be adjusted to suit the solid particulates, their size and thermal conductivity, while the liquid residence time is much shorter, minimizing its over-processing. The particu-

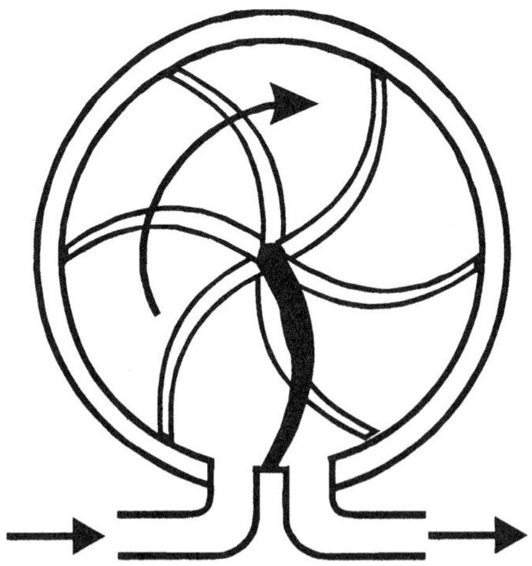

Figure 4–10 The Stork Rota-Hold device. Courtesy of Wilfred Hermans, Stork, Amsterdam.

late residence time in the holding section becomes a definite known minimum value, reducing the uncertainties associated with the particulate residence time distribution outlined earlier in this chapter. In addition, there is a positive liquid flow over and around the solid particles that increases the heat transfer rate from the liquid to the solid and reduces or eliminates the problems outlined in Section 4.15. Under these conditions, it is much easier to calculate the Fo value received by the particulates and severely reduces the chance of producing an unsterile product.

One characteristic of SHS systems is associated with start-up of the process where, for the first few minutes, only liquid is produced, followed by liquid with only one size fraction of particulates if several devices are used. In addition, close-down of processing will result in the process equipment containing particulates but no associated liquid phase. Stork has developed a holding tank to overcome these problems in which the particles collect in a central, perforated inner area that is slowly agitated, while the liquid resides throughout in the tank. Aseptic filling is then a two stage process, the first is of the high particle concentration from the central area and the second is liquid.

Another disadvantage of the Rota-Hold is the increased possibility of contamination due to the inclusion of a rotating device in the holding tube but if the seals are designed aseptically and the device is placed at the start of the holding tube, this risk would be minimized. During cleaning-in-place, the devices are rotated at higher speeds, around 500 rpm.

4.23 OHMIC HEATING OF PARTICULATE FOODS

Ohmic heating is a technique in which electricity is applied directly to the food, which heats up solely due to its electrical resistance (i.e., heat is generated directly within the food). This technique, originally developed at the Electricity Research Council, Capenhurst, United Kingdom, and recently commercialized by APV International, has been the subject of much development and research work in recent years. Although heating using electricity is an expensive method, Ohmic heating has its major advantages when used for the heating of a continuous flow of liquid foodstuff containing solid particles where, if the conditions are correct, the liquid and solid phases will heat at the same rate, reducing the difference between the two; the solid phase also will heat at a much faster rate than in the systems previously described using conventional heat exchangers. The direct conversion of electricity into heat in the product is substantially more efficient than using microwaves or radiofrequency heating, where there is a loss of efficiency in the generation of electromagnetic radiation.

Heating rates of approximately 1°C per second can be achieved in the liquid and, where the electrical conductivity of the liquid and solid phases are similar, within the solid also; that is, both phases can be heated from 50°C to a sterilizing temperature of about 140°C in 90 seconds. For the solid phase, this is much faster than can be achieved by conventional convective heat transfer from a heated liquid for all but the smallest solids, even with an infinite heat transfer coefficient.

The construction of the Ohmic heater is quite simple; there are four, seven, or more electrical elements each connected to one phase of a three-phase, low-frequency (50 or 60 Hz) electrical supply. The use of low-frequency electricity prevents electrolysis of components in the foodstuff. Each electrode is made from a metal alloy that is food compatible and has a high corrosion resistance, supported in a block of PTFE that provides electrical insulation (Figure 4–11). The stainless steel tube between each element is lined with an electrically insulating plastic liner, such as polyvinylidene fluoride (PVDF), polyethylene ether ketone (PEEK), or glass. For electrical safety, the electrical phase connected to the two outer electrodes of the column is electrically earthed.

It is important with three-phase heating to ensure that each phase has a similar current taken from it, and therefore the electrodes are spaced so that there is the same electrical impedance between them. This means that, as the electrical resistance of a foodstuff generally decreases with increasing temperature (the electrical conductance increases), the electrodes must be spaced further apart as the foodstuff progresses toward the outlet.

As the heating effect depends on the electrical conductivity of the foodstuff, itself dependent on the concentration of ions within it, a knowledge of this property is essential but data are rarely available (see Section 1.7.1). For most foodstuffs, the electrical conductivity varies mainly depending on the voltage gradient used and, as mentioned above, temperature. For purely liquid foods, the electrical conductivity increases linearly with temperature but overall falls as the concentration of pulp in it increases (Palaniappan & Sastry, 1991). With solid foods, the situation is more complicated as the electrical conductivity rises less linearly with temperature, especially at low voltage gradients, and may be different in different directions within the solid; for example, the electrical conductivity of carrots is 0.25 S/cm at 25°C across the axis and 0.42 S/cm parallel to the axis, due to its nonhomogeneous internal structure. In addition, the electrical conductivity of many vegetable products has been shown to increase rapidly at about 60°C, due to cell wall breakdown and greater mobility of ions, but may also be due to gelatinization of starches or denaturation of proteins. The electrical con-

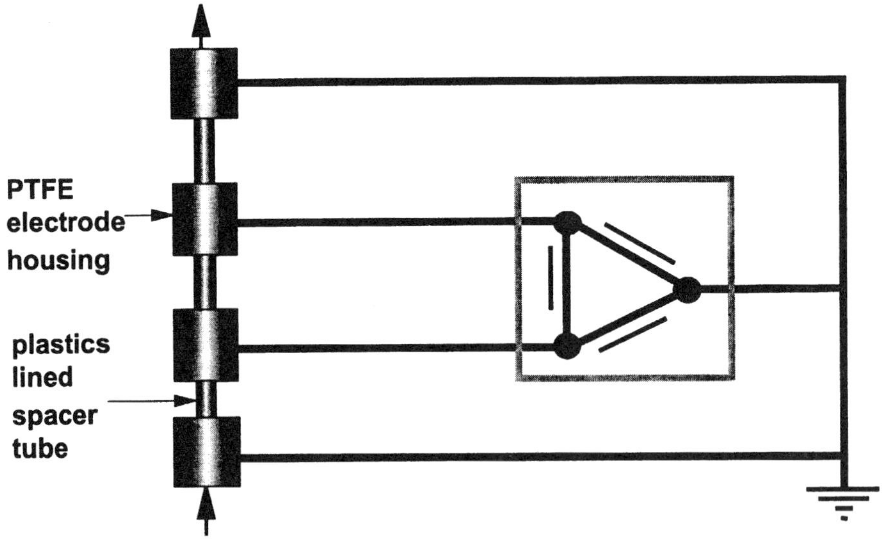

Figure 4–11 Electrical arrangement for an Ohmic heater. Courtesy of APV Company, Crawley, Sussex, United Kingdom.

ductivity of foods may be manipulated by altering its ionic concentration (e.g., for solid foods by leaching or by infusion). This is achieved by placing the solid foodstuff in either water, which causes leaching of salts from the food and a decrease in conductivity, or by immersion in a solution of sodium chloride where the electrical conductivity increases. The increase in conductivity obtained in the latter case can be varied by altering the sodium chloride concentration in the solution or by increasing permeability using vacuum infusion (Sastry, 1994).

4.23.1 Continuous-Flow Thermal Process

The layout for a typical continuous-flow thermal process based on an Ohmic heating unit is given in Figure 4–12. The Ohmic heating column is held either vertically or inclined and the flow is arranged so that it is upward to ensure that the electrical pathway between elements is full of foodstuff. There is also an air vent at the top of the column to ensure this. The product is pumped by a suitable pump (1) through the Ohmic heating column (2) to the holding tube (3), where it is held for the required holding time. Control of flowrate is very important during heating, and the use of a good positive displacement pump is required. Cooling is achieved by a conventional heat exchanger, using any type outlined in Chapter 3 suitable for particulate liquids. Tubular heat exchangers are generally preferred as they give lower structural damage to the heated, fragile solid phase when compared to scraped-surface heat exchangers. Back-pressure arrangements may be provided using methods described in Section 4.10.

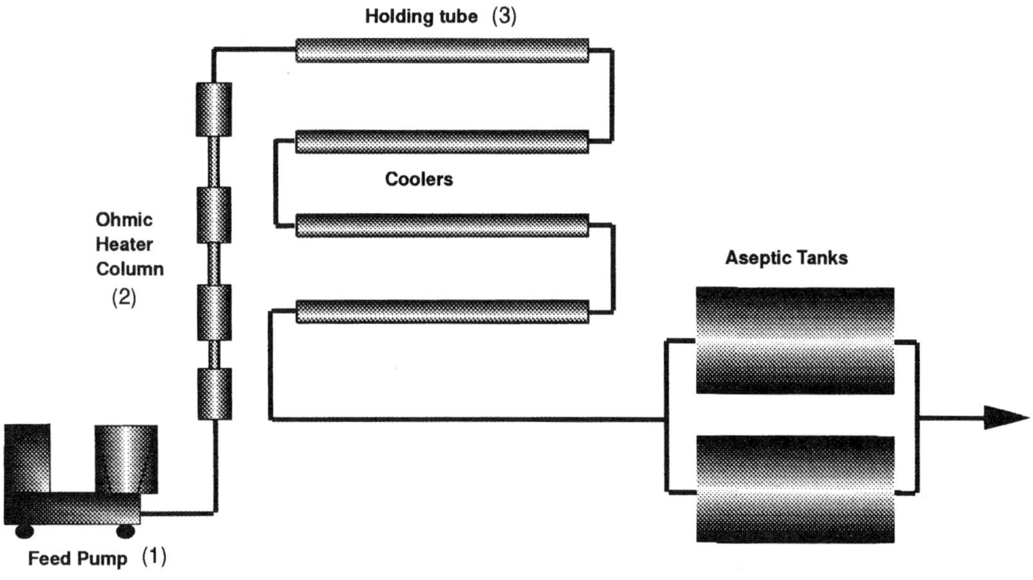

Figure 4–12 Flow diagram for an Ohmic heating process. Courtesy of APV Company, Crawley, Sussex, United Kingdom.

Presterilizing of the equipment is achieved by recirculation of a sodium sulfate solution, which is heated by the Ohmic heater to the sterilization temperature, passed through the rest of the process equipment, and then returned to the feed tank via a poststerilization cooler. During changeover to food product, any sudden change in electrical conductivity would cause a temperature fluctuation that may result in contamination of the process equipment. In order to minimize this, the concentration of the recirculating sodium sulfate solution is adjusted to give an electrical conductivity that approximates to that of the foodstuff to be processed. In addition, during changeover from sodium sulfate to food product, the interface between the two solutions is collected in a catch tank situated just before the aseptic product balance tank.

The process can be readily controlled, but the use of feedback control is less effective due to the long time delay between a change in the inlet conditions and its resultant change in the outlet temperature. Feed forward control is therefore used, where changes in inlet temperature, mass flowrate, and product-specific heat capacity are monitored by a microprocessor; the electrical power required to achieve the outlet temperature is computed and compared to a signal from the power transducer. Feedback monitoring is also used to prevent long-term drift of the outlet temperature. Another control feature is that the power is automatically switched off on loss of pressure from the column.

Another process arrangement has been proposed for products in which the level of particulate solids is relatively low, less than about 40% by volume. The product is split into two streams, a purely liquid phase and a concentrated particulate stream. The liquid stream is treated in a conventional continuous-flow thermal process in which heat recovery is high (using plate heat exchangers where possible) and the particulate stream is heated in the Ohmic heater. The cold liquid stream is injected into the ohmically heated stream at the end of the holding tube of the latter, giving a fast rate of cooling and minimum structural damage to the product. In addition, the better energy economy of the conventional continuous-flow thermal process improves the process economics.

4.23.2 Modeling and Design of the Process

Mathematical modeling of the heating rate of solid particulates in the Ohmic heating system has shown that it is a function of many variables, not just of the particulate shape and size but also the electrical conductivities of the liquid and solid and, in some circumstances, the orientation of the solid in the electric field. The latter has been modeled by de Alwis, Halden, and Fryer (1989) where the electrical conductivities of the solid and liquid phases were different. When the electrical conductivity of the liquid is greater than that of the solid, the latter heats up more slowly than the liquid as the current "bypasses" the solid. This effect would result in a very poor heating rate in the solid if only a single particle were present, but with increasing particle density, the path through the liquid becomes smaller and the liquid conductivity decreases, making the electrical current, and therefore heating rates, in the two phases more even. However, the conductivity of the mixture overall decreases and the heating rate is therefore slower. If a particle in this situation has a long, thin shape, its heating rate if orientated parallel to the current flow is slower than the liquid, but if orientated across the current, it heats faster than the liquid. The reverse happens when the conductivity of the solid is greater than that of the liquid. The effect decreases as the particulate shape becomes closer to a sphere. In addition, nonuniform heating of the liquid next to the solid has been observed. Any uneven heating within the foodstuff as a whole will, however, tend to be evened out by conduction and convection heat transfer within and between the solid and liquid phases, or even by inducing turbulence, as covered previously in this chapter.

It is therefore important to balance the electrical conductivity of the two phases as much as possible to ensure even heating and the absence of cooler parts of the foodstuff, with its chance of compromising the safety of the foodstuff. This may not be as easy as first imagined, as the change in conductivity of both phases may be different as the temperature increases and, although they may be in balance at the feed end, the conductivities and therefore temperatures of each phase may not be balanced at the outlet. The use of sodium chloride to balance the electrical conductivities may give organoleptic problems and may not be possible with some products such as fruit, where other electrolytes should be used. The temperature difference between the phases may be exacerbated by the different residence times of the particles in the Ohmic heating section. De Alwis and Fryer (1992) suggest that if the conductivities of each phase are within ± 5%, even heating will result, but even this may give differences of up to 10°C between the phases, without the effect of convection heating (Fryer et al., 1993).

Other disadvantages of Ohmic heating are the fact that boiling may occur as there is no maximum temperature as found in conventional heat exchangers (defined by the temperature of the heating medium) and that there may be some fouling of the electrodes, and external cooling of them may be necessary (Sastry, 1994).

4.23.3 Verification of the Ohmic Process

The process has been verified using the calcium alginate immobilization technique outlined in Section 4.17. The tests were carried out using alginate analogues of beef and carrot containing spores of *B. stearothermophilus*, as 19-mm cubes suspended in gravy (Skudder, 1993). The whole was processed in a 45 kW Ohmic heating process designed to attain an Fo value of 32 for the liquid. By recovering the particles, determining the number of surviving spores, and relating these to the survival of *Cl. botulinum*, the effective Fo values for the beef and carrot cubes were determined. Values for beef cubes were 23.5 to 30.5 and for the carrot cubes were 28.1 to 38.5.

The Fo values for the center only of the cubes were also determined by repeating the above but including some alginate analogues in which the only *B. stearothermophilus* spores present were held in a 3-mm sphere situated at the center of the analogue cube. The F_0 value at the center of beef cubes was found to range from 34 to 37.5, while for the whole cube it was 29 to 38.5; for the center of the carrot cubes it ranged from 30.8 to 40.2, while for the whole cube it was 23.1 to 44.0. The F_0 value expected for the center of the cube if heat generation within the product had not occurred and the process relied only on heat conduction from the gravy (as would occur in heating by a tubular or scraped-surface heat exchanger) was given as 0.2. These results indicate that the solid and liquid phases received comparable heat treatment over the whole process and that within the solid cubes, the heating rate was uniform.

The electrical conductivity of the cubes or of the gravy were not reported, nor were they compared to natural beef or carrots. A more realistic result might have been obtained if the spore spheres were immobilized at the center of cubes of actual beef or vegetables, a procedure that would be recommended to ensure a safe product in a commercial installation.

REFERENCES

Abdelrahim, K.A., Ramaswamy H.S., Grabowski, S., & Marcotte, M. (1995). Dimensionless correlations for the fastest particle flow in a pilot scale aseptic processing system. *Lebensmittel Wissenschaft und Technologie* **28**(1), 43–49.

Abdelrahim, K.A., Ramaswamy H.S., Marcotte, M., & Toupin, C. (1993a). Mathematical characterization of residence time distribution curves of carrot cubes in a pilot scale aseptic processing system. *Lebensmittel Wissenschaft und Technologie* **26**(6), 498–504.

Abdelrahim, K.A., Ramaswamy H.S., Marcotte, M., & Toupin, C. (1993b). Residence time distribution of carrot cubes in starch solutions in a pilot scale aseptic processing system. *Food Research International* **26**(6), 431–441.

Abdelrahim, K.A., Ramaswamy, H.S., & Marcotte, M. (1997). Residence time distributions of meat and carrot cubes in the holding tube of an aseptic processing system. *Lebensmittel Wissenschaft und Technologie* **30**(1), 9–22.

Adams, J.A., & Rogers, D.F. (1973). *Computer-aided heat transfer analysis*. New York: McGraw-Hill.

Alcairo, E.R., & Zuritz, C.A. (1990). Residence time distributions of spherical-particles suspended in non-Newtonian flow in a scraped-surface heat-exchanger. *Transactions of the ASAE* **33**(5), 1621–1628.

Alhamdan, A., & Sastry, S.K. (1997). Residence time distribution of food and simulated particles in a holding tube. *Journal of Food Engineering* **34**, 271–292.

Alhamdan, A., & Sastry, S.K. (1998). Residence time distribution of foods and simulated particles in a model swept surface heat exchanger. *Journal of Food Process Engineering* **21**(2), 145–180.

Astrom, A., & Bark, G. (1994). Heat transfer between fluid and particles in aseptic processing. *Journal of Food Engineering* **21**, 97–125.

Awuah, G.B., Ramaswamy, H.S., Simpson, B.K., & Smith, J.P. (1996). Fluid-to-particle convective heat transfer coefficient as evaluated in an aseptic processing holding tube simulator. *Journal of Food Process Engineering* **19**(3), 241–267.

Bain, A.G., & Bonnington, S.T. (1970). *The hydraulic transport of solids by pipeline*. Oxford, UK: Pergamon Press Ltd.

Balasubramaniam, V.M., & Sastry, S.K. (1994a). Liquid-to-particle heat transfer in non-Newtonian carrier medium during continuous tube flow. *Journal of Food Engineering* **23**, 169–187.

Balasubramaniam, V.M., & Sastry, S.K. (1994b). Convective heat transfer at particle-liquid interface in continuous tube flow at elevated fluid temperatures. *Journal of Food Science* **59**(3), 675–681.

Balasubramaniam, V.M., & Sastry, S.K. (1994c). Liquid-to-particle heat transfer in continuous flow through a horizontal scraped surface heat exchanger. *Transactions of the Institute of Chemical Engineers* **72**(C4), 189–196.

Balasubramaniam, V.M., & Sastry, S.K. (1995). Use of liquid crystals as temperature sensors in food processing research. *Journal of Food Engineering* **26**, 219–230.

Balasubramaniam, V.M., & Sastry, S.K. (1996a). Liquid-to-particle heat transfer in continuous tube flow: Comparison between experimental techniques. *International Journal of Food Science and Technology* **31**(2), 177–187.

Balasubramaniam, V.M., & Sastry, S.K. (1996b). Estimation of convective heat transfer between fluid and particle in continuous flow using a remote temperature sensor. *Journal of Food Process Engineering* **19**(2), 223–240.

Balasubramaniam, V.M., & Sastry, S.K. (1996c). Fluid-to-particle convective heat transfer coefficient in a horizontal scraped surface heat exchanger determined from relative velocity measurement. *Journal of Food Process Engineering* **19**(1), 75–95.

Baptista, P.N., Oliveira, F.A.R., Cunha, L.M., & Oliveira, J.C. (1995). Influence of large solid spherical particles on the residence time distribution of the fluid in two phase tubular flow. *International Journal of Food Science and Technology* **30**, 625–637.

Bean, P.G., Dallyn, H., & Ranjith, H.M.P. (1978). The use of alginate spore beads in the investigation of ultra high temperature processing. In B. Tabiano (Ed.), *Proceedings of the international meeting on food microbiology and technology*, Parma, Italy, 20–23 April.

Berry, M.R. (1989). Predicting the fastest particle residence time. In J.V. Chambers (Ed.), *1st International Congress on Aseptic Processing Technologies*, Purdue University, USA.

Bird, R.B., Stewart, W.E., & Lightfoot, E.N. (1960). *Transport phenomena*. New York: John Wiley & Sons.

Brown, K.L., Ayres, C.A., Gaze, J.E., & Newman, M.E. (1984). Thermal destruction of bacterial spores immobilized in food/alginate particles. *Food Microbiology* **1**(1), 187–198.

Buchwald, B. (1988). In H. Reuter (Ed.), *Aseptic packaging of food*. Technomic Publishing Co., Ltd.

Cacace, D., Palmieri, L., Pironi, G., Dipollina, G., Masi, P., & Cavella, S. (1994). Biological validation of mathematical modelling of the thermal processing of particulate foods: The influence of heat transfer coefficient determination. *Journal of Food Engineering* **23**, 51–68.

Campos, D.T., Steffe, J.F., & Ofoli, R.Y. (1994). Statistical-method to evaluate the critical Reynolds-number for pseudoplastic fluids in tubes. *Journal of Food Engineering* **23**(1), 21–32.

Carslaw, H.S., & Jaeger, J.C. (1959). *Conduction of heat in solids* (2nd ed.). Oxford, UK: Oxford University Press.

Chandarana, D.I., Gavin, A., III, & Wheaton, F.W. (1989). Simulation of parameters for modelling aseptic processing of foods containing particulates. *Food Technology* **43**(3), 137–143.

Chandarana, D.I., Gavin, A., III, & Wheaton, F.W. (1990). Particle/fluid interface heat transfer under UHT conditions at low particle/fluid relative velocities. *Journal of Food Process Engineering* **13**(3), 191–206.

Chandarana, D.I., & Unverferth, J.A. (1996). Residence time distribution of particulate foods at aseptic processing temperatures. *Journal of Food Engineering* **28**(3/4), 349–360.

Chang, S.Y., & Toledo, R.T. (1989). Heat transfer and simulated sterilization of particulate solids in a continuously flowing system. *Journal of Food Science* **54**(4), 1017–1023, 1030.

Charm, S.E. (1978). *The fundamentals of food engineering* (3rd ed.). Westport, CT: AVI Publishers.

Clement, I., Duquenoy, A., Jung, A., & Morisset, V. (1997). Measuring the coefficient of heat transfer between a liquid and suspended particles. In R. Jowitt (Ed.), *Engineering & food at ICEF-7*. Sheffield, UK: Sheffield Academic Press.

Clift, R. (1978). *Bubbles, drops and particles*. New York: Academic Press.

Coulson, J.M., & Richardson, J.F. (1977). *Chemical engineering*. Oxford, UK: Pergamon Press Ltd.

Dallyn, H., Falloon, W.C., & Bean, P.G. (1977). Method for the immobilization of bacterial spores in alginate gel. *Laboratory Practice* **26**(10), 773–775.

de Alwis, A.A.P., & Fryer, P.J. (1992). Operability of the Ohmic heating process: electrical-conductivity effects. *Journal of Food Engineering* **15**, 21–48.

de Alwis, A.A.P., Halden, K., & Fryer, P.J. (1989). Shape and conductivity effects in the Ohmic heating of foods. *Chemical Engineering Research and Design* **67**, 159–168.

de Cordt, S., Vanhoof, K., Hu, J., Maesmans, G., Hendrickx, M., & Tobback, P. (1992). Thermostability of soluble and immobilized alpha-amylase from *Bacillus licheniformis*. *Biotechnology and Bioengineering* **40**(3), 396–402.

Dutta, B., & Sastry, S.K. (1990a). Velocity distributions of food particle suspensions in holding tube flow: Experimental and modelling studies on average particle velocities. *Journal of Food Science* **55**(5), 1448–1453.

Dutta, B., & Sastry, S.K. (1990b). Velocity distributions of food particle suspensions in holding tube flow: Distribution characteristics and fastest–particle velocities. *Journal of Food Science* **55**(6), 1703–1710.

Fan, K.M., & Wu, W.R. (1996). Residence time distribution of suspended particle in vertical tubular flow. *Journal of Food Science* **61**(5), 982–994, 1067.

Fryer, P.J., de Alwis, A.A.P., Koury, E., Stapely, A.G.F., & Zhang, L. (1993). Ohmic processing of solid-liquid mixtures: Heat generation and convection effects. *Journal of Food Engineering* **18**, 101–125.

Fryer, P.J., Pyle, D.L., & Rielly, C.D. (1997). In P.J. Fryer (Ed.), *Chemical engineering for the food industry*. London: Blackie Academic and Professional.

Ghiron, K., & Litchfield, J.B. (1997). Magnetic thermometry in the aseptic processing of multiphase foods. In R. Jowitt (Ed.), *Engineering & food at ICEF-7*. Sheffield, UK: Sheffield Academic Press.

Grabowski, S., & Ramaswamy, H.S. (1995) Incipient carrier fluid velocity for particulate flow in a holding tube. *Journal of Food Engineering* **24**, 123–136.

Gurney, H.P., & Lurie, J. (1923). Charts for estimating temperature distributions in heating or cooling of solid shapes. *Industrial and Engineering Chemistry* **15**(11), 1170–1172.

Hanks, R.W. (1963). The laminar-turbulent transition for fluids with a yield stress. *AIChemEJ.*, **9**, 306–309.

Härröd, M. (1987). Scraped surface heat exchangers. A literature survey of flow patterns, mixing effects, residence time distribution, heat transfer and power requirements. *Journal of Food Process Engineering* **9**, 1–62.

Härröd, M. (1990a). Modelling of the media-side heat transfer in scraped surface heat exchangers. *Journal of Food Process Engineering* **13**(1), 1–21.

Härröd, M. (1990b). Temperature variations in the outlet from scraped surface heat exchangers. *Journal of Food Process Engineering* **13**(1), 23–38.

Härröd, M. (1990c). Methods to distinguish between laminar and vortical flow in scraped surface heat exchangers. *Journal of Food Process Engineering* **13**(1), 39–58.

Härröd, M. (1990d). Modelling of laminar heat transfer in scraped surface heat exchangers. *Journal of Food Process Engineering* **13**(1), 59–68.

Härröd, M. (1990e). Modelling of vortical heat transfer in scraped surface heat exchangers. *Journal of Food Process Engineering* **13**(1), 69–79.

He, Y.Z. (1995). *Residence time distributions of liquids and particulates in a holding tube.* PhD Thesis. Reading, UK: University of Reading.

Heisler, M.P. (1947). Temperature charts for induction and constant temperature heating. *Transactions of the American Society of Mechanical Engineers* **69**(4), 227–236.

Heppell, N.J. (1985). Comparison of the residence time distributions of water and milk in an experimental UHT sterilizer. *Journal of Food Engineering* **4**, 71–84.

Heppell, N.J. (1990). *Continuous sterilization processes.* PhD Thesis, University of Reading, Reading, UK.

Holdsworth, S.D. (1992). *Aseptic processing and packaging of food products.* London: Elsevier Applied Science.

Holdsworth S.D. (1993). Rheological models used for the prediction of the flow properties of food products. *Transactions of the Institution of Chemical Engineers* **71(C)**, 139–179.

Hong, C.W., Sun Pan, B., Toledo, R.T., & Chiou, K.M. (1991). Measurement of residence time distribution of fluid and particles in turbulent flow. *Journal of Food Science* **56**(1), 255–256, 259.

Hunter, G.M. (1972). Continuous sterilization of liquid media containing suspended particles. *Food Technology in Australia* **4**, 158–165.

Jakob, M. (1949). *Heat transfer.* New York: John Wiley & Sons.

Kelly, B.P., Magee, T.R.A., & Ahmad, M.N. (1995). Convective heat transfer in open channel flow: Effects of geometric shape and flow characteristics. *Food and Bioproducts Processing* **73** (C4), 171–182.

Lareo, C.A., & Fryer, P.J. (1998). Vertical flows of solid-liquid food mixtures. *Journal of Food Engineering* **36**(4), 417–443.

Lareo, C., Fryer P.J., & Barigou, M. (1997). The fluid mechanics of two-phase solid-liquid food flows: A review. *Transactions of the Institution of Chemical Engineers* **75** (C), 73–105.

Lee, J.H., & Singh, R.K. (1993). Residence time distribution characteristics of particle flow in a vertical scraped surface heat exchanger. *Journal of Food Engineering* **18**, 413–424.

Leniger, H.A., & Beverloo, W.A. (1975). *Food process engineering.* Dordrecht, Netherlands: D. Reidal Publishing Co.

Lewis, M.J. (1987). *Physical properties of foods and food processing systems.* Chichester, UK: Ellis Horwood.

Liu, S., Pain, J.P., Proctor, J., de Alwis A.A.P., & Fryer, P.J. (1993). An experimental study of particle flow velocities in solid-liquid food mixtures. *Chemical Engineering Communications* **124**, 97.

Maesmans, G.J., Hendrickx, M.E., de Cordt, S.V., & Tobback, P. (1994). Feasibility of the use of a time-temperature integrator and a mathematical model to determine fluid-to-particle heat transfer coefficients. *Food Research International* **27**(1), 39–51.

Maingonnat, J.F., & Corrieu, G. (1983a). A study of the thermal performance of a scraped surface heat exchanger. Part I. Review of the principal models describing heat transfer and power consumption. *International Chemical Engineering* **26**, 45–54.

Maingonnat, J.F., & Corrieu, G. (1983b). A study of the thermal performance of a scraped surface heat exchanger. Part II. The effect of axial diffusion of heat. *International Chemical Engineering* **26**, 55–68.

Maingonnat, J.F., Leuliet, J.C., & Benezegh, T. (1987). Modelling the apparent state of mixing in a scraped surface heat exchanger with the thermal performance for Newtonian and non-Newtonian products. *Rev. Gen de Therm Francaise*, 306–307, 381–385.

McKenna A.B., & Tucker, G. (1990). Computer modelling for the control of particulate sterilization under dynamic flow conditions. *Food Control* **2**(4), 224–233.

Metzner, A.B., & Gluck, D.F. (1960). Heat transfer to non-Newtonian fluids under laminar flow conditions. *Chemical Engineering Science* **12**, 185–190.

Mishra, P., & Tripathi, G. (1973). Heat and momentum transfer to purely viscous non-Newtonian fluids flowing through tubes. *Transactions of the Institute of Chemical Engineers* **51**, 141–150.

Mutsakis, M., Streiff, F.A., & Schneider, G. (1986). Advances in static mixing technology. *Chemical Engineering Progress* **82**(7), (July 1986).

Mwangi, J.M., Rizvi, S.S.H., & Datta, A.K. (1993). Heat transfer to particles in shear flow: Application in aseptic processing. *Journal of Food Engineering* **19**(1), 55–74.

Oliver, D.R. (1962). Influence of particle rotation on radial migration in Poiseuille flow of suspensions. *Nature* **194**, 1269–1271.

Palaniappan, S., & Sastry, S.K. (1991). Electrical conductivity of selected juices and influences of temperature, solids content, applied voltage and particle size. *Journal of Food Process Engineering* **14**, 247–260.

Palaniappan, S., & Sizer, C.E. (1997). Aseptic process validated for foods containing particles. *Food Technology* **51**(8), 60–62, 64, 66, 68.

Palmer, J.A., & Jones, V.A. (1976). Prediction of holding times for continuous thermal processing of power law fluids. *Journal of Food Science* **41**, 1233–1234.

Palmieri, L., Cacace D., Dipollina, G., Dall'Aglio, G., & Masi, P. (1992). Residence time distribution of food suspensions containing large particles when flowing in tubular systems. *Journal of Food Engineering* **17**(3), 225–239.

Perry, R.H., Green, D.W., & Maloney, J.O. (1984). *Perry's chemical engineers' handbook*. New York: McGraw-Hill.

Ramaswamy, H.S., Abdelrahim, K.A., Marcotte, M., & Toupin, C. (1995). Residence time distribution (rtd) characteristics of meat and carrot cubes in starch solutions in a vertical scraped surface heat exchanger (sshe). *Food Research International* **28**(4), 331–342.

Ramaswamy, H.S., Abdelrahim, K.A., Simpson, B.K., & Smith, J.P. (1995). Residence time distribution in aseptic processing of particulate foods: A review. *Food Research International* **28**(3), 291–310.

Ramaswamy, H.S., Pannu, K., Simpson, B.K., & Smith J.P. (1992). An apparatus for particle-to-fluid relative velocity measurement in tube flow at various temperatures under non-pressurized flow conditions. *Food Research International* **25**(4), 277–284.

Ranz, W.E., & Marshall, W.K., Jr. (1952). Evaporation from drops. *Chemical Engineering Progress* **48**, 141–146.

Rao, M.A., & Anantheswaran, R.C. (1982). Rheology of fluids in food processing. *Food Technology* **36**, 116–126.

Rohsenow, W.M., & Choi, H. (1961). *Heat, mass and momentum transfer*. Englewood Cliffs, NJ: Prentice Hall.

Rowe, P.N., Claxton, K.T., & Lewis, J.B. (1965). Heat and mass transfer from a single sphere in an extensive flowing fluid. *Transactions of the Institution of Chemical Engineers* **43**, T14–T31.

Salengke, S., & Sastry, S.K. (1996). Residence time distribution of cylindrical particles in a curved section of a holding tube: The effect of particle concentration and bend radius of curvature. *Journal of Food Engineering* **27**(2), 159–176.

Sandeep, K.P., & Zuritz, C.A. (1994). Residence time distribution of multiple particles in non-Newtonian holding tube flow: Statistical analysis. *Journal of Food Science* **59**(6), 1314–1317.

Sandeep, K.P., & Zuritz, C.A. (1995). Residence times of multiple particles in non-Newtonian holding tube flow: Effect of process parameters and development of dimensionless correlations. *Journal of Food Engineering* **25**, 31–44.

Sannervik, J., Bolmstedt, U., & Tragardh, C. (1996). Heat transfer in tubular heat exchangers for particulate containing liquid foods. *Journal of Food Engineering* **29**(1), 63–74.

Sastry, S.K. (1994). In R.P. Singh & F.A.R. Oliveira (Eds.), *Minimal processing of foods and process optimization: An interface* (pp. 17–33). Boca Raton, FL: CRC Press.

Sastry, S.K., Heskitt, B.F., & Blaisdell, J.L. (1989). Experimental and modelling studies on convective heat transfer at the particle-liquid interface in aseptic processing systems. *Food Technology* **43**(3), 132–136, 143.

Sastry, S.K., Lima M., Brim J., Brunn T., & Heskitt B.F. (1990). Liquid-to-particle heat transfer during continuous tube flow: Influence of flow rate and particle to tube diameter ratio. *Journal of Food Process Engineering* **13**(3), 239–253.

Sastry, S.K., & Zuritz, C.A. (1987). A review of particle behaviour in tube flow: Applications to aseptic processing. *Journal of Food Process Engineering* **10**, 27–52.

Schneider, P.J. (1963). *Temperature response charts*. New York: Wiley & Sons.

Segré, G., & Silberberg, A. (1961). Radial particle displacement in Poiseuille flow of suspensions. *Nature* **189**, 209–210.

Sieder, E.N., & Tate, G.E. (1936). Heat transfer and pressure drop of liquids in tubes. *Industrial and Engineering Chemistry* **28**, 1429–1435.

Skudder, P.J. (1993). In E.M.A Willhoft (Ed.), *Aseptic processing and packaging of particulate foods*. Glasgow, United Kingdom: Blackie Academic & Professional.

Steffe, J.F. (1992). *Rheological methods in food process engineering*. Freeman Press.

Taeymans, D., Roelans, E., & Lenges, J. (1985). Influence of residue time distribution on the sterilization effect of scraped-surface heat exchanger used for processing liquids containing solid particles. In *Aseptic processing and packaging of foods* (pp. 100–107). Tylosand, Sweden. Sept. 9–12.

Teixeira, A.A., Dixon, J.R., Zahradnik, J.W., & Zinsmeister, G.E. (1969a). Computer determination of spore survival distributions in thermally processed conduction heated foods. *Food Technology* **23**, 137.

Teixeira, A.A., Dixon, J.R., Zahradnik, J.W., & Zinsmeister, G.E. (1969b). Computer determination of spore survival distributions in thermally processed conduction heated foods. *Food Technology* **23**, 137.

Thorne, S. (1989). *Developments in food preservation*, Vol. 3, London: Elsevier Applied Science.

Tucker, G.S., & Withers, P.M. (1994). Determination of residence time distribution of non-settling food particles in viscous food carrier fluids using hall effect sensors. *Journal of Food Process Engineering* **17**(4), 401–422.

van Loey, A., Hendrickx, M., Ludikhuyze, L., Weemaes, C., Haentjens, T., de Cordt, S., & Tobback, P. (1996). Potential *Bacillus subtilis* alpha-amylase-based time-temperature integrators to evaluate pasteurization processes. *Journal of Food Protection* **59**(3), 261–267.

van Loey, A. (1997).

Weng, Z.J., Hendrickx, M., Maesmans, G., & Tobback, P. (1991). Thermostability of soluble and immobilized horseradish peroxidase. *Journal of Food Science* **56**(2), 574–578.

Whitaker, S. (1972). Forced convection heat transfer for flow in pipes, past flat plates, single cylinders, single spheres and for flow in packed beds and tube bundles. *American Institute of Chemical Engineers Journal* **18**(2), 361–371.

Zandi, I. (1971). Hydraulic transport of bulky material. In I. Zandi (Ed.), *Advances in solid-liquid flow in pipes and its application.* Oxford, UK: Pergamon Press Ltd.

Zhang, C.T., Wannenmacher, N., Haider, A., & Levenspiel, O. (1990). How to narrow the residence time distribution of fluids in laminas flow in pipes. *Chemical Engineering Journal and Biochemical Engineering Journal* **45**(12), 64–67.

Zuritz, C.A., McCoy, S.C., & Sastry, S.K. (1990). Convective heat transfer coefficients for irregular particles immersed in non-Newtonian fluid during tube flow. *Journal of Food Engineering* **11**, 159–174.

Chapter 5

Pasteurization

5.1 INTRODUCTION

Pasteurization is a relatively mild form of heat treatment, which is used to inactivate heat-labile microorganisms, such as vegetative bacteria, yeasts, and molds, which may cause food spoilage or food poisoning if left unchecked. For milk and dairy products it has been defined by the International Dairy Federation (IDF, 1986) as follows: "Pasteurisation is a process applied to a product with the object of minimising possible health hazards arising from pathogenic microorganisms associated with milk, by heat treatment which is consistent with minimal chemical, physical and organoleptic changes in the product." Therefore, in this context it is applied to low-acid products to inactivate heat-labile bacterial pathogens and spoilage organisms. It is described as a mild process because the amount of chemical damage caused is small and the changes to the sensory characteristics are minimal.

It is also used with a range of other low-acid products (pH > 4.5), such as cream, ice cream formulations, and eggs for extending shelf life. In addition, temperatures below 100°C are used for a number of acidic products, such as fruit juices, fruit-based drinks, alcoholic beverages, and fermented drinks. In these acidic products, the main concern is inactivation of yeasts and molds, and in some cases heat-resistant enzymes. Generally, the shelf life of such refrigerated products is short, typically from 7 to 10 days, although this can be considerably extended if the precautions for either reducing postprocessing contamination or employing aseptic processing conditions are taken.

In most cases, the heat treatment is best combined with rapid cooling and refrigerated storage. Once pasteurized, it is also crucial to prevent the product from becoming recontaminated. Such contamination is referred to in general terms as postprocessing contamination, but more specifically in this instance as postpasteurization contamination (PPC). To ensure this, care and attention should be paid to the layout of equipment and to plant hygiene and general aspects of cleanliness. As well as producing specific products, pasteurization is an integral part of some other food-processing operations, such as the production of many types of cheese and spray-dried milk powders. In fact, in milk powder production and infant formulations, pasteurization is the critical process in the production of a pathogen (salmonella)-free product. This and other important control points are discussed in more detail by Mettler (1994).

Some of the historical developments of heat treatment are discussed by Cowell (1995), Thorne (1986), and Wilbey (1993). The credit for development of a mild method of processing food, capable of inactivating vegetative organisms and thereby reducing spoilage, has been given to Pasteur, who worked with beer, wine, and vinegar. The realization that milk was a vehicle for pathogens, which cause diseases such as tuberculosis, led to the development of a low-temperature-long-time process, which was of a batch nature. Such a process was first installed by Charles North in 1907 in New York (Wilbey, 1993). Satin (1996) has reviewed the early history of pasteurization. The idea of pasteurizing milk (as opposed to boiling it or drinking it raw) met with some fierce resistance when it was first mooted. Strong arguments were raised against its introduction (Exhibit 5–1), the main objection being that it would destroy the essential quality of milk. Many similar objections are now being targeted at food irradiation. In fact, there is also currently active debate in the United Kingdom about prohibiting the sale of raw milk to the consumer.

The early processes were batch processes; these are labor intensive and involve filling, heating, holding, cooling, emptying, and cleaning. Batch processing is the basis of the current Holder process for milk, which uses a temperature of about 63.0° to 65.6°C for 30 minutes (see Section 5.4.1). They are still widely used, particularly by small-scale producers (see Figure 1–1). They are relatively slow because heating and cooling times are considerable; the total time for one batch may be up to 2 hours. The factors influencing the heating and cooling times (t) arise from equating the rate of heat transfer from the heating medium to the rate at which the fluid absorbs energy. This is summarized in the following unsteady-state equation, which assumes perfect mixing and which can be used to predict heating and cooling times:

$$t = \frac{Mc}{AU} \cdot \ln\left(\frac{\theta_h - \theta_i}{\theta_h - \theta_f}\right)$$

where t = heating time (s); c = specific heat (J kg^{-1} K^{-1}); M = mass batch (kg); A = surface area (m^2); U = overall heat transfer coefficient (W m^{-2} K^{-1}); i = initial; f = final; and h = heating medium temperatures.

The dimensionless temperature ratio represents the ratio of the initial temperature driving force to that of the final approach temperature. The equation illustrates the exponential nature of the heat transfer process. The same dimensionless ratio can be used to evaluate cooling times, which tend to be longer because of the limitations of chilled water temperature and hence the approach temperature. These can be shortened by using glycol systems, but this adds to the complexity. These factors have been discussed in more detail by Lewis (1990).

Advantages of the batch system are that they are flexible in operation (i.e., it is easy to change from one product to another). Also, if the product is well mixed, there is no distribution of residence times.

For larger processing throughputs they have largely been superseded by continuous processes. These processes will form the major part of this chapter. Although some initial trials were performed on a continuous system, the invention and introduction of the plate heat exchanger, together with the development of the flow diversion valves and tubular holding

Exhibit 5–1 Original Objections to Pasteurization

(A) Sanitation

1. Pasteurization may be used to mask low-quality milk.
2. Heat destroys great numbers of bacteria in milk and thus conceals the evidence of dirt.
3. Pasteurization promotes carelessness and discourages the efforts to produce clean milk.
4. Pasteurization would remove the incentive for producers to deliver clean milk.
5. Pasteurization is an excuse for the sale of dirty milk.

(B) Physical and Bacteriological Quality

1. Pasteurization influences the composition of milk.
2. Pasteurization destroys the healthy lactic acid bacteria in milk, and pasteurized milk goes putrid instead of sour.
3. Pasteurization favors the growth of bacteria in milk.
4. Pasteurization destroys beneficent enzymes, antibodies, and hormones, and takes the "life" out of milk.

(C) Economics

1. Pasteurization legalizes the right to sell stale milk.
2. Pasteurization is not necessary in a country where milk goes directly and promptly from producer to consumer.
3. Pasteurization will increase the price of milk.
4. There are always some people who "demand raw milk."
5. If pasteurization is required many small raw milk dealers will either have to go to the expense of buying pasteurizing apparatus or go out of business.

(D) Nutrition

1. Pasteurization impairs the flavor of milk.
2. Pasteurization significantly lowers the nutritive value of milk.
3. Children and invalids thrive better on raw milk.
4. Infants do not develop well on pasteurized milk.
5. Raw milk is better than no milk.

(E) Public Health and Safety

1. Pasteurization fails to destroy bacterial toxins in milk.
2. Imperfectly pasteurized milk is worse than raw milk.
3. Pasteurization, by eliminating tuberculosis of bovine origin in early life, would lead to an increase in pulmonary tuberculosis in adult life.
4. Pasteurization is unnecessary, because raw milk does not give rise to tuberculosis.
5. Pasteurization gives rise to a false sense of security.
6. It is wrong to interfere in any way with Nature's perfect food.
7. Pasteurization would lead to an increase in infant mortality.

Source: Reprinted from M. Satin, *Food Irradiation—A Guidebook*, p. 35, with permission from Technomic Publishing Co., Inc., copyright 1996.

tubes, made continuous processes an attractive proposition. These allow the use of high(er) temperatures for short(er) times (HTST), and conditions of 72°C for 15 seconds are typical for milk and cream in the United Kingdom. It was recognized that this might also improve product quality by reducing degradative chemical reactions, although this is much less of a problem in pasteurization than in sterilization processes. Some developments in plate heat exchangers have been discussed by Trevatt (1994), and for tubular heat exchangers by Daniels (1990).

Some of the advantages of HTST processing over batch processing are as follows:

- It permits more rapid heating and cooling.
- It is more suitable for larger volumes.
- It is better suited to automation.
- It provides scope for improved energy economy, by use of regeneration. The incoming fluid is heated, usually in a countercurrent fashion by the fluid leaving the holding tube. In this way, savings in heating and cooling can be achieved.

Heat treatment regulations also impose extra areas of control, which may be applicable both to batch and continuous systems. These vary from product to product and country to country. The regulations may stipulate temperature/time combinations (usually minimum times), cooling rates and temperatures, temperature recording for continuous processing, flow control, and flow diversion. In addition, chemical, biochemical, and microbiological tests may also be stipulated.

The emphasis with regard to quality assurance has moved to a system based on understanding and controlling the process, rather than one relying on end-product evaluation. The important factors with pasteurization are (1) raw material quality, (2) processing time and temperature, (3) reducing postpasteurization contamination, and (4) storage temperature. These aspects need serious consideration at the design stage.

5.2 HTST PASTEURIZATION

The schematic for the flow of fluid through the heat exchanger and the heat-exchange sections are shown in Figure 5–1. The fluid first enters the regeneration section (A), where it is heated from θ_1 to θ_2 by the fluid leaving the holding tube. It then enters the main heating section, where it is heated to the pasteurization temperature θ_3. It then passes through the holding tube. The tube is constructed such that the minimum residence time exceeds the stipulated residence time. It then passes back into the regeneration section, where it is cooled to θ_4. This is followed by further cooling sections, employing mains water and chilled water. The mains water section is usually dispensed with where it may heat the product rather than cool it, for example, at high regeneration efficiencies or high mains water temperatures.

The regeneration efficiency (RE) is defined as (heat supplied by regeneration/total heat load, assuming no regeneration). It can be determined from the following equation:

$$RE = [(\theta_2 - \theta_1) / (\theta_3 - \theta_1)] \times 100$$

Assuming no heat losses: $(\theta_3 - \theta_4) = (\theta_1 - \theta_2)$. It can also be estimated by measuring the steam consumption; on a small plant this can be done by measuring the overflow from the hot water set. Regeneration efficiencies can be greater than 90%. As RE increases, the size and capital cost of the heat exchanger increases; the heating rate decreases, which may affect quality, but this is more noticeable in ultrahigh-temperature (UHT) sterilization (see Section 6.4.6). Heating profiles tend to become more linear at high RE efficiencies.

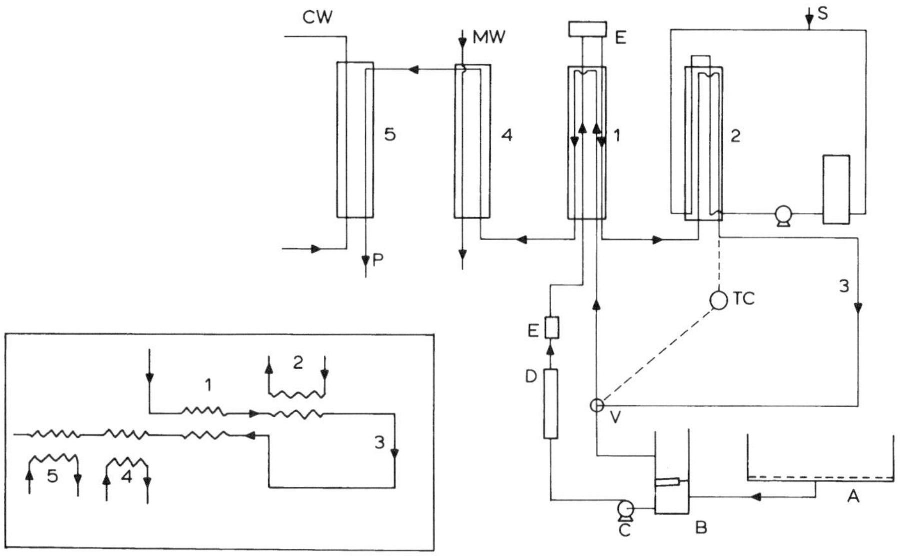

Figure 5–1 Layout of HTST pasteurizer (the insert shows a schematic diagram of the heat exchange sections). (A: feed tank; B: balance tank; C: feed pump; D: flow controller; E: filter; P: product; S: steam injection (hot water section); V: flow diversion valve; MW: mains water coolings; CW: chilled water; TC: temperature controller; 1: regeneration; 2: hot water section; 3: holding tube; 4: mains cooling water; 5: chilled water cooling). *Source:* Reprinted from M.J. Lewis, Physical Properties of Dairy Products, in *Modern Dairy Technology*, Vol. 2, 2nd ed., R.K. Robinson, ed., © 1994, Aspen Publishers, Inc.

5.2.1 Engineering and Control Aspects

A typical layout for an HTST pasteurizer is described in Section 3.5 and shown in Figure 5–2. In the vast majority of cases plate heat exchangers are used (Chapter 3). Other features are a float-controlled balance tank, to ensure a constant head to the feed pump and a range of screens and filters to remove any suspended debris from the material. In most pasteurizers, one pump is used. It is crucial that the flowrate remain constant, despite any disturbances in feed composition temperature, or changes in the system characteristics. The two most common options are a centrifugal pump with a flow controller or a positive displacement pump. If the product is to be homogenized, the homogenizer itself is a positive pump and is sized to control the flow rate. Section 3.1 provides more detail.

In pasteurization literature, the term *holder efficiency* is used; this basically relates the minimum residence time to the average residence time. Values quoted are 50% for streamline flow and 80% to 83% for turbulent flow. Note that the average residence time can be estimated from the holding tube volume and the volumetric flowrate, whereas the minimum residence will depend upon the nature of the flow (i.e., whether it is streamline or turbulent (see Section 3.23). The minimum residence time must be determined experimentally. *The*

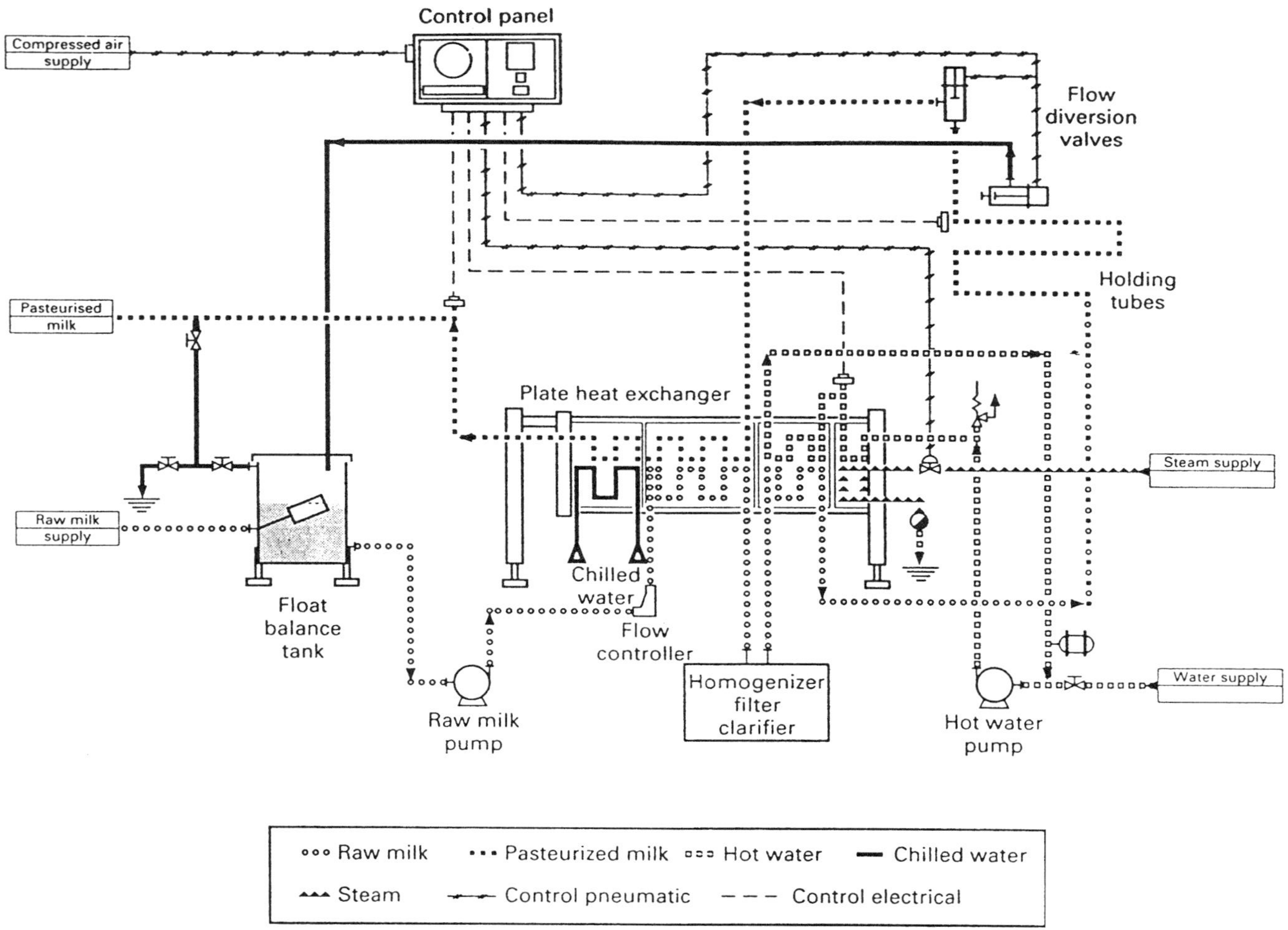

Figure 5–2 Typical milk pasteurization system. *Source:* Reprinted with permission from M.A. Pearse, Pasteurization of Liquid Products, *Encyclopaedia of Food Science and Nutrition*, p. 3445, © 1993, Academic Press.

minimum residence time should always exceed the residence time that is stipulated in any regulations. This can be checked experimentally and is worth doing so at regular intervals.

In the majority of pasteurizers, the final heating process is provided by a hot water set. Steam is used to maintain the temperature of the hot water at a constant value, somewhere between 2°C and 10°C higher than the required pasteurization temperature. There are very few direct-steam-injection pasteurization units, although they are being investigated for extended heat treatments (see Section 5.6.2) and ultrapasteurization.

Electrical heating is used, typically in locations where it would be costly or difficult to install a steam generator (boiler). The author has had experience of a pasteurizer with quartz tubes, with electrical heating elements wound around the tubes.

The holding system is usually a straightforward holding tube, with a temperature recording probe at the beginning and a flow diversion valve at the end. It is not usually insulated. From the holding tube, in normal operation it goes back into regeneration, followed by final cooling. The pasteurized product may be packaged directly or stored in bulk tanks. In addition, there are other valves and fittings; attention should be paid to their hygienic design, especially in terms of minimizing PPC.

In principle, a safe product should be produced at the end of the holding tube. However, there may be an additional contribution to the total lethality of the process from the initial part of the cooling cycle. Thereafter, it is important to prevent recontamination, both from dirty pipes and from any recontamination with raw milk. Failures of pasteurization have resulted from both causes. The most serious incidents have caused food poisoning outbreaks and have arisen where the pasteurized milk has been recontaminated with raw milk, which must have unfortunately contained pathogens. Such contamination may arise for a number of reasons, all of which involve a small fraction of raw milk not going through the holding tube (Figure 5–3).

One explanation lies with pinhole leaks or cracks in the plates, which may appear with time due to corrosion. With plate heat exchangers, the integrity of the plates needs testing for

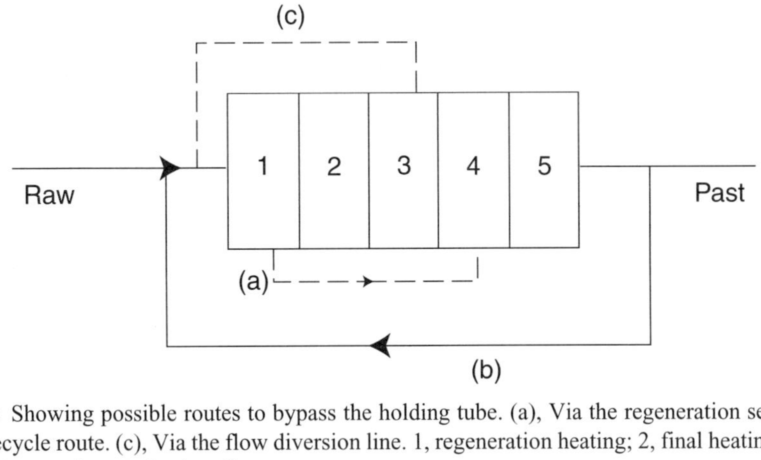

Figure 5–3 Showing possible routes to bypass the holding tube. (a), Via the regeneration section. (b), Via the detergent recycle route. (c), Via the flow diversion line. 1, regeneration heating; 2, final heating; 3, holding tube; 4, regeneration cooling; 5, final cooling.

these. This is most critical in the regeneration section, where there is a possibility of contamination from raw to treated and from high to low pressure. An additional safeguard is the incorporation of an extra pump, to ensure that the pressure on the pasteurized side is higher than that on the raw side, but this further complicates the plant. In some countries this requirement may be incorporated into the heat-treatment regulations. Another approach is the use of double-walled plates, which also increases the heat transfer area by about 15% to 20% due to the air gap. These and other safety aspects have been discussed by Sorensen (1996).

Small holes (pinholes) in plates or tubes in the heating and cooling sections could also lead to product dilution or product contaminating the hot water set, or chilled water; this may result in unexpected microbiological hazards with these water supplies. It is important that plates be regularly pressure-tested and that the product be tested to ensure that it has not been diluted, by measuring depression of the freezing point. Similar problems may arise from leaking valves, either in recycle or detergent lines.

It is now easier to detect whether pasteurized milk has been recontaminated with raw milk since the introduction of more sensitive instrumentation for detecting phosphatase activity. It is claimed that raw milk contamination as low as 0.01% can be detected (Section 5.4.1). Miller Jones (1992) documents a major pasteurization failure in the United States, where over 16,000 people were infected with *Salmonella* and 10 died. The cause was thought to be a section of the plant that was not easy to drain and clean, which led to pasteurized milk's being recontaminated.

One feature of a modern pasteurization plant is the increased level of automation. The accuracy and reliability of sensors and detection methods becomes more important, replacing the role of the experienced process operator, who was more familiar with the properties and behavior of the product he or she was dealing with.

Some further considerations of the engineering aspects are provided by Kessler (IDF, 1986), Society of Dairy Technology (SDT) (1983), and Hasting (1992). Fouling is not considered to be such a problem compared to that found in UHT processing. However, with longer processing times and poorer-quality raw materials it may have to be accounted for (see Section 8.4.2), and some products such as eggs may be more prone to fouling. One important aspect is a reduction in residence time due to fouling in the holding tube (Hasting, 1992).

One interesting question concerns the best position for the temperature detector in the holding tube, in relation to the flow-diversion facility. Usually the flow diverter is situated at the end of the holding tube, whereas the temperature probe may be positioned either at the beginning or toward the end. When positioned at the beginning, there is more time for the control system to respond to underprocessed fluid (i.e., the time it will take for it to pass through the tube), but it will not measure the minimum temperature obtained, as there may be a reduction in temperature due to heat loss as the fluid flows through the tube. This could be reduced by insulating the holding tube, but it is not generally considered to be a major problem on commercial pasteurizers. This is an aspect the purist may wish to consider further. Ideally, the temperature control should be within ± 0.5°C. Note that a temperature error of 1°C will lead to a reduction of about 25% in the process lethality (calculated for $z = 8$°C).

Mechanical stresses due to heating and cooling should also be reduced, to minimize any potential damage to plates. The trend is for HTST pasteurization plants to run for much longer periods (16 to 20 hours), before cleaning and shutdown. Again, monitoring phosphatase activity at regular intervals throughout is useful to ensure uniform pasteurization throughout and for detecting more subtle changes in plant performance, which may lead to a better estimation of when cleaning is required. However, it has also been suggested that there is an increase in thermophilic bacteria due to an accumulation of such bacteria in the regenerative cooling section arising toward the end of such long processing runs.

5.2.2 Tunnel (Spray) Pasteurizers

Tunnel (spray) pasteurizers are widely used in the beverage industry for continuous heating and cooling of products in sealed containers. They are ideal for high-volume throughput. Examples of such products are soft and carbonated drinks, juices, beers, and sauces (Figure 5–4). Using this procedure, PPC should be very much reduced, the major cause being due to defective seams on the containers. There are three main stages in the tunnel—heating, holding, and cooling—in each stage water at the appropriate temperature is sprayed onto the container. Since heating rates are not as high as for plate or tubular heat exchangers, these processes are more suited to longer time/lower temperature processes. The total transit time may be about 1 hour, with holding temperatures between 60° and 70°C for about 20 minutes. Pearse (1993) cites 60°C for 20 minutes as being a proven time/temperature profile. There is some scope for regeneration in these units.

Microwave heating may also be used for pasteurization; one recent example for milk being described by Villamiel, Lopez Fandino, and Olano (1996), who claimed that it was slightly superior in terms of its keeping quality and reducing whey protein denaturation. However, they used 80°C for 15 seconds, which probably would have not complied with the current pasteurization regulations, as it would give a negative lactoperoxidase. Also, microwave processes are much more complex and the benefits that might result have to be weighed against the increased costs compared to well-established HTST processes. Another drawback with microwave pasteurization systems is that of nonuniform heating. This is a critical aspect in pasteurization processes, where it is a requirement that all elements of the food reach a minimum temperature for a minimum time, for example, 70°C for 2 minutes to inactivate *Escherichia coli* O157. If part of the food only reaches 65°C, even though other parts may well be above 70°C, it will lead to serious underprocessing. Identifying with greater certainty where are the slower heating points will help to improve matters. For these reasons, microwaves will not be considered in depth in this chapter. Brody (1992) provides further reading on microwave food pasteurization.

5.3 FACTORS AFFECTING KEEPING QUALITY

Keeping quality is perhaps the most important commercial quality consideration. Since pasteurization only inactivates vegetative spores, the keeping quality will be influenced by a number of factors and may vary considerably. The important control factors are raw material quality, time/temperature conditions, reducing postpasteurization processing, and storage temperatures. It will be demonstrated that keeping quality can be extended by understanding and controlling the overall pasteurization process.

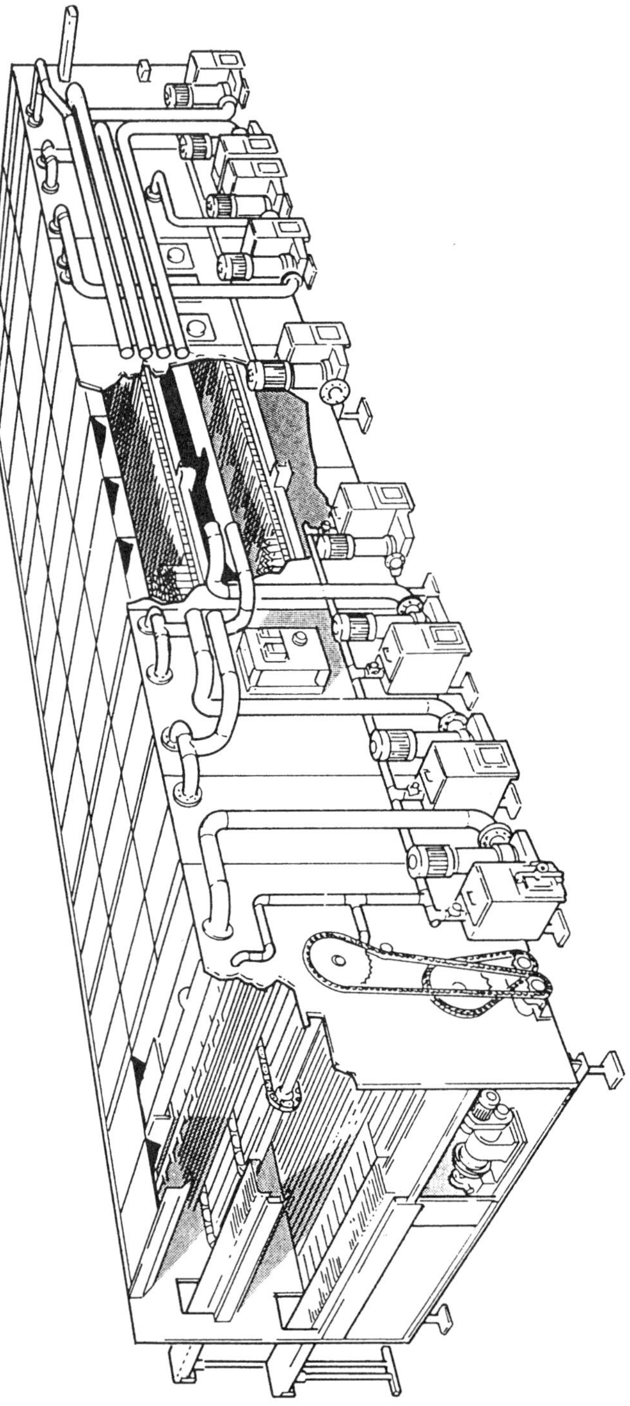

Figure 5–4 Tunnel pasteurizer. *Source:* Reprinted with permission from M.A. Pearse, Pasteurization of Liquid Products, *Encyclopaedia of Food Science and Nutrition*, p. 3449, © 1993, Academic Press.

5.3.1 Raw Material Considerations

It is important to remember that microbial inactivation follows first-order reaction kinetics (Section 2.1). Therefore, the number of microorganisms that will survive the heat treatment and hence the likelihood of spoilage will increase as the initial population increases, which stresses the importance of good microbial raw material quality. However, the situation is complicated by the presence of two distinct populations, one that will be inactivated and the other that will survive pasteurization. Also, some microorganisms that survive the heat treatment (thermoduric and spore formers) may even be activated (heat-shocked) by the process.

The aim in pasteurization is to eliminate vegetative pathogens and reduce the levels of spoilage organisms. A recent survey on *Salmonella* and *Listeria* in bulked milk supplies (O'Donnell, 1995) showed that *Salmonella* were present in 6 out of 1,673 samples tested (0.36%) whereas *Listeria spp.* were present in 310 out of 2,009 samples examined (15.4%). These observations come from a country where the overall quality of the raw milk is considered to be good. Miller Jones (1992) cited the findings of McManus and Lanier (1987), where *Salmonella* was found in 60% of raw milk samples tested, which is much more alarming. Therefore, pasteurization procedures are designed to cope with the worst-case scenario (i.e., that there will be some pathogens present in raw materials).

Another problem arises from the wider range of heat resistances of the microorganisms that may be present. Although pasteurization may achieve 6 to 8 decimal reductions for the major pathogens (see Section 5.3.2), there would be no inactivation of spore-forming bacteria. The source and nature of spore-forming bacteria have been discussed recently in more detail by Lewis (1999).

The standards of hygiene at the production stage are very important. These are summarized in Exhibit 5–2. If raw milk or other raw materials are to be stored before pasteurization, reduce storage temperatures; additional treatments are available to reduce activity of psychrotropic bacteria, such as thermization (Section 5.5.1) or treatment with carbon dioxide (Section 5.6.1).

Generally it is accepted that perishable raw materials should be processed as quickly as possible. Some recent observations have shown that raw milk may be stored for longer periods than was originally thought before the keeping quality of pasteurized milk was affected, according to Griffiths et al. (1989) and Ravanis and Lewis (1995); in fact, the latter authors found that milk pasteurized on the third and fourth days gave a better keeping quality than that processed on the first day. There may be some natural antimicrobial systems operating within the raw materials that can be activated, for example, the lactoperoxidase system (see Section 5.6.1).

5.3.2 Time/Temperature Combinations:

In many countries, heat treatment regulations may apply to specific products. These usually stipulate the minimum heating temperature and time that should be used. As mentioned, pasteurization is generally a mild method of preservation, irrespective of whether the process is batch or continuous. Its main aim is to eliminate those vegetative pathogens and reduce

Exhibit 5–2 Hazards Involved and the Critical Control Points in Raw Milk Handling

1. **Cow**
   Hazards: Milk is obtained from unhealthy animals. The CP (control point) is health control (farmer and the supervising veterinarian).
2. **Milking**
   Hazards: Milking routines may damage tissues, leading to infections of the udder, and contaminate the milk by the environment, milking equipment, etc. The CP is cleaning the udder before and after the milking.
3. **Storage and milk cooler**
   Hazards: High storage temperature and time may create growth of pathogens. CPs are temperature and time.
4. **Collection and transport**
   Hazards: High transport temperature and time as well as unhygienic routines may create growth and contamination of pathogens. CPs are temperature and time.
5. **Delivery**
   Hazards: Seldom associated with the delivery procedure or the milk plant. At this stage CPs can be identified in order to control the hazards.
6. **Storage in silo tanks**
   Hazards: High storage temperature and/or time may create growth of pathogens. CPs are temperature and time. If the milk is to be pasteurized the consumer will be protected by the pasteurization process.

Source: Reprinted with permission from IDF, *Bacteriological Quality of Raw Milk*, Reference S.I. 9601, © 1996, International Dairy Federation.

spoilage bacteria to low levels; this it achieves with some success. However, certain microorganisms will survive the heat treatment and their growth should be restricted during storage. Thermodurics are defined as those organisms that will survive 63° to 64°C for 30 minutes. Those that survive 80°C for 10 minutes are further classified as spore formers. Of major concern are thermodurics and spore formers that are capable of growth at refrigerated temperatures, especially *Bacillus cereus* (see Section 10.2.2).

Some heat-resistant enzymes may survive pasteurization; most notably are proteases and lipases, but many others as well are not totally inactivated (Table 5–1 and Table 5–2). One important feature of pasteurization is that it maintains a high level of protein functionality, which is important for eggs, whipping cream, and other materials.

Inactivation of Pathogens

Pathogens such as *Mycobacterium tuberculosis*, *Campylobacter*, and *Salmonella* are readily inactivated. Brown (1991) has compiled a list of heat-resistance data for a range of vegetative bacteria (Table 5–3). D values are presented at 70°C. This table cites one very high heat resistance value for *Salmonella* in chocolate.

In recent years, there has been more emphasis placed on the effects of pasteurization on the survival of pathogens, particularly *Salmonella* and *Listeria*, although compiled heat resistance data are not readily available. Mackey and Bratchell (1989) suggest that *Listeria* is appreciably more heat resistant than most *Salmonella* serotypes, except *Salmonella senftenberg* 775W. Normal pasteurization procedures will inactivate *L. monocytogenes* in milk, but the margin of safety is greater for vat pasteurization than for HTST treatment. However, reports that the organism can survive heating at 80°C have not been substantiated and are incompatible with carefully determined D and z-values in milk and a range of foods.

Table 5–1 Review of Enzyme Inactivation during Milk Pasteurization

Enzyme	*Temperature (°C)*	*D (s)*	*z (°C)*	*Activity*
Alkaline phosphatase	64	480		
	69.8	15.0	5.1	
Amylase Saccharifying activity	65	224.9	16.2	5.5–31.8
	70	95.9		
Lactoperoxidase	65	1595.4	5.4	0.42–0.93
	70	940.6		
Xanthine oxidase	65	1348.5	6.8	6.6–16.8
	70	1783.7		
Acid phosphatase	65	36744	6.6	0.004–0.37
	70	2552.9		

Cooking food to an internal temperature of 70°C for 2 minutes is adequate to ensure destruction of *L. monocytogenes*, as well as the more recent interest in *E. coli* O157.

Bradshaw, Peeler, and Twedt (1991) reported that the heat resistance of *L. monocytogenes* appeared somewhat greater than that of the other *Listeria spp.* in milks, but the difference was not statistically significant. HTST processing (71.7°C/15 seconds) is adequate for pasteurization of raw milk. Knabel et al. (1990) indicate that, under the conditions of the present study, high levels of *L. monocytogenes* would survive the minimum low-temperature, long-time treatment required by the US Food and Drug Administration (FDA) for pasteuriz-

Table 5–2 Summary of Residual Enzyme Activity after Pasteurization

	% Residual Enzyme Activity	
Enzyme	*Capillary Tubes, Andrews et al. (1987)*	*Plate Heat Exchanger, Griffiths (1986)*
Lipoprotein lipase	1%	
N-Acetyl glucosamidase	19%	
α-*L* -Fucosidase	26%	
γ-Glutamyl transpeptidase	75%	
Xanthine oxidase	78%	95%
α-*D*-Mannosidase	98%	
Acid phosphatase	>95%	95%
Alkaline phosphatase		7%
Catalase		26%
α-Amylase		60%
Lysozyme		82%
β-Amylase		86%
Lactoperoxidase		95%
Ribonuclease		100%

Source: Data from A.T. Andrews, M. Anderson, and P.W. Goodenough, A Study of the Heat Stabilities of a Number of Indigenous Milk Enzymes, *Journal of Dairy Research*, Vol. 54, pp. 237–246, © 1987; and M.W. Griffiths, Use of Milk Enzymes and Indices of Heat Treatments, *Journal of Food Protection*, Vol. 49, pp. 696–705, © 1986.

Table 5–3 Heat Resistance of Some Vegetative Bacteria

Organism	*Heating Medium*	*D value (min) at 70°C*	*z (°C)*	*Reference*
Escherichia coli	Nutrient broth	0.006*	4.9	(Tomlins & Ordal, 1976)
	Milk	0.04*	6.5	(Tomlins & Ordal, 1976)
Lactobacillus casei	Tomato juice	4.0	11.5	(Tomlins & Ordal, 1976)
L. plantarum	Tomato juice	11.0	12.5	(Tomlins & Ordal, 1976)
Listeria monocytogenes	Raw beef	0.15	6.7*	(Mackey & Bratchell, 1989)
L. monocytogenes	Chicken, beef, and carrot homogenates	0.14–0.27	5.98–7.39	(Gaze, Brown, & Banks, 1989)
Salmonella typhimurium	Aqueous sucrose/ glucose (A_w 0.995)	0.03*	17.05	(Tomlins & Ordal, 1976)
Salmonella typhimurium	Milk chocolate	816	19.0	(Tomlins & Ordal, 1976)
Salmonella typhimurium	51% milk	0.35*	6.8	(Tomlins & Ordal, 1976)
Staphylococcus aureus	Milk	0.30*	5.1	(Tomlins & Ordal, 1976)
Streptococcus faecalis var. *zymogenes*	Broth	2.84*	17.0	(Tomlins & Ordal, 1976)
Streptococcus faecium	Broth	0.015*	3.5	(Tomlins & Ordal, 1976)
Streptococcus faecium	Ham	2.57†	6.8–7.5	(Magnus, McCurdy, & Ingeledew, 1988)
Microbacterium lacticum	Skim milk	4.0		(Tomlins & Ordal, 1976)

* Calculated from information presented in the reference.
† D value at 74°C.

Source: Reprinted from J. Reed and J. Bettison, *Processing and Packaging of Heat Preserved Foods*, p. 38, © 1991, Aspen Publishers, Inc.

ing milk. The possible survival of low levels of *L. monocytogenes* during HTST pasteurization and enumeration of injured cells by recovery on selective media under strictly anaerobic conditions are discussed. While *L. monocytogenes* can survive minimum pasteurization treatment (71.7°C/16 seconds) under certain conditions, common methods of handling, processing, and storing fluid milk will provide an adequate margin of safety (Farber et al., 1992).

Linton, Pierson, and Bishop (1990) suggest that there is some evidence that heat-shocking *Listeria* may increase its subsequent heat resistance. Foegeding and Leasor (1990) reported that the heat resistances of all *Listeria* strains (in eggs) were similar. For *L. monocytogenes* F5069, D values ranged from 22.6 minutes at 51°C to 0.20 minute at 66°C. The *z*-value for F5069 was 7.2°C. Minimal pasteurization parameters (60°C for 3.5 minutes) for liquid egg white would result in 99% to 99.9% inactivation (populations reduced 2 to 3 log cycles) for the *L. monocytogenes* strains tested.

It was concluded that the *Yersinia spp*. investigated have a low heat resistance and are rapidly destroyed during milk pasteurization. Heat resistance of *Y. enterocolitica* was on average about 1.5 times higher than resistance in other spp (Pavlov, 1989). The effects of heat resistance in food processing have been discussed by Brown (1991). To summarize, it is considered that the major vegetative pathogens can be controlled effectively by pasteurization and they are not a major determinant of keeping quality.

5.3.3 Process Characterization

A number of parameters have been used to characterize pasteurization processes. One parameter is the pasteurization unit (PU). One pasteurization unit results from a temperature of 60°C (140°F) for 1 minute. The equivalent *z*-value is 10°C (18°F), which is high for vegetative bacteria. Thus the number of pasteurization units is given by:

$$PU = 10^{(T-60)/10} \cdot t\ (\text{min})$$

Thus a temperature of 63°C for 30 minutes would have a value of approximately 60 (Wilbey, 1993), whereas HTST conditions (72°C/15 seconds) would give only 3.96. This discrepancy arises from the large *z*-value: perhaps the lesson to be learned is that it may not be meaningful to extrapolate this to continuous-pasteurization processes. For beer processing, the *z*-value for beer spoilage organisms was 7°C (cited in Holdsworth, 1997). Using this criterion, a satisfactory process for beer processing was 5.6 PU. Holdsworth (1997) points out that an alternative pasteurization unit (P) has also been used, with a reference temperature of 65°C and *z*-value of 10°C. This does not appear to be so widely used, but to avoid any confusion it is good practice to mention the reference temperature and *z*-value used.

Another parameter, introduced by Kessler (1981) is p*. This is based on a reference temperature of 72°C and a *z*-value of 8°C (or Kelvin). Processing conditions of 72°C for 15 seconds are designated as providing a safe pasteurization process for milk and are given an arbitrary p* value of 1.

Its calculation is similar to that for lethality (see Sections 2.3 and 6.1).

$$p^* = (10^{(T-72)/z} . t) / 15$$

where $p = 10^{(T-72)/z}$ T = temperature (°C) and t = time (seconds).

Figure 5–5 shows the time-temperature combinations that correspond to a p* value of 1 (normal pasteurization) as well as other p* values (0.1 to 10).

This simplified equation ignores the contribution of the heating and cooling section. Both these factors provide an additional measure of safety and are further discussed by Kessler (IDF, 1986). Knowledge of the heating and cooling profiles will enable their contribution to be determined. It is based on the same principles as lethality (see Section 6.1). The procedure for this is from the temperature-time profile to plot p against time and determine the area under the curve. Alternatively, the activation energy (285 kJ/mol) can be used. As mentioned earlier, it is important to check the minimum residence time; dye injection can be used to check, calibrate, and certify temperature probes at regular intervals.

Comments on z-values used: It can be seen that the different pasteurization parameters

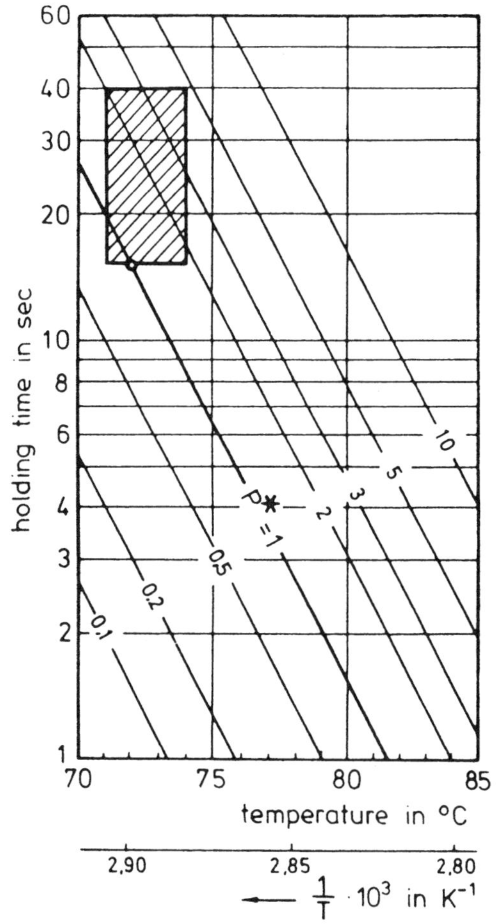

Figure 5–5 Time-temperature combinations that correspond to a p* value of 1 (normal pasteurization) and other p* values. *Source:* Reprinted with permission from H.G. Kessler, Effect of Thermal Processing on Milk, in *Developments of Food Preservation*, Vol. 5, S. Thorne, ed., pp. 91–130, © 1989, Verlag A. Kessler.

make use of different *z*-values: direct comparison of holder (63°C/30 minutes and HTST conditions (72°C/15 seconds) would give a *z*-value of about 4.3°C. It is probably not to be recommended to extrapolate PU from a batch to a continuous process, or p* from continuous to batch processes.

holder/HTST process, z	4.3 °C
p*	8
PU	10

For some acidic products, a standard F value is used to describe the process. In this case both the reference temperature and the *z*-value are stipulated (i.e., an F value is stipulated for a particular process). For example, cucumber pickles (pH 4.1) require a heating process of 36 minutes at 160°F (z = 18°F) and tomatoes (pH 4.5) require a processing time of about 20 minutes at 200°F (z = 16°F) (Pflug & Esselen, 1979).

In general, heat-resistant psychrotrophic bacteria are a major problem in milk pasteurization. Their incidence in raw milk has been reviewed by Phillips and Griffiths (1986). Of these, the most important are *B. cereus*, *B. lichenformis*, and *B. coagulans*. *B. cereus* presents a major problem, since it can withstand pasteurization, is capable of growth under refrigeration conditions, and has been implicated in food poisoning outbreaks, although rarely with milk (Section 10.2.2).

Pasteurization and the criteria for process design have been further discussed by Gaze (1994).

5.3.4 Postpasteurization Contamination

Such processing and microbiological data are useful for determining the level of organisms that survive the pasteurization process. However, it all becomes academic if the product then becomes excessively recontaminated. Therefore, it should be one of the principal objectives in all pasteurization processes to reduce the level of PPC. In fact, most of the major pasteurization failures result from PPC, which will start from the end of the holding tube.

PPC is now considered to be a very important determinant of keeping quality and Muir (1996a, 1996b) describes how this was recognized both for milk and for cream in the early 1980s. In most cases where the keeping quality suddenly deteriorates significantly or slowly declines over a longer period, paying attention to PPC can bring about a significant improvement. Trials conducted under conditions where PPC was changed, where cleaning was both rigorous and inadequate also showed this (Griffiths et al., 1989). Recent HTST pasteurization trials with milk (72°C/15 seconds) conducted at Reading University have shown that when PPC is reduced, keeping qualities (KQ) of 13 to 17 days are obtainable at 10°C and between 25 and 40 days at 4°C. (Borde-Lekona, Lewis, & Harrigan, 1994; Ravanis & Lewis, 1995). Gomez Barroso (1997) has reviewed the literature and showed that keeping qualities in excess of 40 days have been reported by several different research workers, in cases when PPC has been eliminated. PPC may also introduce vegetative pathogens into the final product and compromise its safety.

Controlling PPC relies on special attention being paid to cleaning, disinfecting, and sterilizing. Disinfecting at 90° to 95°C for 30 minutes or in special cases sterilizing at 130°C for

30 minutes may be used. Disinfection in the food industry has been discussed by Lane (1989) and is discussed in more detail in Section 8.8.

PPC is detected by looking for the presence of microorganisms that are readily inactivated by pasteurization (e.g., coliform bacteria or gram-negative psychrotrophs). A wide range of tests for detecting PPC has been compiled by the IDF (1993).

5.3.5 Storage Conditions

The main determinant in this category is storage temperature. It is now possible to control temperature within the food chain more effectively. The weak link in the overall cold chain is usually that indeterminate period after it leaves the retail outlet and reaches the consumer's refrigerator. It is not possible to control the amount of time it spends, for example, in a warm car at elevated temperature. There is a role here to educate consumers that they also have a responsibility to ensure and maintain the safety of heat-treated foods. In general, reducing the storage temperature will increase the keeping quality. Zadow (1989) reports that increasing the temperature by 3°C reduces the shelf life by 50%. Effective refrigeration of milk is considered to be keeping it below 4°C. Allowing milk to go above this temperature is regarded as poor control. Allowing the temperature to increase to 10°C will reduce the keeping quality by more than half. This is illustrated by a number of further examples. Kessler and Horak (1984) found that at 8°C the shelf life of milk ranged between 6 and 11 days, whereas at 5°C it ranged between 10 and 16 days. A linear relationship between log (storage time) and temperature was found and this was used to estimate a *z*-value of approximately 14°C. The keeping quality was taken as the time taken to reach 2×10^6 organisms per milliliter. In these experiments PPC was eliminated. No special mention was made of raw milk quality.

The generation time of psychrotrophic bacteria has been reported by SDT (1989), again showing the beneficial effects of low temperatures. Generation times (hours) were 3 at 10°C, 12 at 4.4°C and 16 at 1.7°C. The time required for the population to increase from 1/ml to 8×10^6/ml was calculated to be 2.9 days at 10°C and 34.5 days at 1.7°C. Griffiths and Phillips (1990), in a survey on growth conditions of *Bacillus spp.*, reported for *Bacillus cereus* no growth at 2°C, and lag times between 28 and 60 hours and generation times between 12 and 23 hours at 6°C.

The dominant flora in cream that had not been subject to PPC was *Bacillus cereus* types. A temperature of 6°C was the critical temperature with regard to growth of *Bacillus spp.* The experimental evidence suggests that keeping the temperature below 5°C would be sufficient to prevent significant growth of *Bacillus cereus*. The question arises whether this level of temperature control is practicable.

Muir (1990) reported on the microbial quality measured at the point of sale. The following percentages for samples with psychrotrophic counts greater than 10^5/ml were found; door-step delivery, 8%; small to medium retailers, 20%; and large supermarkets, 4%. This illustrates the variable quality within the more easily controlled part of the cold chain.

Although it should be possible to achieve a 28-day shelf life, under controlled conditions, it is highly unlikely that it would be attainable, once outside the strictly controlled conditions. Some clear instruction to drink within a short period of purchase, may be required.

Since the weak link in the cold chain is after the point of purchase there is scope for the development of time-temperature indicators (Sections 4.20 and 10.5.6) on the retail package that will register abuse at this stage. There may also be some limited scope for modified atmosphere packaging (MAP) with some products. Flushing raw milk with nitrogen prior to pasteurization was not found to improve significantly the keeping quality of pasteurized milks (Pippas, 1996). In the area of shelf-life prediction, the use of predictive modeling is a useful tool toward helping to predict the shelf life, particularly in abuse situations. A good review article on predictive modeling has been written by Whiting (1995).

5.4 SPECIFIC PRODUCTS

5.4.1 Milk

The biggest volume of pasteurization literature is devoted to milk, including a comprehensive IDF review (1986). Pasteurized milk is widely consumed worldwide and is the most popular form of heat-treated milk in many countries (See Table 1–1). There are many types of milk, namely full-cream and milk standardized to a wide range of fat contents: skim, about 0.1%; semiskim, about 1.5%; and normal milk, about 3.2%. In some countries, semiskim milk has become more popular over the past 10 years; for example, sales of semiskim milk in the United Kingdom have increased from 6.9% to 46.1% of household milk purchases over the period 1985 to 1995 (Dairy Facts and Figures, 1996).

Raw milk quality is of paramount importance: in the United Kingdom, payment schemes based on milk quality have led to a reduction in microbial count. Quality schemes are in force in many other countries and have recently been reviewed by IDF (1996a).

The bacterial count in typical raw milk in the United Kingdom ranges between 10^4 and 10^5/bacteria ml. Even with the best hygiene on farms, it is unlikely that counts below 10^3 bacteria/ml would ever be achieved. Refrigerated storage has also led to gram-negative psychrotrophic bacteria usually being the dominant flora. If the count is greater than 10^6/mL, there may be problems with heat-resistant enzymes or the production of off-flavors (taints). However, these organisms are very heat labile and readily removed by pasteurization; therefore their presence in freshly pasteurized milk is indicative of PPC.

Ideal raw milk would have low levels of psychrotrophic bacteria and thermodurics (Lewis, 1999); there would be no *Bacillus cereus* present. Surprisingly, there is little experimental information available on the effects of raw milk quality on the keeping quality of pasteurized milk.

The Dairy Products (Hygiene) Regulations (SI, 1995) stipulate the following time-temperature combinations for milk pasteurization: (1) retained at a temperature between 62.8°C and 65.6°C for at least 30 minutes and then cooled to below 10°C; and (2) retained at a temperature of not less than 71.7°C for at least 15 seconds and then cooled to below 10°C. In the Scottish regulations an upper limit of 78.1°C has been specified, together with cooling below 6°C. The first set of conditions is known as the holder process and is batch operated. In theory it could be continuous, but the holding tube would be very long. The second set of conditions is known as HTST pasteurization and is amenable to continuous-flow processing.

A third set of conditions is stipulated that permit the use of other temperature-time combinations, as may be specified by the licensing authority. There may be some debate about these conditions, as it would depend upon what are the main criteria used (see Section 5.3.3).

Regulations in most other countries are similar to these, or perhaps more severe. These have been summarized by IDF (1986). Some examples are given by Busse (1981) for Grade A milk in the United States: 63°C for 30 minutes; 77°C for 15 seconds; 89°C for 1 second; 94°C for 0.1 second; 96°C for 0.05 second; 100°C for 0.01 second.

Note the use of extremely short residence times, which may be difficult to control. These conditions produce milk with a shelf life of 18 days or longer at a maximum of 7°C. In central Europe, most dairies use 74°C for 30 to 40 seconds, but these conditions may result in a cooked flavor. However, although this may be appropriate in those particular conditions, there is no general evidence that increasing the severity of the process will increase keeping quality; in fact, in some cases it results in a decrease in the keeping quality. For example, Kessler and Horak (1984) found that more stringent conditions (i.e., 85°C/15 seconds and 40 seconds and 78°C/40 seconds) reduced the shelf life slightly. It should be noted that lactoperoxidase will also be inactivated at these conditions. The main problem with pasteurized milk arises from contaminants arising from PPC and secondly from thermodurics that are capable of growth at refrigeration temperatures.

The main statutory test for ensuring that milk is adequately pasteurized is inactivation of alkaline phosphatase (ALP). It has been demonstrated that if ALP is inactivated, *Mycobacterium tuberculosis* will have been removed satisfactorily, since ALP requires slightly more severe conditions to inactivate it than for the destruction of this particular pathogen. The conventional method (SDT, 1983), involving tintometer standard discs, was capable of detecting as little as 0.2% added raw milk, as well as very slight underprocessing (e.g., 62°C instead of 62.8°C for 30 minutes or 70°C instead of 71.7°C for 15 seconds). Improved methodology for ALP is now available, one example being the Fluorophos test. This is much more sensitive than the original method and can detect down to 0.006% added raw milk; it is now widely used, the legal cutoff for pasteurization being 500 mU of ALP per liter of milk. It is claimed to be useful on high-fat products and chocolate milks and to reduce the incidence of false-positive results arising from phenolics and flavorings, as well as giving no false-negatives (Anon, 1994). It is sensitive enough to be useful for analyzing the performance of a pasteurizer throughout the processing day and picking out any processing deviations (e.g., caused by fouling) very quickly. Painter and Bradley (1996) used the Fluorophos to analyze residual ALP activity of milks with a wide range of fat contents, and chocolate milks subject to batch and HTST pasteurization profiles; all milks subject to the legal minimum heat treatments gave ALP values well below the residual cutoff of 500 mU/L. They also noticed that some underprocessed milks also had values below 500 mU/L. Eckner (1992) found that the inactivation rates of inoculated pathogens (*Salmonella* and *Listeria*) were greater than or equal to that of ALP (measured by Fluorophos) over the temperature range tested. Viable test pathogens were isolated only from milk corresponding to the legal pasteurization, when the initial pathogen inoculum was as high as 10^6/ml.

Mistry (1989) has suggested that removal of lactose during ultrafiltration alters heat inactivation characteristics of ALP; thus, use of ALP to measure the efficiency of pasteurization of highly concentrated milks is questionable.

In some countries, specified microbiological controls have been introduced. For example, European Community regulations state that pasteurized milk should show after incubation at 6°C for 5 days a plate count (at 21°C) of less than 5×10^4/ml; it should also be free of pathogenic microorganisms and coliforms (see Section 10.2).

More recently, United Kingdom and European regulations require that pasteurized milk should show a positive lactoperoxidase activity. This will ensure that the milk is not over–heat-treated. Heating milk to about 80°C/15 seconds would lead to a negative lactoperoxidase and failure of this test, although flash pasteurization conditions may not. Milks that show a negative lactoperoxidase are permitted in some countries, provided they are labeled as "high temperature pasteurized."

The keeping quality of commercial pasteurized milk should be a minimum of 7 to 10 days. There is little published information comparing the keeping qualities of skim, semiskim, and full-cream milks. Comments from the industry suggest that semiskim milk does not keep as well as whole milk. This is despite the fact that microbial inactivation should be reduced in whole milk, due to the protective effect of the fat. This could arise from the additional processing involved and the increased likelihood of post–heat-treatment contamination.

5.4.1.1 Other Changes during Pasteurization

There are some other important changes taking place during pasteurization. In general, as far as chemical reactions are concerned, pasteurization is a mild process; about 5% to 15% of the whey protein is denatured (see Figure 6–8); this is not sufficient to increase significantly the levels of free sulfhydryl group activity or to induce formation of hydrogen sulfide and lead to the development of any cooked flavor.

Whey protein denaturation is higher in skim milk concentrates produced by ultrafiltration, increasing with the increase in the concentration factor (Guney, 1989) (Figure 5–6). There is some suggestion that the holder process may be more slightly more severe than the HTST process in these respects (Painter & Bradley, 1996). There is little change in its rennetting properties and little association of whey proteins with casein, no dephosphorylation, and no significant reduction in ionic calcium. Thus it is possible to make good Cheddar cheese from pasteurized milk, and the majority of milk for cheese making in the United Kingdom is subject to pasteurization (note that significant amounts of cheese are still made in some countries from raw milk). There is also some suggestion that pasteurization may remove some inhibitors that prevent the milk-clotting process.

The effects on heat-sensitive vitamins and other components are very small. Overall, pasteurization results in little change in texture, flavor, and color, compared to raw milk. Wilson, as far back as 1942, reported that it was clear that the majority of people are unable to distinguish between raw and pasteurized milk. Also, the difference in taste between different raw milks appears to be as great as or greater than the difference between raw and pasteurized milks. There is no evidence to suggest that this observation has changed over the past 55 years. Nursten (1995) reports that pasteurization barely alters the flavor of milk and that the volatile flavors responsible for cooked flavor were negligible. It was suggested by a recent academic visitor from the United States that there may be a low level of lipolysis in pasteurized milk, presumably caused by agitation during transportation and processing, which, however, does not bring complaints by most United Kingdom consumers. However, this

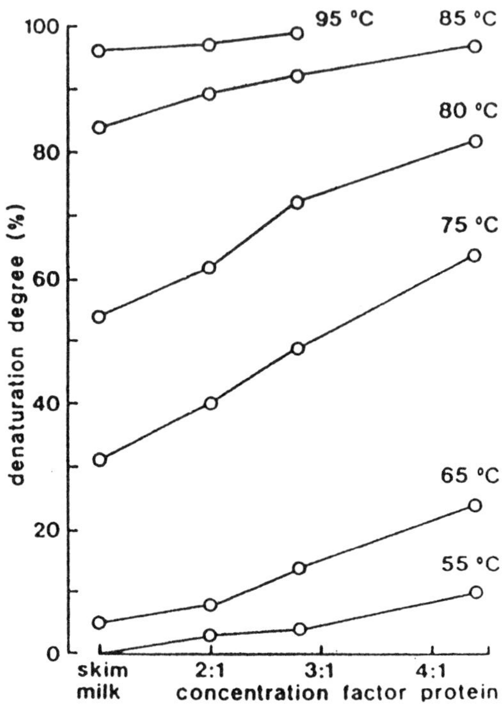

Figure 5–6 Denaturation degree of whey protein in ultrafiltered skim milk retentates depending on the concentration factor and heating temperature (5 minutes heating time). *Source:* Reprinted with permission from E. Renner and M.H. Abd El-Salam, *Application of Ultrafiltration in the Dairy Industry*, © 1989.

---

does raise the issue of different levels of sensitivity arising from exposure to the same product processed in different countries.

5.4.1.2 Enzyme Inactivation

Milk contains an abundance of enzymes and their inactivation in the pasteurization region has been the subject of two major reviews, by Griffiths (1986) and Andrews, Anderson, and Goodenough (1987). Griffiths determined the heat resistance of several indigenous milk enzymes. Some values in the pasteurization region are given in Tables 5–1 and 5–2, starting with the most heat labile. Also recorded is the range of activities determined in different samples of raw milk. Some discrepancies were also noticed between data obtained from capillary tube experiments and those obtained from HTST conditions using plate heat exchangers. Lactoperoxidase activity was thought to provide a useful indicator of overprocessing. Activities determined on a plate heat exchanger (PHE) for 15 seconds were generally lower than expected from the laboratory data. Using a PHE, enzyme activity was almost destroyed at 78°C for 15 seconds and completely destroyed at 80°C for 5 seconds. The enzyme appeared sensitive to temperatures above 75°C, with a *z*-value of 5.4°C. Since that work, ensuring that milk has a positive lactoperoxidase activity has become part of several countries' milk pasteurization regulations to prevent pasteurized milk from being overprocessed. One benefit is that the lactoperoxidose system may contribute to the keeping quality of pasteurized milk (Barrett, Grandison, & Lewis, 1999) (see Section 5.6.1).

Ribonuclease was found to be more heat resistant than lactoperoxidase. Again there were discrepancies between laboratory studies and PHE studies. No loss of activity was observed at 80°C for 15 seconds (laboratory), whereas a 40% loss of activity was found in a PHE at 80°C for 15 seconds.

Andrews et al. (1987) determined the following retention of activities for milk samples heated for precisely 15 seconds at 72°C in glass capillary tubes: acid phosphatase > 95%; α-D-mannosidase 98%; xanthine oxidase 78%; γ-glutamyl transpeptidase 75%; α-L-fucosidase 26%; *N*-acetylglucosamidase, 19%; and lipoprotein lipase 1%. It was recommended that *N*-acetyl-β-glucosamidase could be used for more detailed studies between 65° and 75°C and γ-glutamyl transpeptidase (GGTP) between 70° and 80°C.

Tables 5–1 and 5–2 give a summary of the important experimental conditions found from these studies. Patel and Wilbey (1994) recommend measuring GGTP activity for assessing the severity of HTST heat treatments above the minimum for whole milk, skim milk, sweetened milks, creams, and ice cream mixes. There was a good correlation between the reduction in GGTP activity, destruction of streptococci, and water activity.

Pasteurization is an important process in the production of dried milk and cheese. Both of these processes produce products with a long shelf life. Thus in these cases the level of enzyme activity can affect the product quality. Some work has been reported on phosphatase inactivation during drying by Daeman and Van der Stege (1982). At an inlet temperature of 150°C and an outlet temperature of 95°C, residual phosphatase activity decreased from about 80% to 45%, as the initial total solids increased from 17% to 40%. Note that spray drying conditions are less severe than pasteurization in terms of enzyme inactivation.

Pasteurization will not inactivate indigenous lipases and proteases (see Chapter 6). For indigenous lipases, the kinetics in the range 45° to 75°C are described by Driessen (IDF, 1989). There is some evidence that indigenous lipase is not totally inactivated during HTST pasteurization; this may cause a problem in homogenized milks; it is recommend that 78°C for 10 seconds is needed to inactivate it. Although some of these enzymes may survive heat treatment, their activity during chilled storage will be low, so it is not anticipated that they will cause flavor or texture defects in normal pasteurized milks. This situation may change in extended-shelf-life products, however. General aspects of pasteurization of dairy products have been covered by Kessler (1981) and Wilbey (1993).

Most of the semiskim and full-cream milk that goes into cartons is homogenized, usually at about 60°C. Homogenizers, being positive displacement pumps, will control the overall flowrate. There is some evidence that homogenized milk will give a shorter shelf life (Cox, 1975). One explanation for this is that in nonhomogenized milk, the bacteria will attach to the fat globule and thus will be present in reduced numbers in the bulk, thereby reducing microbial activity there. This needs further investigation.

Pasteurization of recombined milks does not appear to impose any specific problems, although attention should be paid to the water quality. General recommendations on hygienic production of milk and milk products are given by IDF Bulletin No. 292 (1994).

5.4.2 Other Milk-Based Drinks

Milk-based drinks are increasing in popularity. The are defined as those drinks where milk is the main ingredient, usually 80% to 90% of the bulk; the milk may either be fresh or reconstituted. Flavored milks are increasing in popularity, strawberry, banana, and chocolate probably being the most popular. A number of these are based on popular confectionery products (see also Section 5.4.4). Other compounds include the addition of sugars, flavors, emulsifiers, and stabilizers. In general, the total solids content is much higher than in normal milk, giving a slightly reduced water activity, although the pH is not usually much altered. Heat treatment regulations are similar to those for milk. *The main point to bear in mind is that the additional ingredients may adversely affect microbial load, enzyme levels, or stability of milk to heat.* Raw material quality becomes an additional quality control point. Mixing and agitation of raw milk may be required during their formulation, which may give rise to increased lipolytic activity and soapy off-flavors. Some such products may also be acidic in nature, as are fermented milk products.

Milk protein concentrates produced by ultrafiltration may also require pasteurization prior to cheese making or yogurt production. The amount of whey protein will increase as the concentration factor increases (Guney, 1989).

5.4.3 Cream

Many of the general points that have been discussed for milk will also be relevant to cream, such as the four main control points of raw material quality, time/temperature conditions, eliminating PPC, and low-temperature storage. Again, if keeping quality is poor, initial attention should be paid to reducing PPC. Compared to milk, the major difference arises from the higher fat contents of creams and creams are described by their fat contents; the main types of cream are shown in Exhibit 5–3. Fat contents will range from about 9% to 55%: half cream or coffee cream, 9% to 12%; single cream, 18%; whipping cream, 30% to 35%; double cream, 48%; and clotted cream at greater than 50%. In general, cream may be more susceptible to mechanical damage, so positive pumps may be preferred to centrifugal pumps.

Although there is some recognition that fat adds some degree of protection to inactivation of microorganisms by heat, the UK heat treatment regulations for cream are the same as those for milk (a minimum of 72°C/15 seconds) (see Section 5.4.1). This comes as a surprise to most overseas visitors to the United Kingdom. In commercial practice, temperatures a few degrees in excess of 72°C are often used, to compensate for the change in the residence time distribution brought about by the higher viscosity of cream (compared to milk). Regulations in other countries are more severe, although there is no evidence that this improves the overall quality. Also, information from the SDT (1989) suggests that on some specific occasions slightly higher temperatures may be required. Towler (1994) suggests that 80°C may be more appropriate; this should be done with caution, as heat-induced flavors are introduced and, once temperatures reach about 80°C, there may be a deterioration in the keeping quality. Lactoperoxidase inactivation in cream has received less attention than in milk.

Exhibit 5–3 Main Types of Cream and Their Fat Content

Description of Cream	Minimum Butterfat Content % by Weight
Clotted cream	55
Double cream	48
Whipping cream	35
Whipped cream	35
Sterilized cream	23
Cream or single cream	18
Sterilized half cream	12
Half cream	12

Source: Reprinted with permission from National Dairy Council, *Technical Profile on Cream*, p. 5, Copyright © National Dairy Council.

One very important quality attribute is the cream viscosity and its mouthfeel. For any type of cream, it is possible to produce products with a wide range of viscosities. Factors affecting cream viscosity have been reviewed by SDT (1989) and include the separation temperature (lower temperatures giving higher viscosities), the processing temperature (higher processing temperatures will reduce viscosity), and the extent of homogenization. All pasteurized creams need some homogenization to prevent separation taking place, even during the comparatively short shelf life period. In general, increasing the homogenization pressure will reduce fat globule size and lead to an increase in viscosity, caused by some clumping of the fat globules. However, too much homogenization of whipping cream will impair its whipping properties and for double cream may even turn the product solid. Coffee creams and single creams are subject to higher homogenization pressures, and this will lead to an increase in viscosity, creating the impression that the product is richer and creamier than it actually is. The final viscosity will also be affected by the rate of cooling, as this will affect the crystallization behavior of the fat phase. In a continuous process, cooling is fairly rapid; however, since crystallization is not an instantaneous process, further latent heat of crystallization may be given out and may lead to an increase in the temperature on further storage. In general, viscocity can be increased by slow cooling processes; note that this is not the best solution for preventing microbial growth. One solution is to cool partially in a continuous process, to package the product, and then to cool the packaged products further. Another solution is to cool down quickly to below 7°C, package, allow to warm slightly, and then cool again. Further increase in viscosity will occur on cold storage, due to further fat crystallization. There will be a slight reduction in cream viscosity after cream has been packed due to shear in the filling machine, but this is regained on storage.

Cream pasteurization is well covered by the Society of Dairy Technology (SDT, 1989) in their Cream Processing Manual. Cream processing is also reviewed by Towler (1994). The physicochemical properties of cream have been reviewed by Kessler and Fink (1992, IDF Bulletin No. 271).

5.4.4 Ice Cream

Ice cream is a popular frozen dessert that is made by freezing ice cream mix and incorporating air during the freezing process. The mix contains milk solids, not fat (msnf), which is usually derived from skim milk powder, fat (of dairy or vegetable origin), sugar, emulsifiers, stabilizers, flavorings, and colors and water; it is considered to be a dairy product. There are compositional standards: for example, in the United Kingdom, ice cream must contain more than 5% fat and 7.5% milk solids not fat; for it to be called dairy ice cream the fat must be of milk fat; otherwise, hydrogenated vegetable oils are used. Ice cream mixes usually range between 30% and 40% total solids (TS). The amount of air incorporated during freezing is measured by the overrun.

In the United Kingdom and many other countries, it is a legal requirement that all ice cream be subject to pasteurization and this takes place after mixing; homogenization will also be incorporated. There are minimum heat treatment requirements for pasteurization of ice cream. These are as follows:

°C	°F	Time
66	150	30 minutes
71	160	10 minutes
79	175	15 seconds (HTST)

It can be seen that these conditions (both batch and HTST) are more severe than those for milk. It is also necessary to cool the mix to below 7.2°C within 1.5 hours. The mix should not be kept above 7.2°C for more than 1 hour prior to pasteurization.

It is also necessary to record and control temperature with flow control and flow diversion. After pasteurization and cooling, the mix is normally aged for several hours at low temperature before being frozen. Thus there is no requirement for the mix to have a long shelf life and there are no microbiological standards. However, high microbial counts associated with ice cream would most probably have arisen from PPC, as a result of poor hygiene, cleaning, and disinfecting. In this case, coliforms counts can be used to measure the level of PPC. Because these control measures are taken, there are few food poisoning outbreaks associated with ice cream and other frozen desserts.

Holsinger et al. (1992) examined inactivation of *Listeria monocytogenes* in heated ice cream mix and reported that pasteurization guidelines for ice cream mix are adequate to ensure inactivation of *L. monocytogenes*. Increased thermal resistance of *L. monocytogenes* was associated with the common ice cream stabilizer used; the stabilizer contained guar gum and carrageenan. Since results suggest that major ingredients in ice cream, ice milk, and shake mixes increase thermal resistance of *L. monocytogenes*, it is important that every precaution be taken to inactivate the organism. Patel and Wilbey (1994) recommend that GGTP activity will help detect overheat treatment in ice cream mixes.

5.4.5 Eggs

One very convenient form of commercial egg is for it to be supplied as a refrigerated liquid. Shell eggs are usually converted to liquid products, using mechanical breaking machines. Liquid egg, which is widely used in the baking industry as the starting material for many egg products, is very sensitive microbiologically, with a short shelf life. Liquid egg may contain a wide microbiological flora; the main pathogenic microorganism associated with eggs is *Salmonella*. Also of concern is *Enterococcus faecalis*. Otherwise eggs may be frozen, dried, or converted to specialty products. Fortunately, *Salmonellae* are relatively sensitive to heat, with 9D processes accomplished by temperatures of 63° to 65°C for 30 to 60 seconds. In fact, most heat treatment regulations stipulate conditions well in excess of these, introducing an additional margin of safety. In most countries, legislation is based on public health requirements for salmonella-free eggs (Alkskog, 1993). For example, in the United States it is a legal requirement to pasteurize all egg products sold.

A typical heat treatment is based around 64°C for 4 minutes, based on average residence time, calculated from the holding tube volume and the volumetric flowrate. In this case, if the flow were streamline, the minimum residence time would be 2 minutes. It can again be seen that there is a considerable margin of safety. UK heat treatment regulations require that it be treated at 64.4°C for 2.5 minutes, with negative α-amylase activity being a statutory requirement. Note that phosphatase would not be completely inactivated under these conditions. It is important to ensure that other spoilage organisms are inactivated to ensure a reasonable shelf life. Heat treatment regulations in force in other countries are summarized in Table 5–4, (Stadelman et al., 1988). Note that egg white (albumen) will not tolerate such high temperatures as whole egg. Pasteurized liquid egg is kept refrigerated and supplied as such. Even so, its shelf life is short, typically not longer than 6 days.

Use of high-temperature pasteurization, or ultrapasteurization, combined with aseptic packaging has been shown to increase the shelf life without impairing functionality. Alkskog (1993) used a variety of heating regimens, and reported their effects on both the keeping quality and the plant running time. These are summarized in Table 5–5. It can be seen that it was possible to extend the shelf life to 6 weeks or longer. At temperatures above 70°C, egg precipitation on the surface became a major problem, leading to deposition and pressure drop restrictions. There is some scope for increasing run times by replacing the final section of a

Table 5–4 Minimum Pasteurization Requirements for Whole Egg Products in Selected Countries

	Temperature			
Country	*°C*	*°F*	*Time (min)*	*Reference*
Great Britain	64.4	148	2.5	Statutory Instruments (1963)
Poland	66.1–67.8	151–154	3	Cunningham (1973)
China (PRC)	63.3	146	2.5	Cunningham (1973)
Australia	62.5	144.5	2.5	Cunningham (1973)
Denmark	65–69	149–156.5	1.5–3	Cunningham (1973)
United States	60	140	3.5	USDA (1984)

Source: Reprinted with permission from W.J. Stadelman, V.M. Olsen, G.A. Shemwell, S. Pasch, *Egg and Poultry–Meat Processing*, p. 57, © 1988, Wiley-VCH.

Table 5–5 High-Temperature Pasteurization of Egg

(Storage Temperature <5°C)

Process	*Shelf Life*	*Running Time*
PHE 63°C/4 min	6 wk	>8 h
PHE 70°C/1.5 min	12 wk	3 h
Steam injection 70°C/1.5 min	12 w	3 h
Spiraflo 63°C/4 min	—	<8 h
Spiraflo 70°C/1.5 min	12 wk	<7 h

Source: First published by Sterling Publications, Ltd., in *Food Technology International Europe*, Brunel House, 55a North Wharf Road, London, W2 1XR.

PHE with a scraped-surface unit. Low-pressure steam injection also showed promising results as far as shelf life and operating time.

There is now a variety of ready-to-use egg-based products (such as batters) coming onto the retail market, which have a relatively short shelf life and have obviously been subjected to pasteurization.

5.4.6 Fruit Juices and Fruit Drinks

Virtually all fruit fall into the acidic category (pH < 4.5), with most having a pH below 4.0. See Table 1–2 for typical compositions and Table 1–3 for some pH values. The fruit are the starting material for the production of juices, purées, nectars, and fruit-based drinks. Most of these are heat treated, the severity depending upon the pH levels, initial microbial count, and whether the juice is fresh or reconstituted. The main reasons for heat treatment are to inactivate the spoilage microorganisms, which are yeasts, molds, and lactic acid and acetic acid bacteria, as well as some of the enzymes that may pose a problem during processing and storage. The main ones are those responsible for browning and cloud stability. Salunkhe and Kadam (1995) reviewed the production of such products from a wide variety of fruits. In general, acidic juices would have pasteurization conditions within the range 85°C for 15 seconds to 95°C for 2 seconds. However, unlike milk products, there are no heat treatment regulations to comply with, and the conditions used commercially are not often recorded.

Browning may be a considerable practical problem; most of the browning taking place during the production stages will be enzymatic, whereas most that takes place after heat treatment and throughout storage will be nonenzymatic.

The following decimal reduction times at 75°C have been reported: 0.004 second for *Saccharomyces cerivissiae*, 0.053 second for *Lactobacillus fermentum*, and 0.02 second for *Aspergillus niger* spores (Hasselback, Ruholl, & Knorr, 1992). It can easily be seen that these organisms should be well and truly inactivated by normal juice pasteurization conditions. When such products are packaged aseptically, their shelf life should be between 3 and 6 months. Acetic acid bacteria grow well in single-strength juice; temperatures of about 74°C will destroy the major spoilage flora. The conditions required for ensuring cloud stability are more severe; the following combinations were successful in controlling cloud: 99°C for 1.75 seconds, 91°C for 12.75 seconds, and 85° to 88°C for 42.6 seconds, as reported in Nelson and Tressler (1980).

In the United Kingdom, there has been a big increase in fruit juice consumption. Some of this, for example, apple and black currant, is produced in the country. However, for fruit of a Mediterranean and tropical origin, it usually arrives in the United Kingdom as a frozen concentrate. This is then defrosted, diluted back to near its original solids content, pasteurized, and packaged.

Continuous-heat treatment may be used at several stages in the production process and is an important process:

1. It may be used immediately after juice or purée is produced. For example, for apples, flash heating also helps to coagulate those components that make juice difficult to filter (82.2° to 85°C), with final pasteurization at 85° to 90°C.
2. It may be used as a pretreatment to hot filling, or after cooling for the production of frozen concentrates, or as a pretreatment for evaporation or drying.

Although pathogenic bacteria will not grow at these low pH values, they may be present in the raw fruit. Only recently, there was a serious food poisoning outbreak attributed to *E. coli* VTIC 0157 from unpasteurized apple juice in the United States, the source of the *E. coli* being fecal contamination of the apples. Note that the infectious dose is as low as 100 organisms. This bacterium appears to be able to tolerate low pH values. Pasteurization would have effectively inactivated this organism and prevented a serious food poisoning incident.

The key stages in fruit juice technology have been described by Veal (1987) and for soft drink technology by Houghton (1987). In general, they involve cleaning and preparation of the raw materials, and juice extraction, usually by pressing, although extraction has also been investigated. Where enzymatic browning is a problem, it can be controlled by heat treatment, by ascorbic acid addition, or by use of sulfur dioxide. Pectin, which is naturally present, stabilizes the colloidal material responsible for the cloudy appearance of many juices and confers stability. If a cloudy juice is required, it may be necessary to inactivate pectic enzymes or indeed to add some additional pectin. For the production of clear juices, pectic enzymes are added to the juice and some time later the resulting clear juice is filtered from the deposited sediment.

Citrus products involve juice extraction and screening. The juice may also be deoiled; this is done with a small vacuum evaporator, which also degasses the product. If juice is pasteurized at only 65°C, pectinesterases will survive and cause the juice to separate into a clear liquid with a sediment at the bottom. Therefore, higher temperatures (up to 90°C) are required to confer additional stability. Apple juice has been reported to be affected by higher processing temperatures more than orange juice (Hasselback et al., 1992). For orange juice, deaeration of the juice not only improves flavor stability and nutrient retention but improves filling operations and heat exchanger operations. Deaeration is covered in the same operation as deoiling: a certain amount of oil is essential but some need to be removed, as excessive amounts lead to objectionable flavors. Temperatures required to prevent sediment formation by destabilizing pectic enzymes (or loss of cloud) generally range between 86° and 99°C, times may range between 1 and 40 seconds.

For lemons and limes, cloud stabilization is also important. It is achieved by inactivation of pectinesterase activity at a temperature between 69° and 74°C. A higher temperature of 77°C for 30 seconds will provide an adequate margin of safety; the product needs immediate cooling and flashing into a vacuum chamber.

Other juices processed are lemon and grapefruit, where the bitter component naringin is important. Deaeration of mango juice was found to reduce dissolved oxygen from 6.5 to 0.8 ppm during deaeration and to minimize vitamin C loss during processing (Shyu, Hsu, & Hwang, 1996).

Some tropical fruit are sold as purées and may require either tubular or scraped-surface heat exchangers (Jagtiani, Chan, & Sakai). With mango, heat-resistant enzymes become a problem. For purée that is to be frozen, a temperature of 195° to 200°F for 2 minutes is recommended, followed by cooling to 85° to 100°F. This can be accomplished in a PHE and inactivates catalase. Kalra, Tandon, and Singh (1995) report that mango pulp pasteurized for 1 minute at 88° to 92°C by a scraped-surface heat exchanger, filled into sterilized pouches at 88° to 92°C, evacuated and sealed, and then cooled showed no sensory or microbial spoilage when stored at 18° to 20°C for 4 months. Argaiz and Lopez-Malo (1996) showed that pectinesterase was more temperature resistant in mango and papaya nectars than in purées and that this enzyme required more intensive heat than the destruction of deteriorative microorganisms. Cooked flavor development was also studied in these products.

Guava purée is processed at 195°F (90.6°C) for 1 minute, followed by hot filling, using a scraped-surface heat exchanger, or at 93°C for 38 seconds, followed by aseptic packaging. For papaya purée, gelling is prevented by inactivation of pectinesterase enzyme. It is heated at 205°F (96.1°C) and held for 2 minutes, followed by cooling to below 85°F (29.4°C). An aseptic purée is described, which is acidified to pH 3.5 by the addition of citric acid, followed by heating at 93°C for 60 seconds, using a plate heat exchanger. Note that its normal pH is 5.1 to 5.3.

Passion fruit has an extremely sensitive flavor, some of which will be lost during pasteurization. It also contains considerable starch, which produces a gelatinous deposit that is prone to foul heat exchangers. Because of these problems, the usual method of preservation is freezing.

In addition to heat treatment, other compounds that may be added include benzoates. Results suggest that spoilage of thermally processed fruit products can be prevented by the incorporation of low amounts of sorbic acid (Splittstoesser & Churey, 1989).

5.4.7 Tomatoes

Tomatoes have a very high moisture content (94% to 95%). They are processed into a wide range of products, including juices (up to 12% Brix scale), purées (12% to 22% Brix), and pastes (greater than 21% but usually not exceeding 31% Brix). The pH of tomatoes is usually about 4.3 to 4.5, and in general they are treated as an acidic product. Heat treatment is an integral part of these processes, not only in terms of controlling microbial activity but also in terms of its effects on the various pectin enzymes, which in turn controls the viscosity of the product. The desirable processing characteristics of tomatoes are full body, tough to eat (to avoid damage during transportation), high soluble solids, and good color; flavor is also important for juices, but less so for sauces.

The tomatoes are cleaned and crudely chopped and then subject to either a hot-break or a cold-break heating process. The hot break involves heating at 95°C for 5 to 8 minutes and will deactivate the pectic enzymes, which in turn will result in a product with a better viscosity, but flavor loss is higher. This can be done in a semicontinuous fashion. The cold-break process involves a temperature of 65°C for 10 minutes and results in a fresher, more tomatoey product. The material is then screened (hot) to remove skins and seeds, typical screens being about 1.5 to 2.5 mm. Hot-break juice would now be at 70° to 80°C, with a pH of about 4.2 and between 4 and 4.5 Brix. At this stage the titratable acidity and serum viscosity may be measured and citric acid and salt may be added.

Much juice is also evaporated. Tomato evaporation may operate almost continuously throughout the growing season. Following evaporation, the concentrate is subject to a final "sterilization" process (95°C for 2 to 3 minutes) before being packaged. Direct-steam injection processes can be used, or tubular heat exchangers would be suitable. Aseptic packaging systems are becoming more widespread, from 5L to 1 metric ton containers. Some of this concentrate may be used for the production of tomato-based sauces; these would also require a final heat treatment before being packaged.

Practices vary widely worldwide. However, it can be seen that there is scope for continuous-heat exchange in three places: the hot- and cold-break process, final juice or paste sterilization, and the sauce production area. Note the emergence of *Alicyclobacillus acidoterestris*, which can cause spoilage in low-pH products and has recently caused widespread spoilage in processed tomatoes.

5.4.8 Beer

The conditions for pasteurization of beer are about 75°C for 30 seconds. To avoid gas breakout and ensure that about 2.2 volumes of CO_2 are retained up to a temperature of 78°C requires an equilibrium pressure of up to 7.5 bar. Therefore the back-pressure in the plant should be kept at about 0.7 bar higher than this equilibrium pressure. In addition, breweries routinely assume a z-value of 6.9°C for calculation of pasteurization effect at high temperature; the high z-values observed for strain E93 would give greater survival at high temperature than would be predicted from its heat resistance at 60°C. If minimal heat processing of low-alcohol beers is desired, increasing the hopping rate may avoid spoilage problems. Beer can also be pasteurized in cans, in a tunnel pasteurizer (see Section 5.2).

5.5 OTHER HEAT TREATMENTS

5.5.1 Thermization

Thermization is a mild process that has been applied to raw milk that may need to be stored for several days prior to use. The main objective is to reduce the numbers of gram-negative pyschrotrophic microorganisms, which would proliferate during storage of raw milk. When counts reach in excess of 10^6/ml, they will start to produce enzymes such as proteases and lipases, which are very heat resistant. Thermization has now been defined as a heat treatment that uses a temperature of between 57° and 68°C for 15 seconds. It is not widely practiced in the United Kingdom, but is more so in other European countries. It is only effective if thermized milk is kept cool (4°C). Some problems that may be encountered are summarized by Muir (1996a). One arises from the presence of gram-positive cocci (e.g., *Streptococcus thermophilus*), which may build up in the regeneration section of a commercial thermization unit (after about 5 hours) and cause contamination of the thermized product. It may also slightly affect the flavor or texture of cheese, although not its yield. An alternative to thermization is to deep-cool the milk during its storage period, at or below 2°C (see Section 5.3.5).

5.6 STRATEGIES FOR EXTENDED SHELF LIFE OF REFRIGERATED PRODUCTS

Extending the shelf life of pasteurized and low-heat treatment products, without unduly changing their sensory characteristics, is considered to be an important commercial goal. It will also help to build-in additional safety margins into the product to help counteract temperature abuse by the consumer. The focus in this section is on shelf-life extension for products that are to be stored refrigerated (i.e., below 10°C). In this case, the main aim is to control the activity of organisms that are capable of growth at these temperatures. In this respect, psychrotrophic thermoduric and spore-forming bacteria become important; from a safety standpoint *Bacillus cereus* is of interest (Section 10.2.2).

The following comments are directed to examining strategies for extending shelf life of refrigerated products, without unduly affecting sensory characteristics. In these cases, all aspects of control become even more crucial (i.e., raw material quality, inactivating spore-forming pyschrotrophic bacteria, controlling PPC, and effective temperature control. The heat-resistant enzymes may also play a more important role in quality determination as the shelf life is extended.

5.6.1 Raw Material Quality

The highest-quality raw materials should be used (i.e., milk and all other ingredients). These should have a low total psychrotrophic count, which will ensure good quality and a low level of heat-resistant enzymes. The spore count should also be low, with preferably an absence of *Bacillus cereus* (Lewis, 1999). Rapid microbiological tests may be useful for this purpose, although they should be effective at low microbial counts. Poor-quality raw milk may also introduce taints. In the United Kingdom, with the disbanding of the Milk Marketing Board, there is now more opportunity for specifying raw milk quality for specific individual applications. For example, milk destined for cheese needs different quality characteristics compared to milk destined for pasteurized or sterilized milk drinks. Also, the introduction of incentive payments for producing milks with lower bacterial counts has led to an overall improvement in raw milk quality. Other strategies to reduce the initial microbial counts of thermodurics are bactofugation or microfiltration.

Bactofugation involves removing bacteria by using a centrifugal force. It has thus been described as banishing bacteria the centrifugal way and is described by Lehmann (1989). However, reduction of aerobic spores by centrifugation is difficult owing to their size and specific density. Aerobic spores, including *Bacillus cereus*, can be reduced by about 1 log cycle (Fredsted, Rysstad, & Eie, 1996).

Microfiltration, which makes use of a semipermeable membrane that does not allow the passage of microorganisms, has also been found to be useful for reducing microbial counts (Grandison & Finnigan, 1996). The microorganisms are retained by the membrane, whereas the bulk of the milk, which is free of microorganisms, passes through the membrane. Data for skim milk are presented by Bindith, Cordier, and Jost (1996). Values for the filtrate resulting from a typical pass for skim milk at 55°C show typically 2 to 3 decimal reductions of aerobic mesophilic bacteria and 1 to 2 decimal reductions for aerobic spores. The filtered milk may then be subject to further processing. Note that when whole milk is the starting material, there will also be a separated cream phase and a concentrate, with increased microbial count arising from the microfiltration process.

A scheme for producing bactofiltered milk involves mixing the filtrate with pasteurized cream. Note that alkaline phosphatase activities in such milks will still be high. Temperatures up to 70°C are required to inactivate phosphatase (Bindith et al., 1996). Another scheme for extending the shelf life involves mixing the concentrate (enriched in bacteria) with the cream and processing the mixture at 120°C for 2 to 4 seconds, followed by mixing it with the filtrate, homogenizing, and further pasteurizing. All unit operations are done in-line, without any intermediary balance tanks (Larsen, 1996). Improvements in keeping quality are presented, as well as the additional costs involved. Monitoring quality, with the emphasis on rapid testing and increased precision (e.g., Fluorophos test) is also an important part of the overall philosophy.

With perishable raw materials, the normal practice is to process them as quickly as possible. This would certainly apply to milk. However, Ravanis and Lewis (1995) showed that good-quality raw milk could be stored for up to 7 days at 4°C before there was any adverse effect on the keeping quality of pasteurized milk produced from it. A more surprising result was that milk that was pasteurized on both the third and fourth days after production gave a better keeping quality than that processed on the first day after production (about 38 days compared to 25 days). All experiments were performed at standard HTST conditions (72°C/15 seconds) under conditions where PPC was eliminated. One possible explanation of this, which is currently subject to further investigation, is the activity of the lactoperoxidase system, which is one of the natural antimicrobial systems found in milk.

5.6.1.1 Some Special Attributes of Raw Materials

Natural Antimicrobial systems. There are some naturally occurring antimicrobial systems present in raw materials that may be harnessed to improve shelf life. The lactoperoxidase system is considered to provide some natural protection in milk against the growth of spoilage bacteria. It is bactericidal to most gram-negative bacteria and to many pathogens; the antibacterial spectrum is broad and has been summarized by Bjorck (1982). It has been investigated mainly for the preservation of raw milk in warmer climates, where lactic acid bacteria are the main spoilage organisms and also for chilled milk to be used for cheese manufacture, where pyschrotrophic bacteria are the main problem. It is based on the ability of lactoperoxidase to oxidize thiocyanate ions (CNS^-) in the presence of hydrogen peroxide to produce compounds that inhibit growth of lactic acid bacteria and psychrotrophic bacteria. The inhibitory substances are claimed to be short-lived intermediary compounds, such as hypothiocyanite ($OSCN^-$), cyanosulfurous acid (HO_2SCN), and cyanosulfuric acid (HO_3SCN) (Reiter & Harnulv, 1982). It is reported that inhibitory effects are reversed by the addition of reducing agents such as sodium sulfite or cysteine. However, these bacteriocins can be further oxidized to sulfates, ammonium ions, and carbon dioxide, which at the concentrations involved are inert.

Hydrogen peroxide is required for these reactions. Both the hydrogen peroxide and thiocyanate ions required for the reaction are naturally present in raw milk, albeit at low concentrations. Also hydrogen peroxide may be produced by microorganisms present in milk. Lactoperoxidase is present in raw milk (Section 5.4.1). About 70% of its residual activity is retained when it is heated at 72°C for 15 seconds, whereas 80°C for 15 seconds causes almost incomplete inactivation.

It is feasible that the lactoperoxidase system may be exerting some action in pasteurized milk and be contributing to the improved keeping quality of milk heated at 72°C/15 seconds compared to 80°C/15 seconds. This suggestion has been investigated as a possible explanation for the improved keeping quality at the lower heating conditions (Barrett, Grandison, & Lewis, 1998).

There is an extensive literature in both these two areas. Recommendations have been made by the IDF (1988) on activating the system to extend the shelf life of raw milk by addition of hydrogen peroxide and thiocyanate at about concentrations of 10 to 15 ppm. It is interesting that hydrogen peroxide is very quickly used up, with its concentration approaching zero in a short time period and in most situations is likely to be limiting the reaction. However, these procedures have given rise to considerable debate. The main arguments against it are that it would be open to abuse in the sense of disguising poor-quality milk and it would be a dangerous practice to allow farmers to add potentially harmful chemicals to milk.

There are several other antimicrobial systems that are present in milk. Such antimicrobial activity may also be associated with other raw materials, such as lactoferrin, lysozyme, and the bovine immunoglobulins. These have been reviewed recently by the IDF (1994, SI 9404).

Hydrogen Peroxide. Hydrogen peroxide can be added in its own right to preserve raw milk, under conditions where it may be difficult to cool the milk quickly. The concentrations required (300 to 800 ppm) are much higher than those used to activate the lactoperoxidase system. At these levels it will inactivate the lactoperoxidase and adversely affect the clotting properties by rennet. A level of 0.8% by weight combined with a temperature of 49° to 55°C for 30 minutes has been suggested as a substitute for pasteurization. Treatment levels of 0.115% completely inactivated *Mycobacterium tuberculosis*. Gram-positive bacteria are not inactivated to the same extent as gram-negative bacteria (Swart, 1993).

Carbon Dioxide. Carbon dioxide addition has been investigated as a means of preserving raw milk (Mabbit, 1982). It inhibits the growth of gram-negative pyschrotrophic bacteria in raw milk but does not inhibit, and may even stimulate, the growth of gram-positive lactic acid bacteria. It is a naturally present component of raw milk, which may contain up to 5% (v/v) as it leaves the udder. One problem with CO_2 addition is that it will reduce the pH of the milk, which may lead to destabilization of the casein micelle on subsequent processing. Amigo and Calvo (1996) suggested that the pH had to be reduced to about 6.0 for it to be most effective. If required, the gas can be easily removed by degasing, using a vacuum or by sparging with nitrogen.

5.6.2 Extended Heat Treatments

The most obvious way to improve the shelf life would be to increase the severity of the heat treatment, in order to inactivate or control those organisms that survive conventional pasteurization. Of particular interest are the thermoduric bacteria and heat-resistant spores. Again, the main pathogen in this category is *Bacillus cereus*. Two factors that one would wish to prevent are (1) heat shocking of thermoduric and spore-forming bacteria and (2) development of a cooked flavor. In addition, it is important to appreciate that some other subtle changes may take place that may influence the processing characteristics of the milk, for example, how well it makes hard cheese, such as Cheddar.

The simplest approach would be to use conditions slightly more severe than normal HTST conditions of 72°C/15 seconds. This is also practiced in some countries to account for local difficulties and poorer-quality raw milk. Such conditions have been investigated by Schroder and Bland (1984), using 72°, 78° and 83°C for 20 seconds, with storage at 7° and 12°C. At 7°C storage, a temperature of 78°C slightly improved the keeping quality, whereas a temperature of 83°C resulted in a decreased keeping quality. Deterioration was higher when PPC was not eliminated. The KQ of milk stored at 12°C was not significantly affected by processing temperature. The most expected outcome would be that the more severe conditions would give rise to an improvement in the keeping quality. However, this was not always found to be the case.

Kessler and Horak (1984) and others have also arrived at similar findings. The conclusions drawn from such studies have been summarized as follows: it appears that a slight improvement in KQ can be obtained in most practical circumstances by raising the pasteurization temperature from a minimum of 72°C by a few degrees. However, the amount of improvement is likely to be of little practical value and, if the milk is whole milk and unhomogenized, there will be adverse effects on the physical properties of the cream layer. Increase of processing temperature to above 78°C with a holding time of 20 seconds, or an increase of the holding time above 20 seconds at 78°C will reduce KQ sharply, in all practical circumstances (Burton, 1986). Therefore, increasing the severity of the heat treatment does not always lead to an improved keeping quality. The usual explanation offered for these unexpected and seemingly contradictory findings arises from heat shocking of the spores at the harsher processing conditions, although the lactoperoxidase system will become inactive also.

Double processing has been suggested, the principle being that the first heat treatment heat shocks the spores and the second then inactivates them. Double processing has been considered but not found to be beneficial. The problem arises because not all the spores are heat shocked uniformly. Being a double process, some time must elapse between the processes to allow time for the spores to germinate, and this causes operational problems (Brown, Wiles, & Prentice, 1979).

In both extended and double heat treatments, care should be taken not to induce a cooked flavor. An alternative approach is to use higher temperatures for shorter times to inactivate *Bacillus cereus* spores and other more heat-resistant pyschrotrophic bacilli. This takes advantage of the fact that for a given level of spore inactivation, the cooked flavor intensity would be reduced. Conditions of 115°C for 1 to 5 seconds have been investigated, although conclusions have been mixed. Guirguis, Griffiths, and Muir (1983) reported that temperatures of 116°C for 1 second were optimum for spore germination. The implication is that these heat-shocked spores would be readily inactivated by a subsequent process. In double cream, a temperature of 115°C for 5 seconds appears to inhibit spore-forming bacteria. The process was further improved by following this with a further HTST pasteurization process (Griffiths et al., 1986a, 1986b). There is a European patent (1981, EP 0 043 276, B1) on extended-shelf-life milk (120°C for 2 seconds).

Heat-resistance data on *Bacillus cereus* have been summarized by Bergere and Cerf (1992). A wide range of values have been reported, which suggests that a temperature of 116°C for a few seconds would not achieve 2 decimal reductions, in many cases. If this heat-treatment strategy were to be adopted, this would need further experimental investigation. A review of *Bacillus cerus* is included in Section 10.2.2.

Borde-Lekona et al. (1994) reported that milk heated at 116°C for 2 seconds showed a keeping quality (determined by microbial count) of up to 35 days at 10°C. In this case the initial raw milk quality was very good, with a standard plate count of about 2,500/ml. The same milk, processed at 72°C for 15 seconds, kept for only 15 days at 10°C. However, heat treatment at 116°C for 2 seconds did not fully reduce the spore count, but there was no further increase in the spore count during storage. Milks processed at 90°C for 15 seconds also kept extremely well, for over 28 days at 10°C (Wirjantoro & Lewis, 1996). Note that in the United States, ultrapasteurization refers to a process wherein milk is heated at 138°C for 2 seconds, which is close to ultrahigh temperatures (UHT) in Europe and worldwide.

It is also possible to use direct-steam injection methods and flash cooling to improve the flavor further, one example being the Pure-Lak concept. This uses a reduced UHT heat-treatment process to achieve 6 log reductions of pyschrotrophic aerobic spores, typically in the range 130° to 140°C for 2 seconds. The products will have a keeping quality of 10 to 45 days at 10°C, depending upon the filling and packaging conditions (Fredsted et al., 1996). They will also have a built-in quality buffer to prevent spoilage at elevated temperatures. Comparisons are made with pasteurized milks among lactulose, lactoperoxidase, and whey protein denaturation levels and sensory characteristics. Such direct-steam injection products have been reported to have taste characteristics close to normally pasteurized products; there is a slightly perceptible cooked flavor immediately after production, which disappears after about 1 week (Kiesner et al., 1996).

In conclusion, some factors for consideration for extending shelf life are as follows.

Consider whether to use 72°C for 15 seconds or to use conditions that will reduce the count of psychrotrophic spore-forming bacteria. If 72°C for 15 seconds is to be used, one has to face the fact that psychrotrophic spore formers will not be removed; in this case one needs to ensure that they are present in low numbers in the feed, and that they are controlled during storage. Ravanis and Lewis (1995) have shown that milks with a shelf life of over 25 days can be achieved at 4°C, if conditions are controlled.

If 115° to 120°C for 1 to 5 seconds is to be used, one would need to check the flavor, and whether the selected conditions would lead to a reduction in psychrotrophic spore formers. The cooked flavor intensity could also be reduced by using the steam-injection process. Its rapid heating and cooling combined with flash cooling could be beneficial; this will also lower residual oxygen levels. This may also be beneficial in reducing toxin production by any surviving *Bacillus cereus* organisms.

It is also essential to eliminate PPC. One way is to employ plant sterilization conditions used in UHT. Thus the shelf life of low-acid foods can be extended by this type of approach. If PPC is eliminated, spoilage will result only from those organisms surviving the heat treatment. An effective plant sterilization procedure is 130°C for 30 minutes, downstream of the holding tube, combined with good hygiene and cleaning throughout to prevent a buildup of spore-forming bacteria. The use of sensitive tests such as Fluorophos will ensure that pasteurized milk is not being recontaminated with raw milk.

Most sources quote the keeping quality of pasteurized milk to be between 7 and 10 days. In experiments by Borde-Lekona et al. (1994) and Ravanis and Lewis (1995), where PPC was reduced as far as possible, keeping qualities between 15 and 18 days at 10°C and in excess of 35 days at 4°C were regularly achieved.

Consider the importance of aseptic packaging and controlled refrigeration, preferably below 4°C. Dommett (1992) drew attention to spoilage by *Bacillus circulans* in pilot plant investigations with aseptically packaged homogenized milk, cream, and reverse osmosis concentrate. It was established that a monoculture of spore-forming bacilli normally forms, most often *B. circulans* or less frequently *B. cereus*. Eventual shelf life was affected mainly by storage temperature after processing but smaller effects were demonstrated for pasteurization temperature and cycle time. *B. circulans* has important characteristics selecting it for survival and growth, including tolerance to very low temperature, low O_2, and mild acid. These factors and the very high incidence of this organism in this trial suggest that *B. circulans* could be a potential problem in commercial aseptically packaged products.

The emphasis is on control: reducing the initial count, controlling the processing time and temperature, eliminating PPC, and keeping the products cold. All the selected processes should be carefully evaluated, using microbiological and sensory methodology. Quality assurance procedures should emphasize this control.

5.6.3 Alternative Processes

5.6.3.1 Addition of Bacteriocins

Nisin addition has also found some scope (Delves-Broughton, 1990), especially in terms of its control of growth of gram-positive bacteria. It is probably the most well-characterized of the antimicrobial proteins and it is produced by *Lactococcus lactis subsp. lactis*. It is a polypeptide bacteriocin with a molecular weight of 3488. It is bactericidal to many gram-positive organisms, including the vegetative cells of spore formers and is bacteriostatic to spores. Nisin has no effect against gram-negative bacteria, yeasts, and fungi. It functions by molecular damage to cytoplasmic membrane. Nisin has been deemed safe for use and is nonallergenic as a food preservative; it is now used in over 40 countries around the world.

It is well established that spores are more susceptible to nisin than the vegetative bacteria from which they are originate. Although a sporicidal activity has been demonstrated against the spores of a limited number of species, the effect for the majority is sporistatic. Thus nisin provides a means of controlling the activity of gram-positive spore formers. However, for heat-processed foods, it is important to ensure that an effective level of nisin is maintained throughout the shelf life of the food. If the nisin is added prior to heat treatment, higher doses will be required, as some activity is lost as a result of the heat treatment, although this will be reduced in pasteurization compared to sterilization. It shows that loss of activity increases as pH increases, over the pH range 3.0 to 8.0.

Addition of 30 IU/ml nisin was found to extend the shelf life of pasteurized milk stored at 15°C from 2 to 5 days. However, no improvement was observed at 7°C (Delves-Broughton, 1990). This is probably explained by keeping quality being influenced by PPC. Addition of 5 mg/L to whole egg considerably increased its shelf life. Vandenbergh (1993) reported that adding milk in the range of 30 to 50 IU/mL will increase the shelf life of milk by 6 days at 15°C and by 2 days at 20°C. Wirjantoro and Lewis (1996) showed that addition of nisin 40 IU/ml nisin to milk pasteurized at 72°C for 15 seconds increased the shelf life at 10°C by about 7 days. For milk heated at 90°C for 15 seconds, the effect was greater and milk was still acceptable after 28 days of storage at 10°C. Thus, addition of nisin prior to pasteurization provides an opportunity to achieve extended shelf life in regions with poor refrigeration. It was interesting that milk processed at 115°C for 2 seconds had a count of less than 10 organisms per milliliter after 28 days of storage at 10°C. These conditions are being further investigated for extending shelf life at ambient temperature. Phillips, Griffiths, and Muir (1983) found that nisin was effective in increasing the keeping quality of pasteurized double cream, but only when PPC was eliminated.

5.6.3.2 Novel Techniques

Some novel techniques are being investigated for pasteurizing products. Irradiation also offers a great deal of scope, particularly for inactivating pathogens in solid foods such as poultry and seafood, but it still meets consumer resistance. Other techniques under investigation are UHP processing, pulsed light, and high-intensity pulsed electric fields. Some results for this latter method in the continuous processing of apple juice, raw skim milk, beaten eggs, and green pea soup are described by Qin et al. (1995). The lethality mechanism involves permanent loss of cell membrane function as a result of electroporation. These nonthermal methods have been recently reviewed by Barbosa-Canovas et al. (1998).

However, the major problem these novel methods face in gaining widespread acceptance is that thermal processes are so firmly established and are capable of producing foods that are safe and of a high quality and nutritional values in large volumes at very low processing costs. However, some advances are being made, and fruit juices subject to UHP are now commercially available, at a cost.

5.7 EXTENDED SHELF LIFE AT AMBIENT TEMPERATURES

It would be desirable to have a product that combined the keeping qualities of UHT milk at ambient temperature and the flavor characteristics of pasteurized milk. An appreciation of the problems involved in producing such products using reduced heat treatments have been summarized by Muir (1990). These arise from the diverse range of microorganisms found in raw milk. As matters stand at the present, sterilization offers the only realistic method of extending the shelf life of heat-treated products beyond about 7 days at ambient temperature and would be accompanied by heat-induced flavors. An alternative approach currently being investigated for some milk-based products involves a combination of reduced heat treatments (RHT), at 115 to 120°C for 2 seconds, with low additions of nisin (75 and 150 IU/mL). The process is based upon the principle that nisin shows potential for inhibiting gram-positive spore-forming bacteria, which would be the predominant flora surviving in RHT-milk. Experimental trials have shown that such a combination is very effective at inhibiting growth, with most of the samples showing no microbial activity after 100 days at 30°C. However, some spoiled samples were found (about 1 in 50), probably caused by a low incidence of one or more specific microorganisms, so the process is more compatible with the concept of extended shelf life rather than commercial sterilization. It was also found to be important to reduce postprocessing contamination for nisin to be effective, which is in line with the findings of Phillips et al. (1983) (see Section 5.6.3.1). The majority of RHT-nisin samples stored at 10°C showed no signs of spoilage after 1 year, again suggesting scope for extended shelf life refrigerated products.

REFERENCES

Alkskog, L. (1993). High temperature pasteurisation of liquid whole egg. *Food Technology International Europe*, 43–46.

Amigo L., & Calvo, M.M. (1996). Addition of CO_2 to raw milk to increase its useful life during refrigeration. *Revista Espanola de Lechera* **73**, 25–27.

Andrews, A.T., Anderson, M., & Goodenough, P.W. (1987). A study of the heat stabilities of a number of indigenous milk enzymes. *Journal of Dairy Research* **54**, 237–246.

Anon. (1994). Quicker results. *Dairy Field* **177**, 64.

Argaiz, A., & Lopez-Malo, A. (1996). Kinetics of first change on flavour, cooked flavour development and pectinesterase inactivation on mango and papaya nectars and purees. *Fruit Processing* **6**, 145, 148–150.

Barbosa-Canovas, G.V., Pothakamury, U.R., Palou, E., & Swanson, B.G. (1998). *Non-thermal preservation of foods*. New York: Marcel Dekker.

Barrett, N., Grandison, A.S., & Lewis, M.J. (1998). Contribution of lactoperoxidase to the keeping quality of pasteurized milk. *Journal of Dairy Research* **66**, 73–80.

Bergere, J.L., & Cerf, O. (1992). Heat resistance of *Bacillus cereus* spores. In *Bacillus cereus in milk and milk products*, IDF Bulletin No. 275.

Bindith, O., Cordier, J.L., & Jost, R. (1996). Cross-flow microfiltration of skim milk: Germ reduction and effect on alkaline phosphatase and serum proteins: In *Heat treatments and alternative methods*, IDF/FIL No. 9602.

Bjorck, L. (1982). Activation of the lactoperoxidase system as a means of preventing bacterial deterioration of raw milk. *Kieler Milchwirtschaftliche Foschungsberichte* **34**, 5–11.

Borde-Lekona, B., Lewis, M.J., & Harrigan, W.F. (1994). Keeping quality of pasteurised and high pasteurised milk. In A.T. Andrews, & J. Varley (Eds.), *Biochemistry of milk and milk products*. London: Royal Society of Chemistry.

Bradshaw, J.G., Peeler, J.T., & Twedt, R.M. (1991). Thermal resistance of Listeria spp. in milk. *Journal of Food Protection* **54**, 12–14, 19.

Brody, A.L. (1992). Microwave food pasteurisation. *Food Technology International Europe*, 67–72.

Brown, J.V., Wiles, R., & Prentice, G.A. (1979). The effect of a modified Tyndallization process upon the sporeforming bacteria of milk and cream. *Journal of Society of Dairy Technology* **32**, 109–112.

Brown, K.L. (1990). Principles of heat preservation. In J.A.G. Rees & J. Bettison (Eds.), *Processing and packaging of heat preserved foods*. Glasgow, Scotland: Blackie.

Brown K.L. (1991). Effects of heat resistance in food processing. *Food Technology International Europe*, 87–89.

Burton, H. (1986). Microbiological aspects. In *Factors affecting the quality of milk*, IDF Bulletin No. 130.

Burton, H. (1988). *UHT processing of milk and milk products*. London: Elsevier Applied Science Publishers.

Busse, M. (1981). Pasteurized milk, factors of a bacteriological nature. In *Factors affecting the keeping quality of heat treated milk*, IDF Bulletin No. 130.

Cowell, N.D. (1995). Who invented the tin can? A new candidate. *Food Technology* **12**, 61–64.

Cox, A. (1975). Problems associated with bacterial spores in heat-treated milk and dairy products. *Journal of Society of Dairy Technology* **28**, 59–68.

Cunningham, F.E. (1973). Egg product pasteurization. In W.J. Stadelman & O.J. Cotterill (Eds.), *Egg science and technology*, 1st ed. Westport, CT: AVI Publishers.

Daeman, A.L.H., & Van der Stege, H.I. (1982). The destruction of enzymes and bacteria during spray-drying of milk and whey, 2, The effect of the drying conditions. *Netherlands Milk Dairy Journal* **36**, 211–229.

Dairy Facts and Figures. (1996). London: National Dairy Council.

Daniels, A. (1990). Developments in tubular heat exchangers. *Food Technology International Europe*, 60–64.

Delves-Broughton, J. (1990). Nisin and its application as a preservative. *Journal of the Society of Dairy Technology* **43**, 73–76.

Dommett, T.W. (1992). Spoilage of aseptically packaged liquid dairy products by thermoduric psychrotrophs. *Food Australia* **44**, 459–461.

Driessen, F.M. (1989). Inactivation of lipases and proteases, 71-93, In *Heat-induced changes in milk*, IDF Bulletin No. 238.

Eckner, K.F. (1992). Fluorometric analysis of alkaline phosphatase inactivation correlated to Salmonella and Listeria inactivation. *Journal of Food Protection* **55**, 960–963.

European Community Dairy Facts and Figures. (1994). Residuary Milk Marketing Board, Thames Ditton, Surrey, England.

European patent. (1981). EP 0 043 276, B1.

Farber, J.M. Daley, E. Coates, F., Emmons, D.B., & McKellar, R. (1992). Factors influencing survival of *Listeria monocytogenes* in milk in a high-temperature short-time pasteurizer. *Journal of Food Protection* **55**, 946–951.

Foegeding, P.M., & Leasor, S.B. (1990). Heat resistance and growth of *Listeria monocytogenes* in liquid whole egg. *Journal of Food Protection* **53**, 9–14.

Fontana, A.J., Howard, L., Criddle, R.S., Hansen, L.D., & Wilhelmsen, E. (1993). Kinetics of deterioration of pineapple concentrate. *Journal of Food Science* **58**, 1411–1417.

Fredsted, L.B., Rysstad, G. & Eie, T. (1996). Pure-LaC: The new milk with protected freshness and extended shelf-life. In *Heat treatments and alternative methods,* IDF/FIL No. 9602.

Gaze, J.E., Brown, G.D., & Banks, J.G. (1989). *Food Microbiol.*

Gomez Barroso, B. (1997). *Effect of raw milk quality on the keeping quality of pasteurized milk.* MSc dissertation, The University of Reading, Reading, UK.

Grandison, A.S., & Finnigan, T.J.A. (1996). Microfiltration. In *Separation processes in the food and biotechnology industries*, pp. 141–154, Cambridge, UK: Woodhead Publishers.

Griffiths, M.W. (1986). Use of milk enzymes and indices of heat treatments. *Journal of Food Protection* **49**, 696–705.

Griffiths, M.W., Hurvois, Y., Phillips, J.D., & Muir, D.D. (1986a). Elimination of spore-forming bacteria from double cream using sub-UHT temperatures: 1. Processing conditions. *Milchwissenschaft* **41**, 403–405.

Griffiths, M.W., Hurvois, Y., Phillips, J.D., & Muir, D.D. (1986b). Elimination of spore-forming bacteria from double cream using sub-UHT temperatures: 2. Effect of processing conditions on spores. *Milchwissenschaft* **41**, 474–477.

Griffiths, M.W., & Phillips, J.D. (1990). Incidence, source and some properties of psychotrophic *Bacilus* spp found in raw and pasteurized milk. *Journal of the Society of Dairy Technology* **43**, 62–66.

Griffiths, M.W., Phillips, J.D., West, I.G., & Muir, D.D. (1989). The effects of low-temperature storage of raw milk on the quality of pasteurized and UHT milk. *Food Microbiology* **5**(2), 75–87.

Guirguis, A.H., Griffiths, M.W., & Muir, D.D. (1983). Sporeforming bacteria in milk. I. Optimisation of heat treatment for activation of spores of *Bacillus* species. *Milchwissenschaft* **38**, 641–644.

Guney, P. (1989). PhD thesis, University of Giessen, Giessen, Germany.

Hasting, A.P.M. (1992). Practical considerations in the design, operation and control of food pasteurisation processes. *Food Control* **3**, 27–32.

Hasselback, U., Ruholl, T., & Knorr, D. (1992). Fruit juice pasteurisation under reduced thermal load. *Fluessiges-Obst* **59**, 592–593.

Holdsworth, S.D. (1997). *Thermal processing of packaged foods.* London: Blackie Academic and Technical.

Holsinger, V.H., Smith, P.W. Smith, J.L., & Palumbo, S.A. (1992). Thermal destruction of *Listeria monocytogenes* in ice cream mix. *Journal of Food Protection* **55**, 234–237.

Houghton, H.W. (1987). How soft drinks have benefited from new technology. *Food Technology International Europe*, 139–141.

IDF Bulletin. (1981). *Factors affecting the keeping quality of heat treated milk*, No. 130.

IDF Bulletin. (1986). *Monograph on pasteurised milk*, No. 200.

IDF Bulletin. (1988). *Code of practice for the preservation of milk by the lactoperoxidase system*, No. 234.

IDF Bulletin. (1989). *Heat-Induced changes in milk*, No. 238.

IDF Bulletin. (1992). *Monograph on the pasteurization of cream*, No. 271.

IDF Bulletin. (1993). *Catalogue of tests for the detection of post-pasteurisation contamination in milk*, No. 281.

IDF. (1994). Indigenous *antimicrobial agents of milk—recent developments*, IDF Reference SI 9404.

IDF Bulletin. (1994). *Recommendations for hygienic manufacture of milk and milk-based products*, No. 292.

IDF. (1995). *Heat induced changes in milk* (2nd ed.), No. SI 9501.

IDF. (1996a). *Bacteriological quality of raw milk*, Reference SI 9601.

IDF. (1996b). *Heat treatments and alternative methods*, IDF/FIL No. 9602.

Jagtiani, D.J., Chan, H., & Sakai, W. *Tropical fruit processing.* San Diego: Academic Press.

Kalra, S.K., Tandon, D.K., & Singh, B.P. (1995). Mango. In *Handbook of fruit science and technology*. New York: Marcel Dekker.

Kessler, H.G. (1981). *Food engineering and dairy technology*. Freising, Germany: Verlag A Kessler.

Kessler, H.G. (1989). Effect of thermal processing of milk. In S. Thorne (Ed.), *Developments of food preservation*, Vol. 5, (pp. 91–130). London: Elsevier Applied Science.

Kessler, H.G., & Fink, R. (1992). Physical and chemical effects of pasteurisation on cream properties, 11–17, In IDF Bulletin No. 271.

Kessler, H.G., & Horak, F.P. (1984). Effect of heat treatment and storage conditions on keeping quality of pasteurized milk. *Milchwissenschaft* **39**, 451–454.

Kiesner, C., Hoffmann, W., Clawin-Radecker, I., Krusch, U., Neve, H., & Buchheim, W. (1996). Application of direct heating systems for the production of high-pasteurised milks. In IDF (1996), *Heat treatments and alternative methods*, IDF/FIL No. 9602.

Knabel, S.J., Walker, H.W., Hartman, P.A., & Mendonca, A.F. (1990). Effects of growth temperature and strictly anaerobic recovery on the survival of *Listeria monocytogenes* during pasteurization. *Applied and Environmental Microbiology* **56**, 370–376.

Lane, A. (1989). Disinfection in the food industry. *Food Technology International Europe*, 91–99.

Larsen, P.H. (1996). Microfiltration of pasteurised milk. In IDF (1996) *Heat treatments and alternative methods*, IDF/FIL No. 9602.

Lehmann, H. (1989). Bacterial clarification of milk and milk products. *European Dairy Magazine* **3**, 61–65.

Lewis, M.J. (1990). *Physical properties of foods and food processing systems.* Cambridge, UK: Woodhead Publishers.

Lewis, M.J. (1994). Advances in the heat treatment of milk. In R.K. Robinson (Ed.), *Modern dairy technology*, Vol. 1. London: Elsevier Applied Science.

Lewis, M.J. (1999). Microbiological issues associated with heat treated milks. *International Journal of Dairy Technology* **52**(4), 121–125.

Linton, R.H., Pierson, M.D., & Bishop, J.R. (1990). Increase in heat resistance of *Listeria monocytogenes* Scott A by sub-lethal heat shock. *Journal of Food Protection* **53**, 924–927.

Mabbit, L.A. (1982). Preservation of refrigerated milk. *Kieler Milchwirtschaftliche Forschungsberichte* **34**(1), 28–31.

Mackey, B.M., & Bratchell, N. (1989). The heat resistance of *Listeria monocytogenes*, *Letters in Applied Microbiology* **9**, 89–94.

Magnus, C.A., McCurdy, A.R., & Ingeledew, W.M. (1988). *Canadian Institute of Food Science and Technology Journal* 2.

Mettler, A.E. (1994). Present day requirements for effective pathogen control in spray dried milk powder production. *Journal of the Society of Dairy Technology* **47**, 95–107.

Miller Jones, J. (1992). *Food safety*. St. Paul, MN: Egan Press.

Mistry, V.V. (1989). Thermal inactivation characteristics of alkaline phosphatase in ultrafiltered milk. *Journal of Dairy Science* **72**, 1112–1117.

Muir, D.D. (1990). The microbiology of heat treated milks. In R.K. Robinson (Ed.), *Microbiology of milk and milk products*. New York: Elsevier Applied Science.

Muir, D.D. (1996a). The shelf life of dairy products: 1. Factors influencing raw milk and fresh products. *Journal of the Society of Dairy Technology* **49**, 24–32.

Muir, D.D. (1996b). The shelf life of dairy products: 2. Raw milk and fresh products. *Journal of the Society of Dairy Technology* **49**, 44–48.

Nelson, P.E., & Tressler, D.K. (1980). *Fruit and vegetable juice processing technology*. Westport, CT: AVI Publishers.

Nursten, H. (1995). Heat induced changes in the flavour of milk. In IDF (1995) *Heat induced changes in milk* (2nd ed.), No. SI 9501.

O'Donnell, E.T. (1995). The incidence of *Salmonella* and *Listeria* in raw milk from bulk tanks in England and Wales. *Journal of the Society of Dairy Technology* **48**, 25–29.

Painter, C.J., & Bradley, R.L., Jr. (1996). Residual alkaline phosphatase activity in milks subjected to various time/temperature treatments, 396–402. In IDF (1996) *Heat treatments and alternative methods*, IDF/FIL No. 9602.

Patel, S.S., & Wilbey, R.A. (1994). Thermal inactivation of (γ-glutamyltranspeptidase and *Enterococcus faecium* in milk-based systems. *Journal of Dairy Research* **61**, 263–270.

Pavlov, A. (1989). Heat resistance of Yersinia species in milk. *Veterinarna Sbirka* **87**(4), 13–15, 16.

Pearse, M.A. (1993). Pasteurization of liquid products. In *Encyclopedia of food science, food technology and nutrition*, 3441–3450. New York: Academic Press.

Pflug, I.J., & Esselen, W.B. (1979). Heat sterilisation. In J.M. Jackson & B.M. Shinn (Eds.), *Fundamentals of food canning technology*. Westport, CT: AVI Publishers.

Phillips, J.D., & Griffiths, M.W. (1986). Factors contributing to the seasonal variation in *Bacillus spp* in pasteurized dairy products. *Journal of Applied Bacteriology* **61**, 275–285.

Phillips, J.D., Griffiths, M.W., & Muir, D.D. (1983). Effect of nisin on the shelf-life of pasteurized double cream. *Journal of the Society of Dairy Technology* **36**, 17–21.

Pippas, N. (1996). *Dissolved oxygen in milk*, MSc dissertation. Reading, UK: The University of Reading.

Qin, B.L., Pothakamury, U.R., Vega, H., Martin, O., Barborosa-Canovas, G.V., & Swanson, B.G. (1995). Food pasteurization using high intensity pulsed electric fields. *Food Technology* **12**, 55–60.

Ravanis, S., & Lewis, M.J. (1995). Observations on the effect of raw milk quality on the keeping quality of pasteurised milk. *Letters in Applied Microbiology* **20**, 164–167.

Reiter, B., & Harnulv, B.G. (1982). The preservation of refrigerated and uncooled milk by its natural lactoperoxidase system. *Dairy Industries International* **47**, 13–19.

Salunkhe, D.K., Bolin, H.R., & Reddy, N.R. (1991). *Storage, processing and nutritional quality of fruits and vegetables*, Vol. II. Boca Raton, FL: CRC Press.

Salunkhe, D.K., & Kadam, S.S. (Eds.). (1995). *Handbook of fruit science and technology*. New York: Marcel Dekker.

Satin, M. (1996). *Food irradiation—A guidebook*. Lancaster, UK: Technomic Publishing Co, Ltd.

Schroder, M. (1984). Origins and levels of post pasteurization contamination of milk in the dairy and their effects on keeping quality. *Journal of Dairy Research* **51**, 59–67.

Schroder, M.J.A., & Bland, M.A. (1984). Effect of pasteurization temperature on the keeping quality of whole milk. *Journal of Dairy Research* **51**, 569–578.

Shyu, Y.T., Hsu, W.Y., & Hwang, H.S. (1996). Study on the improvements of mango juice processing. *Food Science, Taiwan* **23**, 469–482.

Society of Dairy Technology (SDT). (1983). *Pasteurizing plant manual.* Huntingdon, UK: Author.

Society of Dairy Technology (1989). *Cream processing manual.* Huntingdon, UK: Author.

Sorensen, K.R. (1996). APV Heat Exchanger AS introduces a new security system for pasteuriser installations, 179–183. In *Heat treatments and alternative methods*, SI 9602.

Splittstoesser, D.F., & Churey, J.J. (1989). Effect of low concentrations of sorbic acid on the heat resistance and viable recovery of *Neosartorya fischeri* ascospores. *Journal of Food Protection* **52**, 821–822.

Stadelman, W.J., Olson, V.M., Shemwell, G.A., & Pasch, S. (1988). *Egg and poultry-meat processing.* Chichester, UK: VCH Publishers, Ellis Horwood.

Statutory Instruments (SI). (1963). The liquid egg (pasteurization) regulations 1963. No. 1503. London: HMSO.

Statutory Instruments. (1995). *Food milk and dairies: the dairy products (hygiene) regulations*, No. 1086, London: HMSO.

Swart, G.J. (1993). Other pasteurisation processes. In *Encyclopedia of food science, food technology and nutrition*, 3455–3461. New York: Academic Press.

Thorne, S. (1986). *The history of food preservation.* Kirkby Lonsdale, England: Parthenon Publishing Group Ltd.

Tomlins, R.I., & Ordal, Z.J. (1976). *Inhibition and inactivation of vegetative microbes.* Edited by F.A. Skinner & W.B. Hugo. London: Academic Press.

Towler, C. (1994). Developments in cream separation and processing. In R.K. Robinson (Ed.), *Modern dairy technology.* New York: Elsevier Applied Science.

Trevatt, C. (1994). Developments in plate heat exchangers. *Food Technology International Europe*, 83–85.

USDA. (1984). Regulations governing the inspection of eggs and egg products. 7 CFR Part 59.

Vandenbergh, P.A. (1993). Lactic acid bacteria, their metabolic products and interference with microbial growth. *FEMS Microbiology Reviews* **12**, 221–238.

Veal, K. (1987). Key stages in fruit juice technology. *Food Technology International Europe*, 136–138.

Villamiel, M., Lopez Fandino, R., & Olano, A. (1996). Microwave pasteurisation of milk in continuous flow unit: Shelf life of cows milk. *Milchwissenschaft* **51**, 674–677.

Walstra, P., & Jenness, R. (1984). *Dairy chemistry and physics.* New York: John Wiley.

Whiting, R.C. (1995). Microbial modeling of foods. *Critical Reviews in Food Science and Nutrition* **35**, 467–494.

Wilbey, R.A. (1993). Pasteurisation of foods: Principles of pasteurisation: In *Encyclopedia of food science, food technology and nutrition*, 3437–3441. New York: Academic Press.

Wilson, G.S. (1942). *Pasteurisation of milk* (p. 104). London: Edward Arnold.

Wirjantoro, T.I., & Lewis, M.J. (1996). Effect of nisin and high temperature pasteurisation on the shelf-life of whole milk. *Journal of the Society of Dairy Technology* **4**, 99–102.

Zadow, J.G. (1989). Extending the shelf-life of dairy products. *Food Australia* **41**, 935–937.

Chapter 6

Sterilization

6.1 INTRODUCTION: CRITERIA FOR STERILIZATION

Sterilization of foods by the application of heat can be done either in sealed containers or by continuous-flow techniques. Traditionally it is an in-container process, although there have been many developments in container technology since the process was first commercialized at the beginning of the nineteenth century.

Whatever the process, the main concerns are with food safety and quality. The most heat-resistant pathogenic bacterium is *Clostridium botulinum*; this will not reproduce below a pH of 4.5. On this basis, the simplest classification is to categorize foods as either acid foods (pH < 4.5) or low-acid foods (pH > 4.5). Note that a broader classification has been used for canning: low acid, pH >5.0; medium acid, pH 4.5 to 5.0; acid, pH 4.5 to 3.7; and high acid, below pH 3.7. However, as mentioned earlier, the main concern is with foods where pH is greater than 4.5 (Section 2.4). For such foods the minimum recommended process is to achieve 12 D reductions for *Clostridium botulinum*. This is known as the minimum botulinum cook. This requires heating at 121°C for 3 minutes, measured at the slowest heating point (Section 2.4). The evidence that this provides a safe level of processing for thermally processed foods is provided by the millions of items of commercially processed foods consumed worldwide each year without any botulinum-related problems.

The temperature of 121.1°C (250°F) is taken as a reference temperature for sterilization processes. This is used in conjunction with the *z*-value for *Cl. botulinum*, which is taken as 10°C, to construct the standard lethality tables (Table 6–1). Since lethalities are additive it is possible to sum the lethalities for a process and determine the total integrated lethal effect, which is known as the F_0 value. This is a common method to evaluate and compare the microbiological severity of different processes.

The F_0 values, which are recommended for a wide range of foods, are given in Table 6–2. It can be seen from these values that all foods have in excess of the minimum botulinum cook (i.e., an F_0 of 3), with values ranging between 4 and 18 (Section 2.4). This is because there are some other bacterial spores that are more heat resistant than *Cl. botulinum*; the most heat resistant of these is the thermophile *Bacillus stearothermophilus*, which has a decimal reduc-

Table 6–1 Lethality Tables

Reference Temperature = 121.1°C; Z = 10°C

Processing Temperature (°C)	*Temperature (°F)*	*Lethality* (L)*
110	230	0.078
112	233.6	0.123
114	237.2	0.195
116	240.8	0.309
118	244.4	0.490
120	248	0.776
121	249.8	0.977
121.1	250	1.000
122	251.6	1.230
123	253.4	1.549
124	255.2	1.950
125	257	2.455
126	258.8	3.090
127	260.6	3.890
128	262.4	4.898
129	264.2	6.166
130	266	7.762
131	267.8	9.772
132	269.6	12.300
133	271.4	15.49
134	273.2	19.50
135	275	24.55
136	276.8	30.90
137	278.6	38.90
138	280.4	48.98
139	282.2	61.66
140	284	77.62
141	285.8	97.72
142	287.6	123.0
143	289.4	154.9
144	291.2	195.0
145	293	245.5
146	294.8	309.2
147	296.6	389.0
148	298.4	489.8
149	300.2	616.6
150	302	776.24

* Lethality values are derived from $\log L = t - \theta/Z$, where L = number of minutes at reference temperature equivalent to 1 minute at experimental temperature.

T = experimental temperature, θ = reference temperature (121.1°C)

Table 6–2 F_0 Values That Have Been Successfully Used Commercially for Products on the UK Market

Product	*Can Size*	*F_0 Values*
Baby foods	baby food	3–5
Beans in tomato sauce	All	4–6
Peas in brine	Up to A2	6
	A2 to A10	6–8
Carrots	All	3–4
Green beans in brine	Up to A2	4–6
	A2–A10	6–8
Celery	A2	3–4
Mushrooms in brine	A1	8–10
Mushrooms in butter	Up to A1	6–8
Meats in gravy	All	12–15
Sliced meat in gravy	Ovals	10
Meat pies	Tapered, flat	10
Sausages in fat	Up to 1.lb	4–6
Frankfurters in brine	Up to 16Z	3–4
Curried meats and vegetables	Up to 16Z	8–12
Poultry and game, whole, in brine	A2½ to A10	15–18
Chicken fillets in jelly	Up to 16 oz.	6–10
"Sterile" ham	1 and 2.lb	3–4
Herrings in tomato	Ovals	6–8
Meat soups	Up to 16Z	10
Tomato soup, not cream of	All	3
Cream soups	A1 to 16Z	4–5
	Up to A10	6–10
Milk puddings	Up to 16Z	4–10
Cream	4/6 oz	3–4

Source: Copyright © 1992, J. Brennan.

tion time of about 4 minutes at 121°C. Of recent interest is the mesophilic spore-forming bacteria, *Bacillus sporothermodurans* (Section 10.3.3).

Some values for some other important heat-resistant spores are summarized in Table 6–3. Burton (1988) and Holdsworth (1997) should be consulted for more comprehensive lists. Such heat-resistant spores may cause food spoilage, either through the production of acid (souring) or the production of gas. Again, most of these spores will not grow below a pH of 4.5; for example, *Bacillus stearothermophilus* does not grow below about pH 5.2.

Table 6–3 Reaction Kinetic Parameters for Some Microorganisms, Enzymes, and Chemical Reactions

	$D_{121}(s)$	Z (°C)	Q_{10}
B. stearothermophilus NCDO 1096, milk	181	9.43	11.5
B. stearothermophilus FS 1518, conc. milk	117	9.35	11.8
B. stearothermophilus FS 1518, milk	324	6.7	31.1
B. stearothermophilus NCDO 1096, milk	372	9.3	12
B. subtilis 786, milk	20	6.66	31.7
B. coagulans 604, milk	60	5.98	57.6
B. cereus, milk	3.8	35.9	1.9
Cl. sporogenes PA 3679, conc. milk	43	11.3	7.7
Cl. botulinum NCTC 7272	3.2	36.1	1.89
Cl. botulinum (canning data)	13	10.0	10
Protease inactivation	0.5–27 min at 150°C	32.5–28.5	2.0–2.3
Lipase inactivation	0.5–1.7 min at 150°C	42–25	1.7–2.5
Browning		28.2, 21.3	2.26, 2.95
Total whey protein denaturation, 130°–150°C		30	2.2
Available lysine		30.1	2.15
Thiamin (vitamin B_1) loss		31.4–29.4	2.08–2.19
Lactulose formation		27.7–21.0	2.30–3.00

Source: Reprinted with permission from H. Burton, *Ultra-High-Temperature Processing of Milk and Milk Products*, © 1988.

6.1.1 Brief Review of In-Container Sterilization

The severity of the process (F_0) selected for any food depends upon the nature of the spoilage flora associated with the food, the numbers likely to be present in that food, and to a limited extent on the size of the container, since more organisms will go into a larger container. Such products are termed *commercially sterile*, the target spoilage rate being less than 1 in 10,000 containers (see Section 10.3).

For in-container processes, there is now a much wider range of containers available, including cans, bottles, jars, flexible pouches, and plastic trays. These are high-speed operations, with can seamers now being able to seal up to 2,000 cans a minute on beverage products (less on other foods). It involves the integration of a number of operations, all of which will contribute to the overall effectiveness of the process (Figure 6–1). Since it is not practicable to measure the temperatures in every can, the philosophy for quality assurance involves

Figure 6–1 The canning process. *Source:* Reprinted with permission from J.M. Jackson and B.M. Shinn, Fundamentals of Food Canning Technology, in *Canned Foods*, © 1979.

verifying that the process is initially under control, followed by understanding and controlling the processes that affect the heat transfer process, and preventing recontamination.

Processing conditions such as temperature and time are critical control points. Others are raw material quality (especially counts of heat-resistant spores), and controlling all factors affecting heat penetration. These include filling temperature; size of headspace; ratio of solids to liquid; liquid viscosity; venting procedures; and reducing postprocessing contamination by seal integrity, cooling water chlorination, and avoiding handling wet cans (after processing) and drying them quickly. It is also important to avoid large pressure differentials between inside and outside of container.

The main problem with canning is that there is considerable heat damage to the nutrients and changes in the sensory characteristics; this can be assessed by the cooking value (Holdsworth, 1997), where a reference temperature of 100°C is used and a *z*-value of 20° to 40°C (typically about 33°C). Also summarized are the *z*-values for the sensory characteristics, which range between 25° and 47 °C. Further information on canned food technology is provided by the following excellent reference works: Stumbo (1973), Hersom and Hulland (1980), Jackson and Shinn (1979), Rees and Bettison (1991), Footitt and Lewis (1995), and Holdsworth (1997).

6.2 CONTINUOUS SYSTEMS

Some of the quality-related drawbacks of canning processes result from slow heating and cooling rates and the lower processing temperatures (up to about 130°C) that can be used, due to pressure restrictions within sealed containers. This has led to development of continuous processes, combined with aseptic packaging (see Chapter 1). This gave scope for using higher temperatures and shorter times; for example, for attaining F_0 values of 3 and 8 at higher temperatures, the following conditions should be used (assuming a *z*-value of 10°C):

$F_0 = 3$	$F_0 = 8$
131° C for 0.3 min (18 s)	48s
141° C for 1.8 s	4.8 s

Evidence that data for *Clostridium botulinum* can be extrapolated to 140°C has been provided by Gaze and Brown (1988).

The main advantages that result from using higher temperatures for shorter times (HTST) are that, for a given microbial inactivation, the amount of chemical damage is reduced. This arises because microbial inactivation rates are much more temperature dependent than chemical reaction rates. This can be illustrated by comparing the kinetics for thiamin loss with those conditions required to achieve an F_0 value of 6 (Figure 6–2). This shows that at a temperature of 110°C the time required would be about 75 minutes and the thiamin loss would be about 40%. If the temperature is increased to 140°C, the processing time would be less than 5 seconds and the loss of thiamin would be less than 1%. In general, the reduction in chemical reaction (in the majority of cases) would give rise to an improvement in the quality of the product. Similar data for a wide range of reactions are presented by Kessler (1981, 1989).

There are some situations where this reduction in the amount of chemical damage in ultra-high temperature (UHT) products compared to in-container sterilization may lead to a deterioration in product quality. Some examples are for heat-resistant enzymes (e.g., proteases and lipases) or some antinutritional compounds (see Section 6.8), which may survive UHT

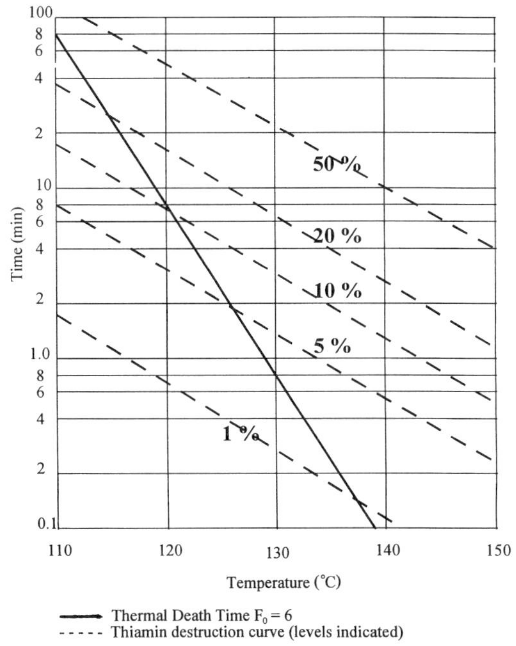

Figure 6–2 Time-temperature sterilization curve

processing and cause problems during storage. Of course, it is important to be aware of these and try to ensure that they are not a problem and that raw materials with high initial concentrations are avoided (Section 10.4.2).

Also, continuous processing allows much faster heating than in-container sterilization processes: this can be further increased by direct steam injection processes. Again, this will reduce the amount of chemical damage taking place. UHT processes can be indirect or direct-processes. With indirect processes there is no contact between the heat transfer fluids (steam, hot water, chilled water) and the product. With direct processes, steam comes in direct contact with the product, with possibilities of product contamination by the steam. These processes are described in much more detail in Chapter 3.

6.2.1 Indirect Heating

Plate heat exchangers or tubular heat exchangers are used for low-viscosity fluids. The choice of heat exchanger can give rise to considerable differences in the heating and cooling rates. Some temperature/time profiles for different UHT plants are shown in both in Figure 3–1 and Figure 6–3. For more viscous products or materials containing particulate matter, scraped-surface heat exchangers or more specialized equipment can be used (Chapters 3 and 4). A typical indirect plant is shown in Figure 6–4.

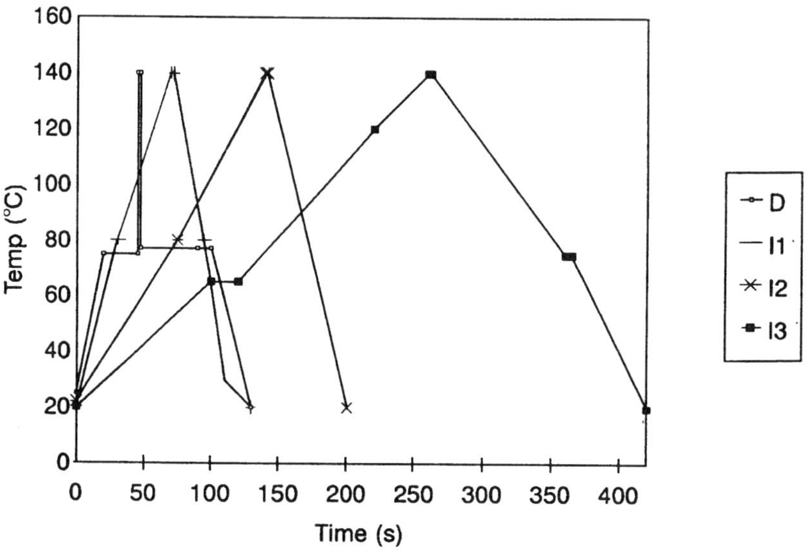

Figure 6–3 Temperature/time profiles for different UHT plants: D, direct; I1, I2, and I3, indirect plants with increasing regeneration efficiencies. *Source:* First published by Sterling Publications, Ltd., in *Food Technology International Europe*, Brunel House, 55a North Wharf Road, London, W2 1XR.

The product is heated by regeneration and finally by steam or pressurized hot water. Cooling is achieved by regeneration and chilled water. An indirect-UHT plant has many similarities to a pasteurization plant. However, there are also some subtle differences, which are briefly summarized below.

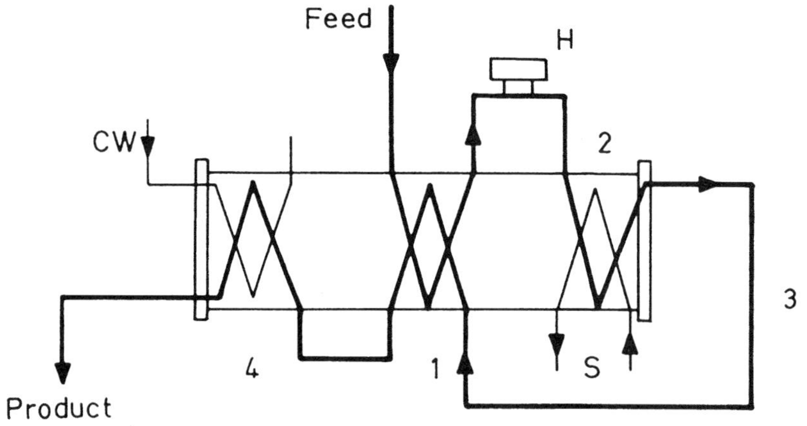

Figure 6–4 Schematic diagram of indirect UHT plant: 1, regeneration section; 2, final heating section; 3, holding tube; 4, cooling section; CW, chilled water; H, homogenizer; S, steam or hot water. *Source:* Reprinted with permission from J. Rothwell, *Cream Processing Manual*, p. 59, © 1989, Society of Dairy Technology.

1. Higher operating pressures are needed to achieve the higher processing temperatures required. The steam tables (see Section 1.4) show the relationship between the saturated vapor pressure for water and its temperature. They can also be used to determine how the boiling point of a material changes with pressure. The usual assumption made is that the food (e.g., milk or fruit juice) behaves like water; that is, there is no elevation of boiling point. The operating pressure must be maintained above the saturated vapor pressure at the desired processing temperature at all times during processing (see Table 1–4). Slightly higher pressures (about 1 bar) are recommended to prevent air from coming out of solution (Table 1–9). Any sudden drop in pressure (e.g., due to a pump failure) will lead to a reduction in temperature, which might compromise product safety.
2. Homogenization is always required with products containing fat. Otherwise this will separate during storage and may crystallize and form a hard plug. It can occur between either the heating or the cooling stages, or in some cases both (see Section 6.3).
3. Before processing commences, the plant needs to be sterilized (in particular, downstream of the holding tube). Conditions should be maintained at 130°C for 30 minutes at all points downstream of the holding tube. During processing sterility should be maintained by not allowing the UHT temperature to fall at any point. This gives rise to the concept of sterile and nonsterile sections of the UHT plant, often referred to as upstream and downstream (of the holding tube).
4. Aseptic storage and packaging of the product are essential (Chapter 7).
5. In general, the residence times are very much shorter, typically 2 to 8 seconds. Times below 1 second are not often used as they become more difficult to control precisely.

Points 2 to 5 also apply to direct processes.

A major practical problem with an indirect-UHT plant is fouling (see Chapter 8). It can be the major constraint in terms of plant operating times. Some indirect plants may also include a deaeration process, employing flash cooling primarily to remove entrained air; this is often the case for reconstituted products, which may be subject to excess aeration during the reconstitution process. Jensen (1996) has reviewed some recent developments in both direct and indirect systems.

6.2.2 Direct Heating

Direct heating processes involve the direct contact of steam with the fluid being processed. A temperature/time profile for direct-UHT is compared with indirect heating processes in Figure 6–3. Usually the product is preheated to about 70° to 80°C, often by regeneration before the direct process itself. Direct contact with steam results in extremely rapid heating and also some dilution of the product (Figure 6–5). Heating rates of between 100 and 200 K/s have been reported. For heating the product from 75°C to 145°C, the dilution would be about 10% to 15%. This increased volume should be taken into account for estimating the holding time. In most applications this excess water is removed by flash cooling. After flash cooling, the product is cooled by regeneration and chilled water. Homogenization nearly

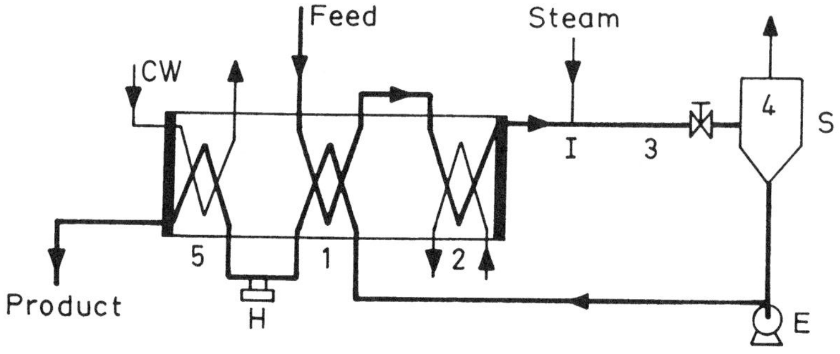

Figure 6–5 Schematic diagram of direct UHT plant: 1, regeneration section; 2, preheating section; 3, holding tube; 4, flash-cooling section; 5, cooling section; CW, chilled water; E, extract pump; H, homogenizer; I, injection point; S, separator. *Source:* Reprinted with permission from J. Rothwell, *Cream Processing Manual*, p. 60, © 1989, Society of Dairy Technology.

always takes place during the cooling section. The quality of steam used needs particular attention, as it should not contaminate the product (see Section 3.15 for more detail). The steam-injection processes themselves warrant further attention. A direct plant is classified as having injection or infusion equipment.

Injection (Figure 6–6) is the process whereby steam is injected into the milk; injectors are single or multiple nozzles; they can be noisy and potentially unstable. Infusion is the process whereby the product is injected in to steam; the steam may be flowing in a pipe or be in an infusion chamber. Infusers require a slightly larger vessel which has to be capable of withstanding UHT temperatures and pressures. The processes are compared and contrasted in much more detail in Chapter 3.

Both types of equipment have been used for processing foods. The heating process is gentler with infusion than injection, as there are fewer shear and noise problems arising from the condensation of small steam bubbles and the associated cavitation. Both give very rapid—but not instantaneous—heating, with estimated times of about 1 second for injectors and 0.1 to 0.5 second for infusers. It has been claimed that injectors may overheat some products because the incoming steam is at a higher pressure and hence a higher temperature than the chamber temperature. The same arguments can be used against infusers, however, because of the rise in pressure in the chamber from the release of dissolved gases. De Jong, Wallewijn, and van der Linden (1996) suggested that infusion resulted in less whey protein denaturation than injection—fouling of the distribution device could be avoided if the initial Reynolds number of the milk jets was over 10,000. The holding time is more predictable when an injector is used, as it is based on flow through a simple tube. Noncondensed steam bubbles may reduce the holding time by reducing the effective volume of the tube. With an infuser, a pool of liquid may collect in the base of the vessel, which will increase the spread of residence times. Also, some of the vessel atmosphere may be carried through the holding tube along with the liquid, thereby reducing the residence time in the same way as for the injection system.

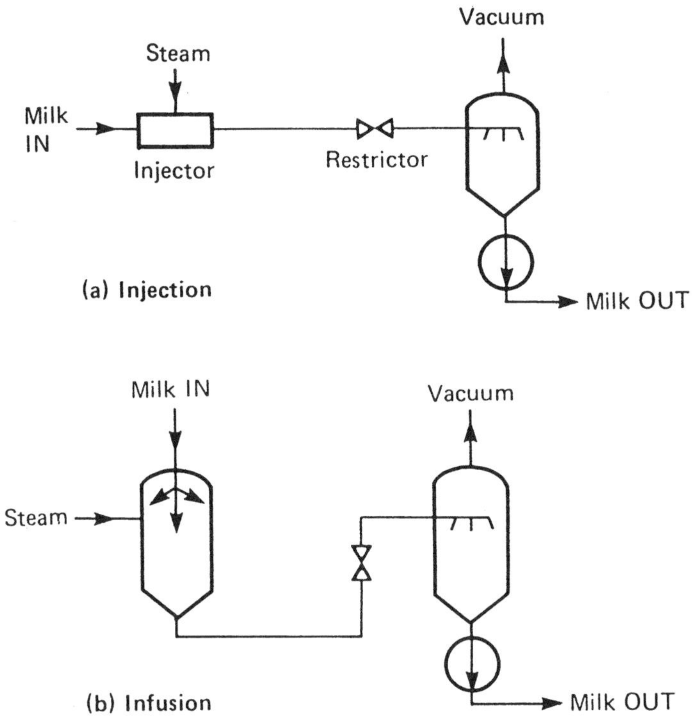

Figure 6–6 Diagram of injection and infusion systems. *Source:* Reprinted with permission from H. Burton, *Ultra-High-Temperature Processing of Milk and Milk Products*, p. 104, © 1988.

After a short period in the holding tube (where the pressure is high), the product is subjected to a flash-cooling process. This is achieved by a sudden reduction in the pressure, across a back-pressure valve, shown in Figure 6–5. This will reduce the boiling point and produce a rapid cooling effect, resulting in the production of some water vapor. The amount of water vapor removed will depend upon the temperature drop across the valve, which in turn is controlled by the flash-cooling temperature and hence the pressure in the separation vessel. The greater the cooling effect, the more water is removed; this is an important control mechanism. This is discussed in more detail in Section 3.14

Some homogenization takes place during the steam injection; it has been stated that steam injection at about 175°C was equivalent to about 1,500 psi in a conventional homogenizer. Three factors accounted for this size reduction:

1. Turbulence from the steam jets produces shearing forces that break globules into smaller units.
2. Streaming occurs from a cavitation of steam bubbles from local turbulence from oscillating bubbles.
3. Collapse of steam bubbles produces a cavitation effect that produces shock waves.

The flash-cooling process involved also results in the loss of some other components, which are more volatile than water; these include some of the dissolved oxygen, low-molecular-weight sulfur components, and other desirable or undesirable flavor components. This may lead to a further change in product quality.

In general, direct-UHT plants are more complex; have higher capital costs; have less regeneration (up to about 50%), but result in a better-quality product; are less susceptible to fouling; result in better quality but higher capital and running costs; and have better quality resulting from more rapid heating and cooling and the flash-cooling process.

Jensen (1996) describes a high-heat-infusion UHT process, which is a compromise between direct and indirect heating. The product is heated to 90°C, held for 30 seconds to stabilize proteins, vacuum cooled to 70°C to remove air, homogenized, and heated to 120°C by a tubular heat exchanger; this is followed by direct-steam infusion to 140° to 150°C. The product is then cooled to the final temperature. Although the process appears complex, the stated advantages are that it gives less chemical damage compared to indirect processing and that it gives better energy economy (75%) than the conventional direct process. More specific differences are covered for individual products, where appropriate.

It is now more widely appreciated that it is misleading to compare processes solely in terms of their holding times and temperatures and that it is important to consider the heating and cooling periods, as these can vary significantly from one plant to another (Chapter 3). Procedures have been established for evaluating the overall extent of microbial inactivation and chemical change from the temperature/time profiles. This has been done for milk to determine B* and C* values by Kessler (1981) (see Section 3.22).

6.3 HOMOGENIZATION

Homogenizers are discussed in more detail in Section 3.1. Homogenization is required for products that contain substantial amounts of fat. It is a process that reduces the fat globule size, usually by mechanical disruption. This will lead to a large increase in the interfacial area. Fat globules in raw milk range between 1 and 10 μm, with an average size of about 3.5 μm. Such milk will be subject to creaming if it is left to stand. Following homogenization, the average size is reduced to below 0.8 μm. Homogenization temperatures should be above 45°C, to ensure that fat is in the liquid form, as this helps their disruption. In UHT processing, homogenization also helps reduce sediment formation.

The nature of the dispersion is affected by the type of homogenizer, pressure, temperature, and nature of the product. Two-stage homogenization is sometimes used, especially to disrupt clumps. Walstra and Jenness (1984) discuss these effects in more detail.

Most pressure homogenizers are positive displacement pumps (see Section 3.1). They work by subjecting the product to a high pressure and forcing it through a narrow restriction (homogenization valve) at high velocity, which produces intense turbulence. It is estimated that the energy density is 10^{10} to 10^{11} W/m^3 and this will determine the size and velocity of the eddies that disrupt the fat globules. Homogenization will also increase product temperature (approximately 1°C/580 psi [40 bar]).

During homogenization, the fat is covered and stabilized by new "fat globular membrane material." In milk this is predominantly casein in nature. It has been estimated that this process takes place within a short time period (~ less than 10 μs). Large micelles are adsorbed in preference to small ones, causing the homogenized fat globules to behave more like large casein micelles. Thus conditions that cause casein micelles to aggregate will also cause homogenized fat globules to aggregate. In other food systems, additional emulsifiers and stabilizers may be required to ensure emulsion stability. Another important reason to homogenize with UHT products is to distribute protein aggregates that may form during heat treatment. Thus, homogenization helps to reduce sediment formation in milk and is best done after heat treatment. Zadow (1975) concluded that homogenization downstream of the holding tube will help reduce such sediment formation. However, care should be taken as excessive homogenization can result in more sediment formation.

One interesting question in UHT processing is what is the best position for the homogenizer (i.e., upstream or downstream)? If it is positioned upstream, it is not in the sterile part of the plant and does not have to operate under aseptic conditions. In terms of reducing potential spoilage, this is the best place to position the homogenizer and removes the requirement to incorporate a sterile block. Typically the product would be heated to about 60° to 80°C before homogenization and final heating to the UHT temperature. However, the drawback of homogenization in this position is that the higher temperatures used later in the process, or the high shear rates found in plate heat exchangers, might destabilize the emulsion and affect its stability during storage. Note that heat is one method for destabilizing emulsions. However, many UHT plants operate with the homogenizer in the upstream position, and this is the best position where stability and sedimentation are not major problems.

An example of where homogenization is necessary downstream is on direct-steam-injection plants; here it is thought that the injection process itself will damage or destabilize the emulsion, hence it is best to do it beforehand. Another example is where there may be excessive destabilization or perhaps even excessive sediment formation. Where downstream homogenization is necessary, it is important to provide a sterile block, which involves the pistons moving through an atmosphere of steam. This position, however, increases the risk of microbial contamination.

Jones (1964) concluded that steam injection is known to reduce fat globule size and that direct-injection processes would require less homogenization pressure to achieve the same degree of fat stabilization. Burton (1988) concluded that downstream homogenization was essential with the direct-UHT process. Perkin (1978) compared injection and infusion methods for sediment formation and found no differences in the amounts, but considerable differences in their appearance.

On a plate UHT heat exchanger, there was evidence for whole milk that fat globule size was influenced by the position of the homogenizer, with downstream homogenization giving the better homogenization effect. The difference was much more marked for direct-steam injection, where upstream homogenization was not at all effective. At equal homogenization pressures, the fat globule size was slightly smaller for indirect compared to direct processes.

With a tubular plant it is possible to have the homogenizer in the upstream position and the homogenization valve downstream. This is made possible because the tubes will withstand the high pressures created by the pumping section of the homogenizer.

Kaw, Singh, and McCarthy (1996) observed that for whole milk, UHT sterilization alone had no effect on particle size distribution (PSD). Homogenization prior to UHT processing resulted in considerable aggregation of fat globules; these aggregates were not easily broken down by a second homogenization process. Homogenization after heating was most effective, reducing the fat globule size considerably, as well as the tendency toward aggregation. Addition of a protein-dissociating medium reduced these aggregates, indicating the role of casein in their formation. The suggestion was that the aggregates were intact fat globules in a protein matrix. Downstream homogenization was also most effective for UHT treatment of reblended concentrated whole milk, made by mixing skim milk and cream (no composition given) and for reverse osmosis (RO) concentrates. The reblended milk behaved differently to whole milk, in that UHT treatment alone caused aggregation; also, the aggregates did not incorporate fat globules.

UHT sterilization of RO concentrates produced an increase in PSD and some considerable aggregation. The RO process alone was also found to have a considerable homogenizing effect caused by the passage through the back-pressure valve; however, UHT sterilization increased PSD considerably, forming large aggregates. These aggregates were not easily broken down by subsequent homogenization.

In general, for concentrates, homogenization will reduce heat stability of concentrated milk, so it is preferable to homogenize after UHT processing; this option is commonly used in the manufacture of heat-evaporated concentrates (Muir, 1984). Homogenization will increase a product's susceptibility to oxidation reactions.

6.4 MILK

Milk is the product that has been subject to most intensive research in terms of sterilization and UHT processing. It has some specific features, such as micelle stability and flavor changes, that are unique to milk. However, much of the work on nutrient losses would be appropriate to other food systems, and much of the knowledge (wisdom) about milk processing could well be applied more generally. There is also a wide range of milk products that are subject to UHT processing: normal (nonstandardized), standardized, skim, semiskim, and fortified milk; and milk with added minerals, vitamins, and flavoring. There is even greater scope for additives: concentrated (evaporated, condensed), with or without sugar, and reconstituted and recombined.

To extend the shelf life of milk much beyond a few days at ambient storage conditions, it must be heat treated more severely than for pasteurization, and postprocessing contamination must be eliminated. Temperatures in excess of 100°C are required. The procedure is known as sterilization. It is much more drastic than pasteurization and can be achieved by two different methods: namely, in-container sterilization or continuous sterilization followed by aseptic packaging. The aim is to produce a product that is commercially sterile. In this sense a target spoilage rate of 1 in 10,000 (10^4) would be the objective. Since it is a low-acid product, the main obvious source of concern is *Clostridium botulinum*; however, this organism is rarely found in raw milk, so the point of interest becomes heat-resistant spores, in particular *Bacillus stearothermophilus*, which is a major problem. More recently, problems arising from a heat resistant mesophilic spore former (HRS) have been identified (Hammer et al., 1996; Meier et al., 1996). Counts as high as 10^5 bacteria per milliliter have been found, without changing the sensory characteristics of the milk (see Section 10.3.3).

When milk is heated, many changes take place. The most important are the following:

- decrease in pH
- precipitation of calcium phosphate and reduction in ionic calcium
- denaturation of whey proteins and interaction with casein
- Maillard browning
- modification of casein; dephosphorylation, hydrolysis of κ-casein and general hydrolysis changes in micellar structure; zeta potential, hydration changes; association and dissociation

These changes are extremely important inasmuch as they affect the sensory characteristics of milk (appearance, flavor, and texture), its nutritional value, and its susceptibility to foul heat exchangers (see Section 8.4) or to form more sediment or even gel during storage. More detailed accounts of the effects of heat on milk are provided by Walstra and Jenness (1984), Kessler (1989), IDF (1981, 1989, 1995), Burton (1988), and Lewis (1994).

6.4.1 Sterilized Milk

Milk that is packaged in airtight containers and subject to temperatures in excess of 100°C is known as sterilized milk. Sterilized milk has a long history and in the United Kingdom is a traditional product, which is still significant, but also declining slightly in popularity. In 1983 it commanded 6.1% of the liquid milk market, whereas in 1993 it was 3.0% (EC Dairy Facts and Figures, 1994). In the European Community it accounted for 9.7% of the milk consumed in 1990 and it is still popular in France, Italy, Belgium, and Spain. It was traditionally sold in glass bottles, sealed with a crown cork. Some is still produced that way, but it is now also available in plastic bottles with a metal foil cap, which will withstand steam sterilization.

The Dairy Products Hygiene regulations (SI, 1995) are much less specific in terms of times and temperatures than those for pasteurization. They state that the milk should be filtered or clarified, homogenized, filled into containers, and maintained at a temperature of above 100°C for sufficient time to ensure that a negative result is obtained when the turbidity test (Section 10.5.7) is performed. The milk should be labeled sterilized milk. In practice, retort conditions range between temperatures of 110° and 116°C for between 20 and 30 minutes, depending upon the degree of caramelization required. Batch or continuous retorts may be used.

Additional regulations for sterilized milk produced by continuous-flow methods state that the milk should be filled into sterile containers using such aseptic precautions that will ensure the protection of the milk from the risk of contamination. Temperature records should be kept for the continuous-heat exchanger and a flow-diversion valve should be provided. The continuous-heat treatment would be in the region of 137°C for 4 seconds, followed by filling and sealing and a subsequent heat treatment in the container, which is substantially lower than that used in the conventional process, but which is sufficient to denature the whey protein fully and hence satisfy a negative turbidity test. This procedure will allow a greater product throughput by reducing the retorting period.

Sterilized milk thus produced must also satisfy the colony count for UHT milk. The turbidity test (Section 10.5.7) measures the extent of whey protein denaturation, a negative turbidity indicating complete denaturation (see Section 10.5.7). Therefore, according to these regulations sterilized milk should contain no undenatured whey protein. On the other hand, one might expect UHT to give a positive turbidity reading.

An interesting phenomenon is the formation of a boiling ring at the interface of the milk and headspace air within the bottle. This will almost certainly have been present (see FAO publication, Milk Sterilization, 1965), but more discerning consumers are now noticing it and suggesting that the bottles have not been properly cleaned, which is not the case. This is a specific problem with glass bottles that are recycled.

6.4.2 UHT MILK

Milk heated under continuous-flow conditions to temperatures in excess of 135°C, held for a minimum of 1 second, and then rapidly cooled and packaged aseptically is known as UHT milk. It can be produced in a variety of heat exchangers, in particular plate and tubular systems, using steam or pressurized hot water. More recently, milk has been heated by direct contact (injection or infusion). The steam condenses and dilutes the product. This excess water is removed by flash cooling. These different processing methods give rise to considerable variation in the quality of the products.

UHT milk differs from sterilized milk in a number of important aspects (Figure 6–7). These can best be seen by plotting reaction kinetic data for the destruction of spores with those of a number of chemical reactions and inactivation of bacterial proteinases. Kessler (1981, 1989) presents such reaction kinetic data for a wide range of chemical reactions, including discoloration of whole milk, evaporated milk, skim milk, and cream containing 10%, 30%, and 40% fat respectively; loss of vitamin B; the formation of hydroxymethylfurfural (HMF); whey protein denaturation; and loss of available lysine. Again, the conclusion is that UHT processing results in a product that is subjected to fewer chemical reactions. Therefore, in comparison to sterilized milk, UHT milk is whiter, tastes less caramelized, and undergoes less whey protein denaturation and loss of heat-sensitive vitamins. Consequently, the two products are different in overall quality and specifically in sensory, chemical, and nutritional terms. Note that changes will also occur during storage (Chapter 9) ; those differences in sensory characteristics are most apparent immediately after processing, but UHT milks will start to be more like sterilized milks in terms of their sensory characteristics after storage at elevated temperatures due to Maillord reactions. One significant problem that may sometimes arise with UHT milks is due to incomplete inactivation of heat-resistant bacterial proteinases (see Section 6.4.10). It also should be appreciated that there will be considerable differences between milk or any other product produced on the different UHT plants available, arising from their different heating and cooling profiles (Reuter, 1984). Students at Reading University have analyzed many UHT milk samples that have given a negative turbidity result, indicating that they have been subject to a fairly harsh UHT processing regime. Mehta (1980) compiled one the most comprehensive reviews on UHT milk, which is still worth reading today.

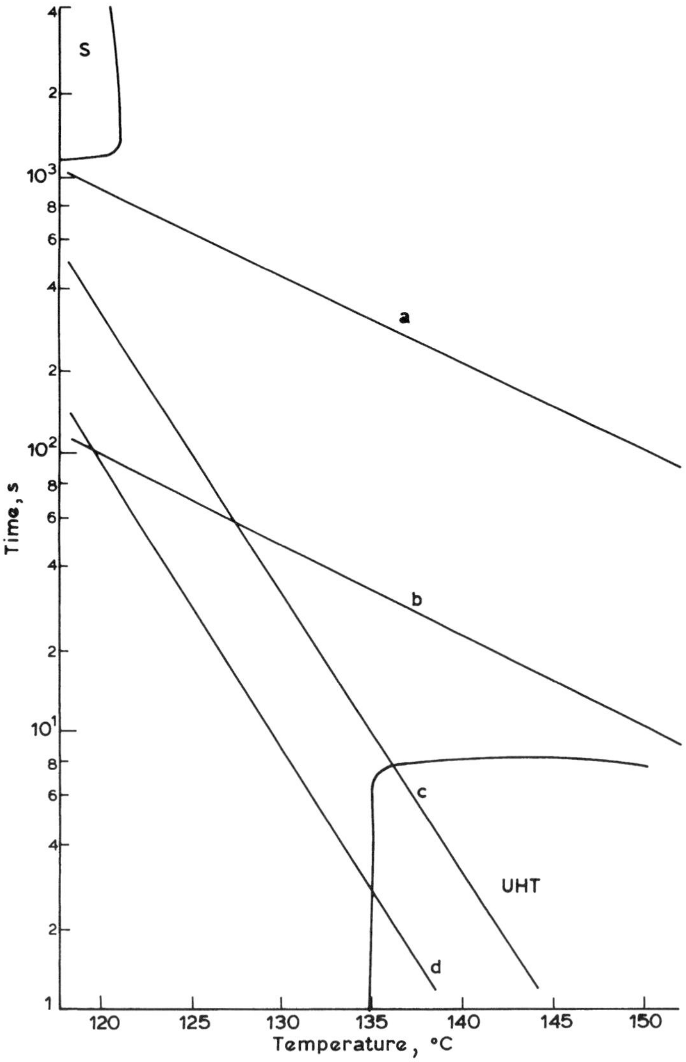

Figure 6–7 Reaction kinetic data for the following reactions: a, Inactivation of 90% of protease from *Pseudomonas* (Cerf, 1981); b, destruction of 3% of thiamin (Kessler, 1981); c, inactivation of thermophilic spores (Kessler, 1981); d, inactivation of mesophilic spores (Kessler, 1981); S, sterilized milk region; UHT, UHT milk region. *Source:* Reprinted from M.J. Lewis, Heat Treatment of Milk, in *Modern Dairy Technology*, Vol. 1, 2nd ed., R.K. Robinson, ed., p. 26, © 1993, Aspen Publishers, Inc.

6.4.3 UHT Milk Regulations

The Food Milk and Dairy Hygiene Regulations (ST, 1995) for UHT milk state the following: UHT milk shall be obtained by applying heat to a continuous flow of milk entailing the application of a high temperature for a short time (not less than 135°C for not less than 1 second), so that all residual microorganisms and their spores are destroyed, but the chemical, physical, and organoleptic changes to the milk are minimal. Note that in the United States, *ultrapasteurization* is described as a process where milk is heated at 138°C for 2 seconds.

Flow-diversion facilities should be provided if the temperature falls below 135°C. Suitable indicating and recording thermometers should be provided and records kept for 3 months. The milk should be packaged aseptically, securely fastened, and labeled ultraheat-treated milk, or UHT milk. A sample of the milk should satisfy the UHT colony count, details of which are provided in Section 6.4.4.

Regulations for allowing UHT milk to be produced by direct methods in the United Kingdom were introduced in 1970. These require that there should be no dilution of milk. How this is best achieved is best described by Perkin and Burton (1970) and Burton (1988). The injected steam should be dry and saturated, free from foreign matter, and readily removable for sampling. The boiler water should be treated only with permitted additives, which are listed (Section 3.15). Heat treatment requirements for skim milk and semiskim milk are the same as for whole milk; however, regulations for flavored milks are different, with 140°C for 2 seconds being the minimum requirement.

6.4.4 Comments on the Regulations

The microbiological severity, corresponding to the minimum conditions of 135°C for 1 second, is not very severe, especially if the spore count in the raw milk is high. The corresponding values are for F_0 (0.41), B* (0.099), and C* (0.033) (Section 3.22). Processing at these conditions may give high spoilage rates due to survival of heat-resistant spores. However, it is an improvement over the earlier heat-treatment requirements, which were 132.5°C for 1 second. Also, they do not account for the contributions of the heating and cooling periods; for indirect processes these may contribute more to lethality than the time spent in the holding tube, whereas in contrast they will be negligible for direct processes.

In general it is not required to label which method of processing (indirect or direct) has been used. The cooked flavor is much less intense for direct-UHT milk. My experience for direct UHT is that there is slightly perceptible cooked flavor immediately after production, which tends to disappear after about 1 week. Caramelized flavors tend to appear on longer-term storage.

Flow diversion is also required. This may suggest that once temperature has returned back above 135°C, it is then safe to resume normal processing. I think that this is debatable with UHT processes. Most processors would favor stopping the process rather than divert and risk the possibility that the plant may have lost its sterility. The best strategy is to try to avoid the problem's occurring by preventing the temperature from falling in the first instance.

There are also microbiological specifications for some products. For example, these stipulate that sterilized and UHT milk, after it has spent 15 days in a closed container at a temperature of 30°C, or where necessary 7 days in a closed container at a temperature of 55°C, should have a plate count (30°C) of less than 100 ml^{-1}. These are thought by many to be very generous for a product that is commercially sterile. However, more recently, UHT milk has been produced in a number of countries that is failing these tests (see Section 10.3.3).

Both the terms *ultrahigh temperature* and *ultraheat treatment* are used. My preference is for *ultrahigh temperature* because it provides a more accurate description of the process. The term *ultraheat treatment* suggests that the product is severely heat treated, which should not normally be the case.

Some interest has also been shown in distinguishing sterilized and UHT milks, principally for legislation purposes. This is not so straightforward because of the wide variety of conditions experienced on a UHT plant. Methods involving HMF values, lactulose, color, and turbidity have been reviewed by Fink and Kessler (1988) (Section 10.5.7).

6.4.5 Overview of Changes Taking Place during UHT Milk Processing

Mehta (1980) provides a good overview of factors affecting the quality of UHT milk to that date. In the literature of the 1960s and 1970s there were large numbers of papers devoted to comparing indirect and direct processes in terms of such factors as spore survival, whey protein denaturation, sediment formation, clotting of enzymes, and nutrient loss. In some of these, the heating profiles were accounted for and in others they were not. Some results from Jelen (1982) (Table 6–4) illustrate that the direct process was the less severe process.

There is also little doubt that direct-UHT milk is less cooked and hence (to many people) superior in flavor to indirect milk. The two reasons for this are the more rapid heating and cooling involved in direct heating, combined with the vacuum used in the flash-cooling process, which reduces some of the heat-induced volatile components that give rise to the cooked flavor. Furthermore, the dissolved oxygen content of direct-UHT milk is also reduced, which would alter the rate of oxidation reactions during storage. Which of these mechanisms contributes most to the improved flavor will probably depend upon how old the milk is when it is drunk (see Chapter 9). It has been claimed that direct-UHT milk cannot be distinguished from pasteurized milk. In general, the direct-UHT milk process is less suscep-

Table 6–4 Concentration of Components after Direct and Indirect Processing

	Direct	*Indirect*
Whey protein nitrogen (mg/100 g)	38.8	27.6
Loss of available lysine (%)	3.8	5.7
Vitamin Loss (%)		
B_{12}	16.8	30.1
Folic acid	19.6	35.2
Vitamin C	17.7	31.6
Hydroxymethylfurfural (μmol L^{-1})	5.3	10.0

Source: Reprinted from M.J. Lewis, in *Modern Dairy Technology*, Vol. 1, 2nd ed., R.K. Robinson, ed., p. 35, © 1994, Aspen Publishers, Inc.

tible to fouling, and it is the best process for poor (microbial) quality raw milk; however, more sediment has been reported in direct milk.

There are many other reactions taking place that would not be influenced by the flash-cooling process but that would be affected by the rate of heating and cooling. Examples are the effects of heat on protein denaturation, mineral loss, and vitamin loss. It is now more well recognized that long heating and cooling periods would significantly increase the extent of chemical reactions in general. Kessler (1981) helped clarify this for milk and shows the reduced amount of chemical damage that results from direct heating and flash cooling (Table 6–5).

6.4.6 B* and C* Values

These are explained in Section 3.22. A process given a B* = 1 value would be sufficient to produce nine decimal reductions of mesophilic spores and would be equivalent to 10.1 seconds at 135°C.

C* is a parameter to measure the amount of chemical damage taking place during the process. A process giving a C* = 1 would cause 3% destruction of thiamin and would be equivalent to 30.5 seconds at 135°C. Again the criteria in most cases is to obtain a high B* and a low C* value. Calculations of B* and C* based on the minimum holding time are given in Table 3–2, as well as data for a number of different UHT plants, both direct and indirect.

Some further results to demonstrate the effects of increasing, heating, and cooling periods on F_0, B*, and C* are shown in Table 6–6. These results are based on heating the product from 80° to 140°C, holding it for 2 seconds, and cooling it down to 80°C. Heating and cooling periods from instantaneously through to 60 seconds are shown. Increasing these periods increases both the chemical and the microbial parameters, with the ratio of chemical/microbial reaction rates increasing with increasing heating period. At a heating period of about 8 seconds, the amount of chemical damage done during heating and cooling exceeds that in the holding tube. It is this considerable increase in chemical damage that will be more noticeable in terms of decreasing the quality of the product. However, this may be required in those circumstances where more chemical damage may be beneficial; that is, for inactivating enzymes or for heat inactivation of natural toxic components (e.g., trypsin inhibitor in soy beans) (see Section 6.4.4) or softening of vegetable tissue (cooking). Differences in temperature/time processes arise due to different heating and cooling rates found for different heat exchangers and also the extent of energy saving by regeneration, with higher regeneration efficiencies usually resulting in slower heating and cooling rates. It should also be remembered that changes will occur during storage (Chapter 9).

6.4.7 Effects on (Milk) Food Components

6.4.7.1 Proteins

Milk proteins can be subdivided into caseins (75% to 80%) and whey proteins (20% to 25%). Casein is in the form of micelles (10 to 100 nm), and the micelles are relatively stable to heat, often needing temperatures of 140°C for 20 minutes to destabilize them. UHT pro-

Table 6–5 Evaluating B* and C* Values

Temperature (°C)	*B**	*C**
120	3.727E–02	0.3328
121	4.641E–02	0.3582
122	5.779E–02	0.3854
123	7.196E–02	0.4147
124	8.961E–02	0.4463
125	0.111	0.4803
126	0.138	0.5168
127	0.173	0.5561
128	0.215	0.5985
129	0.268	0.6440
130	0.334	0.6930
131	0.415	0.7457
132	0.517	0.8025
133	0.644	0.8635
134	0.803	0.9292
135	1	1
136	1.245	1.076
137	1.550	1.157
138	1.930	1.246
139	2.404	1.340
140	2.993	1.442
141	3.727	1.552
142	4.641	1.670
143	5.779	1.797
144	7.196	1.934
145	8.961	2.081
146	11.15	2.240
147	13.89	2.410
148	17.30	2.594
149	21.54	2.791
150	26.82	3.004

$$B = 10^{(T-135)/10.5}: \; B^* = \int B \, dt/10.1$$

$$C = 10^{(T-135)/31.4}: \; C^* = \int C \, dt/30.5$$

Note: B* = 1 corresponds to 9D reductions of thermophilic spores; C* = 1 corresponds to 3% loss of vitamin B_1 (thiamine).

cessing will cause some subtle changes to the micelle, in general leading to a slight increase in size, which may also partly be caused by association with denatured whey protein and some of the mineral components. There will also be some dissociation of casein from the micelle and in general there is a large increase in the number of smaller particles, presumably of aggregated serum proteins and/or dissociated caseins (Nieuwenhuijse, 1995).

Table 6–6 Effects of Increasing, Heating, and Cooling Periods F_0, B*, and C*

Heating and cooling period s	F_0	*B* total*	*C**
0.0	2.59	0.59	0.09
0.1	2.61	0.60	0.10
1	2.77	0.64	0.12
10	4.45	1.04	0.31
30	8.20	1.94	0.73
60	13.8	3.29	1.37

Source: First published by Sterling Publications, Ltd., in *Food Technology International Europe*, Brunel House, 55a North Wharf Road, London, W2 1XR.

Note that fat globules in homogenized milk also tend to behave like large casein micelles. This will alter the light-scattering properties of the milk, making it appear whiter. It will also affect its susceptibility to gelation. Certainly UHT milk is not so easily coagulated by the enzyme rennet compared to raw milk and pasteurized milk. This is attributed to the whey protein denaturation, its association with the micelle, and a reduction in ionic calcium. The whey protein fraction is more heat labile. The major whey protein is β-lactoglobulin, which comprises about 50% of the whey protein fraction. Other major ones are α-lactalbumin, blood serum albumin (BSA), and the immunoglobulin fraction. Note that there is also a proteose-peptone fraction, which represents about 10% of the total whey proteins. This fraction is not denatured by heat. When cheese whey is heated, the whey proteins will unfold and then aggregate, leading to their eventual precipitation. These reactions start at about 70°C. In milk, similar denaturation occurs but the denatured whey proteins will associate with the casein micelle, so the presence of the casein micelles prevents precipitation and has a considerable stabilizing effect. UHT processing will partially denature whey proteins, the range being from about 40% to greater than 90%. In contrast, for sterilized milk, whey protein denaturation should be 100%; this forms the basis of the turbidity test, and sterilized milk must produce a negative turbidity (Section 6.4.1). One proposed distinction between sterilized and UHT milk has been that UHT milk should give a positive turbidity. However, students of the author have analyzed many UHT milks that have also given a negative turbidity, presumably arising from either long heating and cooling periods (Section 10.5.7) or an extensive forewarming process prior to UHT treatment.

There have been a number of detailed studies involving different methods of analysis on whey protein denaturation. Methods of analysis include electrophoresis, immunoelectrophoresis, immunodiffusion, rocket electrophoresis, gel filtration, differential scanning calorimetry (DSC), high-performance liquid chromatography (HPLC), and analysis of SH and S—S groups (Lyster, 1970; Hillier & Lyster, 1979; Kessler, 1989). Some of the these have been reviewed by Burton (1988), Pearce (1989), and Lewis (1994). Perhaps the most detailed study of UHT processing conditions has been made by Kessler (1989), both for the individual whey proteins and for whey proteins in total.

There have also been some studies on the denaturation of β-lactoglobulin A and B (Hillier & Lyster, 1979); their equations show the reaction velocity constant changes with temperature:

$$\log k_A = 4.25 - 1.91\ (1000/T) \quad \text{(range } 100°\text{–}150°\text{C)} \tag{6.1}$$

$$\log k_B = 3.48 - 1.67\ (1000/T) \quad \text{(range } 95°\text{–}150°\text{C)} \tag{6.2}$$

Note T = K; k = reaction velocity constant ($1\ g^{-1}\ s^{-1}$). The reaction follows second-order reaction kinetics. The picture is complicated by the fact that there is a break in the reaction kinetics. Kessler (1989) explains this break by the fact that below 90°C, the unfolding reaction is the rate-determining step for the whole denaturation process, whereas above 90°C, the aggregation process is the rate-determining step. Thermodynamic data are presented to support this. His final calculations (Figure 6–8) show total whey protein denaturation at different times and temperatures, calculated from the data for the individual fractions and their initial concentrations in skim milk. This includes β-lactoglobulin, α-lactalbumin, BSA, immunoglobulin, and proteose peptones. Note that the maximum denaturation level given is 89%, because the proteose peptone fraction is not denatured. These data cover a wide range of heating regimes, from thermization through to in-container sterilization.

Using the data in Equation 6.1, calculations can be done to show how and where β-lactoglobulin is denaturated on heating, as it passes through a tube. In this example, heating is from 75° to 140°C (assuming a linear profile) in time periods of 1, 8, and 30 seconds. It can be seen that the denaturation profile changes completely. At long heating times most denaturation takes place toward the beginning of the tube, at intermediate heating times toward the middle of the tube, and at short heating times toward the end of the tube. Similar differences are observed for different starting concentrations (Figure 6–9); at high concentrations, most denaturation takes place at the beginning, whereas at low concentration it takes place toward the end (see Chapter 8).

Studies have also been performed on denaturation in ultrafiltration (UF) concentrates. Denaturation rates tend to be higher in such systems (see Figure 5–6). This denaturation will not alter their nutritional value (in terms of biological value, digestibility, and available lysine), although sterilized milk has a lower biological value (0.85) compared to UHT milk (0.90) and raw milk (0.91). However, denaturation of whey proteins is responsible for cooked flavor, resulting from formation of free (unmasked) sulfydryl groups and low-molecular-weight sulfur volatiles. Flavor changes are discussed in more detail in Section 9.5. It will also change the stability of the casein micelle to gelation by rennet, as well as age-onset gelation.

6.4.7.2 Minerals

Heating milk will cause the transfer of minerals from the aqueous phase to the casein micelle, which results from the inverse temperature solubility of calcium phosphate; this is partly reversible. It will lead to reduction in ionic calcium (10% to 20%), compared to about 5% in pasteurization. This may be partly responsible for reduced susceptibility of UHT milks

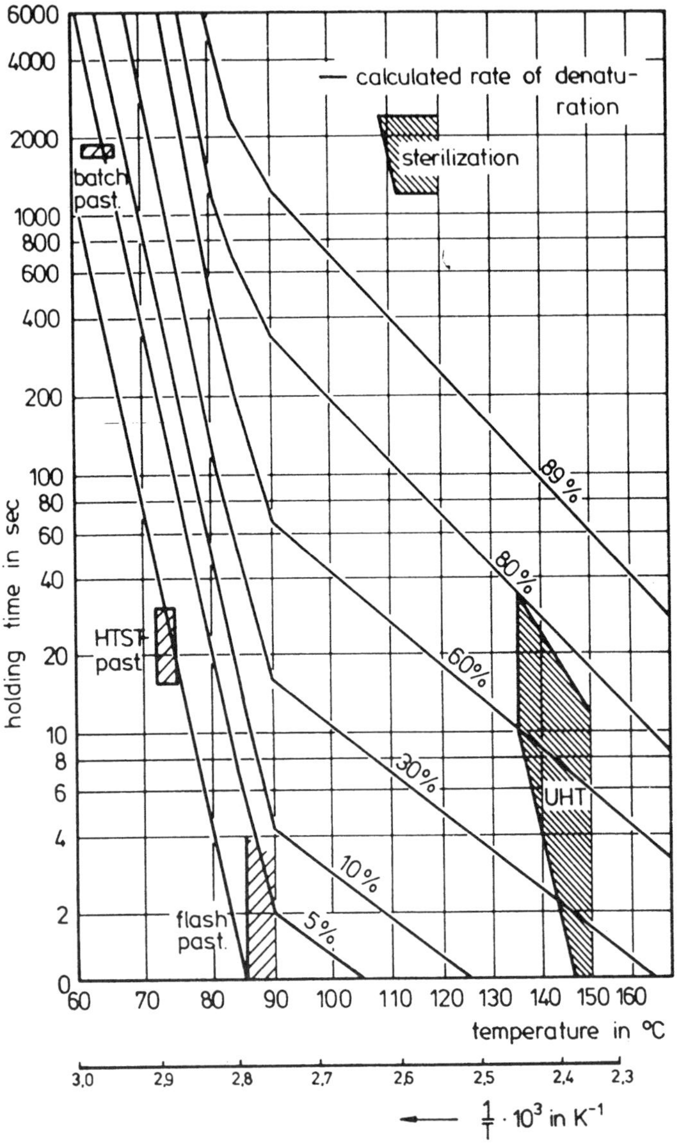

Figure 6–8 Calculated degrees of denaturation of whey proteins (β-lactoglobulin A and B, α-lactalbumin, BSA, immunoglobulin, proteose peptone) in skim milk. *Source:* Reprinted with permission from H.G. Kessler, Effect of Thermal Processing on Milk, in *Developments of Food Preservation*, Vol. 5, S. Thorne, ed., pp. 80–129, © 1989, Verlag A. Kessler.

to coagulation by rennet. This may also help reduce susceptibility to gelation during storage. However, it is generally regarded that there is no mineral loss and that bioavailability of minerals is not affected. More detail is provided by Dalgleish (IDF, 1989) and Holt (IDF, 1995).

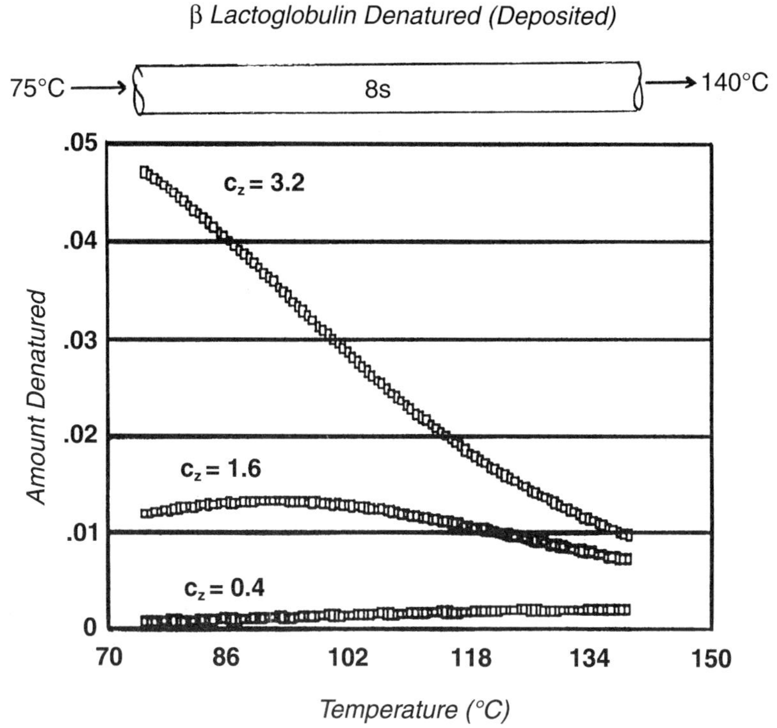

Figure 6–9 Denaturation profile at long, intermediate, and short heating times during passage through a tube.

6.4.7.3 Vitamins

Burton (1988) has reviewed many publications dealing with the effects of UHT processing on vitamins in milk. He pointed out that the published research data are sometimes difficult to interpret. One reason for this is the fact that many of the processes used would have varied in their severity, and this was not fully accounted for in the early work. A second reason is that changes undoubtedtly occur during storage; these are temperature sensitive and some vitamins are also affected by exposure to light and dissolved oxygen. Table 6–7 provides a good summary: it compares the losses of different vitamins during UHT processing of milk with those occurring during HTST pasteurization and in-container sterilization. Burton (1988) concluded that vitamin losses caused by UHT processing are small.

Lewis (1994) summarizes that the fat-soluble vitamins (A, D, and E) and the water-soluble vitamins of the B group (pantothenic acid, nicotinic acid, and biotin) are hardly affected by the UHT processes; riboflavin is stable to heat but not to light. Up to 20% of thiamin and 30% of vitamin B_{12} are lost in UHT processing, whereas losses during in-bottle sterilization are up to 50% for thiamin and 100% for vitamin B_{12}. Considerable work has been done on

Table 6–7 Vitamin Content of Milk, and Typical Percentage Losses Caused by Heat Treatment

Vitamin	Raw Milk Content/100 ml	Loss (%)		
		Pasteurized	Sterilized	UHT
Thiamin	45 μg	<10	30	10
Riboflavin	180 μg	NS	NS	NS
Nicotinic acid	80μg	NS	NS	NS
Vitamin B_6	40 μg	<10	20	10
Vitamin B_{12}	0.3 μg	<10	<90	10
Pantothenic acid	350 μg	NS	NS	NS
Biotin	2.0 μg	NS	NS	NS
Folic acid	5.0 μg	<10	50	15
Ascorbic acid	2.0 mg	20	90	25
Vitamin A	30 μg	NS	NS	NS
Vitamin D	22 μg	NS	NS	NS
Vitamin E	86 μg	NS	NS	NS
β-carotene	17 μg	NS	NS	NS

NS = not significant;
Pasteurized = 72° for 15 seconds;
Sterilized = 115°C for 30 minutes.
Source: Reprinted with permission from H. Burton, *Ultra-High-Temperature Processing of Milk and Milk Products*, p. 259, © 1988 (derived from J.E. Ford and S.Y. Thompson, New Monograph on UHT Milk, IDF Bulletin No. 133, © 1981, International Dairy Federation).

thiamin inactivation; a level of inactivation of 3% would be equivalent to a C* value of 1.0. Again, if the C* for a process is less than 1, thiamin loss is very low (less than 3%). As mentioned, some cases of reported losses of thiamin have exceeded 30%; this suggests that these milks have been very severely treated. Kessler (1989) has reviewed thiamin inactivation in more detail.

One of the more apparently heat labile of the vitamins is vitamin C. This is often used as an index for nutritional loss during processing. Although milk is not a major source of vitamin C in the UK diet, compared to fruit and potatoes (see Table 1–2), it is present in measurable amounts. It will also be present in other foods subject to thermal processes, such as fruit and vegetable juices. Vitamin C is present in two forms, as ascorbic acid (the reduced form) and dehydroascorbic acid (the oxidized form); both forms are active as vitamins. In freshly expelled milk it is present predominantly as ascorbic acid. However, it is readily oxidized to dehydroascorbic acid by dissolved oxygen in the product. The oxidized form is more heat labile, with the product being degraded to 2,3-diketogluconic acid, which has no vitamin activity. Thus, the loss of vitamin C will depend upon the relative proportions of the two forms at the time of processing (i.e., the status of the vitamin C in the raw material). This is not usually known or established in the bulk of the cases. In general, other vitamins may be present in more than one form and present similar problems of analysis (Section 9.6.3).

This would be of crucial importance in foods that are major sources of vitamin C in food processing. For example, fresh raw milk is usually saturated with oxygen. The level of dissolved oxygen falls during storage, consumed mainly by organisms; some may be used to oxidize vitamin C. Bearing this in mind, a summary of vitamin C loss in heat-treated milk is given in Table 6–8 Losses are small compared to those that occur during storage (see Section 9.6.3). Andersson and Oste (1995) provide a more recent review of vitamin loss during pasteurization and UHT processing.

6.4.7.4 Fats

UHT processing causes no physical or chemical changes in fat that are likely to have adverse nutritional consequences. Burton (1988) reported losses of free fatty acids compared to raw milk. Most UHT processes involve homogenization, with creation of new "fat globular membrane" material. Also it is reported that UHT processing will denature agglutin proteins, which form part of the fat globule membrane. Modifications to the fat phase can have a pronounced effect on flavor; most of these occur either during the handling of the raw milk or during the prolonged storage period, not during the UHT process itself.

6.4.8 Other Changes in Milk

A thorough review of the reactions taking place in heated milk that do not involve oxygen has been given by Kessler (1989). These results are presented for a temperature range of 100° to 150°C and a time range of 1 to 6,000 seconds for thiamin inactivation, lysine loss, color changes, and HMF production. A summary of the orders of reaction and activation energies (E_a) for some of these is given below:

	E_a (kJ mol^{-1})
Thiamin (second order)	100.8
Lysine (second order)	100.9
Color changes	116
HMF (zero order)	139

There has always been an interest in indicators that will give a measure of the severity of heat treatment, for quality assurance and legislation purposes, that is for distinguishing be-

Table 6–8 Vitamin C Content of Heat-Treated Milk

Treatment	*Vitamin C (μg/mL)*
Fresh milk	23.3
72°C/16 s	19.4
80°C/16 s	18.3
110°C/3.5 s	17.3
140°C/3.5 s	16.4

Source: Reprinted from M.J. Lewis, in *Modern Dairy Technology*, Vol. 1, 2nd ed., R.K. Robinson, ed., p. 42, © 1994, Aspen Publishers, Inc.

tween UHT and sterilized products or even for distinguishing between direct and indirect UHT processes). It is difficult using any of the naturally occurring components (e.g., vitamin C, see Table 1–2), as one can never be certain of the initial value in the original raw milk. It is better to use components that are not present in raw milk.

One such component of particular interest is lactulose, which is an isomer of lactose (see also Section 10.5.7). It is about four times sweeter than lactose and more soluble. It does not occur naturally in raw milk, but is formed in heated milk (by isomerization). Its importance lies in its increasing reported use as an indicator of the severity of many processes involving milk and milk-based products. In fact, it is appearing in some recent legislation. The detailed chemistry of the reaction and the effect of factors such as protein, citrate, and phosphate on its formation have been reviewed by Andrews (IDF, 1989). It can now be measured by HPLC, which is much simpler than the complex enzyme assays used by Andrews in his pioneering investigations. This may have contributed to its much wider use as a suitable indicator of heat treatment. Typical levels in heat-treated milks are given in Section 10.5.7, as well as further discussion on lactulose as an indicator of the severity of heat treatment. It was not found possible to distinguish between UHT and sterilized milks using lactulose in all cases (Andrews, 1989). Pellegrino (1994) suggested that fat protected various components against heat-induced reactions. For example lactulose formation was 26% less in whole milk (4% fat) subject to direct-steam injection than in skim milk. In the indirect-UHT process, it was only 4% less. There was also less whey protein denaturation and less furosine production. It was concluded that the heat load depends upon the fat content, with the higher-viscosity fat-containing products hindering heat transfer. In contrast, Konna et al. (1990) observed that heat-induced reactions were greater in milk containing 3.5% fat compared to those in 1.0% fat.

Since concern has been shown for inactivation of HRS (see Section 10.3.3), it has been suggested that for HRS inactivation, lactulose levels of greater than 400 mg/kg are to be expected. Surveys have shown that 50% of indirect-UHT milks surpassed a lactulose limit of 400 mg/kg. Tentative proposals for lactulose levels in UHT milk (>100 mg/L) and sterilized milk (>600 mg/L) are being discussed (Wilbey, 1996). Some of the other indicators have been reviewed by Kessler (1989) and IDF 238 (1989).

6.4.9 Sensory Characteristics

6.4.9.1 Flavor

Flavor changes in milk arise because of changes in its chemical constituents. The flavors of particular interest in heated milks are the "cooked" flavor and the bitter and oxidized flavors that may develop during long-term storage. The picture is further complicated because the cooked flavor, which is produced by heating as a result of the production of volatiles such as hydrogen sulfide, methyl sulfide, and others (Section 9.5.1), changes rapidly during the early days of storage. Also the vocabulary used for describing flavors is not straightforward, and terms such as cooked, cabbagey, sulfury, and caramelized are widely used. Flavor changes during processing and storage are discussed in more detail in Chapter 9.

6.4.9.2 Color

UHT milk is often described as whiter than raw milk, arising from a reduction in fat globule size and an increase in casein micelle size, both of which alter its light-scattering properties. This is especially the case with goat's milk. Milk is not subject to any significant browning during normal UHT processing, although changes during storage may be very significant (see Chapter 9). Changes may also become more pronounced as the time/temperature profile is extended.The *z*-value for browning has been taken as 26.2°C and the activation energy as 107 kJ mol^{-1}.

Using these values, Table 6–9 has been compiled to show the relative extents of browning to be expected from different heating rates (2 to 120 seconds) for a final product temperature of 140°C. For a plant where it takes 120 seconds to heat from 80° to 137°C, the amount of browning is 60 times higher, according to the kinetics, than if the process took place in 2 seconds, although in reality the perceived difference might be very small. The amount of browning taking place during this heating period will also be much more extensive than a holding period of several seconds at 137°C. Browning during UHT processing is also compared with both that taking place during in-container sterilization and during storage (Exhibit 9–1). Browning may become significant as C* increases, as pH increases, or if additional sugars are added, glucose being the most reactive. Lactose-hydrolyzed milk will also be more susceptible to browning than normal milk.

6.4.9.3 Texture

UHT processing does not significantly increase the viscosity of milk. The texture of milk is related to its mouthfeel. Blanc and Odet (1981) summarized the factors affecting texture and methods for assessing some of these factors. The main defects are the separation of fat

Table 6–9 Browning at Different Heating Rates and Temperatures

Temperature Range (°C)	*Time (s)*	*Heating,* z = 26.2°C*	*Browning,† Arrhenius 10^{-14}*	*Lethality, F_0*
80–137	2	1.66	1.04	0.11
	10	8.3	5.2	0.53
	30	24.9	15.6	1.58
	60	49.8	31.2	3.16
	90	74.7	46.8	4.74
	120	99.6	62.5	6.33
80–130	90	45.6	29.8	1.07
	120	60.8	39.3	1.43
80–120	90	23.2	15.5	0.13
	120	30.9	20.7	0.18

* Browning expressed as an equivalent time (s) at 121°C.
† Browning expressed as extent of reaction: $\int e^{-E/RT}\,.dt$

and the formation of a sediment that contains protein, fat, lactose, and minerals in varying proportions. The consensus of opinion is that more sediment is produced in milk from a direct heating process, but the use of chemicals that provide additional anions, such as citrate, bicarbonate, or hydrogen phosphate, can be beneficial. The total amount of sediment increases during storage, but the protein content (of the sediment) decreases over the first 5 weeks and then rises again.

6.4.10 Enzyme Inactivation

Enzyme inactivation at pasteurization conditions was dealt with in Section 5.4.1. The main focus in this section is on those enzymes that may survive UHT treatment, namely proteases and lipases. These may be present in the raw milk. Both proteases and lipases are naturally occurring in raw milk (indigenous). Also, both these enzymes can be excreted into raw milk by psychrotrophic bacteria. This normally occurs when their count exceeds 10^6 cfu/ml. There is considerable difference both in their mode of action and their stability to heat between the indigenous and bacterial enzymes.

6.4.10.1 Lipase Activity

Lipases will hydrolyze triglycerides in fats, which will result in the production of free fatty acids. In typical milk at pH 6.7, they will be more dissociated and present in their sodium or potassium forms (soaps). Off-flavors associated with lipolysis are described as rancid or soapy. The fat globular membrane affords some natural protection for the milk fat against lipolysis. Damaging the membrane will give the enzyme access to its substrate. Damage can occur by excessive agitation of raw milk, especially when it is cold, and by aeration or foaming. The presence and action of lipases in raw milk can be demonstrated by homogenizing raw milk at about 35° to 40°C and holding it for less than 30 minutes. Soapy and rancid off-flavors quickly become apparent. The reaction rate is also dependent upon surface area. Therefore, homogenization will drastically increase lipolysis. Levels in raw milk are normally below 0.5 µmol ml^{-1}. Threshold levels have been described as about 1.5 to 2.0 meq/100 g of fat, with trained panelists able to pick up the off-flavor earlier than normal consumers.

Indigenous or native lipases are fairly readily inactivated by pasteurization conditions (see Section 5.4.1). Driessen (1989) reports a D value of 16 seconds at 70°C for lipoprotein lipase. In comparison, bacterial lipases are much more heat resistant. Data are reported at 74°C by Driessen (1989), with D values ranging between 16 and 1,980 seconds for the first inactivation and 985 and 4,000 seconds for the second inactivation. For most of the enzymes there would be over 20% of residual activity following 74°C for 15 seconds. There is little information on their activity at low storage temperatures.

Bacterial lipases appear to be more complicated in their thermal inactivation behavior, showing an initial rapid decrease in activity, followed by a much more gradual decline, suggesting a thermolabile and a thermoresistant conformation. The thermolabile conformation is easily inactivated, but the thermoresistant conformation may well show some residual activity, even after UHT conditions. It is this that is of concern in UHT processing.

Driessen (1989) reported the following D values for the second stage of inactivation of lipase from *Pseudomonas fluorescens* 22F as follows: 110°C (1,800 seconds), 120°C (1,200 seconds), and 130°C (840 seconds), giving a *z*-value of 41.7°C.

Purification of the crude enzyme, as found in milk, has been shown to decrease its heat resistance. Although residual activity may be low, changes caused during prolonged storage may be considerable. Indeed Fox and Stepaniak (1983) isolated a lipase enzyme, which when purified gave a D_{150} value of 22 to 28 seconds.

Bacterial lipases are most active in the temperature range of 30° to 40°C. Some loss of activity does occur during prolonged storage. The pH for optimum activity for bacterial lipases is in the neutral or alkaline range, with suggestions of 6.5 (*P. fluorescens*) and 7.6 (*P. fragi*); another value given is 8.0 (Driessen, 1989). In general, lipase activity is not such a major source of problems as proteolytic activity in UHT products.

6.4.10.2 Protease Activity

Enzymes Exhibiting proteolytic activity are found in raw milk. As for lipases, they can be categorized as indigenous and bacterial. The main class of indigenous proteases are the alkaline proteases, which have their optimum activity about pH 8 and at 37°C. They are much more heat resistant than indigenous lipases and may survive pasteurization conditions. The alkaline protease in bovine milk occurs mainly (greater than 80%) as plasminogen, which is probably identical to plasmin in blood. It is suggested that in fresh raw milk only a small proportion is active, but this becomes more active with storage. It (they) are largely associated with the casein micelle. There are some natural inhibitors present in milk serum that are specific for the proteolysis of casein; these tend to associate with the whey proteins. Denatured β-lactoglobulin is one such inhibitor. There is general agreement that it is a serine proteinase, with trypsinlike activity. It cleaves peptide bonds on the C-terminal side of Arg and Lys residues. It attacks β- and αs-2 caseins at about the same rate. Breakdown products from β-casein are generally called γ-caseins. Proteolysis of αs_2 casein results in positively charged proteins with higher electrolytic mobility. The αs_1 casein is also broken down, but about three times more slowly. Whey proteins and κ-casein are resistant to cleavage by this enzyme. Some activity is lost during cold storage (10% after 4 days at 4°C).

Early studies on heat inactivation were confusing; it was found to be not inactivated at 30 minutes at 60°C, but inactivated at 70°C for 2 minutes. Driessen (1989) found that heat inactivation followed first-order reaction kinetics. However, up to 95°C, residual activity was strongly temperature dependent. Above 110°C, residual activity was much less temperature dependent. Both plasmin and plasminogen activities were determined in UHT milk, directly heated. There were two surprising results. The first was that residual activities were unexpectedly high, for example greater than 40% at 140°C for 4.6 seconds. The second was that residual activity was much more influenced by the holding time than the temperature. The results suggest that inactivation would be much greater in an indirect plant, more so at slower heating rates. This result suggests that enzyme inactivation could be an important quality parameter. Note that, according to the results of the study, milk proteinases would be completely inactivated in milk products sterilized in the bottle.

These data, showing considerable residual activity after UHT processing, contradict earlier data that milk proteinases cannot withstand UHT processing (Nakai, Wilson, & Herreid, 1964; Cheng & Gelden, 1974). Note that proteinase activity cannot be measured accurately from an increase in nonprotein nitrogen (NPN), as the γ-caseins and proteose peptones are precipitated by 12% trichloroacetic acid (TCA) (the method used by Cheng & Gelden). Kohlmann, Nielsen, and Ladisch (1991) showed that addition of low levels of plasmin to UHT milk caused gelation to occur, which further supports a relationship between low levels of plasmin activity and gelation in UHT milk.

6.4.10.3 Bacterial Proteinases

Bacterial proteinases originate from gram-negative bacteria; discrepancies in their optimum activities reinforce the fact that these bacteria produce various extracellular proteinases, hereafter simply called proteinases. They hydrolyse β- and κ-caseins; whey proteins are reported not to be degraded to any extent, although there have been some contrary reports. Note that since bacterial proteinases cause κ-casein hydrolysis, they will also cause gelation to occur; this is a subtle distinguishing difference to the action of indigenous proteases. Their molecular weight is about 40,000 to 50,000. They are metalloproteins, and zinc is important; they are sensitive to chelating agents. They are most active between 37° and 45°C. These proteinases are very thermoresistant and can withstand UHT processing conditions. Most inactivation studies have concentrated on those produced by *P. fluorescens* and *P. fragi*.

Driessen (1989) gives kinetic constants for inactivation of two proteinases, one from *P. fluorescens* 22F and the other from *Achromobacter sp.* 1-10. D values were determined over the temperature range 70° to 130°C. Both enzymes were very stable to heat, with D_{130} values of 660 and 480 seconds, respectively. The *z*-value for both proteinases was about 26°C. Proteinase from *Serratia marascens* D2 showed much more complex behavior. According to the kinetics, bacterial proteinases may survive UHT treatment and even retorting conditions used for sterilized milk. The defects caused in UHT milk by both milk proteinases and bacterial proteinases are summarized in Table 6–10. Burton (1988) reviews some of the heat resistance data for bacterial lipases and proteinases (Table 6–11).

One important quality assurance aspect is to try to avoid problems caused by heat-resistant enzymes. This can be done by avoiding milk containing high enzyme concentrations for UHT treatment. One proposed test for the direct measurement of the activity of bacterial proteases involves incubating them in the presence of luciferase, which catalyzes the reaction of ATP and luciferin to produce light. Luciferase is degraded by the protease and the reduction in light output on addition of luciferin, and ATP can be used as a measure of proteinase activity (Rowe, Pearce, & Crone, 1990). Apart from this, there is no (other) simple direct routine test for measuring enzyme activity and it is made more complex by having two sources of heat-resistant enzymes, which also cause gelation by different mechanisms.

Recio et al. (1996) report that capillary zone electrophoresis was able to differentiate between proteolysis caused by plasmin and by bacterial proteases, by allowing identification of the casein degradation products originating from both types of proteases.

Table 6–10 Survey of Proteolysis in Milk and Milk Products

Proteolysis	*Milk Proteinase*	*Bacterial Proteinase*
Occurrence in	UHT-sterilized milk products	UHT-sterilized milk products
Defects	Bitter off-flavor Transparency of skim milk Glassy appearance of low-fat custard Destabilization	Bitter off-flavor Gelation Separation of whey
Prevention	Sterilization treatment sufficiently high, e.g., 16 s at 142°C	Manufacture of milk of good bacteriological quality Cold storage (<4°C) of the milk for a maximum 3 days at the farm and 1 day at the factory; if not, a thermization treatment of the milk has to be applied Prevention of contamination of the thermized milk Effective cleaning of processing lines

Source: Reprinted with permission from F.M. Driessen, Inactivation of Lipases and Proteases, *Heat Induced Changes in Milk*, IDF Bulletin No. 238, pp. 71–93, © 1989, International Dairy Federation.

Indirectly, bacterial protease activity can be related to psychrotrophic activity for liquid milk. Also, UHT milk itself can be analyzed retrospectively for lipopolysaccharide (LPS) by the limulus assay. LPS is present in the cell wall of gram-negative bacteria and is very stable to heat. Therefore, high assay values will indicate that there were high psychrotrophic counts in the raw milk and may help to further identify the cause of gelation. Another approach is to measure sialic acid in raw milk. It is produced by the action of bacterial proteases on κ-casein. A rapid method of measurement is described by Mottar et al. (1991). Sialic acid was found to be higher in UHT milks compared to pasteurized milks. The levels in UHT milk continued to increase during storage, and it was proposed that the sialic acid content could be used as a quality indicator during storage of UHT milk (Zalazar, Palma, & Cardioti, 1996).

6.4.11 Other Milks

Most of the UHT literature relates to cow's milk; there are some papers on goat's milk and buffalo's milk. Goat's milk presents a real problem for UHT processing; compared to cow's milk it is much more susceptible both to sediment formation and fouling. The level of ionic calcium is much higher in goat's milk, and it also has a much lower alcohol stability (Horne & Parker, 1982). Zadow et al. (1983) studied the stability of goat's milk to UHT processing and concluded that either pH adjustment to well above 7.0 or the addition of 0.2% disodium phosphate before processing was necessary to reduce sedimentation. Both these treatments would reduce ionic calcium. Addition of calcium chloride was found to aggravate the problem. Kastanas, Lewis, and Grandison (1996) compared the fouling behavior of a heat ex-

Table 6–11 Heat-Resistance Data for Some Heat-Resistant Enzymes in the Ultrahigh Temperature Region

Microorganism	*Temperature (°C)*	*Decimal Reduction Time (min)*	Q_{10}	*z (°C)*	*Reference*
Proteases					
P. fluorescens	120	4	3.2	20	1
Pseudomonas	149	1.5	2.0	32.5	2
Pseudomonas	150	1.7	2.1	32	3
Pseudomonas	150	0.5	2.0	32.5	4
P. fluorescens	130	11	1.9	34.5	5
P. fluorescens	150	27	2.3	28.5	6
Lipases					
Pseudomonas	150	1.7	2.5	25	3
Pseudomonas	160	1.25	1.9	37	7
Micrococcus	160	1	1.4	63	7
P. fluorescens	140	3.6	1.9	37	8
Pseudomonas	150	1.0	1.8	38.5	9
P. fluorescens	150	0.5	1.7	42	10

1. Mayerhofer, M.J., Marshall, R.T., White, C.H., & Lu, M. (1973). *Applied Microbiology* **25**, 44.
2. Adams, D.M., Barach, J.T., & Speck, M.L. (1975). *Journal of Dairy Science* **58**, 828.
3. Kishanti, E. (1975). *Bulletin of the IDF*, Document 86, p. 121. Brussels: International Dairy Federation.
4. Barach, J.T., Adams, D.M., & Speck, M.L. (1976). *Journal of Dairy Science* **59**, 391.
5. Driessen, F.M. (1976). *Zuivelzicht*, **68**, 514.
6. Andrews, A.T., Law, B.A., & Sharpe, M.E. Personal communication. See Burton, H. (1977). Journal of the Society of Dairy Technology **30**, 135.
7. Hedlund, B. (1976). *North European Dairy Journal* **42**, 224.
8. Andersson, R.E., Hedlund, C.B., & Jonnson, U. (1979). *Journal of Dairy Science* **62**, 631.
9. Adams, D.M., & Brawley, T.G. (1981). *Journal of Dairy Science* **64**, 1951; *Journal of Food Science* **46**, 673; **46**, 677.
10. Fox, P.F., & Stepaniak, L. (1983). *Journal of Dairy Research* **50**, 77.

Source: Reprinted with permission from H. Burton, *Ultra-High-Temperature Processing of Milk and Milk Products*, p. 58, © 1988 (based on O. Cerf, UHT Treatment, *Factors Affecting the Quality of Heat Treated Milk*, IDF Bulletin No. 130, pp. 42–54, © 1981).

changer with goat's milk and cow's milk and showed that for goat's milk it fouled much faster. They also showed that a number of measures taken to reduce ionic calcium also reduced its susceptibility to fouling. Buffalo's milk is regularly UHT treated in India; it is not susceptible to any specific UHT processing problems, despite its much higher total solids and low alcohol stability, compared to cow's milk (see Section 8.4.3). Changes during storage are reviewed in Chapter 9.

In many parts of the world UHT milk is made from reconstituted or recombined milks. Additional problems may arise from the quality of the milk powder: from the point of view of off-flavors, high bacterial spore count, poor heat stability, and residual enzymes. The problem of detecting enzyme activity also arises when using milk powder for UHT-reconstituted milk.

Cellestino, Iyer, and Roginski (1997) produced whole milk powder from raw milk that was fresh and 2 days old. They also stored the milk powder for up to 8 months at 25°C prior to reconstitution and UHT processing. As the milk powder got older, the amount of sediment in the UHT milk increased, and the fall in pH during storage was larger. Also rates of enzymatic and oxidative reactions were greater, leading to stale, powdery, and other off-flavors, which help distinguish it from UHT milk made from fresh milk. They suggested that the taste of reconstituted milk was affected more by lipolysis than proteolysis. There were also some differences between the UHT milks made from powders from the fresh and the 2-day-old milk. These differences may have been more noticeable if the milks had been kept for longer periods. Other experiments showed that powder made from good-quality raw milk that has been kept for 7 days at 4°C before being made into powder showed no differences to powder produced from the same milk when it was only 1 day old (unpublished results). Suggestions for reducing fouling when using reconstituted milks are given in Section 8.4.8.

Lactose-reduced milks are available in some parts of the world and are particularly suitable for those who are lactose intolerant (this includes cats). They are produced by addition of the enzyme β-galactosidase, which hydrolyzes the lactose to glucose and galactose, making the milk slightly sweeter. Such milks will be subject to slightly more browning both during UHT processing and subsequent storage. The enzyme can be added either before or after heat treatment. In the production of milk for cats, a sterile enzyme solution is aseptically added during the cooling period. The aseptic tank acts as a mixing vessel and an enzyme reactor.

Heat treatment of calcium and magnesium-enriched milks will pose a special problem, as these divalent ions reduce heat stability and increase the susceptibility to fouling and also gelation during storage. Addition of calcium and magnesium has been found to reduce the denaturation temperature of whey proteins (in whey protein concentrates), whereas sodium increased the denaturation temperature (Varunsatian et al., 1983). UHT skim milk with up to 40% extra calcium has been produced by adding sodium citrate as a stabilizer to milk adjusted to pH 7.3 and enriched with calcium lactate. Heat treatment was at 139°C for 4 seconds. (Guamis Lopez & Quevedo Terre, 1996).

6.5 CREAM

Creams come in a range of fat contents, from 9% to over 50% (see Exhibit 5–3). Traditionally cream was sterilized in the container and produced by a process involving preheating, homogenization, filling, seaming, and sterilizing at not less than 108°C for not less than 45 minutes. Further details are provided by the Society of Dairy Technology (SDT) (1989). These have been substantially replaced by UHT cream over the last 20 years. The UK UHT cream regulations stipulate a temperature of 140°C for a time of 2 seconds. Other aspects covered in the regulations are similar to those for milk. UHT cream may also contain certain additives: in the UK it may contain (1) calcium chloride or (2) the sodium or potassium salts of carbonic acid, citric acid, or othophosphoric acid, the maximum concentration of the total of all these ingredients being 0.2%.

Cream is processed on the same equipment that is used for milk, both direct and indirect plants. The IDF (1996) advocates turbulent flow to ensure good mixing and to reduce burn-on, although this is not so easy to achieve compared to milk, because of its higher viscosity.

Additional problems with UHT-sterilized cream arise from the fat phase and are related to stability, plugging, feathering, and disturbance of the milk fat globule (MFG) membrane by homogenization. The problems caused by these fat-related defects become more apparent in UHT cream compared to pasteurized cream because it is generally stored for longer periods.

Homogenization will take place between 50° and 70°C. The IDF (1996) recommends downstream homogenization at about 70°C. It is not easy to find published data on homogenization pressures: low-fat creams will require high homogenization pressures. One problem encountered with low-fat UHT creams is feathering, which occurs when cream is poured on hot acidic beverages, such as coffee; this is the appearance of white floccules or clots that float on the surface. This was investigated by Anderson, Cheeseman, and Wiles (1977), who used a feathering test, using acetate buffer to simulate hot coffee. Resistance to feathering can be increased by a reduction in calcium activity (presumably ionic calcium), for example, by ion exchange or by addition of stabilizers such as sodium citrate or by increasing the amount of casein, which provides more buffering action when cream is added to coffee. Feathering was not found to be influenced by the homogenization pressure (Anderson et al., 1977). Geyer and Kessler (1989) reported that stability to feathering could be improved by ensuring that the fat globules were covered by denatured whey protein; this could be done by an appropriately chosen combination of homogenization and heating conditions.

For a thicker single cream with a good pouring consistency, a pressure of 170 bar would be required. Such a cream may also feather in coffee, so double-stage homogenization at a slightly lower pressure would be used.

Fresh whipping cream is not homogenized but some homogenization is required for UHT whipping cream, A compromise is struck between impairing whipping ability and preventing separation. Pressures used are in the range 35 to 70 bar, with single- or double-stage homogenization. Additives are also permitted. Homogenization efficiency can be assessed by measuring fat globule distribution. One of the major defects during storage is clumping. Some unpublished data collected by the author are presented for cream clumping (Figure 6–10).

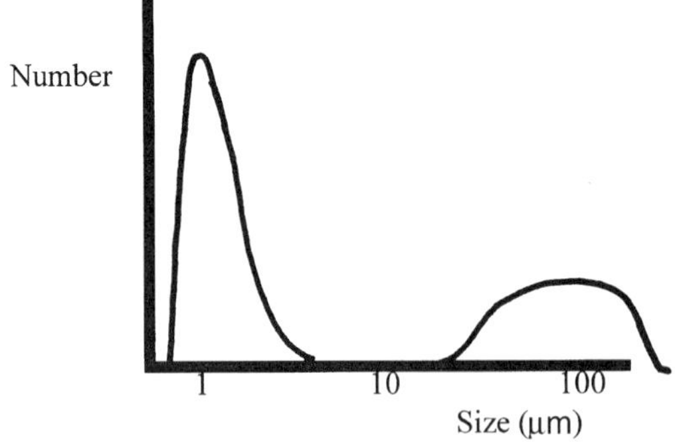

Figure 6–10 Cream clumping

UHT double cream (48% fat) will require very low homogenization pressures. Excessive agitation and homogenization of raw milk will cause release of free fat.

One interesting more recent development is the aerosol can for creams in the fat range 28% to 35%. These creams may also contain a number of additional additives, such as sodium alginate, sodium carboxymethylcellulose, carrageenan, gelatine, sugar, nitrous oxide (as the propellant), and glyceryl monostearate. For special-occasion products, alcohol is also added. One feature of these creams is their high overrun, which can be up to and over 400%.

Ranjith (1995) has looked at improving the stability of UHT cream. He showed that calcium-reduced creams (from about 23 to 17 mM) produced by ion exchange had an improved alcohol stability and resistance to feathering and did not thicken during storage. The method of calcium removal was important, with the best stability resulting when calcium was removed from the permeate, rather than the skim milk fraction.

It is my impression that cooked flavor is less of a problem in cream than it is in milk. Flavor changes in cream during storage are discussed in the *Cream Processing Manual* (SDT, 1989). Corzo, Villamiel, and Martinez Castro (1996) assessed the usefulness of free carbohydrates and whey protein denaturation as thermal treatment indicators in UHT creams (10% to 35% fat). Values (mg/100ml) of lactulose (7 to 177), α-lactalbumin (5.1 to 6.16), and β-lactoglobulin (1.0 to 64.0) were found in the creams and were similar to those levels found in milk.

There has recently been an increase in the sale of synthetic (vegetable) creams, many of which are sold as UHT products. Such whipping creams require less fat (15% to 20%) compared to dairy products. The usual fats are hardened coconut oil or hardened palm kernel oil. Use may be made of sodium caseinate as stabilizer and hydrocolloids for body and flavorings. More information is provided by Bern (1994).

6.6 EVAPORATED MILK AND UHT EVAPORATED MILKS

Concentrated milks are less heat stable then their unconcentrated counterparts, arising from a closer packing of the casein micelles and fat globules. Muir (1984) recommends the following procedures to ensure a good-quality product:

- The bacteriological quality of the raw material must be high, to avoid premature gelation during storage.
- The heat stability of the concentrate must be enhanced by forewarming, to ensure that coagulation does not occur during sterilization. If necessary the pH should also be adjusted.
- Homogenization should be carried out after the heat treatment. If this is not possible, particular attention must be paid to ensure that the stability of the concentrate is adequate to withstand severe heat treatment.
- Finally, to avoid age-gelation, polyphosphates should be added or the resulting concentrate should be further heat treated.

Two manufacturing schemes are proposed based on these recommendations (Exhibit 6–1). Scheme A is designed to produce a product with a minimal heat treatment and a "fresh"

Exhibit 6–1 Potential Routes for the Manufacturing of Sterile Concentrated Milk

A "Fresh" Milk	**B "Cooked" Milk**
Raw milk	Raw milk
↓ Addition of fat or skimmed milk	↓ Addition of fat or skimmed milk
Standardized milk	Standardized milk
↓ Heat 135°–145°C, 15 s	↓ Heat 90° C, 10 min or 135°C–145°C, 15 s
Forewarmed milk	Forewarmed milk
↓ Concentrate × 2.5 to × 3.0	↓ Concentrate × 2.5 to × 3.0
Concentrated milk	Concentrated milk
↓ Add 0.14% sodium hexametaphosphate	↓ Heat 142°C, 15 s; cool
Stabilized concentrate	Sterile concentrate
↓ Heat 142°C, 5 s	↓ Reheat concentrate 115°C, 20 min
Sterile concentrate	Stabilized concentrate
↓ Homogenize 3,000 + 500 psi; 60°–90°C	↓ Homogenize 3,000 + 500 psi; 60°C–90°C
Aseptic homogenization	Aseptic homogenization
↓ Pack in cartons	↓ Pack in cartons
Packed product	Packed product

Source: Reprinted with permission from D.D. Muir, UHT Sterilized Milk Concentrate: View of Practical Methods of Production, *Journal of the Society of Dairy Technology*, Vol. 37, No. 4, p. 40, © 1984, Society of Dairy Technology.

flavor, whereas scheme B results in a product that is close in flavor to conventional evaporated milk. Irrespective of the scheme used, the experimental evidence suggests that the shelf life is in the order of 6 months at 20°C . Longer-term storage may lead to flavor defects.

6.7 ICE CREAM MIX

Heat treatment regulations (UK) for UHT ice cream mix require that it be treated at 149°C for 2 seconds. These conditions are much more severe than for most other milk-based products, and it is difficult to find the experimental basis for how these conditions were derived. However, ice cream mix has higher total solids than most products (except some creams), containing between about 30% and 35% total solids, with up to 12% sugar. If the mix is to be used without being aseptically packed, it should be cooled to below 7.2°C within 1.5 hours. Practical experience has shown that UHT ice cream mix can be very susceptible to browning. This can be reduced by modification of the formulation by removal of those components that are more reactive to browning (e.g., whey powder compared to full-cream milk powder, or glucose compared to sucrose) or by reduction in the processing conditions. Certainly it is possible to produce an acceptable-quality UHT ice cream mix using indirect processing methods, both on plate heat exchangers and on tubular heat exchangers, although the high total solids may favor the use of the latter types. The use of direct steam injection would further reduce the cooked flavor as well as the amount of browning. Again, compositional factors that influence the extent of browning during production will also influence further browning reactions taking place during storage.

6.8 SOY MILK AND OTHER MILK ANALOGUES

Soy milk, which is traditionally consumed in the Far East, has increased in popularity in the United States and other Western countries in the last 10 years. The advent of UHT processing and aseptic packaging has helped improve distribution and contribute to its increasing popularity. Flavored varieties are also available. UHT processing may take place between 135° and 150°C for few seconds, using indirect or direct processing methods.

Basically, soy milk is an aqueous extraction of ground whole soy beans or soy flour. The beans are soaked, ground, and blanched and the soluble components are extracted. The four main extraction techniques are the traditional method, the hot-water grind, the Illinois process, and the rapid hydration hydrothermal cooking (RHHC) process (Exhibit 6–2). These have been compared by Kwok and Niranjan (1995). It is neutral in pH and typical analysis is shown in Table 1–2. The descriptive term *milk* is used because the beverage has approximately the same proximate composition to cow's milk. However, the proteins, fat, and carbohydrates are totally different, as are its sensory characteristics. However, like cow's milk, it is susceptible to spoilage by microorganisms and will undergo acid curdling as well as

Exhibit 6–2 Summary of Soy Milk Process

Traditional	**Hot Water Grind** (Wilkens et al., 1967)	**Illinois** (Nelson et al., 1975)	**RHHC** (Johnson et al., 1981)
SB	SB	SB	SB
↓	↓	↓	↓
Soak overnight	Soak overnight	Soak overnight in 0.5% $NaHCO_3$	Grind SB to flour
↓	↓	↓	↓
Grind	Grind SB with boiling water	Blanch at 100°C, 10–20 min	Slurry in water
↓	↓	↓	↓
Filter	10-min hold at 80–100°C	Grind	Cook at 154°C for 30 s by steam infusion
↓	↓	↓	↓
Cook at 93°–100°C for 30 min	Filter	Heat slurry to 82°C	Cool
		↓	↓
		Homogenize at 3,500 psi and 500 psi	Centrifuge
		↓	
		Add water to give 12% solid	
		↓	
		Neutralize	
		↓	
		Add sugar, salt, and flavoring	
		↓	
		Homogenize	

Note: SB = soy beans; RHHC = rapid hydration hydrothermal cooking.

Source: Reprinted with permission from K.C. Kwok and K. Niranjan, Effect of Thermal Processing on Soymilk, *International Journal of Food Science and Technology*, Vol. 30, p. 266, © 1995, Blackwell Science Ltd.

being subject to proteolysis on standing. These can be controlled by either pasteurization or sterilization.

Heating processes are involved at several stages in the process, including pretreatment of the beans and extraction to produce the soy milk, followed by either pasteurization or sterilization to increase its shelf life. The final quality will depend upon the sum effects of all these heat treatments. Thus for UHT products, it is important to inactivate the heat-resistant spores that may be present, including the mesophilic spore formers (*Clostridium botulinum* and *Clostridium sporogenes*) and the thermophilic spore formers (*Bacillus stearothermophilis*). This latter organism has been isolated from soy milk, and its D value determined as 2.76 minutes (121°C) and 1.39 minutes (125°C), with a *z*-value of 8.36°C (Chung et al., 1988).

Another major problem is the beany flavor, which arises from the action of lipoxygenase on the linoleic and linolenic fractions, resulting in the formation of aldehydes, ketones, and alcohols. This is most likely to occur at the grinding and extraction stage and can be controlled by appropriate heat treatment at that stage. Lipoxygenase inactivation has been reviewed by Kwok and Niranjan (1995). A third problem is to reduce trypsin inhibitors (TI), which are growth inhibitors and are present in raw soy beans. Usually, the aim is to reduce the original activity by at least 90%, as it should not then interfere with the biological value of the protein. The heating conditions to achieve this by UHT are quite severe (more severe than those normally used for UHT cow's milk) and may adversely affect some of the vitamins, as well as increasing the Maillard reaction and denaturing more protein, which will give rise to further heat-induced flavors. These reactions have not been studied anywhere nearly as extensively in soy milk, compared to cow's milk. However, it should also ensure that the UHT soya milk is adequately processed and not subject to an excessive beany flavor.

In general, trypsin inhibitor is more heat labile as pH increases. Kwok, Qin, and Tsang (1993) heated soy milk at high temperatures at three different pH values, namely 2, 6.5 and 7.5. At 95°, 121° and 132°C, trypsin inhibitor activity was considerably reduced as the pH increased. However, at 143°C and 154°C, the effect of pH on thermal inactivation was less pronounced. The kinetic parameters for trypsin inhibitors have been reviewed by Kwok and Niranjan (1995). The data suggest that there are at least two fractions, one that is heat labile and the other heat resistant. Table 6–12 summarizes the data from this and other work on heat inactivation of trypsin inhibitors in soy milk (Kwok & Niranjan, 1995). More recently Kwok (1997) has presented data for quality changes in heat-treated soy milk (Figure 6–11), which show that there is quite a wide range of conditions that will ensure a good color and flavor, over 90% retention of thiamin, and a 90% inactivation of trypsin inhibitor. He also presents a contour plot showing the effects of time and temperature on trypsin inhibitor inactivation (Table 6–12). Pinnock and Arney (1993) observed that there was no difference in sensory responses to flavored soy and cow's milk.

Berger, Bravay, and Berger (1996) describe a patent application for production of an almond milk, based on aqueous dispersions of partially deoiled almond powder, which was subject to UHT treatment. It involves heating and grinding aqueous dispersions (about 8%) at 90°C to obtain solubilization, followed by centrifugation to remove coarse particles, UHT treatment, followed by homogenization during the cooling process. Many other plant extracts have been produced, such as cowpeas and lupins. Few have been subject to UHT processing.

Table 6–12 Reported Kinetic Parameters for the Heat Inactivation of Trypsin Inhibitors in Soy Milk

pH	*Temperature Range (°C)*	E_a *(kJ mol⁻¹)*	*D* (s)*	*Reference*
6.8	93–121	77.4	4,200 (83°C)	Hackler et al. (1965)
6.7	99–154	48.4	3,600 (99°C)	Johnson et al. (1980)
			280 (121°C)	
			165 (132°C)	
			100 (143°C)	
			40 (154°)	
6.5	93–154	60.7	3,600 (93°C)	Kwok et al. (1993)
			360 (121°C)	
			150 (132°C)	
			56 (143°C)	
			23 (154°C)	

* D is the decimal reduction time at the specified temperature, taking into account the holding time only. The coming-up time as well as the cooling time are different in each of the above references.

Source: Reprinted with permission from K.C. Kwok and K. Niranjan, Effect of Thermal Processing on Soymilk, *International Journal of Food Science and Technology*, Vol. 30, p. 279, © 1995, Blackwell Science Ltd.

Chocolate peanut beverages containing 3.0% protein, 5.2% fat, and 9.6% carbohydrate were subject to a UHT process of 137°C for 4 and 20 seconds. The products were commercially sterile, with no bitter flavors. The higher-holding-time sample was slightly darker and less viscous and produced slightly more sediment. Both products were acceptable (Rustom, Lopez-Leiva, & Nair, 1996). Both chocolate- and strawberry-flavored peanut beverages were of good quality after 5 months of storage at 37°C. Changes occurring during storage are discussed in more detail in Chapter 9.

Production of coconut milk is described by Seow and Gwee (1997). Heat-treated coconut milk is available throughout Southeast Asia. Its composition was determined to be solids, 15.6% to 24%; fats, 11% to 18.4 %; protein 0.3% to 0.9%; and carbohydrate, 3.5% to 8.1%. Pasteurized coconut milk (72°C for 20 minutes) has a shelf life of no longer than 5 days at 4°C. Problems arising from sterilization relate to instability during heating and cooling. Instability can arise from curdling or separation, which can be mitigated by selection of appropriate emulsifiers and stabilizers and by two-stage homogenization. It was also helped by preheating and for in-container sterilization products by agitating retorts. It is available as UHT products; it is susceptible to fouling and to chemical and enzymatic deterioration, primarily through lipid oxidation and lipolysis, resulting in strong off-odors and soapy off-flavors.

Schemes for the production of beverages from roasted, partially defatted peanuts are described by Hinds, Beuchat, and Chinnan (1997). These products were more suited to pasteurization rather than sterilization processes, as the latter resulted in more off-flavors, a grittier rather than a smooth texture, and higher viscosities than normally associated with milk-type beverages.

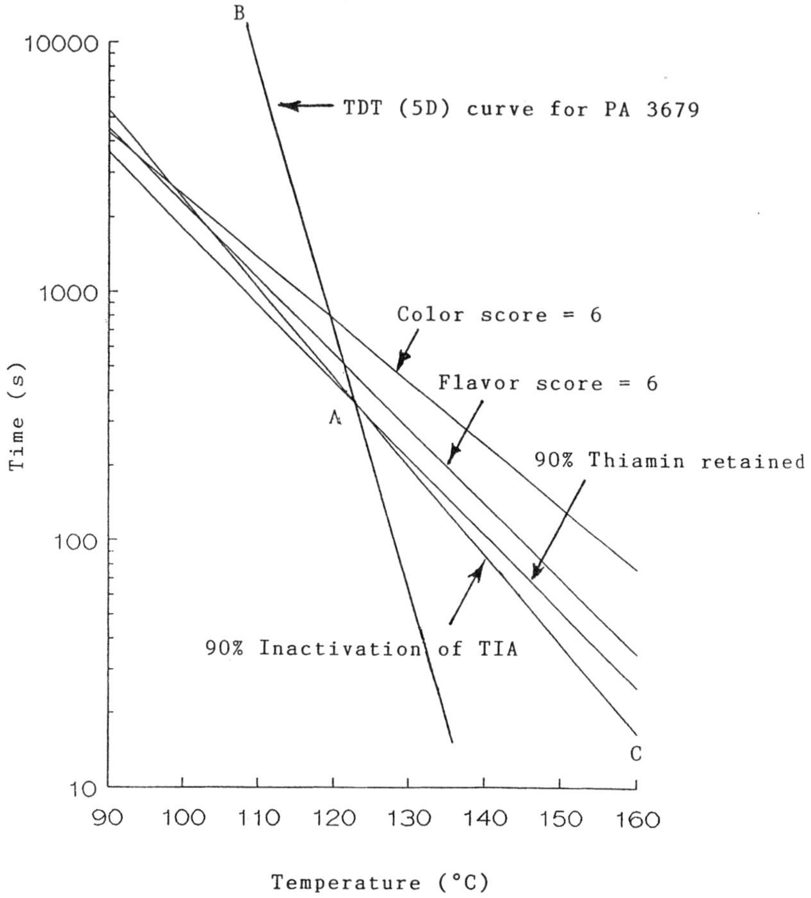

Figure 6–11 Relationship between heating time and temperature in soy milk processing for (A) 5D destruction of the spores of putrefactive anaerobe (PA) 3679, (B) 90% inactivation of trypsin inhibitor activity (TIA), (C) 90% retention of thiamin, and (D) color and flavor corresponding to a score of 6 in a 9-unit hedonic scale. *Source:* Copyright © 1997, K.C. Kwok.

6.9 STARCH-BASED PRODUCTS: SOUPS AND CUSTARDS

Although most soups are either canned or dried, aseptically processed soups are now available. The main functional component in many soups is starch. The use of starches in UHT products has been reviewed by Rapaille (1995). Natural starches can be modified chemically or physically to alter physical and chemical properties. The more important of these for UHT products are viscosity and its stability to heat, acid, and shear; heat penetration; workability; and storage stability. Viscosity stability during heat treatment can be evaluated using a model parameter. It is defined as the ratio of the viscosity before processing to that after processing; the higher the ratio, the better the heat stability. It is usually

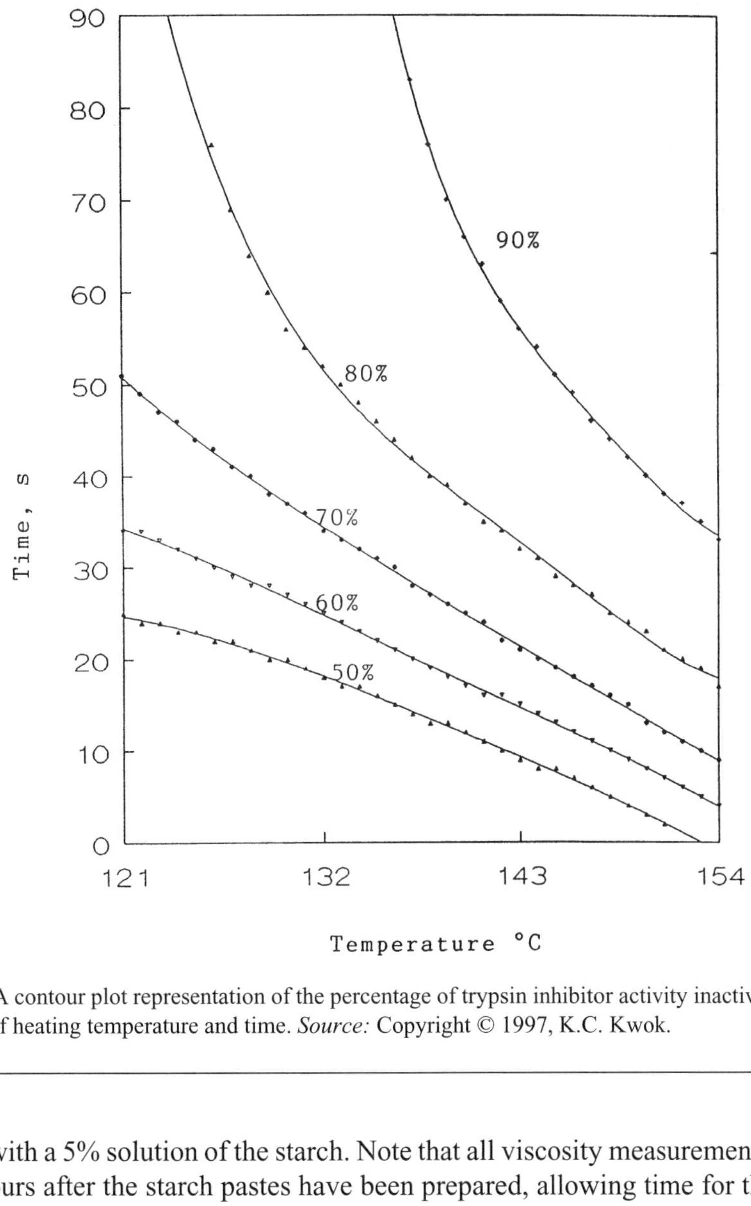

Figure 6–12 A contour plot representation of the percentage of trypsin inhibitor activity inactivated in soy milk as a function of heating temperature and time. *Source:* Copyright © 1997, K.C. Kwok.

performed with a 5% solution of the starch. Note that all viscosity measurements are taken at 20°C, 24 hours after the starch pastes have been prepared, allowing time for the starches to swell fully.

Rapaille (1995) describes some work performed on a creamy dessert, using a number of modified starches, on both indirect (plate) and direct-steam injection systems. In terms of performance, it was obvious that a highly cross-bonded and hydroxypropylated waxy corn starch worked best, followed by a waxy corn starch, having the best stability toward severe heat and shear conditions. Medium cross-bonded starches performed well in plate heat exchangers. However, the use of low cross-bonded starches is not recommended because of their excessive viscosity buildup in the plant, resulting in rapid fouling (Chapter 8).

In studies on scraped-surface heat exchangers, with particulates, slightly different results were found. In general, the viscosity was lower for direct systems than for indirect systems, although the tastes were cleaner.

The choice of stabilizers is also very important with these systems to ensure good textural characteristics. Carrageenans are widely used in flavored milks and thickened milk for their ability to be able to thicken products, to stabilize emulsions, and to prevent suspended particles (for example, in cocoa powder) from separating.

Some recent studies with soups have been involved in the evaluation of different rapid methods for quality evaluation, including direct epifluorescent filter technique (DEFT) and nondestructive testing using ultrasound.

There was some potential for evaluating every package using ultrasound for soups (Wirtanen et al., 1991). However, neither method was suitable without a preincubation step.

6.10 OTHER DRINKS AND BEVERAGES: COFFEES/TEAS

The processing of fruit juices and other acidic products is considered in Chapter 5. These are subject to milder heat treatments than those discussed in this chapter, although aseptic packaging may be required to ensure a long shelf life.

There are also other beverages that are aseptically processed. A wide variety of teas (for example, ice-tea, lemon-tea, and other flavored teas) are available, as well as coffees that are ready to drink. It is difficult to find much technical literature on these products. Chen et al. (1996) noted that the color of oolong tea infusions became darker after UHT treatment. The catechin contents were also changed by UHT processing, the contents of catechin and epicatechin increasing, while epigallocatechin gallate (EGCG) and epigallocatechin (EGC) decreased significantly. Epicatechin gallate was quite stable during UHT processing.

6.11 SPECIALIZED PRODUCTS

Specialized products include infant formulations, nutritional drinks, and other health products, for example, products given to patients with specific medical problems. Particular care needs to be paid to their safety and quality, especially those that are targeted to the more vulnerable in our society, such as the young, elderly, and sick.

REFERENCES

Anderson, M., Cheeseman, G.C. & Wiles, R. (1977). Extending shelf life of UHT creams. *Journal of the Society of Dairy Technology* **30**, 229–232.

Andersson, I., & Oste, R. (1995). Nutritional quality of heat processed milk. In *Heat induced changes in milk* (2nd ed.) pp. 279–307.

Andrews, G.R. (1989). Lactulose in heated milk. In IDF Bulletin, *Heat-induced changes in milk*, No. 238.

Berger J., Bravay, G., & Berger, M. (1996). Method for preparing almond milk and products obtained, PCT International patent application, WO 96/04800 A1.

Bern, M. (1994). Various aspects of producing vegetable whipping cream. *Scandinavian Dairy Information* **8**(3), 42–44.

Blanc, B., & Odet, G. (1981). Appearance, flavour and texture aspects: Recent developments, 25–48. In IDF Bulletin No. 133, *New monograph on UHT milk.*

Burton, H. (1988). *UHT processing of milk and milk products.* London: Elsevier Applied Science.

Cellestino, E.L., Iyer, M., & Roginski, H. (1997). Reconstituted UHT-treated milk: Effects of raw milk quality, powder quality and storage conditions on UHT milk on its physicochemical attributes and flavour. *International Dairy Journal* **7**, 129–140.

Cerf, O. (1981). UHT treatment, 42–54. In IDF (1981) Bulletin No. 130, *Factors affecting the quality of heat treated milk.*

Cheng, W.S., & Gelden, C.S. (1974). Proteolysis in ultra-high-temperature treated and canned 10% cream. *Journal Dairy of Science* **57**, 1502–1504.

Chin Chuang Chen, Shin Hwei Yuo, Ji Bin Sun, & Chu Chin Chen. (1996). Effect of firing on the quality of oolong tea infusion. *Food Science Taiwan* **23**, 308–319.

Chung, J.B., Lee, K.H., John, H.S., & Kim, S.M. (1988). Studies on thermophilic flat-sour bacteria in soymilk: Isolation, identification and determination of heat resistance. *Korean Journal of Food Science and Technology* **20**, 218–224.

Corzo, N, Villamiel, M., & Martinez Castro, I. (1995). Lactulose, monosaccharides and undenatured serum protein contents of commercial UHT creams and their usefulness for thermal treatment assessment. *Food Chemistry* **56**, 429–432.

De Jong, P., Wallewijn R., & van der Linden, H.J.L.J. (1996). Direct heating of milk by infusion, 51–72. In IDF (1996), *Heat treatment and alternative methods*, IDF/FIL No. 9602.

Driessen, F.M. (1989). Inactivation of lipases and proteases, 71–93. In IDF (1989), Bulletin No. 238, *Heat-induced changes in milk.*

European Community Dairy Facts and Figures. (1994). Surrey, UK: Residuary Milk Marketing Board.

Fink, R., & Kessler, H.G. (1988). Comparison of methods for distinguishing between UHT treatment and sterilization of milk. *Milchwissenschaft* **43**, 275–279.

Food and Agriculture Organisation. (1965). *Milk sterilisation.* FAO Agricultural Studies, No. 65.

Footitt R.I., & Lewis, A.S. (Eds.) (1995). *The canning of fish and meat.* London: Blackie Academic and Professional.

Fox, P.F., & Stepaniak, L. (1983). Isolation and some properties of extracelluler heat stable lipases from *Pseudomonas fluorescens* strain AFT 36. *Journal of Dairy Research* **50**, 171–184.

Gaze, J.E., & Brown, K.L. (1988). The heat resistance of spores of *Clostridium botulinum* 213B over the temperature range 120 to 140°C. *International Journal of Food Science and Technology* **23**, 373–378.

Geyer, S., & Kessler, H.G. (1989). Effect of manufacturing methods on the stability to feathering of homogenised UHT coffee cream. *Milchwissenschaft* **44**, 423–427.

Guamis Lopez, B., & Quevedo Terre, J.M. (1996). Calcium enrichment of skim milk subjected to UHT treatment. *Alimentaria* **34**, 79–82.

Hackler, L.R., Van Buren, J.P., Streinkraus, K.H., El-Rawi, I., & Hand, D.B. (1965). Effect of heat treatment on nutritive value of soymilk protein fed to weanling rats. *Journal of Food Science* **30**, 723–728.

Hammer, P., Lembke, F., Suhren, G., & Heeschen, W. (1996). Characterisation of heat resistant mesophilic *Bacillus* species affecting the quality of UHT milk. In IDF (1996), *Heat treatments and alternative methods*, IDF/FIL No. 9602.

Hersom, A.C., & Hulland, E.D. (1980). *Canned foods.* Edinburgh, Scotland: Churchill Livingstone.

Hillier, R.M., & Lyster, R.I.J. (1979). Whey protein denaturation in heated milk and cheese whey. *Journal of Dairy Research* **46**, 95–102.

Hinds, M.J., Beuchat, L.R., & Chinnan, M.S. (1997). Properties of a thermal processed beverage prepared from roasted partially defatted peanuts. *International Journal of Food Science and Technology* **32**, 203–211.

Holdsworth, S.D. (1997). *Thermal processing of packaged foods.* London: Blackie Academic and Professional.

Horne, D.S., & Parker, T.G. (1982). Some aspects of ethanol stability of caprine milk. *Journal of Dairy Research*, **49** 459–468.

IDF Bulletin. (1981). *New monograph on UHT milk*, No. 133.

IDF Bulletin. (1989). *Heat-induced changes in milk*, No 238.

IDF. (1995). *Heat induced changes in milk* (2nd ed.), No. SI 9501.

IDF. (1996). *Heat treatments and alternative methods*, IDF/FIL No. 9602.

IDF Bulletin. (1996). *UHT cream*, No. 315.

Jackson, J.M., & Shinn, B.M. (Eds.). (1979). *Fundamentals of food canning technology: Canned foods.* Westport, CT: AVI Publishers.

Jelen, P. (1982). Experience with direct and indirect UHT processing of milk: A Canadian viewpoint. *Journal of Food Protection* **45**, 873–883.

Jensen, J.A. (1996). Recent developments in direct and indirect UHT systems, 44–50. In IDF (1996), *Heat treatment and alternative methods.*

Johnson, L.A., Hoover, W.J., Deyoe, C.W., Erickson, L.E., Johson, W.H., & Schwenke, J.R. (1980). Modeling of the kinetics of heat inactivation of trypsin inhibitors during steam-infusion cooking of soymilk. *Transactions of the American Society of Agricultural Engineers* **23**, 1326–1329.

Johnson, L.A., Deyoe, C.W., & Hoover, W.J. (1981). Yield and quality of soymilk processed by steam-infusion cooking. *Journal of Food Science* **46**, 239–243.

Kastanas, P., Lewis, M.J., & Grandison, A.S. (1996). Comparison of heat exchanger performance for goat and cow milk, 221–230. In Publication No. 9602, *Production and utilization of ewe and goat milk.*

Kaw, A.K., Singh, H., & McCarthy, O.J. (1996). The effects of UHT sterilization on milks concentrated by reverse osmosis, 299–314. In IDF (1996) *Heat treatments and alternative methods*, IDF/FIL No. 9602.

Kessler, H.G. (1981). *Food engineering and dairy technology*. Freising, Germany: Verlag A Kessler.

Kessler, H.G. (1989). Effect of thermal processing on milk. In S. Thorne (Ed.), *Food preservation*, Vol. 5 (pp. 80–129). New York: Elsevier Applied Science.

Kohlmann, K.L., Nielsen, S.S., & Ladisch, M.R. (1991). Effects of a low concentration of added plasmin on ultra-high temperature processed milk. *Journal of Dairy Science* **74**, 1151–1156.

Kwok, K.C. (1997). *Reaction kinetics of heat-induced changes in soymilk.* PhD thesis, Reading, UK: University of Reading.

Kwok, K.C, & Niranjan, K. (1995). Review: Effect of thermal processing on soymilk. *International Journal of Food Science and Technology* **30**, 263–295.

Kwok, K.C., Qin, W.H., Tsang, J.C. (1993). Heat inactivation of trypsin inhibitors in soymilk at ultra-high temperatures. *Journal of Food Science* **58**, 859–862.

Lewis, M.J. (1994). Advances in the heat treatment of milk. In R.K. Robinson (Ed.), *Modern dairy technology*, Vol. 1, 1–60. London: Elsevier Applied Science.

Lyster, R.L. J. (1970). The denaturation of α-lactalbumin and β-lactoglobulin in heated milk. *Journal of Dairy Research* **37**, 233–243.

Mehta, R.S. (1980). Milk processed at ultra-high-temperatures: A review. *Journal of Food Protection* **43**, 212–225.

Meier, J., Rademacher, B., Walenta, W., & Kessler, H.B. (1996). Heat resistant spores under UHT treatment, 17–25. In IDF (1996) *Heat treatments and alternative methods*, IDF/FIL No. 9602.

Mottar, J., Merchiers, M., Vantomme, K., & Moermans, R. (1991). A colorimetric assay for the enzymic determination of sialic acid in milk. *Overdruk uit Zeitschrift fur Lebensmitteluntersuchung und-Forschung*, **192**, 36–39.

Muir, D.D. (1984). UHT sterilized milk concentrate: A view of practical methods of production. *Journal of the Society of Dairy Technology* **37**, 135–141.

Nakai, S., Wilson, H.K., & Herreid, E.O. (1964). Assaying sterile concentrated milks for native proteolytic enzymes. *Journal of Dairy Science* **47**, 754–757.

Nelson, A.I., Steinberg, M.P., & Wei, L.S. (1975). Soybean beverage and process. *US patent, 3,901,978.*

Nieuwenhuijse, J.A. (1995). Changes in heat-treated milk products during storage. In IDF (1995) *Heat induced changes in milk* (2nd ed.), No. SI 9501.

Pearce, R.J. (1989). Thermal denaturation of whey proteins. In IDF (1989) Bulletin No. 238, *Heat-induced changes in milk.*

Pellegrino, L. (1994). Influence of fat content on heat-induced changes in milk and cream. *Netherlands Milk and Dairy Journal* **48**, 71–80.

Perkin, A.G. (1978). Comparison of milks homogenised before and after UHT sterilisation, 20th International Dairy Congress, 709.

Perkin, A.G., & Burton, H. (1970). The control of the water content of milk during UHT sterilization by a steam injection method. *Journal of the Society of Dairy Technology* **23**, 147–154.

Pinnock, C.B., & Arney, W.K. (1993). The milk mucus belief: Sensory analysis comparing cow's milk and a soy placebo. *Appetite* **20**, 61–70.

Ranjith, H.M.P. (1995). *Assessment of some properties of calcium-reduced milk and milk products from heat treatment and other processes*, PhD thesis. Reading, UK: University of Reading.

Rapaille, A. (1995). Use of starches in heat processed foods. *Food Technology International Europe*, 73–76.

Recio, I., Frutos, de M., Olano, A., & Ramos, M. (1996). Protein changes in stored ultra-high-temperature-treated milks studied by capillary electrophoresis and high-performance liquid chromatography. *Journal of Agricultural and Food Chemistry* **44**, 3955–3959.

Rees, J.A.G., & Bettison, J. (1991). *Processing and packaging of heat preserved foods*. Glasgow, Scotland: Blackie.

Reuter, H. (1984). UHT plants: State of technological development, 651–658. In B.M. McKenna (Ed.), *Engineering and food*, Vol. 2. London: Elsevier Applied Science.

Rowe, M., Pearce, J., & Crone, L. (1990). New assay for psychrotroph proteases. *Dairy Industries International* **55**(12), 35, 37.

Rustom, I.Y.S., Lopez-Leiva, M.H. & Nair, B.M. (1996). Nutritional, sensory and physicochemical properties of peanut beverage sterilised under two different UHT conditions. *Food Chemistry* **56**(1), 45–53.

Seow, C.C., & Gwee, C.N. (1997). Coconut milk: Chemistry and technology. *International Journal of Food Science and Technology* **32**, 189–201.

Society of Dairy Technology (SDT). (1989). In J. Rothwell (Ed.), *Cream processing manual*. Huntingdon, UK: Author.

Statutory Instruments (SI). (1995). *Food milk and dairies, the dairy products (hygiene) regulations*, No. 1086. London: HMSO.

Stumbo, C.R. (1973). *Thermobacteriology in food processing*, 2nd ed. New York: Academic Press.

Varunsatian, S., Watanabe, K., Hayakawa, S., & Nakamura, R. (1983). Effects of Ca^{++}, Mg^{++} and Na^{+} on heat aggregation of whey protein concentrates. *Journal of Food Science* **48**, 42–46.

Walstra P., & Jenness, R. (1984). *Dairy chemistry and physics*. New York: John Wiley.

Wilbey, R.A. (1996). Estimating the degree of heat treatment given to milk. *Journal of the Society of Dairy Technology* **49**(4), 19–112.

Wilkens, W.F., Mattick, L.R., & Hand, D.B. (1967). Effect of processing method on oxidative off-flavours of soybean milk. *Food Technology* **21**, 1630–1633.

Wirtanen, G., Matilla-Sandholm, T., Manninen, M., Ahvenainen, R., & Roenner, U. (1991). Application of rapid methods and ultrasound imaging in the assessment of the microbiological quality of aseptically packed starch soup. *International Journal of Food Science and Technology* **26**, 313–324.

Zadow, J.G. (1975). Ultra-treatment of dairy products. *CSIRO Food Research Quarterly* **35**, 41–47.

Zadow, J.G., Hardham, J.F., Kocak, H.R., & Mayes, J.J. (1983). The stability of goats milk to UHT processing. *Australian Journal of Dairy Technology* **38**, 20–23.

Zalazar, C., Palma, S., & Candioti, M. (1996). Increase of free sialic acid and gelation in UHT milk. *Australian Journal of Dairy Technology* **51**, 22–23.

CHAPTER 7

Packaging Systems

The aim of the processing plant is to give the product a sufficient heat treatment to achieve the desired effect, that is, pasteurization, sterilization, or a degree of heat treatment in between. From the moment that the product has cooled below a lethal temperature in the process, it must be considered vulnerable to recontamination by microorganisms, and the aim of all subsequent downstream operations is to allow the product to be packaged into clean or aseptic packaging without this happening. Recontamination may occur from process equipment surfaces in the cooling section, balance tanks, or filling equipment or by ingress of nonsterile air, water, or soil. The hygienic design of the equipment, correct sterilization procedures for process plant contact surfaces, and sterilization of the food-contact surfaces of packaging materials are obviously vital aspects of maintaining product integrity and must all be carefully considered during process design as well as process operation. Preprocessing sterilization of the process equipment has already been covered (see Chapter 3), and the latter is now considered.

7.1 CONNECTION OF PACKAGING EQUIPMENT TO PROCESS PLANT

The way in which the process plant is connected to the packaging equipment must generally be arranged so as to allow for one or more of the following:

- to ensure the most economical use of each item of equipment to reduce processing costs
- to allow some flexibility in case of short interruptions in packaging equipment operation
- to allow a single heat-treatment plant to be connected to more than one type of filler so that different products can, if necessary, be processed and packed into different types and sizes of container

This must be possible while preventing microbial recontamination of the product and even to allow cleaning and resterilization of the process equipment or of each filler independently.

The simplest systems have the processing plant and the filler connected directly, with no intervening equipment. More complex systems use aseptic balance tanks to build up a stock of sterile product between processing plant and filler. Even with such an aseptic balance tank, the system used must be as simple as possible, as the risk of bacterial contamination increases with the number of components in the aseptic product line, and the difficulty of tracking the source of contamination that might occur becomes correspondingly greater. If heterogeneous products are being processed, the balance tank must also keep the particulate solids in suspension in the required proportions, further increasing the complexity.

7.1.1 Direct Connection of Filler to Process Plant

The simplest possible system is the interconnection of a single filler with a single-process ultrahigh-temperature (UHT) plant (Figure 7–1). It can only be used with certain types of filler that accept product continuously, and is not suitable with positive displacement fillers that have a discontinuous demand.

The output of the process plant must be matched to the product requirements of the filler. With most fillers, a low, constant, positive pressure must be maintained at the filler inlet by means of a back-pressure valve. For this to operate satisfactorily, about 5% more product must be provided by the process plant than is required by the filler. In the early stages of development of commercial aseptic filling processes, the excess was returned to the inlet balance tank of the process plant and reprocessed. This may still sometimes happen, but there is increasing recognition that recirculation of product leads to overprocessing and a drop in quality. The excess may therefore be returned to another storage tank for use elsewhere. Direct connection of a filler to a process plant is suitable for establishing an operation and acquiring experience in its operation, as the risks of bacterial contamination are reduced to a minimum. However, it is inflexible because of the firm link between processing plant output and filler requirements. If there is a fault with either plant or filler, causing it to be shut down, then the other must also stop. The full sequence of cleaning and resterilization for the processing plant and the filler must then be repeated (see Section 3.3) before any more product can be produced. With a single filler, only one container type can be produced and often only a single-container volume.

The output capacity is relatively low, as it is determined by the capacity of a single filler. This is low in relation to processing plant capacity unless large-bulk containers are being filled, and may make processing costs relatively high. Increased throughput is possible by

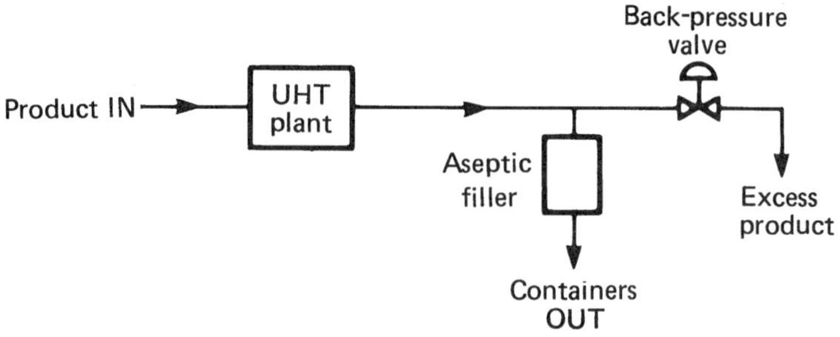

Figure 7–1 The interconnection of a single filler with a single-process UHT plant. *Source:* Reprinted with permission from H. Burton, *Ultra-High-Temperature Processing of Milk and Milk Products*, p. 244, © 1988.

using more than one filler, and different container volumes can then also be produced, but there is little improvement in flexibility. If one of the fillers ceases operation, the resulting excess of processed product can be returned to the inlet of the processing plant, but since half the product will then be recirculated, there will be severe adverse effects on chemical quality. Attempts have been made to deal with this problem by varying the throughput of the processing plant by the use of a variable-speed homogenizer. In practice a reduced flow through the plant without some change in the flow system causes overprocessing because the time spent by the product at high temperatures is increased, and again there is a loss of product quality, although some systems may cut out several passes of the holding tube by aseptic valves to compensate.

7.1.2 Use of a Small Sterile Balance Tank Integral with the Filler

When any filler operates discontinuously, dispensing product at a high rate for a short time, a balance tank is needed to isolate the filling system from the product supply, which is at a constant rate determined by the processing plant. This balance tank need only be of small volume, sufficient to smooth out the demand made on the supply by the filler. It will normally be fitted with a level control to maintain a constant product head at the filler. For the level control to operate, there will be from time to time an excess of product that can be either recirculated or discarded for use elsewhere. If the filler is aseptic, the balance tank has to be constructed so as to be sterilizable, and any headspace above the product in the tank must be maintained sterile, usually by being supplied with sterile air through a bacterial filter. The use of a small, individual sterile balance tank effectively converts a discontinuous filler into one taking a continuous flow of product from the processing plant. The advantages and disadvantages of such a system have been described in the previous section.

7.1.3 Use of a Sterile Balance Tank of Large Capacity

Greatly increased flexibility of operation is possible if a sterile balance tank is used that holds a volume of product sufficient to supply the filler installation for an hour or more, independently of the heat treatment plant. By such complete separation of the filling system from the processing plant, many advantages can be obtained, for example, the ability to shut down and clean the processing plant while keeping the fillers in operation, or the ability to use one process plant for more than one product successively without shutting down either the process plant or fillers, so improving its utilization. The use of large sterile tanks of this type has become common in aseptic processing/aseptic filling systems. The range of tank sizes available is from about 4,000 to 30,000 L or more.

There are several alternative ways of associating a sterile balance tank with a heat treatment plant and with one or more fillers. The simplest system is shown in Figure 7–2. The balance tank is connected to a tee from the pipeline between the processing plant and the filler. It is pressurized with oil-free air, supplied at a pressure of about 5 bar, and sterilized by passing it through bacterial filters: the filters are themselves sterilizable with steam, as described later. The air pressure in the tank during operation is controlled by a pressure control

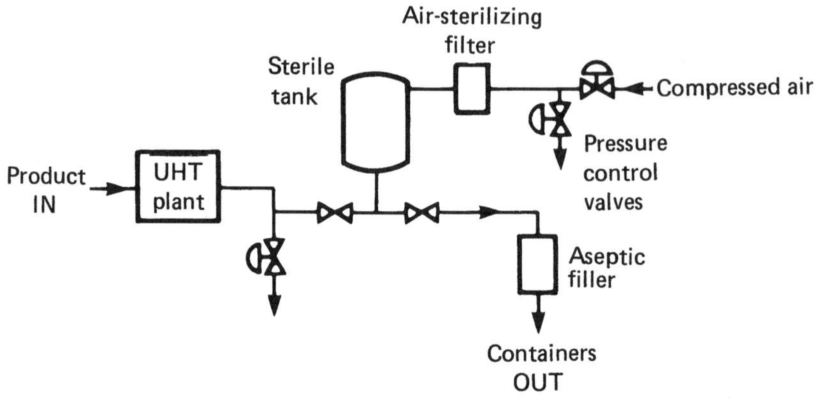

Figure 7–2 One alternate way of associating a sterile balance tank with a heat treatment plant with one or more fillers. *Source:* Reprinted with permission from H. Burton, *Ultra-High-Temperature Processing of Milk and Milk Products*, p. 246, © 1988.

system, to give the product pressure required at the filler for satisfactory operation. With the operating pressure provided in this way, there need be no excess of product passing to the filler. One aspect of air pressure, however, is the incorporation of oxygen into the product, which may be a problem during storage of the final package.

During normal operation, the pressure of product from the processing plant is controlled by a back-pressure valve, and is delivered at a rate above that required by the filler, so that the tank slowly fills while the filler is in operation. It is then possible to continue to process if the filler has to be shut down, the full output of the processing plant being stored in the tank until it can be supplied to the filler later. Alternatively, the processing plant can be shut down, for example, for intermediate cleaning, and the filler operated in the meantime using the product held in the balance tank.

Up to four aseptic fillers can in general be operated in parallel from a single product supply line. When more than one filler is being supplied through an aseptic balance tank, any of them can be shut down without interfering with the operation of the others, and without any change in the operation of the heat treatment plant: the excess sterile product is accommodated in the balance tank. There is then no need to change the flow rate of the heat treatment plant, or to recycle product, both of which have an adverse effect on the product quality.

An even more versatile system of processing plant, fillers, and sterile tank is shown in Figure 7–3. Here, the tank is associated with up to four fillers as in the system just described. However, additional fillers can be installed in the process heat exchanger circuit to operate independently of the balance tank. This system retains the advantages of that in Figure 7–2 but has additional operational flexibility in that two separate products can be processed and filled without the heat treatment plant's being shut down and perhaps having to be cleaned and resterilized.

During processing of the first product, the heat treatment plant can supply the fillers (A), with the surplus product being stored in the sterile tank; fillers (B) are not operating at this

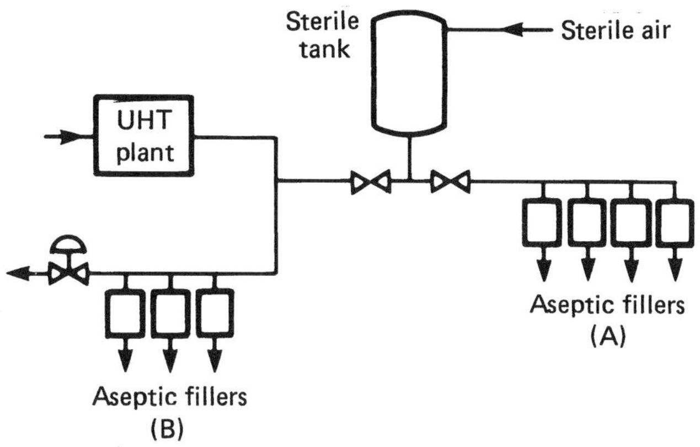

Figure 7–3 A more versatile system of processing plant, fillers, and sterile tank. The tank is associated with up to four fillers, but additional fillers can be installed in the process heat exchanger circuit to operate independently of the balance tank. *Source:* Reprinted with permission from H. Burton, *Ultra-High-Temperature Processing of Milk and Milk Products*, p. 247, © 1988.

stage. When processing of the first product is complete, the sterile tank and fillers (A) are isolated from the heat treatment plant but filling continues from the product stored in the tank.

If necessary, the process plant can be cleaned without being shut down and without losing sterility before the next product is processed. When the cleaning is complete, processing of a second product can begin, this product being aseptically filled using fillers (B) supplied directly from the process plant in the way shown in Figure 7–1, while fillers (A) continue to operate on the first product stored in the sterile tank.

With compatible product combinations, no cleaning of the processing plant will be needed between the two processing periods. For example, if two similar products differing in composition are to be processed and packaged, one product can be processed first and filled using fillers (A) in combination with the sterile tank. At the end of the first product run, the tank and fillers (A) can be isolated from the process plant and the second product processed and filled using fillers (B). By sequencing the products in this way, it may be possible to eliminate compositional problems, for example, the processing of milks or creams of different fat contents or different recipes of orange juices.

If products to be processed are more fundamentally different in nature, then an intermediate clean of the heat exchanger will be needed to prevent residues of one product contaminating the next undesirably.

7.1.4 Construction of Sterile Balance Tanks

Sterile balance tanks are made of stainless steel, and are designed to be cleaned in-place and sterilized with steam under pressure. The shape is normally that of a vertical cylinder, with its height about 1.5 times its diameter: this shape is suitable for cleaning by spraying, and for pressurization.

In addition to the entries for product, normally in the center of the base, and for sterile air near the top of the tank, a spray ball or nozzle for cleaning fluid is mounted in the center of the tank roof. To help to cool the tank after sterilization, it may be surrounded by a jacket through which cooling water can be circulated.

The tank must be designed as a pressure vessel, according to appropriate rules, so as to withstand the steam pressure used for sterilization. Although the tank control system, and particularly the supply of sterile air, should prevent it, the vacuum drawn as the steam in the tank condenses during cooling after sterilization gives a risk of implosion. The tank should be designed not only to resist pressure, but also to prevent implosion under vacuum. If a vacuum is drawn within the tank at any stage during cooling, there is a serious risk of contaminated air being drawn in from outside, and this must be guarded against by ensuring that sufficient sterile air is available to maintain a positive internal pressure during cooling. A product level indication system normally will be provided, and this will also prevent overfilling of the tank by cutting off product supply from the processing plant to the sterile tank and the aseptic fillers associated with it if the level rises too high. A pressure sensor will be installed to control the sterile air supply; this may be incorporated within the level indication and control system. For products containing particulates, the tank must maintain the particles in suspension, which may require an agitator and baffles. In this case, where the agitator shaft enters the vessel wall, a gland seal is required that should be continuously purged with steam to prevent contamination through it. Extra care is required in presterilization of the tank, especially around the gland seal area and where the baffles are attached to the vessel wall. One commercial design of tank specifically for particulate liquids consists of a horizontal short cylinder that rotates around its central axis; internal baffles ensure that the particulates remain in suspension without the need for a separate agitator. Steam, sterile air, and product inlets and outlets are by way of a gland seal on the central axis. In West Germany, guidelines have been established for the official type testing of sterile tanks for use in UHT systems (Reuter, Biewendt, & Klobes, 1982a). These cover constructional requirements, as well as cleaning and sterilization, and bacteriological standards that are the same as those set for aseptic fillers.

7.2 CLEANING AND STERILIZATION OF STERILE TANKS

As with the equipment for heat treatment and filling, cleaning of a sterile tank and its associated pipework and valves after one processing run, and effective sterilization immediately before the start of another, are of vital importance for satisfactory operation. The sterile tank may be supplied with its own self-contained cleaning-in-place (CIP) system, or it may be cleaned with the aseptic fillers associated with the tank using a common CIP system. Exceptionally, the tank may be cleaned with the heat exchanger, but this is unusual. The cleaning requirements for the heat exchanger are different and more severe than those for the tank and aseptic fillers, which operate with relatively cool product and so are relatively easy to clean. Furthermore, the cleaning cycles of the heat exchanger and the sterile tank are not likely to coincide, as the heat exchanger may need to be cleaned more than once during a day.

The cleaning cycle is conventional, consisting of prerinsing, detergent circulation, and final rinsing. The detergent circulation may be single stage with a suitable alkaline detergent at high temperature followed by a water rinse, followed by nitric acid, generally as described for heat exchangers in Chapter 3, with concentrations and temperatures as recommended by the plant and detergent manufacturers.

Figure 7–4 shows one form of cleaning circuit, in which a separate CIP unit is used to clean a sterile tank and the pipework up to the associated aseptic fillers. The rinse water and the detergent solutions are pumped in sequence from the CIP balance tank, heated by steam injection, and sprayed into the sterile tank. They flow from the tank, along the product line to the fillers, and through a return line back to the CIP balance tank. Circulation times to give satisfactory cleaning are determined empirically, and the timing of each circulation phase starts when the required circulation temperature is reached at a sensor placed between the aseptic fillers and the return to the CIP balance tank.

For sterilization of the tank and the associated pipework before processing begins, steam is introduced into the system through the steam injector in the CIP line and the tank sprayball. A valve at the end of the product line, after the fillers, is closed to maintain the sterilizing steam pressure while discharging condensate through a steam trap. A valve on the nonsterile side of the bacterial air filter (Figure 7–2) also closes, so that the filter is pressurized with steam for sterilization. The sterilization time starts when temperature sensors placed in parts of the pipe system remote from the tank and steam supply register the preset minimum sterilization temperature, (e.g., 135° to 140°C).

At the end of the required sterilization time, about 30 minutes, the steam is shut off and exhausted from the system, and compressed air is supplied through the bacterial filter, which is now sterile, in order to maintain a pressure of sterile air in the tank and pipework as the steam condenses. If a water jacket is fitted to the sterile tank, water can be applied to speed the final stages of cooling when most of the steam has been exhausted. Cooling water must not be applied too early in the cooling cycle, or a vacuum may be created that may cause mechanical damage or bacterial contamination from outside.

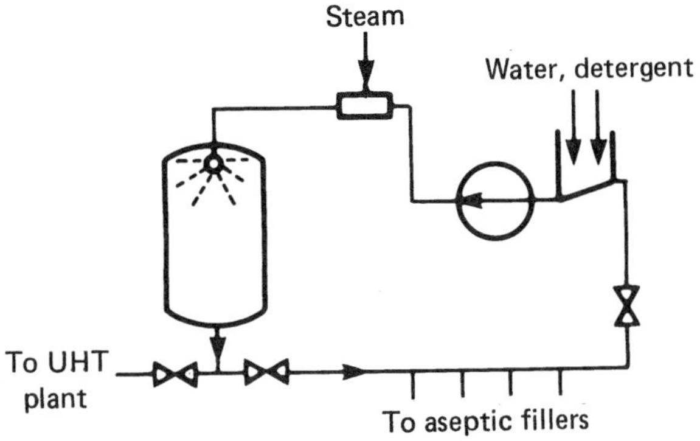

Figure 7–4 A separate CIP unit is used to clean a sterile tank and the pipework up to the aseptic fillers. *Source:* Reprinted with permission from H. Burton, *Ultra-High-Temperature Processing of Milk and Milk Products*, p. 250, © 1988.

7.3 USE OF STERILE BARRIERS

At various times during operation, parts of the circuit are sterile and it is important that they remain so, while adjacent parts of the circuit are not sterile. For example, in the circuit shown in Figure 7–2, the process plant may be cleaned while the tank is still supplying sterile product to the fillers. In the circuit shown in Figure 7–4, after cleaning and during aseptic filling, the line from the tank to the fillers contains sterile product, while the return line from the fillers to the CIP balance tank is not sterile. An aseptic filler is connected to the sterile product line through a valve: it may be necessary, particularly in multiple filler installations, to shut down and either clean or repair a filler while the supply line still contains sterile product. Although a single valve may be shown isolating the sterile from the potentially nonsterile parts of the system, in practice a single valve is insufficient to eliminate the possibility of cross-contamination by bacteria passing across the valve seat. Where it is essential to avoid contamination, a sterile barrier is used.

In a sterile barrier, two valves are connected in series in the line. In the closed position, both valves are closed, and the volume between them is filled with steam under pressure through additional valves. One form of this system is shown in Figure 7–5. Contamination from one side of the barrier to the other is then prevented not only by the two closed valves, but by the sterile zone between them. The group of valves is operated together through the automatic control system for the process. A temperature sensor may be mounted within the barrier zone, to activate an alarm if for any reason the correct temperature is not reached through failure of the steam supply or improper valve operation. Barriers of this kind can be employed wherever in the circuit there is a risk of contamination of a sterile area from an adjacent nonsterile area.

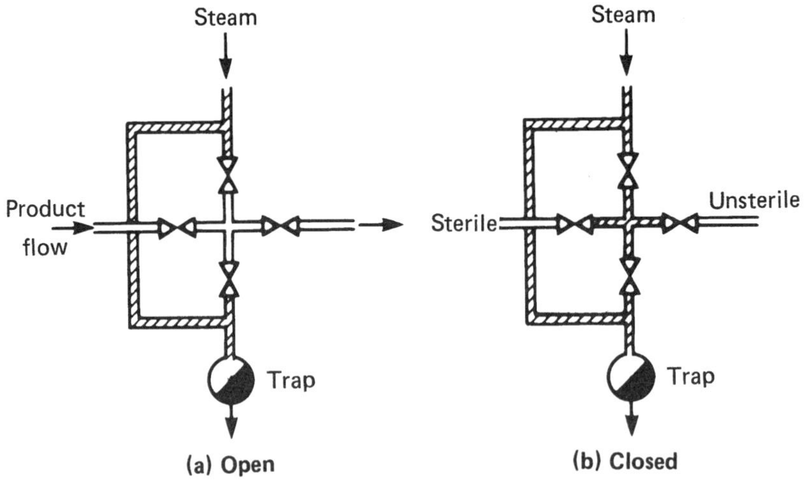

Figure 7–5 A sterile barrier. (a), Both valves are open; (b), valves are closed to prevent contamination from one side of the barrier to the other. *Source:* Reprinted with permission from H. Burton, *Ultra-High-Temperature Processing of Milk and Milk Products*, p. 252, © 1988.

7.4 GENERAL MECHANICAL PRINCIPLES

In sterile circuits, it has become standard practice to use fully welded construction. Most forms of pipe union have gaskets to form the liquid seal, located in recesses, and these are difficult to clean and sterilize because cleaning and sterilizing fluids have only limited access to them. The ability to dismantle pipework carrying product has become much less important with the development of effective CIP systems, so that orbital welding has become the standard method of connection. However, welding standards need to be high so that the inner weld bead is sound and free from porous areas or cavities, and does not project into the tube bore, as these defects will reduce the effectiveness of both cleaning and sterilization. The subject of hygienic welding is covered in more detail by Henon (1997) and EHEDG (1993a).

The principles of pipe layout that are accepted as important for effective cleaning in normal systems are even more important in sterile systems, since they influence the effectiveness of plant sterilization as well as cleaning. Dead ends at tee junctions and instrument mounting points must be eliminated where possible or, if unavoidable, must be as short as practical, and flow of product must be arranged into the dead end, not past it. Not only will they make cleaning difficult, but when water circulation is used for plant sterilization, any dead volumes will only be heated poorly as the circulating water will not penetrate into them or, if steam under pressure is used for sterilization, air is likely to be trapped in the dead volume and interfere with heat transfer; in both cases the specified sterilization temperature may not be reached.

Care is also needed in designing pipe runs. If steam is to be used for sterilization, pipes must be laid to a gradient (minimum 1:100) and be well supported without sagging, with steam traps at low points, so that condensate does not lie in pools within the pipes and prevent direct contact with the high-temperature steam, but flows away and out of the system. If circulating water is used, pipe layouts must be suitably designed, and the water flow rates made high enough, for all air pockets to be displaced.

Valves should be of suitable design for aseptic operation. They should be crevice-free, and diaphragm or similar seals must be included to isolate product from sliding shaft seals, which are difficult to clean and sterilize. For particulate liquids, valves must be eliminated wherever possible, but where found to be necessary, high-lift designs are desirable to allow easy passage of the solids; preferably they should be fitted with a sharp-edged plug to cut particulates rather than allow them to be trapped as the valve closes. Instruments must be cleanable and sterilizable and not intrude further into the pipeline than necessary, with special attention paid to the gap between the sensor shaft and pipe wall. For example, pressure sensors preferably should be the electronic strain-gauge type but if a gauge is required, it should be of the tubular diaphragm type. For particulate liquids, noninvasive instrumentation should be used wherever possible.

Sterile product systems should always be made as simple as possible to meet the specified processing needs, because added complication always means greater problems in obtaining plant sterility initially and maintaining it later. The subject of hygienic design for continuous-flow thermal processes and guidelines for the processes have been generally covered by the European Hygienic Equipment Design Group (EHEDG) (EHEDG, 1992, 1993b, 1993c,

1994) as well as other aspects of hygienic design published in a discontinuous series of articles in the same journal.

7.5 PACKAGING SYSTEMS FOR CONTINUOUS-FLOW THERMALLY PROCESSED PRODUCTS

Thermal processing systems as described in Chapters 3 and 4 produce a continuous stream of product that has been thermally treated to some predetermined degree between thermization, pasteurization, and full sterility. This product must now be packaged into a container in which the internal surfaces, in contact with the product, are of a comparable or higher microbiological standard than the product. The packaging must also conform to the usual packaging requirements for any food product, that is, it must provide an appropriate barrier to microorganisms, water vapor, and oxygen, be physically strong with a surface to support printing, but be easy to open and with a tamper-evident seal. Highly desirable characteristics, but less essential would be ease of opening (a criticism of earlier aseptic cartons and some current pasteurized-product cartons), be resealable, and easily stackable and with an option to come in packs of multiple units. Attractive appearance, recyclability, and, of course, low cost are also important.

In practice, two standards of packaging are used: fully aseptic containers for commercially sterile products with ambient storage of 3 to 12 months and clean containers for lesser-heat-treated products designed for refrigerated storage with a short shelf life (up to 3 weeks). The latter is well established and requirements are relatively slight; the short shelf life and refrigerated storage mean that barriers are relatively unimportant and that nonsterile, but clean, container surfaces can be tolerated. The requirements for aseptic packaging are very much more severe and must be satisfied before the whole aseptic processing system can be considered satisfactory.

The remainder of this chapter will concentrate on aseptic filling systems rather than systems for pasteurized products. However, there is one other system between these two extremes that is worth consideration. Hot filling may be used to extend the refrigerated shelf life of nonsterile products or alternatively to sterilize packaging surfaces for high-acid products to give a commercially sterile product.

7.6 HOT FILLING

In hot filling, the product is thermally processed as normal, but not cooled to any extent, and then filled hot into the container, which is then sealed and agitated or inverted to ensure that the hot product contacts all internal surfaces. The container and its contents are then cooled as rapidly as possible. The temperature the container surface achieves is obviously limited to 100°C or less at atmospheric pressure and so cannot fully sterilize the packaging (for example, it cannot eliminate *Clostridium botulinum* spores). However, hot filling is used for high-acid products (pH < 3.7) and will give a fully commercially sterile product with a long shelf life at ambient temperatures. The process has, alternatively, been applied to low-acid foods, such as chilled fresh soups and sauces. For these types of products, two cases may be considered:

1. The nature of the product will not support the growth of *Cl. botulinum* even at a pH value greater than 4.5, and the level of thermal processing of the product in this case should be sufficient to inactivate non–spore-forming food poisoning microorganisms.
2. The product can be maintained at a sufficiently low temperature after processing and through the distribution chain to suppress the growth of *Cl. botulinum*, and therefore toxin production. A thermal process sufficient to inactivate *Cl. botulinum* spores (an F_0 value of 3 minutes) would normally be applied to the product to reduce reliance on the chilled temperature control of the product, but contamination with *Cl. botulinum* spores may occur through the inherently low temperature achieved by the packaging surface during hot filling. There is an improved product quality, as the thermal process applied does not give full commercial sterility, but great care must be taken in subsequent temperature control of the product.

One major aspect that has a bearing on the final product quality is the rate at which the product and package can be cooled after it has been filled. Cooling can be achieved by blast chilling with air or by immersion in, or spraying with, a cold liquid such as water or brine. The rate of cooling will be greater the lower the temperature of the cooling medium, but it must not be low enough to allow freezing of the product to occur. The rate of cooling can also be maximized by agitating the container to ensure increased internal heat transfer by convection. As with conventional in-container processes, it must be ensured that any leakage of the cooling medium into the package will not cause contamination of the product, and therefore chlorinated water or ice would be required.

7.7 ASEPTIC PACKAGING

The rest of this chapter concentrates on aseptic packaging and the packaging surface sterilization systems that may be used.

The term *aseptic packaging* is applied to systems in which the packaging material is fully sterilized; that is, it receives a treatment that removes all spores and vegetative bacteria that will grow in the product under normal storage conditions. The final package can therefore be considered commercially sterile and to have a storage life at ambient temperatures of at least several months.

An aseptic filling system must meet a series of requirements, each of which must be satisfied individually before the whole system can be considered satisfactory.

1. The container and the method of closure must be suitable for aseptic filling, and must not allow the passage of organisms into the sealed container during storage and distribution. It is also desirable for the container to have certain physical properties that will help minimize chemical change in the product during storage. These are discussed in more detail in Chapter 9. Here we are concerned only with the need for the container to prevent the entry of contaminating organisms.
2. The container, or that part of it that comes into contact with the product, must be sterilized after it is formed and before being filled. The level of sterilizing effect will be related to the probable initial contamination of the container surface. The container sterilization process may be single stage, either within the aseptic filler as part of its

operation or as a preliminary process, or it may be two-stage, with the second stage forming part of the filler operation.

3. The container must be filled without contamination by organisms either from the equipment surfaces or from the atmosphere that surrounds the filler. Filling is usually done in an enclosed area that is supplied with a sterile atmosphere (e.g., air sterilized by heating or filtration).
4. If any closure is needed, it must be sterilized immediately before it is applied.
5. The closure must be applied and sealed in place to prevent the passage of contaminating organisms while the container is still within a sterile zone.

There are many possible ways of meeting these general requirements, even for a single basic type of container. It is not practicable to describe all the different methods adopted by manufacturers, and the systems described here are intended to illustrate the variety of approaches that can be used in attempts to meet this common set of requirements.

7.7.1 Methods of Container Sterilization

Many physical methods and chemical reagents can be used for container sterilization (von Bockelmann, 1985; Cerny, 1985). The sterilization performance needed for different containers depends on the probable number of organisms existing in the container before sterilization. For example, for plastic pots freshly removed from their external packaging, an average count of 0.7 organism per container with a maximum of 10 organisms per container has been quoted, with less than 10% of these being spores. Swartling and Lindgren (1962) reported that polythene-coated paper used for containers could have surface contamination of only 0.02 organism per square centimeter. von Bockelmann (1982) has given figures of 0.02 to 0.05 organism per square centimeter, corresponding to 27 organisms for a 0.5-L carton and 39 for a 1-L carton: micrococci were the largest group (44%), and spores represented about 3%. Franklin and Clegg (1956) estimated that a well-washed returnable glass bottle contained about 0.5 spore per container.

All of these levels of contamination are similar, and imply that a sterilizing process giving about 3 to 4 decimal reductions of resistant spores is adequate to give a single survivor in about 5,000 containers of 0.5- to 1-L capacity. Smaller containers with a smaller area of container material will have correspondingly less initial contamination according to the ratio of surface areas, and a lesser sterilization effect will be needed to give a satisfactory commercial spoilage level. Conversely, larger containers will require a higher sterilization effect to give satisfactory performance, in accordance with the larger area of container material to be sterilized. However, as container volume varies with the cube of linear dimensions, while surface area varies only with the square of dimensions, the variation in required surface sterilizing effect with container size is less than might be expected.

7.7.1.1 Saturated Steam

The most reliable sterilant without doubt is wet heat in the form of saturated steam. The kinetics of sterilization with saturated steam are well understood and reproducible for different organisms, and are those covered in Chapter 2. However, saturated steam presents difficulties in application. For example, to give steam temperatures high enough for the sterilization time to be short and practical in a commercial process, the steam must be under pressure. A pressure chamber must therefore be used, with the container or container material to be sterilized entering and leaving the chamber through suitable valves. Air in a steam pressure chamber interferes with the transfer of heat from the steam to the container surface, so that any air entering with the containers must be removed and not allowed to accumulate. Condensation of the steam during heating of the container surface produces condensate that may remain in the container and dilute the product.

Some early aseptic filling systems for cans (Ball & Olson, 1957) and for glass bottles (Hansen, 1966) used saturated steam as the container sterilant. Fillers have also been designed in which polystyrene-thermoformed cups (Cerny, 1982) and polypropylene-preformed cups (Cerny, 1983) are sterilized by steam under pressure: sterilizing effects against *Bacillus subtilis* spores of 5 to 7 decimal reductions have been demonstrated.

7.7.1.2 ***Dry Heat***

Dry heat can be applied either in the form of a hot gas or as a hot, nonaqueous liquid such as glycol. Dry heat has the advantage that high temperatures can be reached at atmospheric pressure, thus simplifying the mechanical design problems for a container sterilization system.

However, the resistance of microorganisms to dry heat is much greater than to wet heat. Much higher temperatures are therefore needed for thermal inactivation by dry heat than for inactivation in a similar time by wet heat (Pflug, 1960), and it may be that temperatures of the order of 200°C may be needed for inactivation in a reasonable length of time. The kinetic information given in Chapter 2 is not valid under these conditions.

Superheated steam (i.e., steam that has been heated to a temperature well in excess of the boiling point of water at the prevailing pressure) also behaves in the same manner as a dry gas if the relative humidity is less than about 0.5% (Han, 1977). Collier and Townsend (1956) recommend minimum time-temperature conditions in superheated steam of 1 minute at 177°C or 10 minutes at 160°C to give satisfactory container sterilization in an aseptic canner. Quast, Leitao, and Kato (1977) found that 6.3 minutes at 170°C in similar equipment gave 10 decimal reductions of *B. stearothermophilus* spores; estimates from the data in Table 2–2 suggest that, in saturated steam, less than 0.5 second at 170°C would be needed for this sporicidal effect. Similarly, nonaqueous liquids, such as ethylene glycol, at high temperature provide dry heat, as do infrared heating systems, and their sterilizing effect is relatively low. Infrared radiation has no specific sterilizing effect other than that arising from the temperature, and acts merely as a form of heating of the container material surface.

The high temperatures needed for satisfactory sterilization by dry heat mean that containers that are heat sensitive cannot be used. Dry heat has been commercially applied on a wide scale in aseptic canning, but even here at one time the melting of solder in the seams of the cans was found to be a problem if the specified sterilization temperature (210° to 220°C) is exceeded (Hersom, 1985b). Dry heat has been used in some systems for glass bottles (Burton, 1970; Schreyer, 1985), but the high container sterilization temperature has caused breakage problems when a relatively cool liquid is then filled. Some plastic materials can withstand temperatures of 200°C but again, as this is very much an upper limit, temperature control is very important.

7.7.1.3 Hydrogen Peroxide

It has been known for many years that hydrogen peroxide (H_2O_2) is lethal to microorganisms, including heat-resistant spores, and has the advantage that it degrades to water and oxygen on heating, leaving very little residue, unlike other chemical sterilants. The first successful aseptic filling system for cartoning, the aseptic Tetra Pak of 1961, used a combination of hydrogen peroxide and heat for the sterilization of the surface of the container material, and this method has been used by many manufacturers for different types of container ever since. There have been many studies on the death of resistant spores in suspension in hydrogen peroxide solution. The resistance of spores to H_2O_2 bears no relation to their thermal resistance; for example, Toledo, Escher, and Ayres (1973) found the survival of spores at 24°C in 25.8% H_2O_2 to be in the order *B. subtilis* > *B. subtilis* var. *globigii* > *B. coagulans* > *B. stearothermophilus* > *Clostridium PA 3697*, with D values of 7.3, 2, 1.8, 1.5, and 0.8 minutes, respectively. *Staphylococcus aureus* was more sensitive than any of these, but not as much as might be expected from relative thermal resistances, with a D value of 0.2 minute. Cerny (1976) found that *B. stearothermophilus* spores were more resistant than *B. subtilis* spores, and that yeasts and molds were more resistant than would be expected from their thermal resistances. In practice, the sterilizing effectiveness of H_2O_2 in relation to aseptic filling is usually assessed against spores of either *B. subtilis* or *B. subtilis* var. *globigii* because of their relatively high resistances.

Sterilizing performance increases with both peroxide concentration and temperature. Swartling and Lindgren (1968) and Cerny (1976) found for *B. subtilis* spores a linear decrease of log (number of surviving spores) with time, similar to that which is normally found in thermal sterilization (see Chapter 2). Swartling and Lindgren (1968) found that 4 decimal reductions (which they considered adequate for the sterilization of carton material in aseptic filling) were obtained after suspension in 20% H_2O_2 at 80°C for 15 seconds. Reduction of the peroxide concentration to 15% increased the time required by about 50%, and a decrease of temperature of 10°C increased the time by about 70%. Some of the test organisms investigated by Toledo et al. (1973) were more resistant than this, but they still obtained 4 decimal reductions of spores of *B. subtilis* var. *globigii* after about 12 seconds in 25.8% peroxide solution at 76°C. The spores of *B. subtilis* studied by Cerny (1976) were also more resistant than those studied by Swartling and Lindgren (1968), but 4 decimal reductions were obtained at about 40 seconds in 20% H_2O_2 at 80°C; the time was about 20 seconds in 30% H_2O_2 at 70°C. There are many uncertainties in the use of H_2O_2 for surface sterilization. Some workers have found that the resistances of "wet" and "dry" spores differ (Toledo et al., 1973; Smith & Brown, 1980), although it is difficult to see what this means when the spores are in contact with an aqueous solution of peroxide. Cerf and Hermier (1972) and Cerf and Metro (1977) have found survival curves for some spores of bacilli that differ greatly from the classic semilogarithmic form, with a very resistant subpopulation. We have seen in Chapter 2 that a similar resistant population can lead to "tails" in the thermal death curves of some spores, but the proportion involved is not as high as that found in sterilization by hydrogen peroxide. Cerny (1976) found that spores apparently inactivated by peroxide treatment could be reactivated by subsequent heat treatment. It is therefore difficult to predict the sterilizing effect that any specific combination of peroxide concentration and temperature is likely to have.

However, one of the principal factors that prevents a container sterilization process using H_2O_2 from being designed in the way that a thermal sterilization process can be designed in terms of times and temperatures is a practical one. Some aseptic filling systems use a bath of hot peroxide to sterilize the container or the container material, so that the process can be based on data of the kind summarized above. However, the majority of systems apply the peroxide solution (usually at 30% to 35% concentration) to the surface of the container material by dipping or in a finely dispersed spray. The surface is then subjected to heat, either from radiant heating elements or from hot air jets. The peroxide solution on the surface is therefore heated and evaporated, to sterilize the surface and at the same time remove the peroxide solution to prevent it contaminating the product after filling. The sterilization conditions are consequently very complex: the temperature of the peroxide solution on the surface rises during the process, and at the same time the concentration of the solution increases as water is evaporated, until a concentration is reached when peroxide itself will be evaporated. There seems to be no data that will allow the sterilizing effect of such a complex sequence of events to be predicted.

In most fillers using H_2O_2 for container sterilization, therefore, the sterilization conditions have been determined empirically. These conditions include peroxide solution concentration, quantity applied to the container material per unit area, intensity of radiant heat or temperature and quantity of the drying air, and the time for which it is applied (Huber, 1979). Hydrogen peroxide is a poison, and in some countries there are strict limits to the concentrations that may remain in the filled container and that are allowed in the atmosphere surrounding an aseptic filler in operation, which could be inhaled by operators. For example, in some countries the H_2O_2 concentration in the atmosphere must not exceed 1 part in 10^6.

The Food and Drug Administration (FDA) of the United States has ruled that the "level of H_2O_2 that may be present in milk packaged in material, sterilized by H_2O_2 must be no greater than 100 parts per billion (100 parts in 10^9) at the time of filling and must fall to approx. 1 part per billion within 24 h" (FDA, 1981). In fact, peroxide cannot be measured accurately in a heated milk product, since it is rapidly eliminated by the reducing compounds that are present. In practice, therefore, the test of the initial level must be made on packs filled with water. Colorimetric tests are available for measuring H_2O_2 concentrations to confirm the required initial level in the water, although more sensitive methods can detect concentrations in filled packs as low as 1 part in 10^9 (Perkin, 1982).

7.7.1.4 Other Chemical Sterilants

Two other chemicals that are used as sterilants in association with aseptic filling systems are ethylene oxide and peracetic acid.

Ethylene oxide is a toxic gas. It may be used for the presterilization of paperboard-based packaging materials, particularly preformed carton blanks that are to be assembled in the aseptic filler. As a gas, it can penetrate porous materials and sterilize the interior of the board layer. Because of its toxicity, the stacked blanks are presterilized in a special facility and the gas is allowed to disperse from the stacks before they are shipped to the filling plant in sealed boxes.

Peracetic acid (sometimes called Oxygal) is a liquid sterilant that is particularly effective against spores of aerobic and anaerobic bacteria (Mrozek, 1985), and is effective at lower temperatures than hydrogen peroxide. Because of its toxicity, it is less suitable for the sterilization of container surfaces but is more commonly used in sprays for the presterilization of the internal surfaces of aseptic fillers before filling begins. However, one packaging system using peracetic acid with a wetting agent claims a 5 to 6 log reduction in microorganism load after exposure to peracetic acid followed by drying at 45°C and rinsing with sterile water (Binet & Gutter, 1994).

In practice, peracetic acid is used in a solution that also contains hydrogen peroxide, since peracetic acid is produced by the oxidation of acetic acid by hydrogen peroxide in an equilibrium solution. The end result of the decomposition of peracetic acid is acetic acid. The solution containing peracetic acid and hydrogen peroxide is effective against resistant bacterial spores even at 20°C; for example, a 1% solution will eliminate 10^7 to 10^8 of most resistant spore strains in 5 minutes at 20°C, and the most resistant strains in 60 minutes. The maximum usable temperature is 40°C, when the sterilization times are about five times shorter.

Ethyl alcohol was used in some early aseptic filling systems for the treatment of plastic packaging films. At 80% concentration, ethyl alcohol is a very effective sterilant against vegetative organisms. However, it is ineffective against bacterial spores and so is no longer used.

7.7.1.5 Ultraviolet Irradiation

It has been known for many years that ultraviolet (UV) irradiation can inactivate microorganisms, and that the optimum wavelength is about 250 nm with a rapid falling away of effectiveness at shorter and longer wavelengths (e.g., Bernard & Morgan, 1903). The effect seems to arise from direct absorption of the radiation by the deoxyribonucleic acid (DNA) of the bacterial cell.

Attempts have been made to use this effect for the pasteurization of milk (Burton, 1951), but they have not been successful. The turbidity of milk makes it difficult to ensure that all parts are exposed to the radiation, but there are more fundamental difficulties, as it appears that microorganisms are very variable in their reaction to irradiation. Nevertheless, UV irradiation has been successfully applied to the sterilization of air, and of water that is free from suspended matter.

With the development of more powerful sources of UV irradiation (Bachmann, 1975), attempts have been made to use it for surface sterilization. However, the results have not been entirely satisfactory, and it is possible to achieve a much greater sterilizing effect with hydrogen peroxide. There are practical problems, also, in ensuring that the radiation intensity is uniform and adequate for sterilization over the whole of a container that may be of a complex shape, and that bacteria are not protected from the radiation by particles of dust and dirt (Cerny, 1977, 1984).

7.7.1.6 Combination of Hydrogen Peroxide with Ultraviolet Irradiation

The mechanism of the sterilization effect of H_2O_2 is not fully understood. It has been suggested that spores are inactivated by the hydroxyl radicals produced by the decomposition of the peroxide, so that the lethal action is a function not of the peroxide itself but of its decomposition. On this assumption, the effect of heat in improving the lethal effect of peroxide is not only one of increasing the reaction rate, but also involves promoting breakdown.

The breakdown of H_2O_2 can also be promoted by UV irradiation. It has been found that the lethal action of peroxide solutions with and without the application of heat is increased by simultaneous UV irradiation, and that the overall lethal effect is greater than the sum of the effects of the peroxide and irradiation alone (Bayliss & Waites, 1979, 1982). The effect is optimum at a relatively low peroxide concentration, between 0.5% and 5% (Figure 7–6); at higher concentrations the peroxide appears to have a protective effect, and the greater the UV intensity the higher is the optimum peroxide concentration.

Four decimal reductions of *B. subtilis* spores of a strain very resistant to H_2O_2 were easily obtained by irradiation with a UV dose of 1.8 W/cm^2 in 2.5% peroxide (Bayliss & Waites, 1982). Subsequent heating to 80°C increased the kill further.

Experiments with cartons artificially contaminated with *B. subtilis* spores sprayed with 1% H_2O_2 and then irradiated for 10 seconds with a high-intensity UV source above the carton showed 5 decimal reductions with polyethylene-lined material and 3.5 decimal reductions with a polyethylene/aluminium foil laminate (Stannard, Abbiss, & Wood, 1983). No heat was applied. This sterilization system has potential advantages over the use of peroxide and heat alone. Since a lower peroxide concentration can be used (less than 5% as against 30% to 35% in most applications of peroxide and heat), the problems of atmospheric contamination

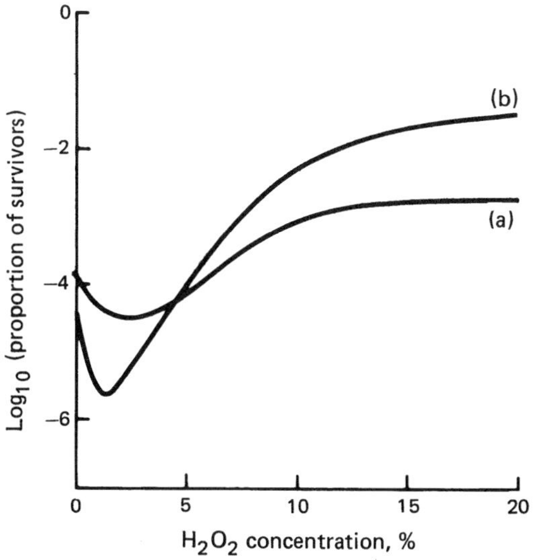

Figure 7–6 Lethal effect on *B. subtilis* spores of simultaneous UV and H_2O_2 treatment at room temperature. (a), *B. subtilis* NCDO 2130. (b), *B. subtilis* NCDO 2129. *Source:* Reprinted with permission from H. Burton, *Ultra-High-Temperature Processing of Milk and Milk Products*, p. 190, © 1988.

and of peroxide remaining in the filled product are reduced. However, too high a peroxide concentration reduces the sterilization effectiveness, so that strict control of concentration is essential. This combination of H_2O_2 with UV has now been applied commercially to carton sterilization during aseptic filling.

7.7.1.7 Ionizing Radiation

Ionizing radiations that have penetrating power (e.g., gamma rays) have been used to sterilize the interior of sealed but empty containers, particularly those made of materials that cannot withstand the temperatures needed for thermal sterilization. The bags, made of plastic laminates for use in aseptic bag-in-box systems, are generally sterilized by gamma irradiation after manufacture and before dispatch to the filling plant. They are treated in a specialized irradiation plant, and given a radiation dose of 25 kGy (2.5 Mrad) or more, which is sufficient to ensure sterility. As the bags remain sealed, they remain internally sterile until they are opened under protected conditions in the aseptic filler.

7.7.2 Types of Aseptic Filling Systems

Aseptic filling systems vary widely in the ways in which they are designed to meet the basic requirements we have set out earlier. The principal controlling factor in design is the type of container that is to be filled, with subsidiary factors being the container material and whether the container is supplied preformed or whether it is formed during the aseptic filling process.

7.7.2.1 Cans

One of the earliest aseptic fillers to be widely used commercially was designed to use cans. This type of filler is still successfully used, but cans are now seen as having certain shortcomings. They are relatively expensive, particularly for a low-cost product such as milk, and they are bulky to ship and store before use. Aseptic canning may also have suffered for psychological and marketing reasons. It is not possible to distinguish at the point of sale between a can that has been conventionally retorted and one in which the product has undergone a continuous-flow thermal process and then been aseptically filled. The processors and marketers of an aseptic product generally want to emphasize the newness of the process and the improved quality of the product, and for this reason they look for a type of container that is clearly distinguishable from one that is associated with an older type of process.

Aseptic canners are available to handle can sizes from 4.5 US fl oz up to 5 US gal (18.8 L), with operating speeds up to 450 cans per minute for the smaller sizes. The cans may be of tinplate or drawn aluminium; the solder in tinplate cans may have to be of higher melting point than normal to withstand the can sterilization temperatures. The processes take place as the cans pass along a conveyor belt within a continuous tunnel. The cans are sterilized in the tunnel at atmospheric pressure by steam at 200° to 220°C, superheated with gas flames; the sterilizing time is about 40 seconds. The hot flue gas may be mixed with the superheated steam to fill the can sterilizing and filling chambers and prevent bacterial contamination from the external atmosphere; the proportion of flue gas to steam controls the vacuum developed in the headspace of the filled cans after sealing and cooling.

When the cans have passed through the sterilizing tunnel, they continue through the filling chamber, where they are filled, often using a simple in-line filler of the slit or multiport type. The filling level is determined by the product flow rate and the speed of travel of the can under the filler.

The can lids are sterilized, again by superheated steam, in a separate unit. They are then placed in position and seamed by a conventional seamer operating in a sterile chamber attached to the end of the filling chamber, kept sterile by the atmosphere of superheated steam and flue gas.

Before filling begins, the can sterilizing, filling, and sealing zones are sterilized by the hot mixture of superheated steam and flue gas that fills them during normal operation. A modification of this system has been developed to use "cans" formed from a paper and foil composite, sterilized by hot air. The limitations on the sterilizing temperature that can be withstood by this type of "can" mean that the filler is not truly aseptic, but may be used with high-acid products such as fruit juices where relatively low process temperatures will give commercial sterility (see Chapter 2).

7.7.2.2 Cartons

Paperboard cartons are commonly used in aseptic filling systems for milk, cream, soy-based milks, fruit juices, and recently for soups and sauces, with and without solid particulates. The filling systems are of two different kinds: those in which the carton is formed within the filler from a continuous reel of material; and those in which the cartons are supplied as preformed blanks, folded flat, which are assembled into cartons in the filler.

The carton material is normally a laminate of paperboard coated internally and externally with polyethylene, which makes the carton impermeable to liquids and allows thermal sealing of both the internal and external surfaces. For the packaging of most products that are intended to have a long storage life, an oxygen barrier is desirable for reasons that are discussed in a later chapter. This is normally provided by a thin aluminium foil incorporated in the laminate. The carton material structure is shown in Figure 7–7. Plastic oxygen-barrier materials are being developed, and these may replace the aluminium foil layer in the laminate structure at some time in the future.

7.7.2.3 Paperboard Cartons Formed from the Reel

In the fillers most generally used, the container material moves continuously downward in a strip, and is formed by shaping rolls into a cylinder (Figure 7–8). An overlapping longitudinal seal is formed by heat sealing. At the same time, an additional thin polyethylene strip is heat-bonded along the inside of the longitudinal seam; the purpose of this is to seal the edge of the laminate, which is inside the cylinder, and so prevent filled product from penetrating the paper layer, or organisms in the paper contaminating the product.

As the continuous cylinder moves downward, a series of transverse heat seals is made by jaws that move down at the same speed as the cylinder. These seals have the effect of closing

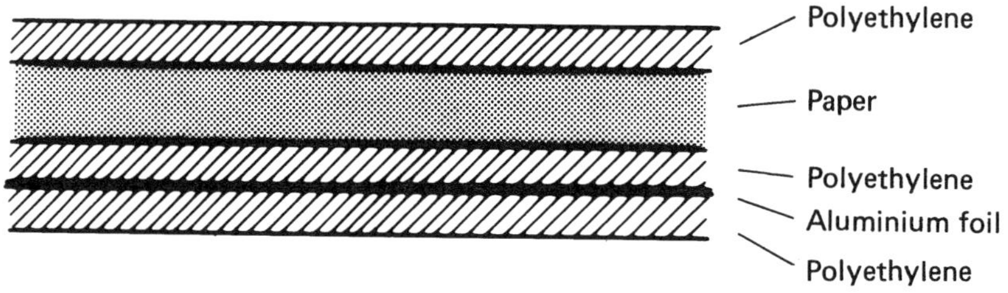

Figure 7–7 Carton material structure. *Source:* Reprinted with permission from H. Burton, *Ultra-High-Temperature Processing of Milk and Milk Products*, p. 193, © 1988.

the bottom of the cylinder, so that it can be filled with product. This is done through a filling tube from above, and a float-operated filling valve at the outlet of the tube maintains the liquid level above the sealing level (Figure 7–9). The seals are made through the liquid, so that considerable pressure is needed to expel the liquid from the sealing zone and make good contact between the plastic layers that have to be fused together. In the latest designs, the heat to melt and fuse the layers in contact in the overlapping joint is provided by induction heating of the aluminium foil component of the laminate, from coils in the sealing jaws.

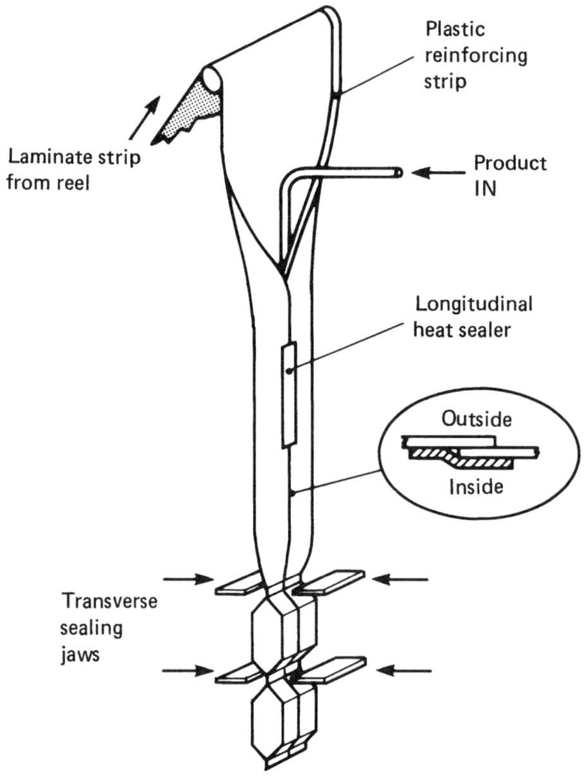

Figure 7–8 Paperboard carton formed from the reel. It is formed by shaping rolls into a cylinder. *Source:* Reprinted with permission from H. Burton, *Ultra-High-Temperature Processing of Milk and Milk Products*, p. 194, © 1988.

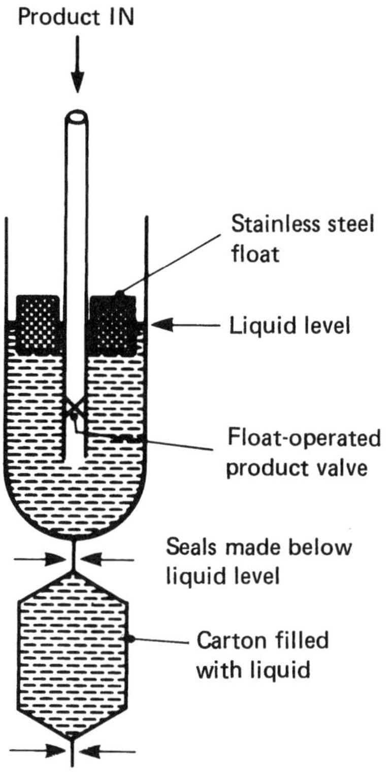

Figure 7–9 As the continuous cylinder moves downward, transverse heat seals are made, which close the bottom of the cylinder for filling. *Source:* Reprinted with permission from H. Burton, *Ultra-High Temperature Processing of Milk and Milk Products*, p. 195, © 1988.

In the earliest forms of this system, alternate transverse seals were made at right angles to each other, so that the cylinder was transformed into a series of tetrahedrals, each filled entirely with liquid. The volume of each was determined by the diameter of the formed tube and the longitudinal distance between each seal. After sealing, the individual tetrahedra were divided by knives to form the separate packs.

In the later forms, the transverse seals are all in the same plane, and the filled volumes between adjacent seals are pressed by molds into rectangular blocks. After the individual blocks are separated by knives, the "wings" at each end are bent down and heat sealed to make a true block form. The volume of each block is determined as above, by the diameter of the cylindrical tube that is first formed and by the longitudinal distance between adjacent transverse seals. Although the most common method of operation involves cartons that are completely filled with product, it is possible to fill cartons to include a headspace by arranging the float-controlled filling valve to give a product level below the top transverse seal. The filled volume is then determined by the carton dimensions and the filling level as set by the valve.

Before the cartons are formed and filled in this way, the packaging material is sterilized with hydrogen peroxide and heat. Two alternative methods are used (Figure 7–10). In the first (Figure 7–10(a)), the strip of laminate, after the reinforcing polyethylene strip has been bonded to one edge, passes through a bath of H_2O_2 at 35% concentration. A wetting agent is added to the peroxide to improve the formation of a liquid film on the laminate, which then passes through a pair of rollers to remove excess liquid. After the container material has been formed and sealed into a cylinder, it passes a tubular electric heater that heats the inside surface. The heater raises the temperature of the laminate surface that will be in contact with the product to about 120°C to sterilize the surface and remove the H_2O_2. The sterilizing effect of this process is 4 to 6 decimal reductions of spores of *B. stearothermophilus*.

In the second system (Figure 7–10(b)), the laminate strip with the longitudinal seam-reinforcing strip sealed to one edge passes through a deep bath of hot H_2O_2 (35%) at a temperature of 78°C; the time spent in the liquid is 6 seconds. Squeezer rollers remove much of the peroxide from the packaging material and return it to the bath. Air at 125°C is directed through nozzles onto both sides of the laminate, to heat it for increased sterilizing effect and to evaporate the peroxide. This process gives a sterilizing effect of 5 to 7 decimal reductions of spores of *B. stearothermophilus*.

Because of its high temperature, the peroxide solution becomes concentrated by evaporation of water. As a routine, the concentration of H_2O_2 should be checked each day, and when

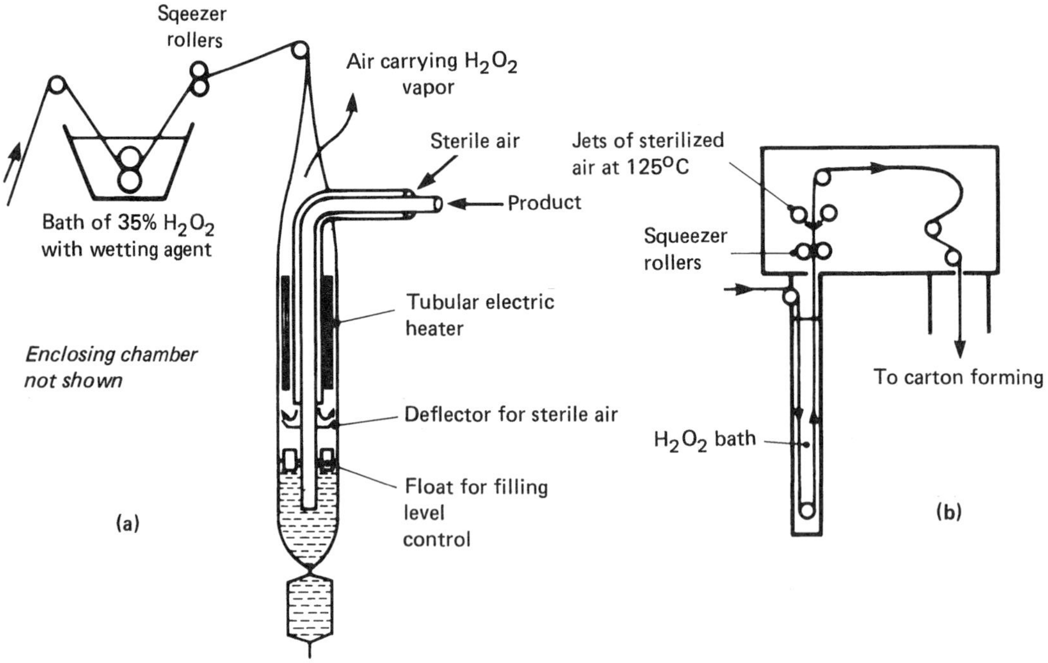

Figure 7–10 Before the cartons are formed and filled, the packaging material is sterilized with hydrogen peroxide and heat by one of two alternative methods, (a) and (b). See text for descriptions of each method. *Source:* Reprinted with permission from H. Burton, *Ultra-High-Temperature Processing of Milk and Milk Products*, p. 197, © 1988.

it reaches 40%, water should be added to restore the original concentration. Since the chemical stability of H_2O_2 solutions depends on added stabilizers, which are diluted by water addition, solution in the bath should in any event be discarded after 120 hours of operation.

To protect the sterilized carton material and the product before and during filling and sealing, both the alternative systems are contained in a chamber maintained at an overpressure of 0.5 bar with sterile air. This air is circulated in a closed circuit using a water-ring compressor. It is heated to 380°C for sterilization and then cooled to about 55°C. With the first material sterilization method described above, the air is supplied through an annular tube outside the product filling tube to a point immediately below the tube heater, where it is deflected upward to carry away the peroxide residues evaporated from the surface of the carton material (Figure 7–10(a)). In the second sterilization system, the fully cooled air for pressurization is again applied through an annular tube around the product tube to a point just above the filling level. However, a part of the sterile air is cooled only to 125°C for supply to the nozzles, which remove the peroxide from the cartoning material after the peroxide bath and which complete the material sterilization by final heating (Figure 7–10(b)). In both cases, the pressurizing air in the sealed chamber contains peroxide residues evaporated from the cartoning material. These are removed by recirculation of the air through the water-ring compressor, where the peroxide is dissolved in the sealing water: to conserve water, this is then normally used in the air cooler before discharge.

Aseptic filling systems of this kind can give carton sizes from 125 ml to 1,000 ml, with outputs of 33 to 100 cartons per minute, depending on the type of machine and carton size. A reel of carton material may provide 1.5 hours of production of larger cartons, but up to 4.5 hours of production of small cartons. Replacement reels can be spliced during operation, automatically with some machines.

In an alternative system, the polyethylene-paperboard-aluminium laminate is taken horizontally from the reel and formed into a continuous cylinder by butt-welding of the longitudinal seam, with inner and outer supporting plastic strips, using induction heating of the aluminium layer. After the continuous cylinder is cut into appropriate lengths for the required carton size, each section is fitted with a laminated base formed from continuous-strip material, heat sealed in place. The carton is then erected and sterilized internally, filled, and sealed with a lid, again made from strip laminate. These stages of the process are similar to those used with preformed paperboard cartons and described in the next section. This type of filler can produce cartons of 150 to 1,000 ml capacity, and operate at a rate of up to 200 packages per minute in a four-lane sterilizing, filling, and sealing system.

7.7.2.4 ***Preformed Paperboard Cartons Assembled from Blanks on the Filler***

In systems of this type, the cartons are manufactured and partly assembled elsewhere, with the longitudinal seams completed and creases applied for folding, but with the cartons in lay-flat form ready to be finally shaped in the filler and the tops and bottoms formed and bonded. The partly assembled lay-flat blanks are shipped in boxes in stacks, ready for loading into the filler. For aseptic operation, the boxed blanks are frequently presterilized with gas (e.g., ethylene oxide) before the boxes are sealed to maintain sterility during shipment.

Stacks of blanks are loaded into the aseptic filling machine. They are individually extracted by suction, opened, and placed on a mandrel, and the base is folded and heat sealed.

The carton is transferred to a pocket on a conveyor, and the top is prefolded. All these operations take place under nonsterile conditions, although steps are taken to avoid recontamination. The empty cartons are then carried by the conveyor into the aseptic area of the filler.

In all aseptic fillers of this type, the aseptic area consists of several separate functional zones where operations are carried out in sequence as the carton is carried by the conveyor step-by-step through them (Figure 7–11). Sterility is maintained in each zone by a slight overpressure of sterile air, obtained by the high-performance filtration of outside air. Careful distribution of the sterile air is needed to prevent bacteria from entering the aseptic area from the surrounding atmosphere through the entry and exit ports and any other openings. Bacteria may also be carried into the first sterile zone on the unsterile outer surface of the erected carton, and it is important that these bacteria should not be detached and carried by the air flow into the headspace of the filled cartons further along the conveyor.

The first functional zone (Figure 7–11) is that in which the inside surface of the carton is sterilized. First, the inner surface of the carton is coated with a film of H_2O_2. In the most common method, a 35% solution without a wetting agent is applied while the carton is stationary beneath a jet, either as a fine spray or as peroxide vapor in hot air so that the vapor condenses as liquid peroxide on the container surface. The spray is more economical in its use of peroxide solution, 0.1 to 0.2 ml of solution being used per 1-L carton. It is, however, claimed that the vapor method gives more uniform coverage of the surface, and this is important for consistent sterilization.

The surface sterilization is completed and the H_2O_2 is removed by the jetting of the cartons with hot air at 170° to 200°C. This air is obtained by heating, usually with gas, part of the air sterilized by filtration for use in the aseptic area of the filler. Several consecutive jetting stations are used, usually between four and eight, with hot air applied while the carton is stationary under each.

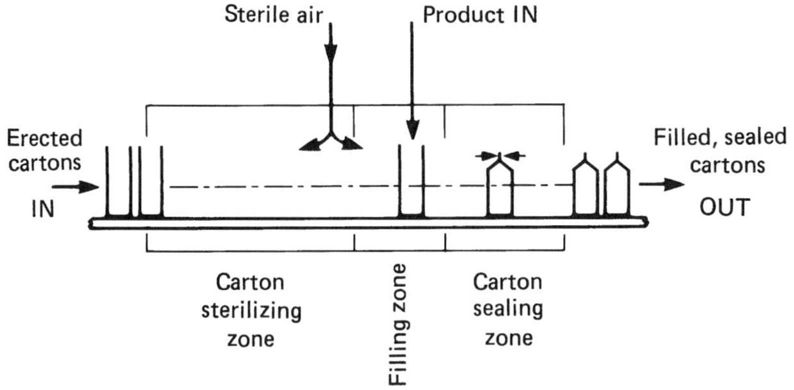

Figure 7–11 The carton is carried by the conveyor step-by-step through several separate functional zones. *Source:* Reprinted with permission from H. Burton, *Ultra-High-Temperature Processing of Milk and Milk Products*, p. 199, © 1988.

Application of the peroxide normally takes 2 to 4 seconds, and its removal by hot air for 4 to 8 seconds. In an alternative system, the interior surface of the carton is sprayed uniformly with 0.1 ml of a much weaker solution of peroxide, 1% to 2% concentration, and is then irradiated for about 10 seconds with high-intensity UV irradiation. The peroxide is then heated and removed by hot-air jets as above. The complete process has a sterilizing effect similar to that of the two-stage peroxide-hot air process. Because the total amount of peroxide used is some 20 to 30 times less, the problems of residual peroxide in the carton (Stannard & Wood, 1983) and of peroxide contamination in the surrounding atmosphere are more easily dealt with. In any case, the peroxide-contaminated air from the sterilizing zone must not be allowed to discharge freely into the space surrounding the filler, because of undesirable effects on the operators, but it must be piped away to the open air.

Uniform cover of the carton surface with peroxide solution is essential for good and uniform sterilization. Since the inner surface is invariably a hydrophobic plastic material such as polyethylene, the solution will not of itself spread to give good coverage. The smaller the particle size in the spray, the more complete will be the surface cover. Good design of spray systems is therefore important, and vapor systems have been used as alternatives. However, vapor systems waste more peroxide in the general flow of filtered air, and make the disposal problem more difficult.

After the peroxide has been removed completely from the carton surfaces, the carton passes to the second aseptic area, where it is filled (Figure 7–11). The types of filler used vary, but they are all of a type that dispenses a fixed volume of product into the carton, leaving a headspace. Different types of piston fillers are commonly used, modified to eliminate or reduce crevices that might be difficult to clean and sterilize, and constructed of materials that are sterilizable. It is claimed that one type of aseptic filler for cartons can accept suspended particles up to 15-mm size.

Foaming can be a problem with piston fillers operating on some products. Foam can contaminate the sealing surfaces of the carton, making it difficult to obtain a good, bacteria-tight seal. It can also soil the filling area, causing problems in keeping it clean and sterile. A foam-removing system may therefore be included in the filler, and this must be within the sterile zone, with the product removed being stored in a sealed container until the end of the filling run.

Finally, the filled carton passes to the final aseptic zone, where its top is closed and heat sealed (Figure 7–11). This is an important process, because the carton must be resistant to leaking and bacterial contamination. As an alternative to simple heat sealing, ultrasonic welding may be used.

Conventional cartons are of the "gable-top" type. However, there is nowadays a demand for cartons of block type, which can be packed one above another. After the sealed cartons have left the sterile area, they can be further shaped into a block shape if required.

Aseptic fillers of this type can be used with carton sizes from 200-ml capacity to 0.5 US gal (1.9 L). With carton sizes up to 1 L, throughputs up to 105 cartons per minute are possible in twin-line machines. With larger cartons, 65 cartons per minute can be filled in single-line machines.

7.7.2.5 Plastic Pouches

A simple, cheap form of container for liquids is the pouch. Heat-sealable plastic film is used, fed continuously from a roll. The film is folded longitudinally, and the edges heat sealed to make a flat vertical tube. Transverse heat seals divide the tube, so that measured volumes of liquid product can be filled into the space above a seal. A further seal above the liquid produces a pillow-shaped flexible pouch or sachet that can be separated from the tube as an individual pack.

The process is similar to that used to form cartons from the reel, but the containers are flexible and not self-supporting, and in general a volumetric filler is used. The system has been used widely as a low-cost process for pasteurized milk and other nonsterile products. It was developed in modified form for aseptic filling, but was initially unsatisfactory. The simplest type of film—polyethylene—was used, which, because of its transparency and oxygen permeability, is unsuitable for the prolonged storage of liquid milk or milk products. Furthermore, unsuitable film-sterilizing systems were applied. More recent development in materials and methods has now made aseptic filling into sachets commercially practicable.

The packaging material is fed as a continuous strip from a roll. Different film materials may be used, according to the distribution system and expected shelf life of the filled sachets. For short shelf life (up to 2 weeks), a black/white polyethylene coextruded film can be used, which excludes light but not oxygen. If the pouch is to be packed in a separate opaque outer container, white polyethylene can be used alone for this short life.

For longer shelf life (up to 3 months), a film is needed that provides an oxygen barrier as well as light protection. A coextrusion of polyvinylidene chloride (PVDC) or ethylene vinyl alcohol (EVOH) with black/white polyethylene has suitable properties. A coextrusion with white polyethylene alone is suitable if an opaque outer is to be used.

In a typical filler (Figure 7–12), the pouch-forming, filling, and sealing operations take place within a sealed chamber, which is supplied with air sterilized by a bacterial filter that is capable of removing 99.9% of particles of 0.3-μm diameter. The film enters the chamber through a bath of hydrogen peroxide solution where it remains for a minimum of 20 seconds when 1-L pouches are being filled; with smaller pouches the film feed rate is lower, so that, for example, with 250-ml pouches the sterilizing time is almost 50 seconds. With a peroxide concentration of 34% and a temperature of 44°C, a sterilizing effect against bacterial spores of at least 5 decimal reductions is claimed. At the outlet of the bath, the film passes through mechanical scrapers that remove surplus liquid.

The film then passes through the flow of sterile air being supplied to the filling chamber at a temperature of 45°C. The air removes the peroxide from the film, so that its concentration in the filled pouch is less than 1 part in 10^6.

The film is moved forward by grippers discontinuously in lengths sufficient to produce a single pouch. It is folded longitudinally and heat sealed by vertical sealing jaws into a flat tube over the length of a pouch. The transverse seal produced by the forming of the previous pouch closes the bottom of the tube. Product is filled in fixed and controllable volume through a vertical tube and nonfoaming filling nozzle.

The filler is of the positive-displacement, piston type with the filled volume determined by the travel of the piston. For low-viscosity products, the filler is supplied by gravity from a

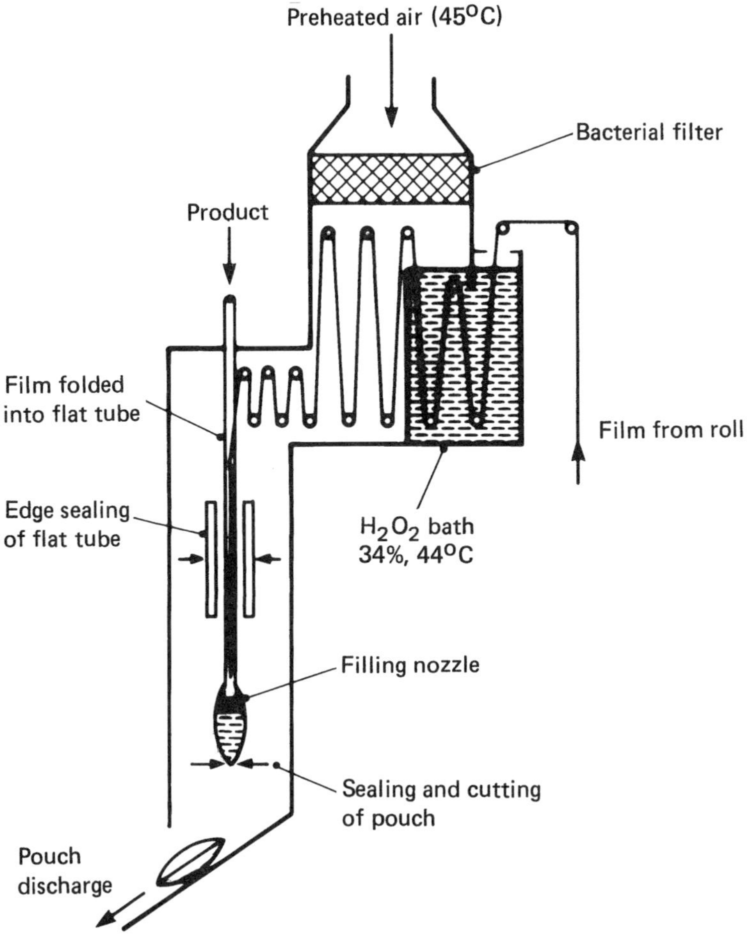

Figure 7–12 A typical filler in which pouch-forming, filling, and sealing operations take place within a sealed chamber. *Source:* Reprinted with permission from H. Burton, *Ultra-High-Temperature Processing of Milk and Milk Products*, p. 203, © 1988.

balance tank; with more viscous products a positive-displacement pump can be used. When filling is complete, the tube is moved down by the length of a pouch, and the top is sealed and cut, so that the filled pouch is detached and the bottom of the next is made at the same time.

In commercial form, the filler consists of two filling heads within a single aseptic chamber, and each head can operate independently of the other. If both are filling the same size of pouch, the throughput lies between 83 pouches per minute for a 200-ml volume and 44 pouches per minute for a 1.5-L volume. The two filling heads can fill different sizes of pouch at the same time.

Filling volumes can easily be varied on the machine without necessarily changing the roll of film, although a wider roll (e.g., 380 mm) is preferred for larger pouches of 1 L and above, and a narrower roll (e.g., 320 mm) is preferred for smaller pouches. The filling accuracy is claimed to be within 1.5% for the smallest pouches and within 0.5% for the largest.

7.7.2.6 Blow-Molded Plastic Bottles

Blow-molded plastic bottles have been used for many years as a cheaper alternative to glass for nonreturnable containers. They have been used for a variety of liquid foodstuffs, including pasteurized and in-container sterilized milks and flavored milks. For in-container sterilized milks they have the advantages, in addition to cheapness, of good heat transfer rates because of their thinness, and of immunity from thermal shock breakage during heating and particularly cooling.

For ease of molding and cheapness, polyethylene or polypropylene has been used for milk. The transparency and oxygen permeability of these materials can cause oxidation problems in milk and milk products intended for a long storage life. Pigments have been used to overcome transparency, but the resulting color of the bottle has not been well accepted by the consumer. However, developments in materials and blow-molding techniques have now made it possible, at somewhat higher cost, to mold bottles from multilayered material that has satisfactory light and oxygen barrier properties. These materials are little used as yet in comparison with polyethylene and polypropylene, but they will provide blow-molded containers that will be more satisfactory, technically and scientifically, for long-shelf-life products (Mottar, 1986).

Aseptic filling systems using blow-molded bottles are of three different types:

1. A standard, nonsterile bottle is sterilized and then filled and sealed aseptically.
2. A bottle is first blown in such a way that it is sterile, and then filled, and sealed aseptically.
3. A bottle is blown aseptically, filled, and sealed consecutively at the same station so that sterility is maintained.

7.7.2.7 Using Nonsterile Bottles

A bottle is blown and trimmed conventionally, usually as the first stage of the aseptic process but possibly separately from the filling line. Polyethylene or polypropylene can be used, or a more complex material for improved keeping quality. The bottles are then conveyed (Figure 7–13) into a sterile chamber, which is kept at a slight overpressure with air sterilized by a high-performance (Class 100) filter. The bottles are inverted and sprayed inside and outside with a solution of H_2O_2. The bottles are erected and then pass down a hot air tunnel in which the peroxide is evaporated and the vapors ducted away. They are then reinverted, rinsed inside and out with sterile water, and erected once more. The bottles are filled using a rotary volumetric filler, and the headspace can be filled with an inert gas if necessary. The bottles are then heat sealed with chemically sterilized plastic film or other heat-sealable closure, and an outer plug or screw cap can be applied. Sterile air is supplied from above in a laminar flow system to the parts of the sterile chamber where the bottles are finally rinsed, filled, and sealed. The bottles then leave the sterile chamber and can be handled for distribution.

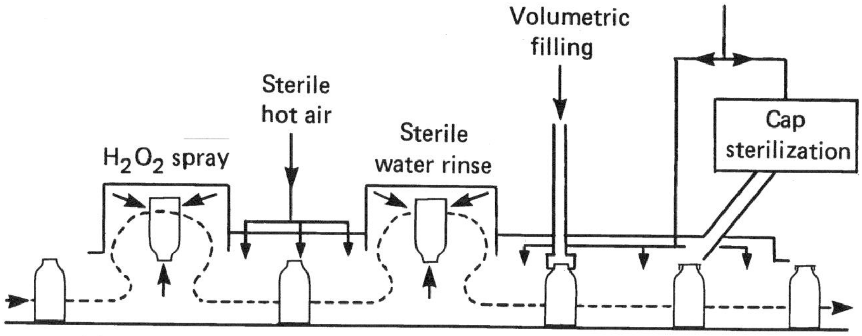

Figure 7–13 Nonsterile bottles are conveyed into a sterile chamber, which is kept at a slight overpressure with air sterilized by a high-performance filter. *Source:* Reprinted with permission from H. Burton, *Ultra-High-Temperature Processing of Milk and Milk Products*, p. 205, © 1988.

It is recommended that the filling system should be installed in a separate room, supplied with sterile, filtered air and with the floor, walls, and atmosphere disinfected regularly. It is also recommended that the operating staff should enter the room through a chamber for washing, disinfecting, and changing clothing.

Typical container sizes range from 250 ml to 1 US gallon (3.8 L). For the smallest size of container, the largest filler can fill and seal 500 bottles per minute; for the largest container the maximum capacity is 106 bottles per minute.

7.7.2.8 Using Sterile Blown Bottles

In this system the bottles are first blown with sterile air in a completely sealed form. The high temperature of the thermoplastic material during molding, and the sterile air, ensure that the inner surfaces are sterile, and since the bottle leaves the mold completely sealed there can be no recontamination before the aseptic filler.

All the operations are performed within a single, large chamber with glass walls, supplied with sterile, filtered air in laminar flow from high-performance bacterial filters in the top of the chamber (Figure 7–14). The sealed bottles are carried into the chamber, and the outsides are first sterilized with H_2O_2 sprays. The closed top of the bottle is cut away and the neck trimmed. The bottle is then filled using a rotary volumetric filler; the headspace can be filled with inert gas if required.

Foil caps or any other heat-sealable closures are chemically sterilized outside the chamber, fed under the protection of sterile air into the aseptic chamber, and heat sealed to the bottles. If necessary, any suitable screw- or plug-type overcap can be applied if required. The bottles are then carried from the chamber ready for subsequent handling.

As with the previous system using nonsterile bottles, it is recommended that the filler should be installed in a room that can be disinfected and supplied with sterile air, and having a separate chamber for the operating staff to enter after personal cleaning. The container sizes that can be used, and the system capacities, are the same as those for the previous system.

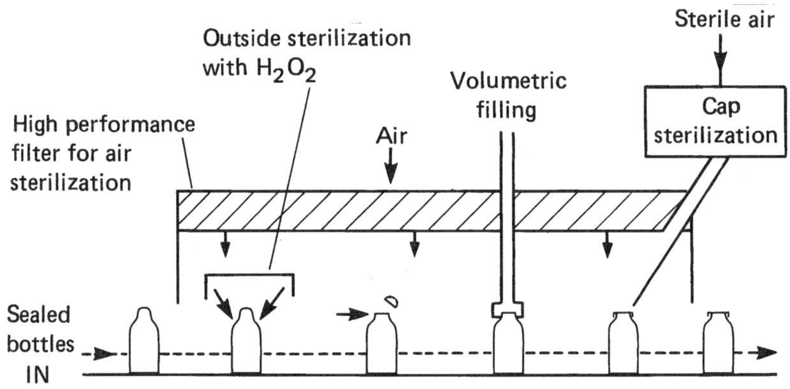

Figure 7–14 Sterile blown bottles. All operations are performed within a single, large chamber with glass walls, supplied with sterile, filtered air in laminar flow from high-performance bacterial filters in the top of the chamber. *Source:* Reprinted with permission from H. Burton, *Ultra-High-Temperature processing of Milk and Milk Products*, p. 206, © 1988.

7.7.2.9 Bottles Blown, Filled, and Sealed at a Single Station

This system is unique, being offered by a single manufacturer. It is mechanically complex, with the separate operations of parison extrusion, blow molding, bottle filling, and sealing all taking place in sequence in a single head. It is used in the pharmaceutical industry for the packaging of sterile liquids, and is used in the food industry for edible oils and vinegar as well as for fruit juices, milk, and milk products.

These combined bottle-forming and filling operations are at present only practicable with polyethylene or polypropylene. Light transmission through the container can be reduced by the incorporation into the molding material of titanium oxide, aluminium powder, or dyes acceptable for foodstuffs, but the oxygen permeability remains high, with the consequent risk of adverse effects on the product during storage.

The sequence of operations is as follows:

1. The plastics granules are worked and softened in an adiabatic extruder. A temperature of about 200°C is reached and held for about 3 minutes, which ensures sterility of the material. The material is then extruded in the conventional way for blow molding in a vertical cylindrical parison of suitable length and diameter for the container to be produced.
2. The mold closes on the parison, sealing the bottom, and the container is blown with sterile air conventionally through the top of the parison.
3. The product filling tube is concentric with the air-blowing tube, and after blowing is complete the container is filled using a volumetric piston-type system; the air is discharged from the container during filling through a vent. The filling accuracy is stated to be within 0.5%.
4. When filling is complete, the blowing and filling nozzles retract, and jaws in the mold head close to seal the bottle, form the neck, and form a suitable opening device.

For machines of relatively low throughput, all the operations take place within a single mold, although, depending on container size, more than one container can be formed and filled in the single mold.

In machines of larger throughput, lines of half-molds circulate in a vertical plane about the axis of the parison, which is being continuously extruded. Each pair of molds comes together at the top as the parison is first extruded. As the molds and parison move downward, the container is first blown and then filled through pipes that extend down the axis of the parison. At the bottom, after the container has been blown, filled, and sealed, it is cut from the parison and released as an individual pack while the half-molds separate and recirculate to the top of the parison.

In this system, sterility at all stages is ensured by the method of operation, and no special arrangements are required such as aseptic chambers protected by sterile air. Sterility of the inside surface of the container is ensured by the high temperature of the plastics material during extrusion of the parison and the use of sterile air for blowing. There is then no access to the surrounding air while the container is filled and sealed, still within the closed mold. Machines suitable for the aseptic filling of milk and milk products have container sizes from 200 ml up to 1 L, with filling rates up to 100 bottles per minute in the smaller sizes and 67 bottles per minute in the 1-L size.

7.7.2.10 Preformed Plastic Cups

There are perhaps more types of aseptic fillers for plastic cups than for any other type of container, but many are similar in their operating principles and differ only in detail. The cups are normally of polystyrene or polypropylene, but there is no reason why more sophisticated materials with better barrier properties should not be used. Products aseptically filled into cups range from milk and cream to formulated products such as custards and multilayer desserts in which several products have to be filled in succession into the same pot. All the aseptic cup fillers conform in general to the sequence of operations of the fillers for cartons made from preformed blanks, as described above and shown in Figure 7–11. The first, nonaseptic stage is the restocking of the cups and feed into pockets on the conveyor, instead of the assembly and initial sealing of the cartons. A cup also needs a separate lid or other closure that needs to be sterilized and heat sealed into place after the filling stage.

The operations take place within a sterile tunnel, supplied with sterile air from a high-performance filter, normally arranged in laminar flow from above the sterile area. The air-supply arrangements should be such that bacterial contaminants, for example, any that may be carried into the sterile area on the outside of a cup and therefore not actively sterilized, are carried away from the filling area and not toward it. An extraction system may be fitted below the conveyor so that any contaminants are carried through the conveyor and away from the cups.

The first operation is cup sterilization. Spraying the inside of the cup with hydrogen peroxide solution, and then removing the peroxide with hot air, may be considered the conventional method, established during many years of use not only with cup fillers but also with carton fillers. In one typical and good example, 35% H_2O_2 solution is sprayed into the cup (35 mg for a 250-ml cup size) and remains for about 3 seconds before the application of hot, drying air. The drying air passes through a high-efficiency filter bank to give not more than 100 particles of >1 μm size per cubic foot (=0.028 m^3) of air. The air is then compressed and heated to a maximum of 400°C, depending on the material of which the cups are made. It is injected into the cups at a series of stations to raise the surface temperature of the inside of the cup to about 70°C, to complete the surface sterilization and reduce the peroxide residues to acceptable levels. Higher surface temperatures can be used with polypropylene than with polystyrene. The sterilization performance is stated to be a minimum of 3 decimal reductions of spores of *B. globigii*. If packs of cups are opened and used immediately, the average contamination level has been found to be about 0.7 organism per cup, with a maximum of 10 organisms per cup. About 10% of these are spores. On average, therefore, the number of nonsterile cups after this process will be significantly less than 1 in 10^4. Opened packs of cups should not be stored before use, as static electricity on the plastics material will attract dust and bacteria and greatly increase the difficulty of sterilization.

The H_2O_2 vapor evaporated from the cups is extracted with the air from below the conveyor and ducted away, so that a peroxide concentration of 1 part in 10^6 is not exceeded in the atmosphere near the filler.

Modifications to this "conventional" system are also used. The effectiveness of coverage of the container surface with peroxide solution depends on the size of the spray particles; the smaller the particles, the greater the proportion of area that can be covered with liquid. According to Cerny (1984), a conventional spray gives drops of over 30 μm diameter on the surface, and only 30% to 40% of the surface area is covered.

An ultrasonic system can be used to give particle sizes of only 3-μm diameter (Anon., 1986), which will give an average surface cover of about 60%. A small amount of 30% to 35% H_2O_2 solution is held in a vessel with an ultrasonic transducer in its base. With a suitable ultrasonic frequency (about 1.7 MHz) and energy, standing waves are developed in the solution that project a mist of 3-μm particles into the space above it. The mist is carried by air into the cups. To give uniform and effective cover with such small particles, and to prevent them from coalescing into less-effective larger ones, they are given a negative electric charge by corona discharge from a point source maintained at 20 to 50 kV before the mist is injected. At the injection point, the cup is surrounded by an earthed electrode so that the droplets are attracted outward to the beaker surface and deposited uniformly.

With this system, and a peroxide contact time of 2.4 seconds or more before drying with hot air, a sterilizing effect against *B. subtilis* spores of about 4 decimal reductions is claimed. The amount of peroxide used is 40 mg/200-ml cup (i.e., similar to that used in the conventional sterilization process described above).

Yet another way of applying peroxide solution is to use a vapor in hot air, which condenses on the container surface. It is said that the condensed particles are an order of magnitude smaller than those given by a conventional spray. However, there will be some loss of peroxide in forming the vapor and through incomplete condensation.

Cups can also be sterilized by carrying them through a peroxide bath before heating. Such a system is shown in Figure 7–15. Polystyrene cups are loaded onto a conveyor and carried through a bath of 35% H_2O_2 at a temperature of 85° to 90°C with a transit time stated to be about 2 seconds. The bath forms a seal between the outside atmosphere and a heating chamber in which the wetted cups are said to be kept at the temperature of the hot peroxide solu-

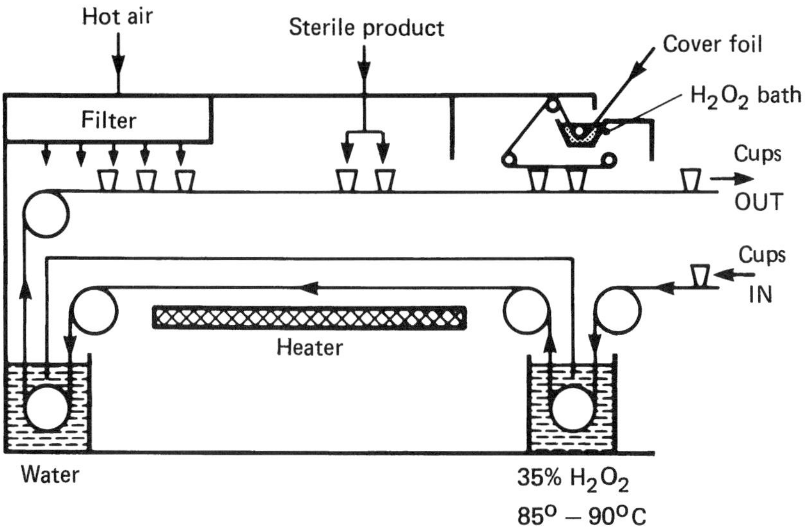

Figure 7–15 Cups are sterilized by carrying them through a peroxide bath before heating. *Source:* Reprinted with permission from H. Burton, *Ultra-High-Temperature Processing of Milk and Milk Products*, p. 211, © 1988.

tion for about 5 seconds. The cups are carried by the conveyor out of the heating chamber, through a water bath, which acts as a further seal for the chamber and removes some of the peroxide film, into a separate upper chamber that is pressurized with sterile air.

The upper chamber corresponds to the sterile tunnel in which all the operations take place as in other aseptic cup fillers. In this case, when the cups are carried into the sterile upper chamber, their sterilization is completed by spraying with sterile water and drying with hot air. There are many theoretical arguments used to support the alternative methods of peroxide sterilization. Ultimately, as we have seen earlier, the effectiveness of sterilization can only be determined empirically, and the relative merits of the alternatives established by practical performance.

In one cup filler, polypropylene cups are sterilized by saturated steam under pressure (Cerny, 1983). Thermal sterilization by saturated steam is more predictable than chemical sterilization, and it leaves no residues other than sterile condensate.

The nested cups are divided into groups of 18 to 20, inverted, and loaded into a vertical, cylindrical sterilizing chamber. The chamber is sealed top and bottom, evacuated to remove air that would otherwise interfere with heat transfer to the cups, and then pressurized with saturated steam at 3 to 3.5 bar, corresponding to 140° to 147°C. The vacuum is applied for 3 seconds and the steam for 3 seconds, by which time the full pressure and temperature have been reached. The sterilizing conditions are held for about 1.2 seconds and the chamber is then vented to atmospheric pressure in about 3 seconds. Finally the cups are cooled with cold, sterile air and discharged from the bottom of the cylinder into the sterile chamber of the filler. They are separated individually from the group and inverted, to be placed into pockets in the filler conveyor.

The cups are inverted for sterilization, so that condensate does not remain in them to dilute the product but is ducted away with the vented steam at the end of sterilization. The sterilizing effect of this process against spores of *B. subtilis* is 5.5 decimal reductions with 3-bar steam pressure and 7 decimal reductions with 3.5 bar. After cup sterilization, whether by peroxide or steam, in all types of filler the cups are carried by the conveyor into the filling zone. To prevent contamination during filling, this area may be supplied with a separate sterile air supply obtained by filtration through an absolute bacterial filter. The headspace in the top of the filled cup will then also be filled with air of the highest bacteriological quality. Alternatively, a sterile inert gas can be supplied to fill the container headspace.

The filler is of the positive-displacement type, normally a piston filler designed and made of such materials as to be cleanable in-place, and sterilizable by steam under pressure.

The filling section will sometimes be divided into several parts, so that layers of different products can be filled one after another. Each section will be a separate aseptic filler. Finally, the cup is sealed with a suitable closure, which, like the cup itself, must be sterilized effectively. The conventional sterilization method with coated aluminium foil is to pass the foil strip through a solution of H_2O_2 (a 35% solution containing wetting agent is typical), which also acts to isolate the sterile sealing zone from the surrounding zones. The peroxide solution is then removed from the foil by radiant heat, by hot sterile air ducted in countercurrent flow along the foil surface, or by passing the foil over a heated roller. The sterilizing effect obtained is similar to that for the cup sterilization process. In some systems, UV irradiation may be used either alone or in conjunction with peroxide; UV alone may be more effective with closures than with cups because a flat surface is being irradiated, which can be brought close to the UV source to increase the radiation intensity. After sterilization, the closures are stamped from the foil and heat sealed onto the cups within the sterile zone. The cups can then be discharged from the filler. Other types of closure (e.g., aluminium tops with pull tags) can be sterilized with peroxide and heat and applied in a similar way.

The cup-filling system that uses pressure steam sterilization of the cups, as described above, also uses steam sterilization of the coated aluminium capping strip. The conveyor system links every part of the aseptic filler, and may be a contamination risk since in most systems it is not actively sterilized with the cups or the caps. In some systems, therefore, the conveyor belt is cleaned with water and high-pressure steam and sterilized with an H_2O_2 spray as it returns from the outlet of the filler to the inlet. Where a peroxide bath is used for cup sterilization (Figure 7–15), the conveyor is automatically resterilized with the cups. Typical cup-filling systems will accept cup capacities between about 90 and 600 ml, and fill to an accuracy of about ± 0.5%. Most fillers will accept liquids such as milk, semiliquids such as sauces and custards, and liquids containing particles, provided that the particle size is not too great. For a single lane of the filler, the throughput may be up to 60 cups per minute, depending on the time required for filling, which itself depends on the viscosity of the product. Fillers of up to 10 lanes are manufactured.

7.7.2.11 Form-Fill-Seal Systems

In form-fill-seal fillers, the packaging material is a web of plastic material fed from a roll. The containers are first produced by thermoforming with most materials but by mechanical forming with others, to give multiple containers still in web form. The containers are then filled, and then heat sealed with a plastic or plastic-coated aluminium closure. Finally, the individual containers are separated from the web.

The simplest packaging material is polystyrene, which is commonly used because it is easily thermoformed. However, more complex, coextruded, multilayer materials can be used, such as polystyrene/PVDC/polyethylene or polystyrene/EVOH/polyethylene, to reduce oxygen penetration and light transmission and so prevent chemical deterioration of the filled product during storage.

A multilayer film containing an aluminium foil as a light and oxygen barrier can also be used for the container, in which case mechanical forming rather than thermoforming is used. The container-forming process has limitations on the dimensions of the container produced, mainly because of the restricted depth that is obtainable from the deformation of a flat roll-fed sheet. Form-fill-seal systems are therefore used more commonly for the smaller sizes of container.

The basic form-fill-seal systems have been developed for aseptic operation by incorporating sterilization of the container material after it leaves the roll, and similarly of the closure material. The sterilization, forming, filling and closing operations take place within a tunnel, previously sterilized and supplied with sterile, filtered air as in many of the systems we have previously considered. The sterilization of the packaging material normally involves combinations of hydrogen peroxide and heat, as we have described previously.

In a typical aseptic form-fill-seal machine, the web of container material, of a thickness up to 1.6 mm and up to 570-mm wide, leaves the roll and enters the sterile machine chamber through an air lock formed by a bath of 35% H_2O_2 at room temperature. The peroxide is pumped from a supply tank, which also feeds the sterilizing bath for the lid foil, so that the peroxide in contact with the film is constantly agitated to give some mechanical cleansing. The container material takes about 15 seconds to pass through the peroxide bath, and surplus liquid is removed by air knives as the web leaves the bath vertically.

As the web moves horizontally through the sterile tunnel, it is first heated by hot platens, which are brought into contact with its upper and lower surfaces to remove peroxide residues and to preheat the web material to 130° to 150°C to prepare it for thermoforming. In other types of machine, radiant heat is used to heat the web after it has left the peroxide bath. The containers are then formed within the web, into a water-cooled mold below the web, usually by a combination of mechanical forming and compressed air; the compressed air is that used to pressurize the sterile chamber, and is sterilized by an absolute filter excluding all particles larger than 0.5 μm. The containers are then filled with a suitable design of positive filler, usually piston operated.

The closing film is fed into the sterile tunnel through a bath of peroxide solution forming an air lock, the peroxide being removed by a combination of air knives and heating as with the main container web. After it is laid over the web of containers, the film is heat sealed along both sides of the web to protect the filled containers from bacterial recontamination. The covered web of containers then leaves the sterile chamber; the individual containers have the closure material heat sealed in place, and are cut and separated from the web under nonsterile conditions.

In one type of form-fill-seal machine, the plastic material from which the containers are formed is not sterilized by a combination of H_2O_2 and heat before the thermoforming stage, and the lidding material is not sterilized by peroxide. In this machine, the unsterile film (e.g., polystyrene) for the containers is first preheated and thermoformed, and the formed containers are then sterilized by saturated steam under pressure, at 3 to 6 bar (135° to 165°C) for about 1.5 second (Cerny, 1982). The aluminium/plastic sealing foil is similarly sterilized by saturated steam under pressure. Experiments have shown a surface sterilizing effect of 5 to 6 decimal reductions of bacillus spores with this method. The other operations are as described above, and take place within a sterile chamber protected by sterile, filtered air as before.

In form-fill-seal systems, the mold must be designed for the shape and size of container required, and the positive displacement filler designed and set to dispense the correct volume into that container. Containers with capacities up to about 250 ml can be formed and filled, but most aseptic form-fill-seal machines are used to fill small, so-called "portion packs" of UHT milk or cream with a content of up to about 15 ml. Output depends on the width of the container web material, the dimensions of the formed cup, and the cycle time of the machine (usually about 20 cycles per minute). With small cups, outputs up to 175 per minute are common; outputs as high as 400 per minute have been quoted.

Systems exist in which the container web and the lid foil are disinfected by dry, radiant heat. As we have seen previously, this form of heating does not give effective surface sterilization unless the temperatures reached are high and the treatment times long. Although the method may give acceptable results when filling high-acid products such as fruit juices, it will not be acceptable with low-acid products such as milk.

An alternative version of the form-fill-seal system avoids the need to sterilize actively the surfaces of the container and lid materials. Coextruded multilayer films are used in which one outer layer of polypropylene can be peeled away as the first operating step within the sterile chamber. The coextrusion is sterile throughout because of the high temperatures reached during the extrusion process. When the outer layer is peeled away, a sterile inner surface is exposed (Figure 7–16) for both container forming and cap application, so that no further sterilization of the product contact surfaces is needed. After the outer layer is removed from the container film, the remaining coextrusion is heated by radiation and then thermoformed conventionally and filled. The film to close the containers is then applied and heat sealed in place. Finally the web of filled and sealed containers leaves the sterile chamber, and the containers are cut from the web under nonsterile conditions. Ideal container capacities for this system are said to be 500 ml or less, and filling rates up to 800 containers per minute are possible.

7.7.2.12 Glass Bottles

Aseptic filling into glass bottles was one of the first processes to be attempted after aseptic filling into cans, and before aseptic filling systems suitable for cartons and plastics containers of different types were available. At that time, chemical surface-sterilization systems had not been established, and the bottles were sterilized either by saturated steam under pressure or by dry heat (Burton, 1970). When dry heat was used, the high temperatures needed for bottle sterilization led to a high risk of thermal-shock bottle breakage when cool product was filled, unless extended cooling with sterile air was used. With saturated steam, air had to be

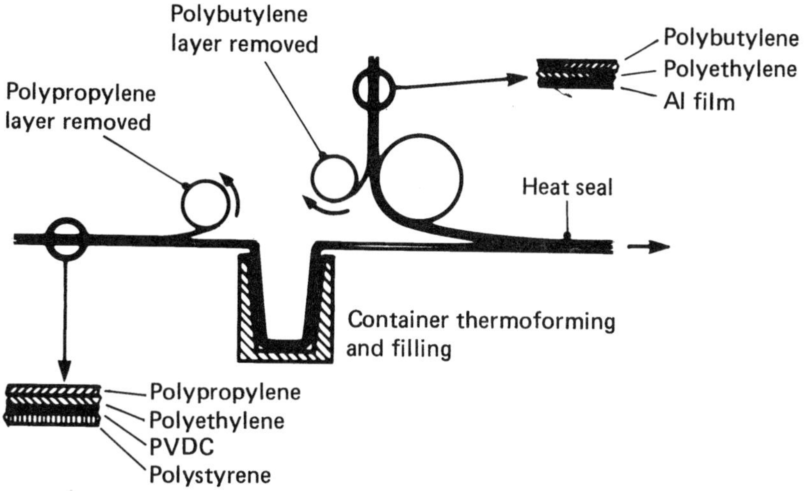

Figure 7–16 An alternative version of the form-fill-seal system. *Source:* Reprinted with permission from H. Burton, *Ultra-High-Temperature Processing of Milk and Milk Products*, p. 216, © 1988.

removed by special means before the steam was applied, in order to give good heat transfer to the bottle surface. Reduction of condensate levels was also a problem.

None of the prototype aseptic fillers for glass bottles reached commercial operation, and the successful development of systems for cartons and plastic containers led to a reduction in interest in glass containers. Development has continued at a low level, however, and several systems for aseptic filling into glass bottles have been developed (IZG, 1984). One of these uses the earlier method of dry heat sterilization, followed by cooling and filling in a sterile area. Others use hydrogen peroxide sterilization in a bath or with spraying, followed by drying with hot air, filling, and sealing, all within a sterile chamber protected by a supply of sterile, filtered air. This latter method is identical in principle to that used for blow-molded plastics bottles, one form of which is illustrated in Figure 7–13. It remains to be seen whether these new systems find a place in industry. The extra cost of a nonreturnable glass bottle as compared with some form of carton may have to be absorbed by the use of bottles for higher-value products that can sustain a higher price.

7.7.2.13 Aerosols

A small-scale and specialized type of aseptic filling is the filling of UHT-sterilized whipping cream (about 30% fat content) into aerosol cans. This form of aseptic canning is very different from the system for conventional cans considered earlier.

The empty aerosol cans are carried on a conveyor, and inverted and cleaned by spraying with potable water at room temperature for 5 seconds, and then with 20% hydrogen peroxide at 20°C. The cans are then erected and filled with 20% H_2O_2 at 20°C. After 10 seconds they are inverted again and allowed to drain for 10 seconds. They are then dried in a hot air tunnel for 6 minutes, using air at 150°C that has first passed through a high-efficiency bacterial filter; a minimum can temperature of 135°C should be reached.

The dispensing valves are presterilized by irradiation, and are applied in a sterile laminar-flow chamber supplied with air sterilized by filtration.

The aerosol cans must be able to withstand H_2O_2 and temperatures that may reach 150°C, as well as the normal conditions for an aerosol.

7.7.2.14 Bulk Filling Systems

A bulk pack of product is one that is larger than is suitable for domestic use, but is convenient for higher-volume uses such as catering. Most of the aseptic filling systems we have considered so far have a maximum container capacity of about 1 L (i.e., suitable for domestic use). A bulk system may be defined as one able to fill aseptically, and maintain sterile, volumes above about 5 L suitable for catering use or for wholesale marketing. The maximum volume for which aseptic bulk systems are used is at present 1,000 L (1 tonne weight).

The bulk shipment of even larger quantities is possible. In the 1950s, UHT milk aseptically filled into a bulk tank was shipped by rail from Bern, Switzerland, to Assisi, Italy. An aseptic road tanker has been used for the shipment of sterile milk between England and France for equipment test purposes. However, these movements of very large volumes have been purely experimental, and it seems unlikely that anything of the sort will be attempted commercially in the near future because of the economic risks involved in the possible spoilage of these quantities of product.

A bulk aseptic filling system for 55-gal (208-L) metal drums has been available for many years. The drum is placed in a small vertical retort and sterilized by steam under pressure. After the sterilizing steam is vented, a vacuum is drawn in the sealed retort to remove condensate, and the drum is filled and closed while still in the retort. This system has never been used widely for dairy products, but mainly for fruit juice concentrates.

A more recent development, and one that has been adapted to the aseptic filling of dairy products as well as other liquids into containers between 1- to 1,000-L capacity, has been the so-called "bag-in-box." In this system, of which several forms exist, the product is filled into a plastic bag that is put when full into, or in the larger sizes is filled when inside, a suitable outer protective case, either a metal drum or more commonly a rectangular paperboard box.

The bag is supplied to the filler premade and in lay-flat form. As it fills, it assumes the shape of the outer case in which it is to be shipped. The bags can be fed to the filler individually by hand, or, with some of the machines filling smaller volumes, they can be supplied as a continuous web in which the bags are mechanically fed to the filler consecutively and separated at the time of filling.

For aseptic applications, the bag material is a laminate of three or four layers, of which one will be a barrier material such as metalized polyester (a polyester with a coating of aluminium particles) or EVOH. The composition of the outer layers will depend on the required sealing properties of the laminate. The filling valve and its construction materials depend on the filling, sealing, and sterilizing methods used by the individual manufacturers, but the valve body is welded to the bag as it is manufactured and some form of bacterial seal is incorporated.

The bag and its valve or connector assembly are sterilized by gamma irradiation after manufacture and before shipping to the processing plant, with a radiation dose of about 25 kGy (2.5 Mrad), which is sufficient to destroy all microorganisms, including resistant spores. As the bag remains sealed until the actual filling operation, the internal product contact surfaces remain sterile. It is only necessary during filling to sterilize those exposed surfaces in the filling valve that come into contact with the product.

The details of the filling system depend on the type of connector used, and on the method of sterilization of the connector surfaces before filling. In some systems, the operations take place entirely within a sterile cabinet, protected with sterile, filtered air in laminar flow. The valve assembly of the empty bag is connected, either by hand or mechanically, to the product outlet in the filler (Figure 7–17) and all the further operations follow automatically. The product outlet and the outside surfaces of the connector are sterilized with an intense spray of a suitable chemical disinfectant (e.g., hydrogen peroxide and peracetic acid solution). The connector cover is removed, the interconnection to the product filler made, and filling proceeds. The quantity filled is determined either by bag weight or by an electromagnetic product flow meter. When filling is complete, the filling nozzle is cleaned with a hot air blast, the interconnection is broken, and the cover for the interconnector on the bag is automatically replaced. The filled bag is then placed in the outer box, which is sealed for distribution.

Other filling systems use saturated steam under pressure for sterilization of the interconnection area, and enclose the whole of the area in a sealed unit comprising the steam and product supply arrangements. All of this unit is resterilized with each bag. The sequence of operations for one such filler is shown diagrammatically in Figure 7–18. The connector attached to the bag consists in this case of a simple body with a completely heat-sealed outer membrane that preserves the sterility of the interior of the empty bag, and an inner membrane that is only partly attached to the body. The initial sterilizing process with saturated steam at

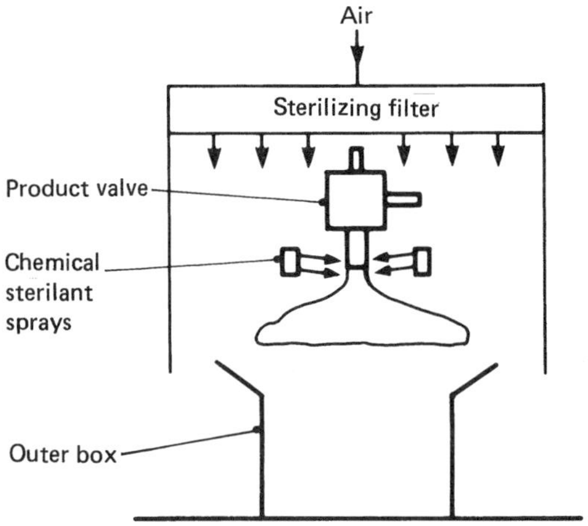

Figure 7–17 Bag-in-box filling system. *Source:* Reprinted with permission from H. Burton, *Ultra-High-Temperature Processing of Milk and Milk Products*, p. 220, © 1988.

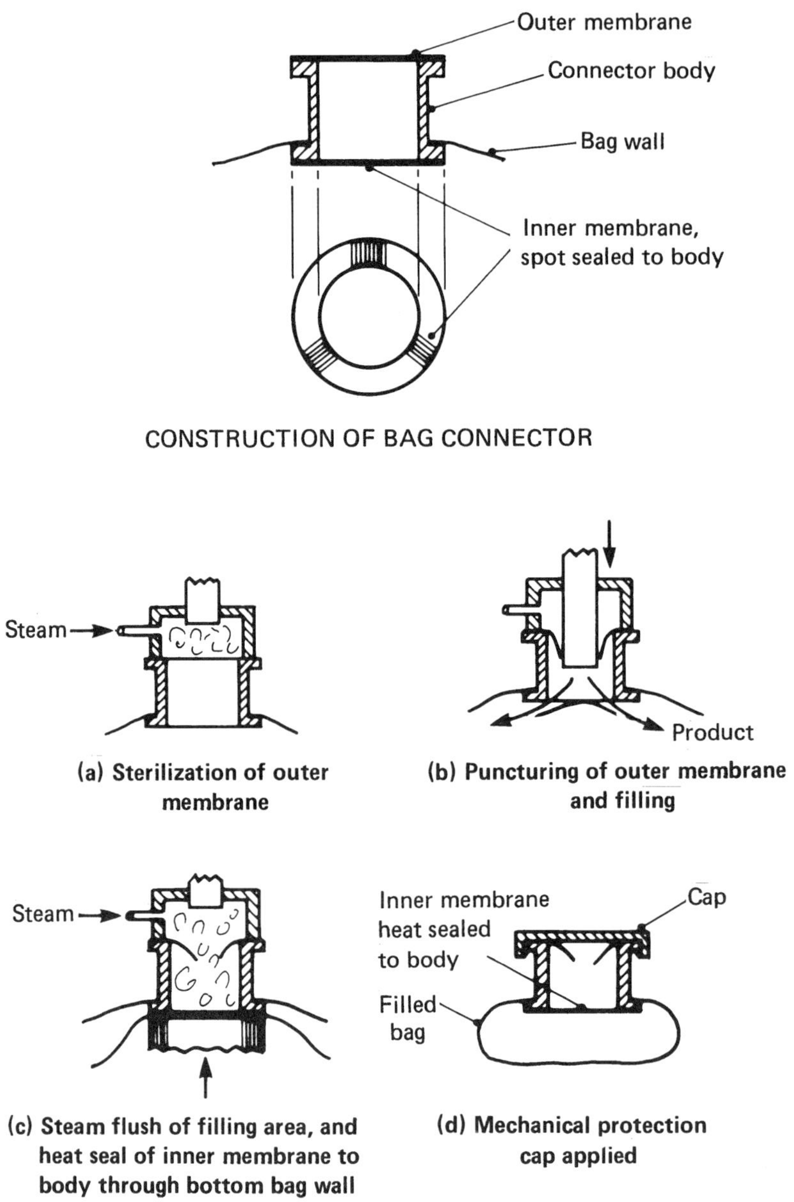

Figure 7–18 Sequence of operations for a system using saturated steam under pressure for sterilization of the interconnection area. *Source:* Reprinted with permission from H. Burton, *Ultra-High-Temperature Processing of Milk and Milk Products*, p. 221, © 1988.

a temperature of 150°C, applied for about 3 seconds, sterilizes the surface of the outer membrane and of the filling system.

The outer membrane is then pierced, and the bag is filled through the spaces between the inner membrane and the connector body. The volume filled is determined by a turbine or electromagnetic flowmeter in the product line. When the bag is full, the filling area is flushed with steam, and the inner membrane is heat sealed to the inner face of the connector body, the heat and pressure being applied from outside the bag through the underneath wall. The membrane and inner layer of the bag are made from materials chosen so that the heat and pressure do not form a seal between them, but only between the membrane and the connector body.

The inner membrane forms the bacterial seal that maintains sterility. A tap or other mechanical protection can then be applied to the connector body. To take product from the bag, the specially designed tap is operated first to penetrate the inner membrane and then to control the product flow.

The short time required for the initial steam sterilization means that it is practical to fill relatively small bags without too large a proportion of the operating time being occupied with processes other than filling. Volumes as low as 1 L are practicable, but commercially a 10-L bag size appears to be the most popular.

An alternative type of filler using steam sterilization of the filling zone has a much more complex valve system, which is used for both filling and dispensing. A rotary, threaded valve attached to the bag interlocks with the filling head, which during operation flushes the valve with steam for 20 seconds and then sterilizes it with steam at 140°C for 2.5 minutes; the relatively long time is determined by the need to sterilize the internal parts of the valve system. The filling head then unscrews the valve and fills the bag with a volume of product determined by an electromagnetic flowmeter, flushes the valve seats with steam, and then reseals the valve by rotation.

The long initial sterilization time means that the process is most economical when filling large volumes of product so that the preparation time is a small proportion of the total cycle time, and this system is exclusively used to fill 1,000-L bags. To compensate for the large bag size, this is the only bag-in-box system that allows part of the contents of a bag to be extracted aseptically, leaving the balance of the contents still sterile for further extended storage. The discharge system, like the filling system, is totally enclosed and purged with steam during operation, so that product can be taken from the bag without recontamination.

7.8 CONTROLS AND SAFEGUARDS IN ASEPTIC FILLING SYSTEMS

We have seen that aseptic filling systems are usually very complex, needing the satisfactory operation of many different functions if the overall performance of the filler is to be consistently of a high standard. The filler must initially be effectively cleaned and sterilized before operation, but subsequently there are still many factors that need to be controlled, for example, those that determine the effectiveness of sterilization of the containers, the maintenance of sterile atmospheres in the filler, and the correct sequence of events. If any of these pass outside the range required for satisfactory operation, there will be a significant risk of producing nonsterile containers.

With early filler designs, many of these factors were left to manual control. It is perhaps surprising that performances were so good under these conditions. Nowadays, with improvements in measuring methods and the development of microprocessor control and logic systems, complex controls can be designed that ensure that every factor identified as important and measurable is monitored throughout the processing plant, and any deviation from specified conditions either gives a warning to the operator or, if necessary, shuts down the filler.

Each type of aseptic filler has its own critical points that will be taken into account. For example, a filler that uses hot peroxide solutions for sterilization may have a system to monitor concentration and temperature of the peroxide, one that sprays peroxide solution into the container may monitor concentration and the volume of solution applied, and one that depends on heat for sterilization may detect the temperature reached at the critical point and allow the process to continue only if the temperature is adequate. A filler that depends on air from a bacterial filter to protect the filling zone may have a system for detecting the level of particles in the filtered air, indicating a defect in the filtration system if the level is too high, and may detect the air overpressure in the filling zone.

The system used for each type of filler will depend on what is needed to ensure the safe performance of that filler. Normally, a monitoring and control system is designed exclusively for one type of filler. However, independent systems are marketed that can be fitted to any filler, allowing any of a range of critical factors to be surveyed and used as the basis of control.

7.9 PERFORMANCE OF ASEPTIC FILLERS

Hersom (1985a, 1985b) has suggested 1 non-sterile container in 5,000 as an acceptable and attainable standard for commercial aseptic filling. Whereas, Wiles (1985) has gone further and suggested that while 1 in 1,000 may be an acceptable commercial level for primary dairy products, other more sophisticated products may require 1 in 10,000 for acceptability.

These performance levels are difficult to obtain, and many processors with specific products and distribution systems may find that higher levels of potential spoilage are commercially satisfactory for them, and that an aseptic filling system that does not claim to be capable of such high-performance levels meets their needs.

Few manufacturers are prepared to give a guarantee of performance in practice, and few are even prepared to estimate a spoilage level. They argue that even the best aseptic filler can be made ineffective by poor operating conditions. However, some manufacturers are prepared to be specific about the performance of their fillers, and these give some indication of the spoilage levels (presumably optimum spoilage levels) that might be achieved.

A British manufacturer of fillers for plastic cups up to 420-ml capacity is prepared to guarantee a spoilage level of not more than 1 in 5,000, for cups filled with separated milk fortified with 0.1% sucrose. Acceptance tests for the filler after installation require four test runs, each of 10,000 cups, with all being incubated. In any one run, there shall be no more than 3 nonsterile cups, and in total there shall be no more than 7 nonsterile cups. Statistically, this gives a high probability of the spoilage level of 1 in 5,000 being reached. A West German manufacturer of form-fill-seal machines claims a spoilage level of 1 in 10,000 for small (up to 15 ml) packs. Low spoilage levels are much easier to obtain in smaller packs than in larger, as we have shown in Chapter 3 and earlier in this chapter, and this fact must be remembered in making direct comparisons. A French manufacturer of aseptic pouch fillers is prepared to guarantee not more than 1 in 500 nonsterile pouches, in sizes from 250 ml to 2 L.

Finally, a West German manufacturer of a cup-filling machine guarantees a sterilizing effect of 3 decimal reductions for the packaging material and the machine itself. Of course, this cannot easily be translated into a spoilage level.

These few examples, and the wide range of spoilage levels that they show, demonstrate how difficult it is to define the performance of an aseptic filler, and how difficult it is to obtain data on such a sensitive commercial subject. It is known that many commercial aseptic filling systems are able to meet or approach the stringent levels set by Hersom (1985a, 1985b). All the systems that have found a place in the market have presumably done so because they meet the requirements of some processors, taking into account product characteristics, available containers and their price, types of distribution system, and the level of spoilage acceptable to the producer.

In West Germany, processing equipment is officially tested and approved for use. Guidelines for the testing of aseptic fillers have been adopted (Reuter, Biewendt, & Klobes, 1982b) that cover the assessment of construction, operation, cleaning, and so forth. These guidelines require 304 samples to be taken on each of 3 days, 4 samples immediately and the rest after storage at room temperature for the stated shelf life. Not more than 1 sample should be nonsterile, which demonstrates <0.1% spoilage at a 5% statistical probability.

REFERENCES

Anon. (1986). Filter uses sound to save sterilant. *Food Engineering International* **11**(1), 48.

Bachmann, R. (1975). *Brown Boveri-Mitteilungen* **62**, 206.

Ball, C.O., & Olson, F.C.W. (1957). *Sterilization in food technology*. New York: McGraw-Hill.

Bayliss, C.E,. & Waites, W.M. (1979). The combined effect of hydrogen peroxide and ultraviolet irradiation on bacterial spores. *Journal of Applied Bacteriology* **47**, 263–269.

Bayliss, C.E., & Waites, W.M. (1982). The effect of simultaneous high intensity ultraviolet radiation and hydrogen peroxide on bacterial spores. *Journal of Food Technology* **17**, 467–470.

Bernard, G.E., & Morgan, H. (1903). *Proceedings of the Royal Society* **72**, 126.

Burton, H. (1951). *Dairy science abstracts* **13**, 229.

Burton, H. (1970). In *Ultra-high-temperature processing of dairy products*. Huntingdon, UK: Society of Dairy Technology.

Cerf, O., & Hermier, J. (1972). *Le Lait* **52**, 1–10.

Cerf, O., & Metro, F. (1977). Tailing of survival curves of *Bacillus licheniformis* spores treated with hydrogen peroxide. *Journal of Applied Bacteriology* **42**, 405.

Cerny, G. (1976). Sterilization of packaging materials for aseptic packaging. I. The microbiological activity of concentrated hydrogen peroxide solutions. *Verpackungs-Rundschau* **27**(4); *Technischwissenschaftliche Beilage*, 27–32.

Cerny, G. (1977). Sterilization of packaging materials for aseptic packaging. II. Studies on the microbicidal effects of C-band UV irradiation. *Verpackungs-Rundschau* **28**(10); *Technischwissenschaftliche Beilage*, 77–82.

Cerny, G. (1982). Sterilization of packaging materials for aseptic filling. IV. Sterilization of packaging material surfaces by means of saturated steam, using the Hassia thermoforming and sealing machine THM 13/37-AS. *Verpackungs-Rundschau* **33**(8); *Technischwissenschaftliche Beilage*, 47–50.

Cerny, G. (1983). *Verpackungs-Rundschau* **34**(8); *Technischwissenschaftliche Beilage*, 55.

Cerny, G. (1984). Basic aspects of sterilization of packaging materials for aseptic packaging. *Zeitschriftfiir Lebensmittel-Technologie und Verfahrenstechnik* **35**, 242-245.

Cerny, G. (1985). In *Symposium on Aseptic Processing and Packaging of Foods, Tylosand, Sweden, Sept. 1985*. Lund, Sweden: Institute of Technology, p. 166.

Collier, C.P., & Townsend, C.T. (1956). The resistance of bacterial spores to superheated steam. *Food Technology* **10**, 477–481.

EHEDG (1992). Microbiologically safe continuous pasteurization of liquid foods. *Trends in Food Science & Technology* **3**(11), 303–307.

EHEDG (1993a). Welding stainless steel to meet hygienic requirements. *Trends in Food Science & Technology* **4**(9), 306–310.

EHEDG (1993b). The microbiologically safe continuous-flow thermal sterilization of liquid foods. *Trends in Food Science & Technology* **4**(4), 115–147.

EHEDG (1993c). Hygienic design of closed equipment for the processing of liquid food. *Trends in Food Science & Technology* **4**(11), 375–379.

EHEDG (1994). The continuous or semicontinuous flow thermal treatment of particulate foods. *Trends in Food Science & Technology* **5**(3), 88–95.

FDA (1981). Food & Drug Administration Federal Register. 21 CFR, Part 178, Vol. 46, No. 6. Washington DC: FDA.

Franklin, J.G., & Clegg, L.F.L. (1956). Problems affecting the sterilization of milk bottles. *Dairy Industries* **21**, 454–460.

Han, B.H. (1977). Bulletin of the Korean Fisheries Society **10**, 145. (1978). In *Food Science and Technology Abstracts* **10**, 12B100.

Hansen, W.F. (1966). *Milk Industry* **58**(6), 48.

Henon, B.K. (1997). *Proceedings of the 7th Annual Symposium of the European Hygienic Equipment Design Group*. Brighton, UK.

Hersom, A.C. (1985a). In *Symposium on Aseptic Processing and Packaging of Foods, Tylosand, Sweden, Sept. 1985*. Lund, Sweden: Institute of Technology, p. 9.

Hersom, A.C. (1985b). Aseptic processing and packaging of food. *Food Reviews International* **1**(2), 215–270.

Huber, J. (1979). Removal of bacteria from materials for aseptic packaging. III. Treatments of preformed blanks for UHT packaging by spraying with concentrated hydrogen peroxide solutions. *Verpackungs-Rundschau* **30**(5); *Technischwissenschaftliche Beilage*, 33–37.

IZG (1984). Report über die Verpackung aus Glas. Glas-Info 3/84, November. Düsseldorf: IZG Informations-Zentrum Glas GmbH.

Mottar, J. (1986). The usefulness of co-extruded high-density polyethylene for packaging UHT milk. *Milchwissenschaft* **41**, 573–577.

Mrozek, H. (1985). Detergency and disinfection. *Journal of the Society of Dairy Technology* **38**, 119–121.

Perkin, A.G. (1982). Chemiluminescent determination of hydrogen peroxide in water. *Journal of the Society of Dairy Technology* **35**, 147–149.

Pflug, I.J. (1960). Thermal resistance of microorganisms to dry heat. Design of apparatus, operational problems and preliminary results. *Food Technology* **14**, 483–487.

Quast, D.G., Leitao, M.F.F., & Kato, K. (1977). Death of *Bacillus stearothermophilus* 1518 spores on can covers exposed to superheated steam in a Dole aseptic canning system. *Lebensmittel-Wissenschaft und Technologie* **10**, 198.

Reuter, H., Biewendt, H.-G., & Klobes, R.H. (1982a). Type testing for the purpose of official approval of sterile tanks for intermediate storage of UHT milk. *Kieler Milchwirtschaftliche Forschungsberichte* **34**, 401–407.

Reuter, H., Biewendt, H.-G., & Klobes, R.H. (1982b). Type testing for the purpose of official approval of machines for sterile packaging of UHT milk. *Kieler Milchwissenschaftliche Forschungsberichte* **34**, 409–414.

Schreyer, G. (1985). Environmentally favourable alternative—UHT milk in glass bottles. *Deutsche Molkerei-Zeitung* **106**, 482–484, 486.

Smith, Q.J., & Brown, K.L. (1980). The resistance of dry spores of *Bacillus globigii* (NCIB 8058) to solutions of hydrogen peroxide in relation to aseptic packaging. *Journal of Food Technology* **15**, 169–179.

Stannard, C.J., Abbiss, J.S., & Wood, J.M. (1983). Combined treatment with hydrogen peroxide and ultraviolet radiation to reduce microbial contamination levels in preformed food packaging cartons. *Journal of Food Protection* **46**, 1060–1064.

Stannard, C.J., & Wood, J.M. (1983). Measurement of residual hydrogen peroxide in preformed food cartons decontaminated with hydrogen peroxide and ultraviolet irradiation. *Journal of Food Protection* **46**, 1074–1077.

Swartling, P., & Lindgren, B. (1962). Report No. 66. Alnarp, Sweden: Milk and Dairy Research Institute.

Swartling, P., & Lindgren, B. (1968). The sterilizing effect against *Bacillus subtilis* spores of hydrogen peroxide at different temperatures. *Journal of Dairy Research* **35**, 423–428.

Toledo, R.T., Escher, F.E., & Ayres, J.C. (1973). Sporicidal properties of hydrogen peroxide against food spoilage organisms. *Applied Microbiology* **26**, 592–597.

von Bockelmann, B. (1982). Collection of Lectures from Tetra Pak Seminars. Lund, Sweden: Tetra Pak International AB.

von Bockelmann, B. (1985). In *Symposium on Aseptic Processing and Packaging of Foods, Tylosand, Sweden, Sept. 1985.* Lund, Sweden: Institute of Technology, p. 150.

Wiles, R. (1985). Aseptic filling into containers other than cartons, with special reference to sterile bulk packs. *Journal of the Society of Dairy Technology* **38**, 73–75.

CHAPTER 8

Fouling, Cleaning, and Disinfecting

8.1 INTRODUCTION

When foods are heated, reactions may take place that will give rise to the formation of deposits on the surface of the heat exchanger; this is termed *fouling*. If the aggregates responsible for such deposits do not attach to the surface of heat exchangers, or become dislodged, they may end up in the final product and be perceived as sediment. Some interesting observations are made by Swartzel (1983) relating sediment formation during storage in aseptic packages to the severity of heat treatment and the extent of fouling.

The problem of fouling was first described in the literature in the 1940s; it is also referred to as milk scale, milk stone, or just deposit formation. Where fouling occurs to a significant extent, these deposits must be subsequently removed by the process of cleaning. Surfaces must then be disinfected or sterilized. Cleaning and disinfecting are two separate but very important procedures (Section 8.7 and 8.8).

Fouling occurs in many continuous processes involving heat transfer, in particular ultra-high temperature (UHT) sterilization and evaporation. When it occurs it may result in a fall in the overall heat transfer coefficient (OHTC) and a drop in the product outlet temperature. This will have a marked influence on product safety and quality, especially on the microbiological flora surviving. This situation is often found in tubular heat exchangers. Also the flow channels will become narrower and in extreme cases be completely blocked. Depending upon the pumping arrangement, this may cause a drop in flow rate for centrifugal pumps or an increased pressure and higher pressure drop for positive pumps. Problems related to blockage and pressure buildup tend to be the more usual situation in plate heat exchangers. To ensure a constant flowrate and residence time may require higher pressures to be developed.

The deposit itself may also interfere with product, again with possible safety and quality implications. The higher wall temperatures may cause more local overheating, which in turn may influence the sensory characteristics of the product; this aspect has not been so fully researched.

Therefore there are several important economic implications of fouling. These have been summarized as follows:

- The reduction in heat transfer efficiency means that the heat exchanger capacity is reduced. If fouling is not controlled properly, an oversized heat exchanger is required for a particular heating duty.
- The energy costs are increased to overcome the thermal resistance of the fouling layer.
- Cleaning costs are increased in terms of the amounts of detergents used. This leads to an escalation of detergent costs. Effluent disposal costs are also increased.
- Processing times are reduced and downtime is increased.

If fouling leads to a compromise in product safety (i.e., due to underprocessing), the cost of a product recall and the adverse publicity can be very high and seriously damage a branded product.

There are a number of other safety implications with regard to process operators; in plate heat exchangers the additional pressures may lead to gasket leakage or even failure, with release of hot and dangerous chemicals, under considerable pressure. Quality implications may result from overheating of some elements of fluid (those in the thermal boundary layer), which may influence the color and flavor.

If not properly removed, the fouling layer may act as a source of nutrients or breeding ground for microorganisms. Although most fouling deposits are chemical in nature, the term *biofilm* is becoming more widespread to describe a film that comprises primarily microorganisms and having microbial activity associated with it.

In general, the severity of fouling increases as the processing temperature increases, or as the total solids of the feed increases. For example, in milk pasteurization, fouling is rarely a major problem, whereas in UHT processing, it can be a very serious problem and affect both the selection of the most suitable equipment and the length of processing time that can be achieved. Also, fouling is very product dependent and is influenced by the composition of the food (see Table 1–2). For any particular raw material, changes will occur in composition. For example, as raw milk gets older, its titratable acidity increases and its pH falls; this also gives rise to an increase in ionic calcium. It is interesting that UHT direct plants are capable of dealing with poor-quality raw materials much better than indirect plants. However, some other foods may foul so excessively at normal UHT temperatures that they cannot be processed. Examples are those containing substantial protein (e.g., cheese whey and eggs) or starch (e.g., desserts, soups, and batters). Thus the amount of fouling may be influenced by processing conditions and even by the choice of heat exchanger, together with factors such as flowrate, viscosity (turbulence), heating and cooling rates, pretreatment, and processing temperature.

It would be very useful to have a simple indicator for each product that could be used to predict whether fouling was likely to be a problem. In cases where it is possible, unsuitable raw material could be diverted to other products or processes where fouling is not likely to be encountered. For formulated products, such as flavored drinks or ice cream mix, there is also the opportunity to manipulate composition to reduce fouling. Also, with reconstituted products, water quality may play an important role, especially with regard to the amount of calcium and magnesium addition that may be present in hard water (Section 8–7).

The main product where fouling has been investigated is liquid milk. However other products where fouling has been investigated are

- foods consisting of milk or milk constituents (e.g., whey concentrates, creams, milk-based drinks, sauces, spreads, and desserts)
- nondairy foods, such as fruit and vegetable juices and fluids in the brewing industry.

Fouling can also be a problem with water that is to be used for heating purposes, due to calcium and magnesium salts, which contribute to water hardness. In such areas, boiler water needs to be treated to remove these salts, as will water that is to be used for plant sterilization. Water used for cooling needs to be treated to remove microflora, which might give rise to biofilms in the cooling section. There has been some success in controlling water scale, using magnetic (electronic) water-descaling systems (Environmental Concepts, Ltd.), although the mechanisms are far from clear. These methods are currently being evaluated with milk and other products (Jungro-Yoon & Lund, 1994).

With some products, some fouling will always occur. However, it is important to understand the processes taking place to reduce the problems caused by fouling. A recent overview of fouling is given by Grandison (1996).

8.1.1 Terms Used in Fouling

Fouling is the buildup of deposits on the surface of a heat exchanger. When deposits attach to surfaces, they will block the flow passages and adversely affect the performance of the heat exchanger. The deposits will also have a low thermal conductivity and there may be a considerable temperature gradient set up over the deposit, with the wall temperature being considerably higher than the bulk product temperature (Exhibit 8–1).

In general, a number of terms have been used to describe and explain the fouling process. Since many fouling experiments have shown two distinct periods, reference is often made to an *induction period,* where the measured fouling parameter does not change with time, and a *fouling period*, where the measured property does change with time (Figure 8–1). In a tubu-

---

Exhibit 8–1 Fouling and Resistances

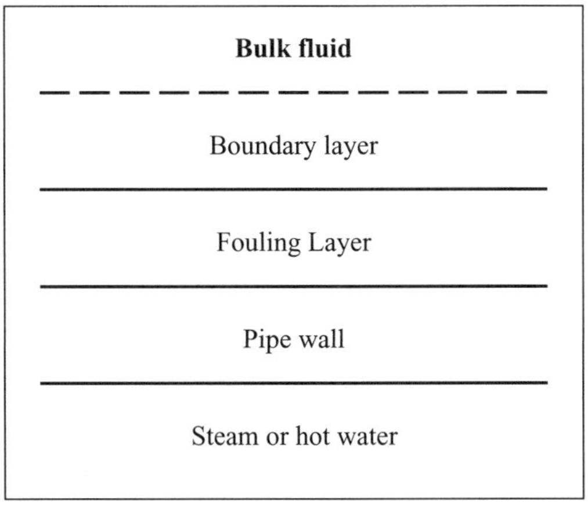

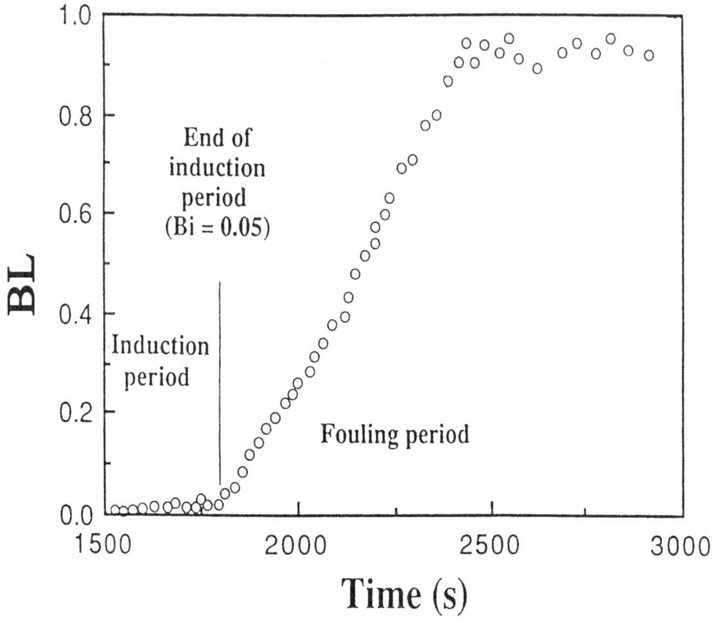

Figure 8–1 Decrease in heat transfer during operation of a tubular test section, showing induction and fouling stages. *Source:* Reprinted with permission from P. Fryer et al., The Uses of Fouling Methods in the Design of Food Processing Plants, *Journal of the Society of Dairy Technology*, Vol. 42, No. 1, p. 25, © 1989, Society of Dairy Technology.

lar system, the end of the induction period was found to coincide with a Biot number of 0.05 (see Section 8.2.3 for definition of Biot number). Fryer (1989) reported that sometimes these periods were obvious and sometimes not. For example, experiments in tubes gave generally obvious induction periods, whereas experiments with plate configurations gave no obvious induction periods, although fouling increased with time. Kastanas (1996) also found that the induction period was very short (or even absent) in some fouling situations.

Fouling is a dynamic process, and the rate of fouling is a balance between deposition and erosion. Fouling may be mass-transfer controlled, where the rate of movement of the foulant to the wall is rate controlling (probably less temperature dependent), more shear dependent, or reaction rate controlled, where the fouling is more influenced by either the bulk or the wall temperature. Surface reaction is concerned with what is happening at the surface and will be influenced by the condition at the surface, especially the effect of temperature on the substances responsible for fouling. If the controlling reaction for fouling takes place in the fluid bulk itself, this could be in the thermal boundary layer or the bulk of the fluid (turbulent core). Thus, fouling may be temperature, concentration, and shear dependent. Some recent experiments with tubular heat exchanger Kastanas (1996) showed that fouling manifests itself in a number of ways. The extremes are as follows:

- No change in OHTC but a buildup of pressure, which in the end limits the extent of the process.

- A regular decrease in OHTC, without any substantial increase in pressure; in such cases, the rate of fall of OHTC is temperature dependent.

There is often an intermediate situation, with an induction period where OHTC remains constant, followed by a period where the OHTC decreases with time; during this decline phase, the pressure may or may not increase. The length of the induction period represents the time taken to condition the surface, so that heavier fouling can take place. For the first layers, a mechanism that involves the interaction among proteins, phosphates, and calcium is thought to play a key role for milk-based products.

8.2 MEASUREMENT OF FOULING

To summarize, when fouling occurs, the two main effects are a reduction in the size (dimensions) of the flow passage and a decrease in the overall rate of heat transfer.

A number of methods have been described in the literature to measure the extent of fouling.

A wide range of experimental equipment and techniques have also been used, including production-scale plant, pilot plant, heat-flux meters, and rigs for measuring the formation of deposits on heated surfaces. Some of the methods for measuring fouling are now reviewed.

8.2.1 Deposit Formation

The amount of fouled deposit formed during the process is measured. For achieving this, it is necessary to get access to the heat exchanger surface or, where this is not possible, to solubilize the deposit and determine the amount of material that is dissolved. Also, the experimental product must usually be flushed from the heat exchanger, prior to any analysis, without removing any of the deposit.

One method is to weigh the total deposit produced; it could be on a wet or dry weight basis. However, it is also possible to measure how the amount of deposit varies at different locations within the heat exchanger or how the deposit composition changes with location. It is also possible to determine how the composition of the deposit changes through the thickness of the deposit. It is interesting to note that the amount of any component in the deposit is usually a minute fraction of the total amount of that component that has passed through the heat exchanger in the duration of the processing period. The thickness of the deposit can also be measured directly, using a microscope or other optical amplifying equipment, or indirectly by selecting some physical characteristic of the deposit that is different from that of the metal wall, for example, electrical conductivity. The difference between the final and the initial value is used to calculate the thickness.

Correlations between the thickness of the deposit (e, m) and the fouling resistance (R_f, $W^{-1}\ m^2\ K$) have been established by Corrieu, Lalande, and Ferret (1981), as follows:

$$R_f = 1.378\ e + 1.603$$

Fiber optics may be used to inspect more inaccessible locations to determine whether deposits have been removed effectively. This is an important function of the cleaning process.

8.2.2 Temperature Measurement

In a heat exchanger operating at steady state, the outlet temperature will be constant. If the heating medium conditions remain constant (temperature and flowrate), the outlet temperature will fall as fouling proceeds (Figure 8–2). Eventually the temperature will fall to below that which is considered to be safe, signifying the end of the run, and the plant must be cleaned.

In commercial practice, it is more normal to keep the product outlet temperature constant, using a temperature controller. In this case, when fouling occurs, it will be accompanied by an increased demand for steam or pressurized hot water, in order to maintain the same rate of heat transfer. The steam temperature and pressure will rise or the hot water set temperature will increase. Thus changes in these temperatures and pressures will indicate that fouling is taking place.

There will also be a considerable temperature gradient across the fouled layer, and in the heating section the temperature at the wall will be considerably higher than the temperature in the bulk fluid. In theory, measuring this temperature gradient should provide an efficient way to measure fouling, but achieving this presents many practical difficulties. The higher wall temperatures may also affect product quality.

8.2.3 Decrease in the Overall Heat Transfer Coefficient

Fouling can be monitored by measuring a decrease in the overall heat transfer coefficient. Some typical data are shown for fouling of goat's milk, at different temperatures, in Figure 8–3.

Traditionally heat exchangers have been designed, sized, and costed by adding a fouling resistance (R_f) to the initial or clean overall heat transfer coefficient.

For a clean heat exchanger:

$$1/U = 1/h_1 + 1/h_2 + L/k.$$

For a fouled heat exchanger:

$$1/U_f = 1/U + R_f$$

where R_f is termed the fouling resistance, which is equivalent to L_f/k_f; f refers to fouling layer.

For modeling purposes a Biot number (Bi) has been proposed, where Bi = (R_f U) (see Figure 8–2) (Fryer, 1989). Thus as the fouling deposit covers the heat transfer area or increases in thickness, there will be a reduction in the overall heat transfer coefficient. This will particularly affect the overall heat transfer coefficient when the fouling resistance becomes the limiting resistance, more so when the overall heat transfer coefficient is high. The overall heat transfer coefficient can be measured at a specific location, or over the entire heat exchange section, which will give an average value for that particular heating or cooling section.

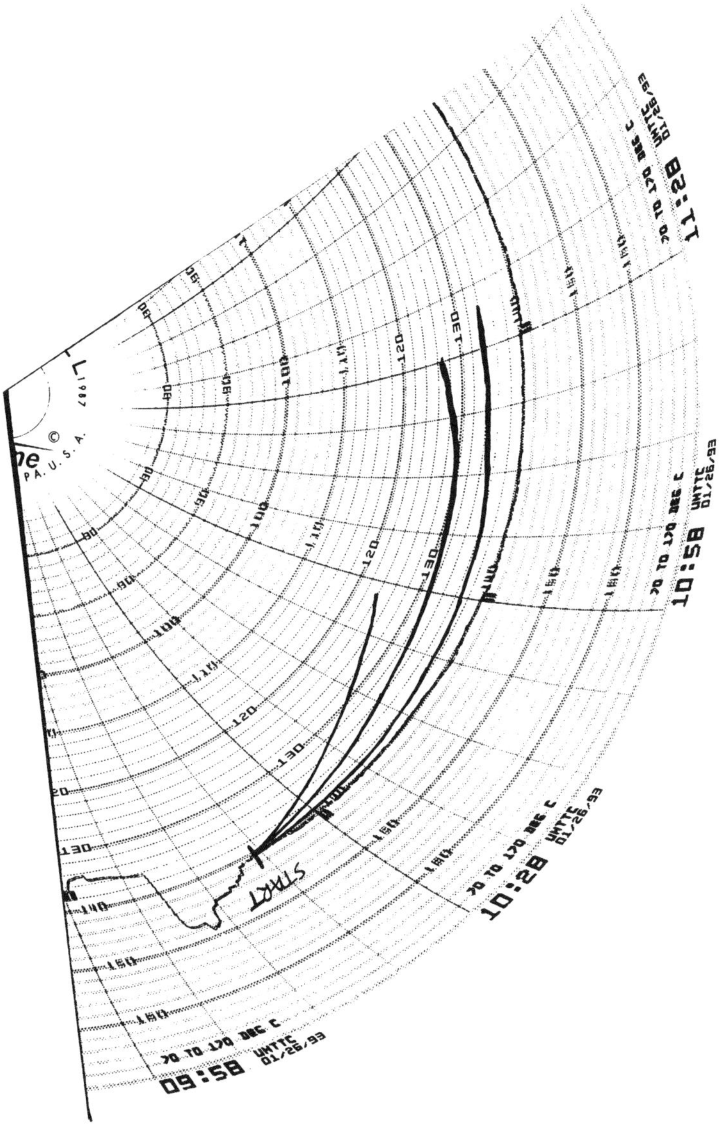

Figure 8–2 Temperature measurement, showing constant temperature and three traces for decline in temperature. *Source:* Copyright 1987 Honeywell International Inc. Truline is a registered trademark of Honeywell. Reprinted with permission of Honeywell International Inc.

8.2.4 Pressure Monitoring

Pressure monitorings are useful as they provide an indirect indication that the dimensions of the flow passage have altered. Some typical data are shown in Figures 8–4 and 8–5. As fouling deposits accumulate, the flow passage will become narrower. If the same flow rate is

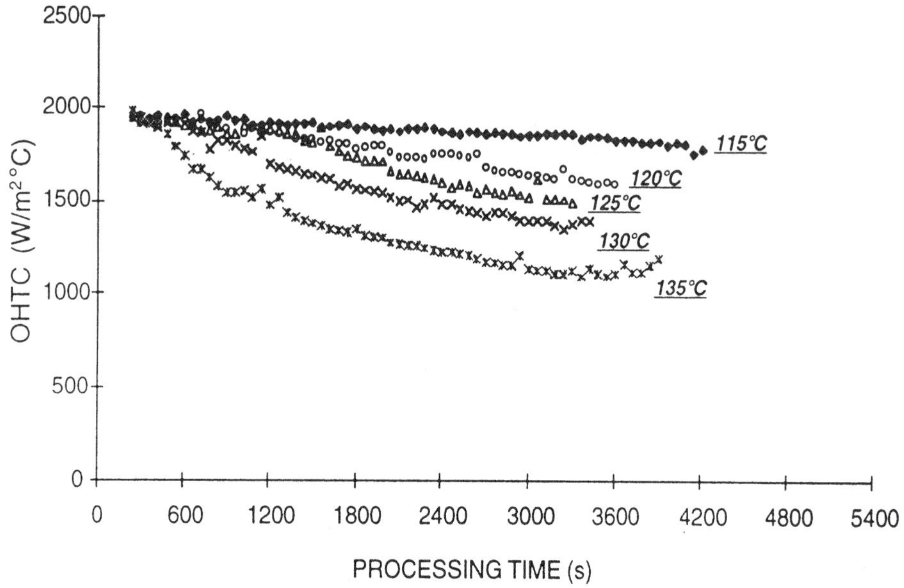

Figure 8–3 The change of the OHTC with processing time at various processing temperatures (goat's milk). *Source:* Reprinted with permission from IDF, *Production and Utilization of Ewe and Goat Milk*, IDF/FIL Publication No. 9602, © 1996, International Dairy Federation.

maintained, then the pressure drop over the fouled section will increase. Thus the extent of fouling can be determined by monitoring both the inlet pressure and the pressure drop over various sections of the heat exchanger. In some cases, processes are terminated when the pressure reaches a preset-limiting pressure. Pressure buildup is more of a problem in plate heat exchangers, as it may cause gaskets to fail. The more voluminous protein deposits make a major contribution to the buildup of pressure. Thus, observations of temperatures and pressures can provide useful information on the development of fouling on commercially operated UHT equipment.

8.2.5 Fouling Rigs

A wide variety of experimental methods have been used to investigate fouling. Some involve test rigs that are connected to the main plant. The test rigs can be more easily dismantled than the main plant, and the temperature regimes and flowrates are well established. Others involve the use of laboratory scale or pilot-scale equipment. Some of these are summarized in Exhibit 8–2.

8.2.6 Fouling Sensors

The concept of an on-line sensor that indicates that fouling has occurred and is impairing the efficiency of the heating process would be useful to indicate exactly when to start the

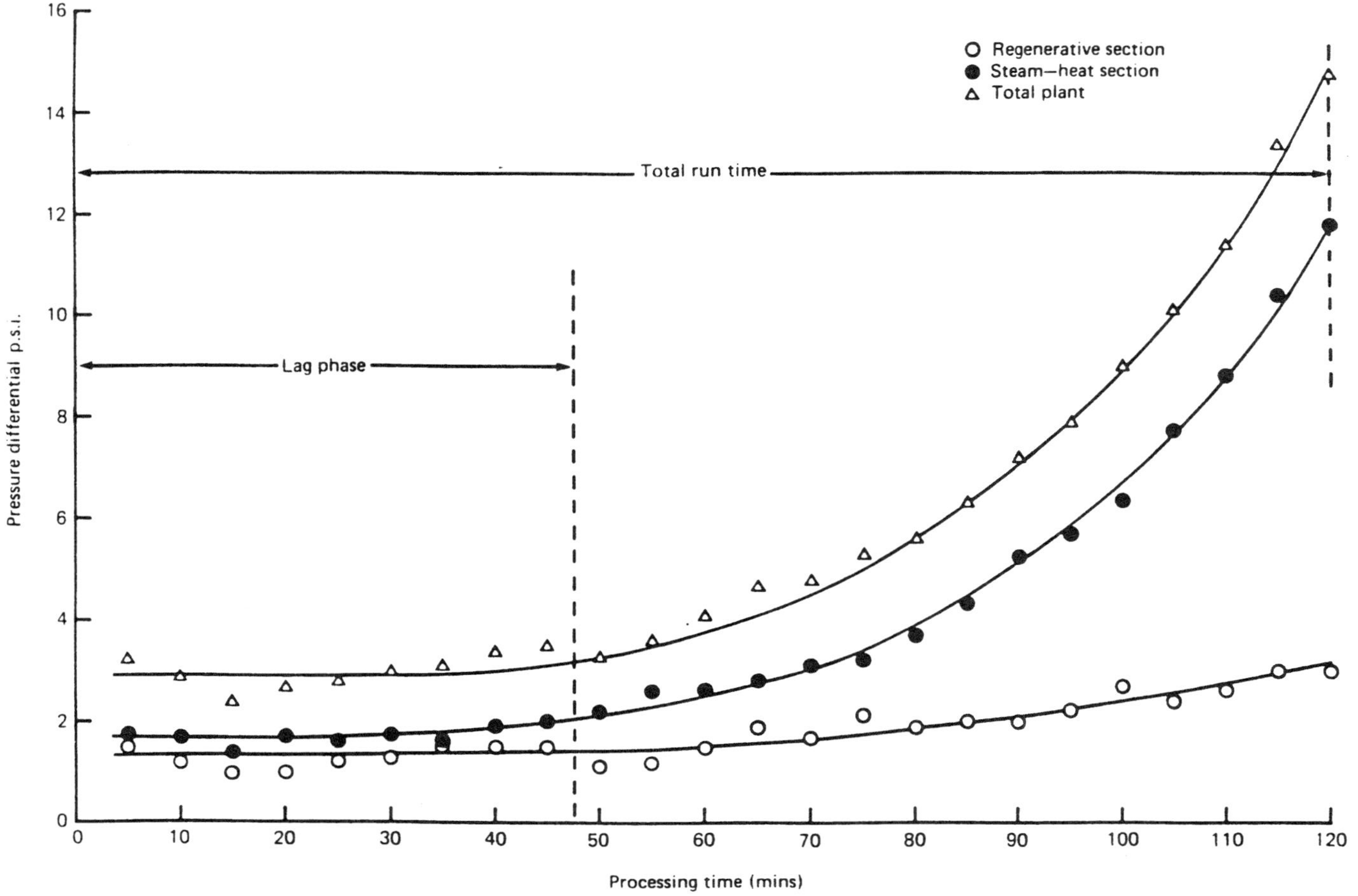

Figure 8–4 Pressure differentials across the regenerative and steam-heated sections during a typical UHT processing run. *Source:* Reprinted with permission from A.S. Grandiscon, UHT Processing of Milk: Seasonal Variation in Deposit Formation in Heat Exchangers, *Journal of the Society of Dairy Technology*, Vol. 41, No. 2, pp. 43–49, © 1988, Society of Dairy Technology.

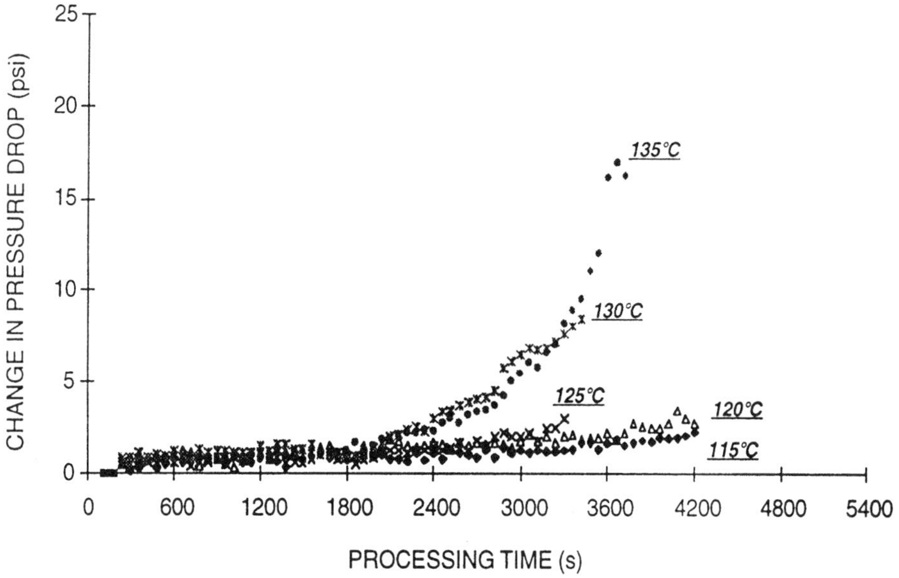

Figure 8–5 Change in heat exchanger pressure drop when goat's milk is heated at various temperatures. *Source:* Reprinted with permission from IDF, *Production and Utilization of Ewe and Goat Milk*, IDF/FIL Publication No. 9602, © 1996, International Dairy Federation.

cleaning procedure. A number of principles have been investigated to try to solve this problem.

Jones, Ward, and Fryer (1994) developed a fouling monitoring device based on a heat flux sensor, which monitored the change of heat flux between a heated copper block and the product. The change in OHTC can be measured from the change in heat flux. Davies et al. (1997) used a heat flux sensor to measure the thermal resistance (density × thermal conductivity) of whey protein deposits in situ. This method gave values of 470 W kg m^{-4} K^{-1} for deposits exposed to wall temperatures of less than 85°C. Deposits formed at higher wall temperatures showed the effects of aging and gave larger thermal resistance values.

Withers et al. (1994) have developed an ultrasonic sensor for detecting fouling during UHT processing. This is a noninvasive technique based on the time required for ultrasound to travel through deposits as they accumulate on a surface. This still needs further develop-

Exhibit 8–2 Summary of Some Experimental Methods To Investigate Fouling and Cleaning

Electrically heated wires	Burton (1964, 1966)
Radial flow cell	Fryer et al. (1984)
Heated plates	Shilton et al. (1992)
Miniature tubular UHT plant	Kastanas et al. (1995); Wadsworth and Bassette (1985)
Pilot-scale fouling rigs	Schreier et al. (1994); Delplace et al. (1994)
Commercial plant	Timperley, Hasting, and De Goederen (1994)

ment to measure accurately the change in deposit at UHT conditions. General principles of ultrasonics are discussed by Withers (1994).

Daufin, Kerherve, and Quemerais (1985) studied the effect of electrochemical potential on metal surfaces and concluded that neither electrochemical potential nor current intensity can be used as sensors for fouling, as there was no correlation between these parameters and fouling deposits.

In a personal communication, Lalande described an on-line fouling sensor based on evaluating disturbances in hydrodynamic functioning resulting from fouling. It had been used in the heat-regeneration section (98° to 120°C) and the high-temperature section (120° to 140°C) on a 10-m^3/h plant.

8.2.7 Indirect Indicators

Some other properties of products, which can be measured easily, can be used to determine the susceptibility of that product to fouling. Examples are heat stability, alcohol stability, sedimentation, gelation, pH, and titratable acidity.

8.3 REVIEW OF FACTORS AFFECTING FOULING

In general, fouling is influenced by a number of factors, which can be categorized as process or product dependent. *Process-dependent* variables include operating conditions, such as flowrate, turbulence, shear rate, temperature, and pressure. *Product-dependent* variables are related to the composition of the product and the effects of heat and shear on reaction rates. Some of these aspects are discussed in more detail for specific products in Section 8.4.

8.3.1 Plant Construction, Design, and Operation

There are a number of process-related parameters that affect fouling. Wherever possible, these should be considered at the design stage. However, once installed, there may not be much scope for changing them to reduce fouling (especially temperatures or flowrates), as they will be predetermined by safety and quality issues, as well as residence times and pressure drops.

8.3.2 Material of Construction and Surface Finish

Since the surface is one of the major participants of the fouling process, it is considered to play an important role in the fouling process. The normal material of construction is stainless steel. It is a common belief that a rough surface is likely to increase fouling by providing more surface to which material may adsorb and by allowing better "keying" of the deposit on the surface. However, the evidence for this is conflicting (Burton, 1988; Kastanas, 1996) and is probably more true for mineral fouling. One reason for these conflicts could be that once an initial deposit has formed, its subsequent development will then be dependent upon the deposit composition and not the nature of the surface.

Stainless steel has a lower thermal conductivity than copper and aluminium; lower thermal conductivities give rise to higher temperature gradients, which result in increased fouling. Covering the surface with polytetrafluoroethylene (PTFE) was found not to eliminate fouling, but cleaning was easier, as its adhesion was poorer. This PTFE layer will also decrease the thermal conductivity of the plate. Jungro-Yoon and Lund (1994) compared the fouling rates on three stainless steel surfaces (titanium plate, standard 304, and electropolished stainless steel) and found very little difference. However, when the stainless steel was coated with polysiloxane and Teflon, there was about a 14% and 20% increase in the fouling rates, respectively. Stainless steel is a hydrophobic surface. The adsorption of whey proteins onto hydrophilic and hydrophobic surfaces has been studied by a number of workers, who found that greater masses of adsorbed fouling mass are found on hydrophobic surfaces as opposed to hydrophilic surfaces. There is evidence that highly polished surfaces appear to improve cleaning, but there is no evidence that it gives a decreased resistance to fouling. More work is being done on the role of surface hydrophobicity on the fouling and cleaning processes.

8.3.3 Temperature

Temperature (see Section 8.2.2) is considered to be a very important operating variable to affect fouling. However, it is not totally clear whether this temperature is the bulk product temperature, the surface temperature, the temperature of the heating medium (steam or hot water), or the temperature difference between the product and the heating medium.

In most practical applications, attention is normally paid to the bulk product temperature, as this can most easily be measured. In a continuous-heat exchanger, the bulk product temperature will change along the length of the heat exchanger. There is also no direct relationship between the bulk product temperature and the amount of deposit formed at that location (Figures 8–6 and 8–7). The nature and location of deposits may also be influenced by the rate at which the product is heated. The most common method used to study the effect of product temperature on fouling involves heating the product at UHT conditions up to a predetermined temperature and studying the deposit formation at the various sections of the plant; this procedure can then be repeated at different temperatures, but it is time consuming. Some results for fouling of goat's milk when heated to different temperatures are shown in Figure 8–5 (Kastanas, Lewis, & Grandison, 1996).

Fouling is no doubt related to the processing temperature, and the severity of overall fouling process increases as the processing temperature increases. Some products are more susceptible to changes in temperature than others (see Section 8.4.8).

8.3.4 Flowrate

Product flowrate was among the first parameters studied for its effect on fouling, but its effect is still not clear. In general, fouling is high at low fluid velocities and Reynolds numbers (Re) and is considerably reduced at high flowrates and Reynolds numbers. However, this trend has not been quantified successfully and the transition point from extensive fouling

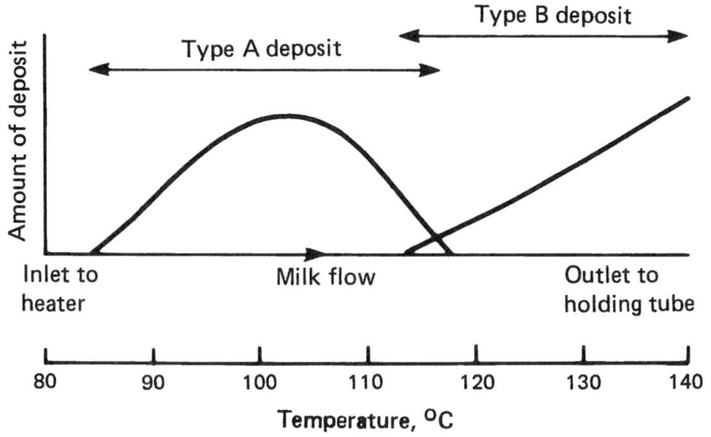

Figure 8–6 Fouling distribution in an indirect heat exchanger operating on raw milk. (See text for description of Type A and Type B deposits.) *Source:* Reprinted with permission from H. Burton, *Ultra-High-Temperature Processing of Milk and Milk Products*, Fig. 10.2, © 1988.

to reduced fouling remains unknown. Also, visual inspection of fouled plates prior to cleaning shows that more deposit forms where flow is relatively stagnant. However, some reports have shown that fouling rate was influenced comparatively little by flowrate, compared to other factors (Lalande and Corrieu, 1981).

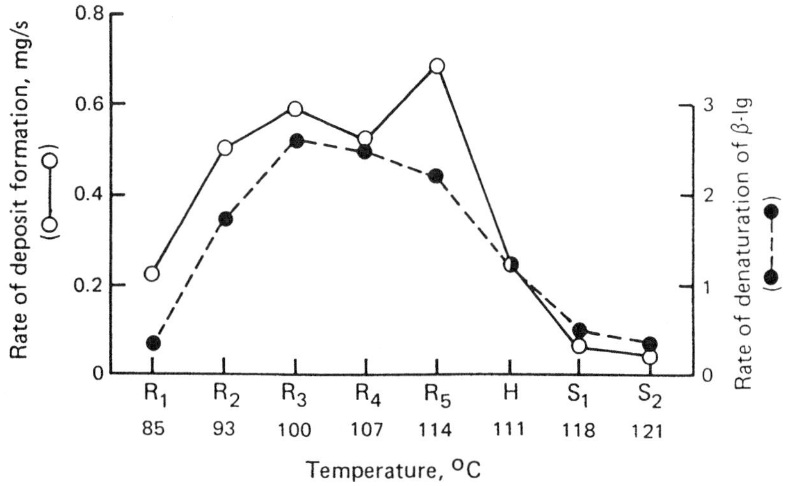

Figure 8–7 Relationship between rate of deposit formation and rate of denaturation of β-lactoglobulin in different sections of a heat treatment plant. *Source:* Reprinted with permission from H. Burton, *Ultra-High-Temperature Processing of Milk and Milk Products*, p. 297, © 1988 (derived from J. Hiddink et al., Heat Treatment of Whipping Cream, I. Fouling of the Pasteurisation Equipment, *Milchwissenschaft*, Vol. 41, pp. 542–546, © 1986).

8.3.5 Other Factors

The use of pressurized hot water rather than steam and the reduction of temperature differentials between the heating medium and the process fluid will also help to reduce fouling rates in general. Again, Hiddink et al. (1986), suggest that temperature differentials only affect fouling once they start to exceed 10° to 15°C. However, reduced temperature differentials will mean that the heat exchange surface is larger and there will be a bigger area for the deposit to be spread over, thereby diluting the overall effect. It is recommended to apply a reasonable back-pressure to avoid air coming out of solution; this is reported to help reduce fouling.

8.3.6 Fouling Models

A number of models have been proposed for the fouling process, related to changes in some of the measured characteristics described earlier. Modeling fouling processes is complex because the fouling rate at any point is a function of temperature, but temperatures change as a result of fouling. There are some extensive models describing fouling dynamics. These can be classified into three groups, according to the parameter they predict: (1) predicting the amount of deposit, (2) predicting the thickness of the deposit, and (3) predicting changes of the overall heat transfer coefficient. These models have been summarized by Kastanas (1996, Table 8–1).

All rely on simplifying assumptions and are semianalytical models. Few of these apply to heating food fluids. Even fewer include one or more parameters that represent the composition of the product, and none has found acceptance for prediciting the fouling behavior of food systems. Most of these comments will also apply to models proposed for cleaning.

8.4 FOULING OF SPECIFIC PRODUCTS

The factors affecting fouling are discussed as related to specific products.

8.4.1 Milk

Fouling of milk has been widely studied and the mechanisms are still not clearly fully understood. In this section, studies relating specifically to investigations on milk will be discussed. There has also been considerable fouling studies on whey proteins and whey protein concentrates. This work is also discussed in Section 8.4.9.

Briefly stated, the proteins in milk comprise caseins (75%) and whey proteins (25%). When milk is heated to above 70°C, the whey proteins start to denature and will associate with the casein micelle; there is no noticeable precipitation of whey proteins. Thus good-quality raw milk is stable (in the sense that there is no noticeable deposit formation or coagulation), even at temperatures of 140°C. Despite this, milk is still susceptible to fouling during UHT processing, so studies have been undertaken on the factors affecting fouling. There is also considerable variation between milks from different species. UHT trials have shown that goat's milk will foul a plate heat exchanger much more quickly than cow's milk (see Section 8.4.5).

Table 8–1 A Summary of the Models Describing Milk or Whey Fouling

Characteristic Parameter	*Author*	*Predicts*
Deposit weight	Lalande & Corrieu (1981)	Rate of change of the deposit weight from processing temperature and milk ammoniac concentration
	Sandu & Lund (1985)	Rate of deposition from Re and processing temperature
	Mottar & Moermans (1988)	Amount of deposit from processing temperature and holding time
	de Jong et al. (1992)	The increase of deposit weight with processing time
	Toyoda et al. (1994)	Amount of deposit with processing temperature
Thickness of deposit or change in hydraulic diameter	Kern & Seaton (1959)	Thickness of deposit from mass flowrate and pressure drop
	Hiddink et al. (1986)	The fouled fraction of the hydraulic diameter for whipping cream
	Belmer-Beiny & Fryer (1993)	Distribution of the deposit in a tubular heat exchanger from wall shear stress and dynamic viscosity
Changes of the heat transfer coefficient	Lund & Bixby (1974)	Change of the product film heat transfer coefficient from Re
	Fryer et al. (1984)	Changes of the overall heat transfer coefficient from wall and bulk temperature and the Re

Source: Copyright © 1996, P. Kastanas.

In milk, calcium phosphate is saturated and solubility decreases with increase in temperature. When milk is heated, the calcium phosphate becomes insoluble and can associate with the casein micelle, remain suspended in the bulk solution, or deposit on the wall of the heat exchanger. It is also known and stated that when whey protein is denatured, it will also associate with the casein. However, the amount associated is reported to decrease as the pH decreases. Obvious questions are what happens both to this denatured whey protein and to the calcium phosphate and are these the clues to the initiation of the fouling process? Also, in the absence of casein, whey proteins are much more susceptible to coagulation. However, since fouling also occurs in the absence of proteins, for example, in milk permeate from ultrafiltration, the situation is more complex and more than one fouling mechanism may be involved.

Burton's review articles (1968 and 1988) provide a good summary of the state of knowledge at those times. The nature of the deposits formed during UHT treatment of milk was investigated by Lyster (1965) and this, together with the work of Burton (1968), led to the recognition of two distinct types of deposit, which were termed type A and type B deposits (Figure 8–6). The major area where deposits occurred was the preheating section. Maximum deposit formation is in the temperature range 95° to 110°C. The deposit that forms between 80° and 105°C (in the early part of the final heating section) is a white voluminous deposit, which has a higher protein content (50% to 70%) and a significant mineral content (30% to 40%) and tends to block the flow passages. This is known as Type A deposit. At the lower end of the temperature spectrum, the protein is predominantly denatured β–lactoglobulin but toward the top end of the range it is predominantly casein.

Tissier, Lalande, and Corrieu (1983) also found two major deposit peaks, the first at 90°C (predominantly protein [50%] and the second at 130°C, which was predominantly mineral [75%]. The major protein contributing to the lower temperature peak was β-lactoglobulin (62%) while in the second peak β-casein (50%) and α_{s1}-casein (27%) were predominant. Work by Lalande et al. (1984) was in general agreement with this. Figure 8–7 (Hiddink et al., 1986) shows the distribution of deposits (superimposed on this are the relative amounts of β-lactoglobulin denatured from the author's calculated data at different temperatures along the heat exchanger). It can be seen that deposit formation does correlate well with the denaturation profile.

Type B deposits are those that form at higher temperatures (in the latter section of the final heating section). They are finer, more granular, and predominantly mineral in origin (70% to 80%), with only small amounts of protein (10% to 20%); because of this and their formation toward the end of the heating section, they tend to reduce the efficiency of the heat transfer process and make it difficult to reach the product outlet temperature without raising the temperature of the heating medium to an undesirable extent. The mineral content is probably β-tricalcium phosphate.

It is interesting that fat did not feature significantly in any of the deposits, usually less than 5%, despite its being present in equal concentrations to protein and in greater concentration to minerals. It appears to play little part in the fouling mechanism and is present due to the fact that a small proportion gets trapped by the other fouling constituents. Fouling matter in creams, where the fat might represent up to 75% of the dry matter, is also reported to be predominantly protein and mineral in character, with the fat playing an insignificant part in the fouling story.

However, when the product was forewarmed (4 to 6 minutes at 85°C), sufficient to denature most of the whey proteins, the nature of the deposit changed to that of a composition similar to the Type B deposit described earlier, throughout the heat exchanger.

This type of deposit produced much less buildup of pressure and on this basis, forewarming could be recommended where pressure buildup was the limiting factor. Most of the discussion focused on fouling in the second heating stage. Some light fouling was found to occur in the preheater, and fouling could also be significant in the holding tube and in the early stages of the cooling section.

Fouling in direct plant is much reduced, due to the reduced heat transfer surface at high temperatures and the very quick rise in temperature; this leads to more deposit formation in the final product compared to indirect plant. Deposits may still form in the holding tube, the back-pressure valve, and the beginning of the cooling section. Higher proportions of fat (30% to 35%) were found in deposits from direct plants, attributed to fat destabilization during the injection process; high-temperature homogenization (138°C) also produced deposits with a higher proportion of fat.

8.4.2 Effects of Raw Milk Quality on Fouling

Raw milk quality and composition will affect the fouling behavior during UHT processing.

8.4.2.1 Acidity

The role of pH is not straightforward. For cow's milk the natural pH range is about 0.15 unit, but for freshly drawn milk of good quality the pH alone is not a reliable indicator of susceptibility to fouling, as fresh milks with the same pH showed considerable variation in their susceptibility to fouling (Burton, 1988). However, for any individual batch of milk, in general a fall in pH of about 0.15 unit during storage, from 6.67 to 6.52, would significantly increase fouling (Figures 8–8 and 8–9). Reducing pH was also found to increase the amount of fat within the deposit. The addition of sodium hydroxide to increase the pH by about 0.1 unit prior to processing had little effect, whereas the addition of trisodium citrate (4 mM) increased the pH by 0.7 unit and increased both processing times and product quality by slightly increasing its viscosity.

As raw milk quality deteriorates, its pH will decrease. Early investigations focused upon the role of decreasing pH as the cause of fouling, although it is now well established that a decrease in pH will also give rise to an increase in ionic calcium. Note that in milk, the total calcium concentration is about 30 mM, whereas the ionic calcium concentration is about 2 mM (less than 10%). Ionic calcium increases significantly as pH is reduced, probably brought about by the dissociation of calcium phosphate from the casein micelle and its increased solubility. These two factors combined will increase susceptibility to fouling. Also, reduction in pH tends to reduce the association of denatured whey protein with the casein micelle.

Burton (1968) showed that for the same milk sample, the amount of deposit formed as the pH was reduced (with hydrochloric acid) was increased, with significant changes taking place below a pH of 6.5. Kastanas et al. (1995) showed that fouling increased as pH was reduced at 140°C. This was measured by a decrease in OHTC and an increase in pressure drop in a miniature tubular UHT plant.

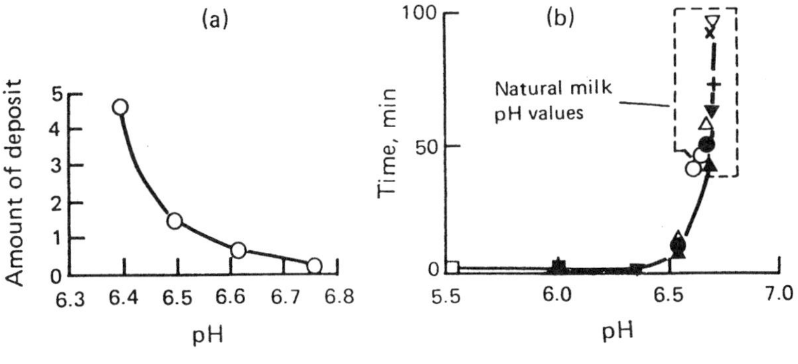

Figure 8–8 Effect of pH on deposit formation. (a), Variation of amount of deposit with pH; (b), Time to reach limiting pressure in a plate heat exchanger. *Source:* Data from (a) H. Burton, A Method for Studying the Factors in Milk which Influence the Deposition of Milk Solids on a Heated Surface, *Journal of Dairy Research*, Vol. 32, pp. 65–78, © 1965; and (b) P.J. Skudder and A.D. Bonsey, The Effect of Milk pH and Citrate Concentration on the Formation of Deposit during UHT Processing, in *Fouling and Cleaning in Food Processing*, D. Lund, A. Plett, and C. Sandu, eds., © 1985, University of Wisconsin, Extension Duplicating.

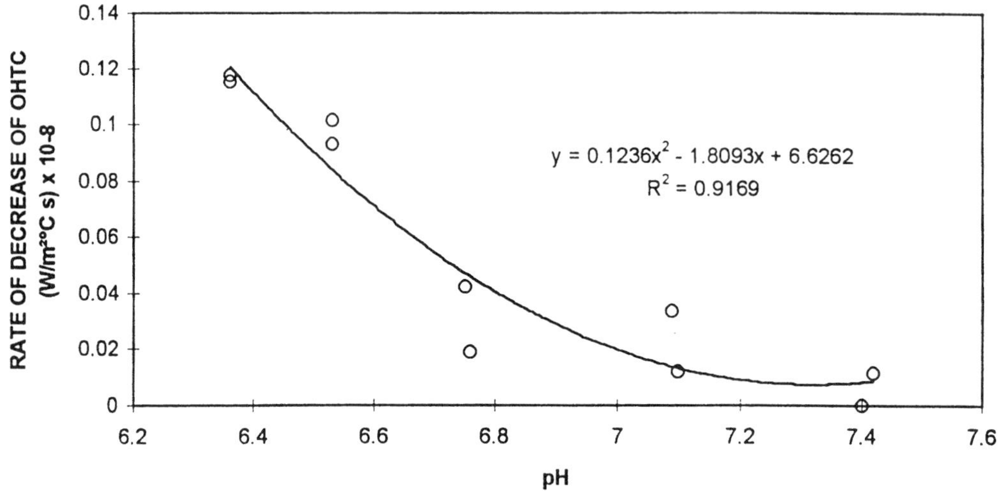

Figure 8–9 The rate of increase of the OHTC with time when milk with varying pH is heated at 140°C. The correlation coefficient (R^2) of the regression line is high. *Source:* Reprinted with permission from P. Kastanas et al., Design and Development of a Miniature UHT Plant, *Transactions of the Institute of Chemical Engineers*, Vol. 73, Part C, June 1995, © 1995, Institute of Chemical Engineers.

Skudder et al. (1986) observed that it was extremely useful to measure pH, as a slight increase in pH improves processing times. Similar results were obtained for reconstituted milks (Zadow & Hardham, 1978). Grandison (1988), from experiments with a wide number of raw milk samples, suggested that pH alone was not a reliable indicator by itself, concluding that there were other factors in milk that would make a significant contribution toward fouling. Experiments designed to look at seasonal variations in milk composition showed that there were significant differences between fouling behavior throughout a complete year (Figure 8–10). However, it was not possible to correlate fouling with any physical and chemical parameters, so the reasons for this observation were not clearly established. Milks with high levels of κ-casein were also more prone to deposit formation, but the addition of pyrophosphates to milk brings about a significant improvement.

It should be noted that during heating to over 100°C, the pH of milk falls to well below 6.0, but the pH recovers as the product cools down.

8.4.2.2 Role of Ammonia and Urea

Lalande and Corrieu (1981) have shown for different raw milks that there is a strong positive correlation between the rate of deposit formation and the concentration of ammonia, measured by an ion-specific electrode; an increase in ammonia concentration from 3.7 to 7.2 ppm is associated with a doubling of the fouling rate constant. One suggestion is that ammonia concentration may be associated with urea content of milk. Added urea was also found to reduce deposit formation. One explanation is that natural urease will break down urea in milk, thereby increasing its susceptibility to fouling and producing ammonia. Burton (1988)

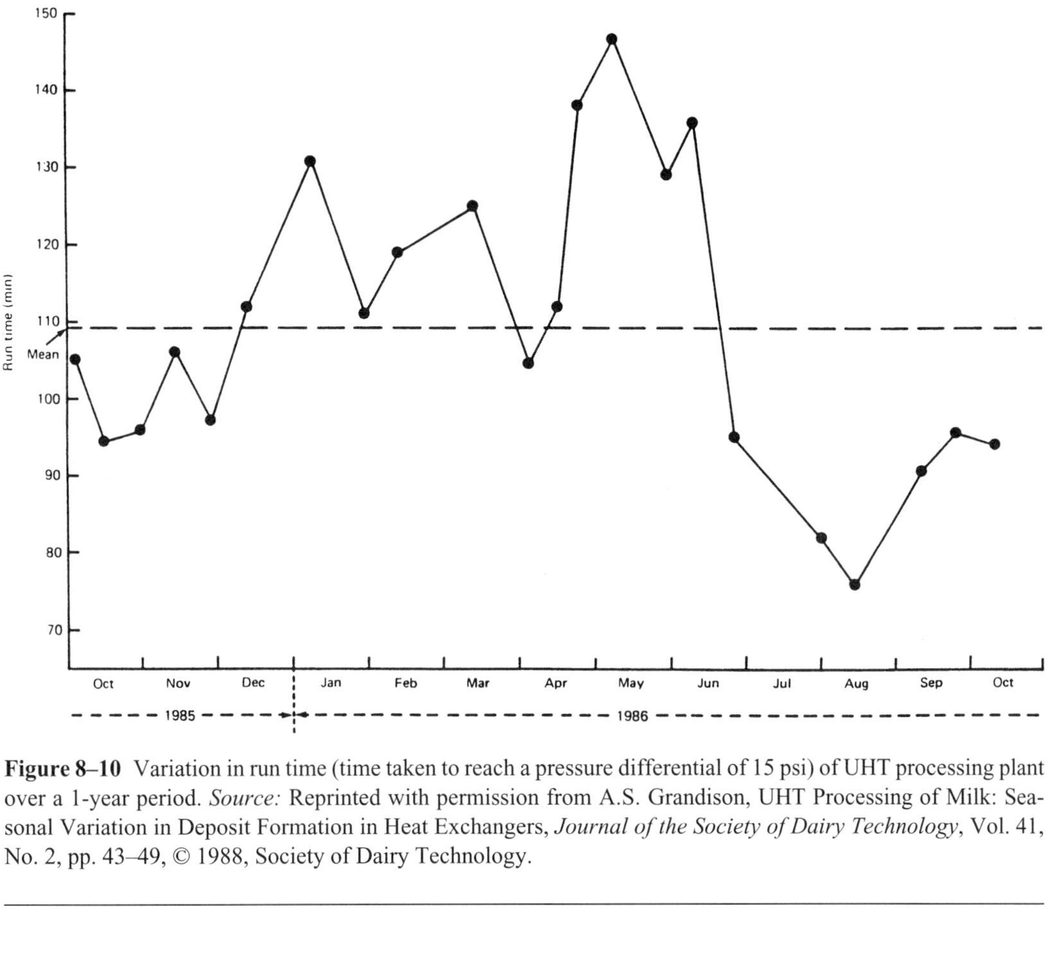

Figure 8–10 Variation in run time (time taken to reach a pressure differential of 15 psi) of UHT processing plant over a 1-year period. *Source:* Reprinted with permission from A.S. Grandison, UHT Processing of Milk: Seasonal Variation in Deposit Formation in Heat Exchangers, *Journal of the Society of Dairy Technology*, Vol. 41, No. 2, pp. 43–49, © 1988, Society of Dairy Technology.

suggested that an ammonia electrode may be a useful quality assurance measurement, but no further work appears to have been reported since that time.

8.4.3 Methods To Reduce Fouling

Although many of the following observations are specific to milk, the general principles and approaches may be relevant to and help provide explanations for the fouling of other products. The first approach is to avoid the use of raw materials that may be susceptible to fouling and to divert them to other products or processes, where fouling is not a major issue. For this, suitable test methods are required.

8.4.3.1 Alcohol Stability Test

The alcohol stability test may be useful in this respect to avoid raw milk that will be susceptible to fouling. In general, there are no quick and simple direct (involving heating) tests to establish whether a batch of milk will be susceptible to fouling. The heat coagulation test has been extensively investigated and could be used, but it is not quick and the results are not easy to interpret directly; it involves heating milk at 140°C and observing how long it takes for coagulation and gelation to occur (Walstra & Jenness, 1984). A review of the factors affecting heat stability is given by Fox and Morrisey (1977). One test that has been investigated because of its simplicity and speed of operation is the alcohol stability test. Equal volumes of milk and alcohol are mixed together to see whether the milk flocculates. If a number of different alcohol concentrations are used, the concentration that fails to coagulate the milk can be determined, and this is termed the alcohol stability for that milk. In general, the higher the alcohol stability the less likely is the milk to be susceptible to fouling. Virtually all milk will coagulate in 100% alcohol. Most good-quality raw milk in the United Kingdom is stable in at least 80% alcohol. The minimum stability to be suitable for UHT processing is recommended at 74% (IDF, 1981). In contrast, goat's milk flocculates in lower concentrations of alcohol, typically 50% to 60% alcohol. This is attributed to different proportions of the individual caseins present, especially the absence of α_{s1}-casein (Horne & Parker, 1982). The alcohol stability for buffalo's milk has been reported as 72% to 88% for bulked milk and 64% to 92% for samples from individual goats (Laxminarayana & Dastur, 1968).

Compositional factors affecting alcohol stability in cow's milk have been reviewed by Horne (1984), Donnelly and Horne (1986), and for goat's milk by Horne and Parker (1982) (see Section 6.4.11).

Burton (1988) also mentions that the alcohol stability test is a useful indirect test for assessing whether UHT-treated milks are microbiologically sound: it is recommended that such sound UHT milks should be stable in 68% alcohol (Section 10.5.5).

8.4.3.2 Aging Raw Milk

Aging raw milk at 4°C for 12 to 24 hours, without change of pH, was found to reduce its susceptibility to fouling. This was thought to be due to lipolysis and the production of fatty acids capric and later stearic acids. (Al-Roubaie, 1977; Al-Roubaie & Burton, 1979). Capric acid was thought to associate with the casein micelle and prevent interactions which would lead to the build-up of the deposit. After about 36 hours, significant amounts of stearic acid were produced, which increased susceptibility to fouling. Experiments with good-quality cow's milk showed that it could be kept for more than 14 days at 2°C before it started to foul UHT plant significantly. This was also observed for goat's milk (Kastanas, 1996).

8.4.3.3 Forewarming

One method to reduce fouling is by forewarming; however, this will inevitably increase the overall intensity of the heat treatment (as measured by whey protein denaturation or lactulose formation) (Section 10.5.7). However, this may provide a reasonable practical solution where milk quality is poor and fouling is known to be a severe problem. A wide range of conditions can be used from 85°C for 10 minutes, up to 90°C for 15 minutes, and for some very poor supplies, up to more than 110°C. A reasonable compromise is about 90° to 100°C for 2 to 5 minutes. However, it is not easy to hold milk for these periods in a continuous process, and batch processes may have to be used. The main change is to the denaturation of β-lactoglobulin and its association with the casein micelle, which will reduce the amount of Type A deposit. A second effect is the reduction in ionic calcium, brought about by the precipitation of calcium phosphate onto the casein micelle after heat treatment. Calcium phosphate itself has a negative temperature solubility. In milk at room temperature it is supersaturated; further increase in temperature will cause it to precipitate and associate with the casein micelle; this in turn will reduce the level of ionic calcium.

8.4.3.4 Dissolved Air

Gynning, Thome, and Samuelsson (1958) showed that removal of air from milk reduced the total amount of deposit produced by between 50% and 75% in a laboratory pasteurizer operating at 85°C. On a commercial UHT plant this can be achieved by using a deaerator, but there is no reported evidence that this will reduce fouling. Fouling is believed to result from the presence of bubbles as the air becomes less soluble at increased temperatures and increases deposit formation. The application of a reasonable back-pressure (about 1 bar) above that required to prevent boiling should ensure that air stays in solution.

8.4.3.5 Use of Additives

Various additives can be used to reduce fouling, but they may not be legally acceptable in all countries. Any component that increases pH and/or reduces ionic calcium should help reduce fouling (see Section 6.4.11). This will include alkali addition to increase pH and substances such as phosphates and citrates, which bind with ionic calcium. Care should be taken with the choice of phosphate salts, as their effects on protein stability is mixed. Sodium and potassium pyrophosphates, added at 100 ppm, were both effective and doubled the running time in the plate heat exchanger (Burdett, 1974). Skudder et al. (1981) found that the addition of iodate extends running time considerably by interfering with the formation of Type A deposit. Addition of only 10 ppm can double the running time; unfortunately, 20 ppm causes bitterness due to proteolysis during the storage of UHT milk (Skudder, 1981).

Some concluding thoughts on milk processing are as follows: fouling occurs under conditions where the casein micelle is less able to associate with fouling precursors formed by the action of heat in the bulk solution, or the boundary layer. Such precursors then move toward the wall of the heat exchanger and deposit there. The extent of the deposit, whether localized or extensive, will further depend upon environmental conditions, such as pH, ionic calcium, and whey protein concentration. This will also influence changes in OHTC and pressure drop.

8.4.4 The Mechanisms for Milk Fouling

The fouled material is predominantly a protein deposit early in the heat exchanger and a mineral deposit later. Is this a reflection of the order of the fouling reactions or a reflection of a more advanced stage of fouling in the sections at higher temperature? A full explanation of the fouling behavior of milk is still far from clear.

Sometimes fouling affects the OHTC, whereas on other occasions it influences the pressure drop. The severity of fouling in goat's milk is very much higher than that in cow's milk, despite many similarities in their gross composition. Fouling also occurs in permeate from milk ultrafiltration, despite the absence of proteins. Fouling has also been observed to be initiated by the use of hard water for plant sterilization (130°C for 30 minutes), prior to UHT processing.

The following explanation is based upon experimental evidence from a number of workers. Fouling appears to be initiated by denatured whey protein's adsorbing onto the heated surface. This fact has been established by a number of workers (Belmer-Beiny & Fryer, 1993). Factors influencing this process are bulk temperature and the level of undenatured whey protein present. This process is believed to take place very quickly. The next stage involves the association of casein micelles from the bulk of the solution with the denatured whey protein. This can lead to a substantial increase in the thickness of the deposit and will be that stage that gives rise to most processing problems. This aggregation is also affected by pH and calcium ion concentration. Decreasing pH and increasing calcium ion concentration will promote aggregation. Such deposit will lead to either an increase in OHTC or an increase in pressure drop, depending upon the conditions. There may not be an even coverage. Toward the end of the heating section, undenatured whey protein may become depleted; at this point, calcium phosphate may attach to the surface. It can be shown from reaction kinetic data for β-lactoglobulin that location of deposits may change as heating profile and initial concentration increases (see Figure 6–9). Also, on more prolonged contact with the hot surface, the whey proteins will further decompose, and, depending upon when it is analyzed, whey proteins may not predominate as the main deposit in immediate contact with the surface. The deposit will become drier and its thermal conductivity will fall. The number of contact points will decrease and, over a period of time, the deposit becomes more patchy, but this may be well after the time that normal processing has ceased.

8.4.5 Goat's Milk

From personal experience, goat's milk is a very difficult product to UHT process. It is both very susceptible to fouling, and the UHT product can produce a very significant deposit in the container after only a few hours of storage. Its alcohol stability is also well below that of cow's milk, between 50% and 60%. Zadow et al. (1983) studied the problem of UHT processing and concluded that either pH adjustment to well above 7.0 or addition of 0.2% disodium phosphate before sedimentation was necessary to reduce sedimentation. The higher levels of ionic calcium, in part due to its lower levels of citrate, were thought to be responsible for its instability.

Kastanas et al. (1996) found that goat's milk fouling lead to a rapid decrease in OHTC, with no initiation period; one explanation is that it represents a uniform coverage of the heat transfer surface with deposit. It seems to occur at relatively high ionic calcium levels, which suggest that calcium ions may accelerate the accumulation of deposit, probably by encouraging the interaction between casein and the denatured whey protein on the surface. Reconstituted goat's milk, made from freeze-dried powder, had a much lower ionic calcium level and was less susceptible to fouling than fresh goat's milk. Kastanas et al. (1996) showed that a number of treatments, such as citrate addition, forewarming, and pH adjustment, all reduced fouling in goat's milk (Figure 8–11). On the other hand, buffalo's milk is widely produced in some countries (e.g., Egypt and India), some of which is UHT processed, often mixed with cow's milk. There are no reports that this provides a severe fouling problem, despite its higher total solids (16% to 18%). In fact its alcohol stability and heat stability (heat coagula-

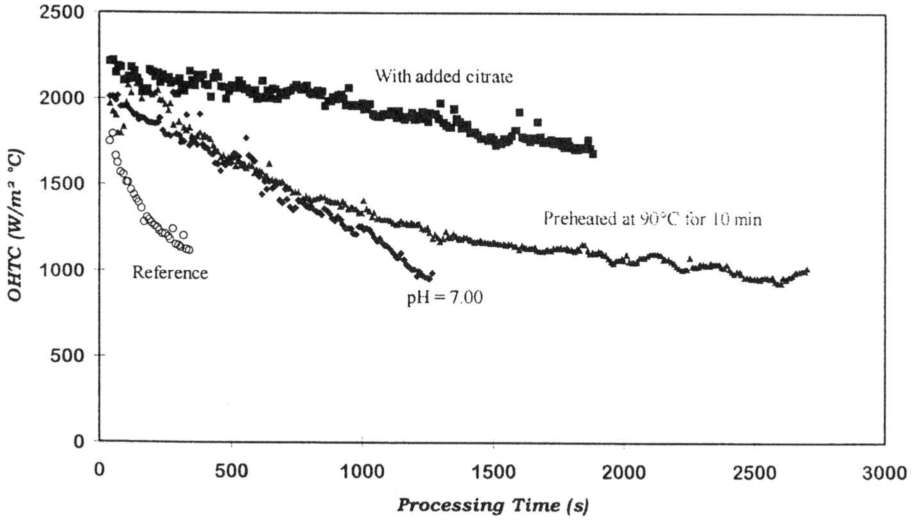

Figure 8–11 Modification to goat's milk composition to minimize fouling. *Source:* Reprinted with permission from IDF, *Production and Utilization of Ewe and Goat Milk*, IDF/FIL Publication No. 9602, © 1996, International Dairy Federation.

tion time) compares favorably with cow's milk (Section 8.4.3.1) (Laxminarayana & Dastur, 1968).

8.4.6 Milk Permeate

When milk is subject to ultrafiltration, the permeate is the fraction that passes through an ultrafiltration membrane. It contains all the low-molecular-weight components in milk and about one third of the calcium (~ 10 mM). Milk permate is very susceptible to fouling. In fact, when it is heated, it will start to go cloudy at about 60°C, due to calcium phosphate precipitation (as pH is decreased, the temperature required to cause precipitation increases). Kastanas et al. (1994) showed that on a tubular heat exchanger permeate fouling was seen as a rapid reduction in OHTC. However, on a plate pilot plant, no fall in the outlet temperature was seen over 20 minutes, although the pressure drop was showing signs of increasing. It should be noted that the permeate used in these experiments was produced from good-quality milk. Also, the ionic calcium in permeate is much lower than that in milk (about 1.3 mM, compared to 2.0 mM). However, it is not entirely clear why these differences do occur. Since permeate contains none of the proteins present in milk or whey, there is obviously a different fouling mechanism taking place to that in milk or whey. This again shows the role of casein in milk in terms of stabilizing those reactions that will lead to fouling (i.e., whey protein denaturation and precipitation of calcium phosphate). My view is that calcium phosphate and other minerals could also be a good nucleator and initiator of fouling, providing the binding sites for denatured whey protein and casein. However, Lund and Bixby (1974) suggest that is not the case, as prefouling of a heat exchanger with calcium phosphate was not found to affect its performance over a 3-hour period. (Note, however, that the OHTC of the prefouled heat exchanger was half that of the clean-plate heat exchanger.) If this is so, it would suggest that efficient removal of mineral scale is an important aspect of the cleaning process. Barton, Chapman, and Lund (1985) looked in more detail at fouling by model solutions of calcium phosphate.

8.4.7 Cream/Concentrates

In general, fouling of cream is not reported to be a major problem, again suggesting that the fat phase does not play an important part in fouling processes. Cream products are not produced in such abundance as milk products, so processing runs would be shorter and this may be a significant factor to explain the shortage of literature on fouling during cream production. Preholding was found to be effective in reducing fouling in cream, recommended conditions being 95°C for a few minutes (Hiddink et al., 1986). Also, fouling deposits from whipping cream contained little fat and were predominantly protein and minerals, Therefore, cream fouling may be influenced by factors similar to those for milk fouling. However, with cream being more viscous, the thermal boundary layer will be thicker and cream may thus be more susceptible than milk, at similar flowrates.

8.4.8 Reconstituted Milk

Some interesting work has been reported by Newstead (1994) on the fouling of reconstituted milks. The work was performed on a plate heat exchanger and fouling was monitored by measuring the rate of change of approach temperature with time. Experiments were performed over 2 hours and results were presented in terms of the run time possible. The following provides a summary of the results:

- *Process temperature* had a considerable effect; increasing it from 135° to 142°C resulted in a 45% increase for each °C rise; the increase was 70% from 142° to 145°C.
- *Pretreatments*: Lower pretreatment tended to favor lower fouling rates.
- *Type of water*: Water in minerals was found to influence fouling significantly, with harder waters producing more fouling. However, other factors, as yet unidentified, were also affecting the fouling rates.
- *Cold storage*, for up to 48 hours, was found to improve fouling rates significantly.
- *Polyphosphates* can reduce fouling significantly. It is interesting that the point of addition also had a significant effect; it was found best to add it to the milk after it had been recombined.
- *Homogenization* during recombining was found to increase fouling significantly; in this sense, recombined milks behave differently to fresh milks. It is probably best to avoid homogenization prior to heat treatment.

Again, these results raise the question of the role of water and water quality, especially the amounts of calcium. Kastanas (1996) looked at the fouling behavior of skim milk powder, which was reconstituted with hard water and soft water, and found that using soft water reduced its susceptibility to fouling.

8.4.9 Whey Proteins

Cheese whey contains about 20% to 25% of the protein present in whey. Whey proteins will start to denature and coagulate at temperatures in excess of 75°C. It is very difficult to UHT-treat whey, as it will result in almost complete blockage in a very short time period (author's painful experience). Considerable research work has been done on whey protein systems and in some cases this has been used as a model system for understanding fouling in milk. However, although there are some similarities, there are also some subtle differences between the two, arising from the stabilizing effect of casein and its ability to prevent denatured β-lactoglobulin's precipitating in milk. Fryer (1989) and Gotham, Fryer, and Pritchard (1992) have done considerable work on whey protein denaturation.

Delplace, Leuliet, and Tissier (1994) examined whey protein fouling in a plate heat exchanger over a 6-hour period using an inlet temperature of 70°C to an outlet of 96°C. They measured deposit formation along the plate and found an induction period, followed by a decrease in OHTC, followed by leveling. There was no significant increase in the pressure over the period where the OHTC was decreasing. The maximum amount of deposit occurred in the middle section, which coincided with the predicted β-lactoglobulin denaturation profile. It was estimated that about 3.6% of the denatured β-lactoglobulin was involved in deposit formation. It was also calculated that the fraction of the hydraulic diameter occupied by the thermal boundary layer volume was about 3%, so it was proposed that only proteins denatured in the hydraulic boundary layer were involved in deposit formation. Therefore, strategies for reducing the hydraulic layer would therefore reduce fouling.

With whey protein systems, what is the pH dependence of fouling? This will depend upon pH effects on whey protein denaturation and on the solubility characteristics. Barlow, Hardham, and Zadow (1984) found that whey protein concentrates were unstable to UHT processing, with plant blockage and sedimentation being observed. Stability to UHT processing was improved by addition of 0.25% dihydrogen orthophosphate, combined with preheat treatment at 85°C for 5 minutes.

8.4.10 Tomato Products

A wide variety of tomatoes is used for processing. Their main characteristics are full body; tough to eat (withstand transportation), high-soluble solids, and good color. Flavor characteristics are not so important, as many are later blended for sauces. The composition of tomatoes is given in Table 1–2. They can be processed at high rates, up to 300 tonnes per hour, 24 hours per day, through the season, which can be up to 120 days. The main tomato products are juice up to 12% Brix, purée 12% to 22% Brix, and pastes (> 21%, but not exceeding 31%, above which there is a quality loss). Tomato processing was summarized in Chapter 5.

Fouling may occur in a number of places: (1) hot break system, (2) evaporation, (3) sterilization at high total solids), and (4) sauce production.

Adams, Nelson, and Legault (1955) found that addition of pectic enzymes to tomato juice completely inhibited fouling, suggesting that pectin contributes to fouling as well as protein. Tomato juices that had not been heated during the break process and so contained natural pectic enzymes, when concentrated produced a reduced amount of fouling. However, this method of controlling fouling will produce a product with a greatly decreased viscosity.

Morgan and Carlson (1960) found that temperature was the most important factor that affected fouling of tomato products.

8.4.11 Eggs/Batters

Ling and Lund (1978) found that PTFE-coated surfaces did not foul when egg albumen was heated for 2 hours at temperatures between 67° and 72°C. Also, surface temperature affected greatly the fouling rate; increasing it from 70.5° to 74°C increased the fouling rate by 100%.

8.5 RINSING, CLEANING, DISINFECTING, AND STERILIZING

8.5.1 Introduction

At the end of the processing run, all surfaces, but especially the heat-transfer surfaces, need to be cleaned and disinfected to bring them back to an acceptable starting condition. Cleaning involves the removal of fouling deposits from the surface and may take place in one or more stages.

It should be pointed out that food processing equipment should be designed with cleanability in mind, as this is fundamental to the concept of producing safe products. The hygienic design of food-processing equipment has been discussed in more detail by Jowitt (1980) and Romney (1990). Particular care and attention should be taken to avoid deadspaces in pipelines, valves, and other fittings.

After processing has finished, the normal procedure is first to rinse or flush out the product with water before starting the cleaning procedure, so rinsing operations are an integral part of the cleaning procedure. Detergent solutions are used for cleaning. After or between detergents, the plant also needs to be rinsed with water. These rinsing processes are necessary to avoid any cross-contamination, but are unproductive and are regarded as downtime. They also consume valuable resources, for example, water for rinsing, detergents for cleaning, and energy, as well as producing effluent, which will also need to be treated. Thus, these processes involve issues that impinge on the environment. Following cleaning, water is used to flush detergent from the plant, and the surfaces may then either be disinfected (e.g., for pasteurization processes) or sterilized for UHT processes, prior to shut-down. This may not always be necessary at the end of the process, but it is essential prior to processing. Disinfecting involves reducing the microbial count on the surface to predetermined low levels, whereas sterilization involves their complete removal.

There is also scope for optimization of the entire process. For example, the simplistic approach is to ensure that the processing run is as long as possible. However, this may create a more difficult deposit to remove, thereby increasing the time for cleaning and the costs for detergents and effluent treatment. At this time, modeling is thought not to be very helpful for such optimization, because of the dearth of satisfactory models that fully describe both the fouling and the cleaning processes.

8.6 RINSING

Rinsing is the first step following processing and involves flushing out the product, usually with water at ambient temperature. Rinsing may be done at the end of a day's production, or when changing from one product to another. It can be very wasteful of water, as well as producing significant effluent. It is particularly important to rinse correctly, particularly between batches of different products, to avoid cross-contamination. Rinsing has been studied less than fouling and cleaning.

Changeover processes of any kind (e.g., between product and water or detergent and water) can be monitored by conductivity, turbidity, or fluorescence measurements; conductivity is most widely used.

For pipeline flow, the transition from product to water can be envisaged as a step change in input function (provided the flowrate of the two fluids remains the same). If the concentration is measured against time at the outlet, the curve shown in Figure 8–12 (Loncin & Merson, 1979) results.

The ideal situation for maximum washout of the material is plug flow. Any item of equipment that extends the distribution of residence time will make rinsing more difficult. One way to analyze the rinsing process is to plot $\log(c/c_o)$ against time at the outlet, following a step change from c_o to c at the inlet. There are three distinct phases when $\log(c/c_o)$ is plotted against time, which have been modeled by Loncin and Merson (see Figure 8–12):

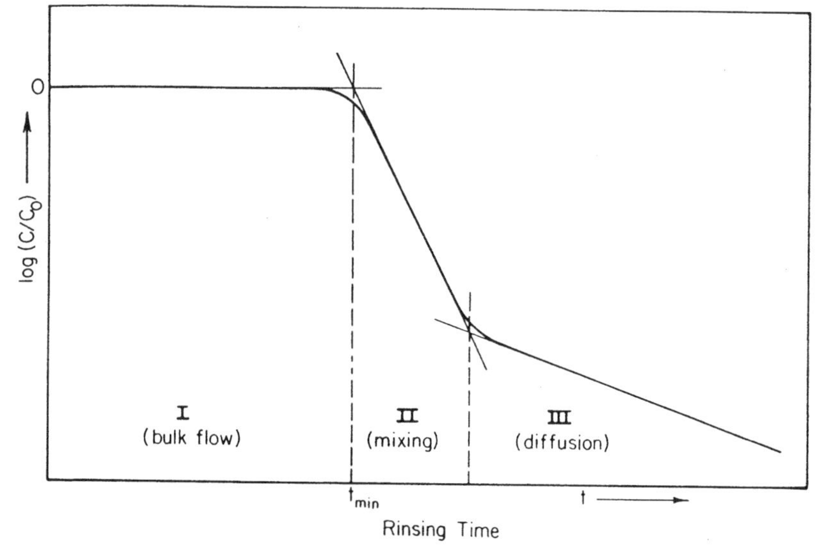

Figure 8–12 Three stages of mass transfer during rinsing. (I) Bulk-flow removal of detergent solution: $\log(C/C_o)$ = 0. (II) Boundary layer mixing; $\log(C/C_o = A_1 - k_1 t$. (III) Boundary layer diffusion; $\log(C/C_o = A_2 - k_2 t$. C_o is the initial detergent concentration; A_1, A_2, k_1, and k_2 are experimental constants; t_{min} is the minimal residence time of the apparatus. *Source:* Reprinted with permission from M. Loncin and R.L. Merson, *Food Engineering*, p. 315, © 1979, Academic Press.

1. Phase I: No change in composition. This ends at a time corresponding to the minimum residence time through the equipment.
2. Phase II: Removal of the bulk of the fluid by displacement of water. There is mixing of the water with the material.
3. Phase III: A slower removal of material from the boundary layer, by molecular diffusion. This is very much the rate-limiting process.

8.6.1 Water/Product Changeover

This can be considered to be the reverse of the rinsing process. After the plant has been sterilized with water at the start of the process, and the appropriate temperature and flow conditions have been established, there is a changeover from water to product. It may take some time for steady state to be reestablished. From water to milk, there is little difference in terms of flow characteristics, but from water to a viscous product there may be some large differences in product viscosity.

If the product is collected too early, then it may be diluted. If it is not collected early enough, there may be substantial waste. The problems are almost the same as those encountered in rinsing, the slow depletion of material from the boundary layers.

8.7 CLEANING

Cleaning is the removal of deposits from a fouled surface. This section focuses upon cleaning-in-place (CIP) methods, whereby deposits are removed by contacting them with hot detergent solutions that eliminate the fouled deposits, thereby largely removing the need for manual cleaning. In the past 30 years, this has largely superceded cleaning-out-of-place, where pipe unions were taken apart by hand and lengths of pipeline were manually cleaned.

Cleaning can be regarded as the combination of three energy sources: thermal energy, chemical energy, and mechanical energy. It involves the transfer of the cleaning agent from the bulk of solution to the surface of the deposit, followed by diffusion into the deposit and its reaction and breaking down of the deposit. This is then followed by the transfer of the dispersed or dissolved deposit back into the bulk solution. Cleaning is achieved by a progressive soaking and dispersion of the soil layer and not by simple dissolution (Loncin & Merson, 1979). The aim is to achieve a surface that is both clean and hygienic. Thus the surface must be free of soil, pathogenic microorganisms, and food-spoilage organisms, and also detergents and disinfecting agents (Plett, 1985). The types of deposit have been summarized by Plett and are presented in Table 8–2.

A detergent has been defined as any substance that either alone or in a mixture reduces the work requirement of a cleaning process. A wide range of detergents is available. Simple ones are based on either caustic soda or nitric acid. These form the basis of the classic two-stage cleaning process, involving alkali followed by acid washes, to remove the types of deposit listed in Table 8–2. An intermediate water rinse would also be used. Typical concentrations would be up to about 2.0% for caustic soda and 1.0% for nitric acid. Other alkalis (e.g., sodium orthosilicate, sodium carbonate) or acid (phosphoric, hydroxyacetic) materials may be incorporated. Caustic soda and nitric acid are also effective bactericides.

Table 8–2 Typical Components in Food Process Fouled Layers

Component	*Solubility*	*Removal*	*Heat Alterations*
Sugar	Water: soluble	Easy	Caramelization
Fat	Water: insoluble Alkali: poor Acid: poor	Difficult (good with surfactants)	Polymerization
Protein	Water: poor Alkali: good Acid: medium	Difficult Good Difficult	Denaturation
Mineral salts, monovalent	Water: soluble Acid: soluble	Easy	
Mineral salts, polyvalent	Water: insoluble Acid: soluble	Difficult Good	Precipitation

Source: Reprinted with permission from E.A. Plett, Cleaning of Fouled Surfaces, in *Fouling and Cleaning in Food Processing*, D. Lund, A. Plett, and C. Sandu, eds., p. 288, © 1985, University of Wisconsin-Madison, Extension Duplicating.

However, it is claimed that in some circumstances, such caustic and acid processes may not remove all the soil. Other additives may be used to improve the performance of a detergent. These include surface-active agents to improve wetting, emulsification, and dispersion, and sequestering agents to prevent water scale from developing and to give good rinsing properties. Table 8–3 lists the properties of some of the main components of detergents.

This has led on to the introduction of single-stage commercial cleaners, usually detergent based, that are more complex and contain many of the additives listed in Table 8–3, to improve their performance. These are more expensive. However, they should be more efficient and save time and energy and produce less effluent. The complexity of modern detergents is illustrated by the large numbers of formulations (in some cases over 200) available from some detergent suppliers.

8.7.1 Methods To Measure Cleaning Effectiveness

The effectiveness of a cleaning program can be evaluated by determining the amount of soil remaining after cleaning. This can be done directly by weighing or by use of some property that is dependent upon the amount of soil still present, for example, ranging from visual or microscopical observations, optical, microbiological, radiological, or chemical methods. However, such methods can only be used if the surface is accessible and would be used for laboratory testing procedures. Having said this, it is also important that production-scale plant is accessible, as it is a useful to be able to inspect internal surfaces periodically, should the need arise. Such testing procedures and equipment have been reviewed by Romney (1990) and Shapton and Shapton (1991).

Table 8–3 Properties of Detergent Raw Materials

Detergent Raw Material	*Chemical Formulation*	*pH 1% Solution*	*Chemical Reactivity*	*Wetting*	*Soil-Lifting Power*	*Dispersion*	*Emulsification*	*Water Softening*	*Threshold Effect*	*Foam*
Caustic soda	NaOH	13.3	5	1	1	1	1	2	—	—
Sodium orthosilicate, monohydrate	$2Na_2O:SiO_2:H_2O$	12.8	4	2	3	3	2	2	—	—
Sodium metasilicate, pentahydrate	$Na_2SiO_3:5H_2O$	12.3	3	2	3	3	2	2	—	—
Trisodium orthophos-phate, crystalline	$Na_3PO_4:12H_2O$	11.95	3	2	3	3	2	2	—	—
Soda ash										
Sodium	Na_2Co_3	11.5	1	1	1	1	1	2	—	—
hexametaphosphate	$Na(PO_3)_6$	7.6	1	1	4	4	2	5	4	—
(Calgon)										
Sodium tripolyphos-phate	$Na_5P_3O_{10}$	9.6	1	1	4	4	2	5	4	—
Sodium gluconate										
Tetra sodium salt of		7.1	1	1	4	4	2	5	—	—
EDTA*		11.2	1	1	4	4	2	5	—	—
Sodium dodecyl benzene sulphonate		6.9	1	5	3	4	4	1	—	5
Nonyl phenol/9EO*										
Disodium salt of		7.1	1	5	3	4	5	1	—	2
acetodiphosphonic acid		8.5	1	1	1	5	2	3	5	—

* 9EO = 9 moles of ethylene oxide; EDTA = ethylenediaminetetraacetic acid.

Source: Reprinted with permission from A.J.D. Romney, *Cleaning in Place Manual*, p. 27, © 1990, Society of Dairy Technology.

Galotte-Lavallee, Lalande, and Corrieu (1981) developed a method for control of CIP systems, based on an optical fiber, which measured the changes in optical density of the cleaning solution due to dissolution of deposits during cleaning.

8.7.2 Kinetics—Dynamics of Cleaning

There has been considerable interest in the dynamics of the cleaning process and how quickly the soil is removed. Again there are a number of ways of doing this, by monitoring properties that are dependent upon the amount of soil that is solubilized, such as optical density, or the concentrations of protein fat or minerals.

Loncin and Merson (1979) have reviewed the kinetics of cleaning. Within certain limits soil removal can be described by a first-order reaction. Thus if m represents the mass of soil per unit area, then the rate of removal is directly proportional to the mass;

$$dm/dt = -km$$

or

$$\ln(m_i/m) = k\,t$$

where m_i = initial mass, m = mass after time (t), and k is a rate constant.

The rate constant depends upon several factors, as follows:

- The kind of soil and its state. There is considerable difference between soils from different foods and their ease of removal (see Table 8–3).
- The nature of the support material. This will affect the forces of adhesion between the deposit and the surface. For example, highly polished surfaces may be assumed to clean more easily, although they may also be more susceptible to corrosion. The use of coatings such as Teflon may also ease cleaning.
- The nature and concentrations of the detergents used. For simple caustic soda solutions, k was found to vary in a linear fashion with concentration. However, this finding is not likely to apply to surface-active components. Note also, in some cases a decline in cleaning rates has been observed when the concentration is too high, attributed to transport inhibition, perhaps caused by a very rapid swelling of the outer soil layers at high concentrations.
- The temperature of cleaning. Note that the rate constant follows Arrhenius kinetics (straight-line relationship between k and 1/T). However (as for concentration), there may be an optimum cleaning temperature beyond which cleaning rates decline due to heat-induced alterations of some soils.
- Mechanical action and the amount of shear. All researchers have found an improvement of cleaning rates with mechanical action; thus increasing the flowrate (hence the flow velocity and Reynolds number) will improve cleaning action but will also increase the pumping and energy costs. There has been some controversy on how to characterize this effect better (i.e., by wall shear stress or by Reynolds number). Note the wall shear stress (τ_w) is given by $D\,\Delta P/4\,L$.

Plett (1985) reviews the mathematical models to correlate the effect of mechanical action on cleaning rates. He concludes that the tendency goes clearly toward correlations involving wall shear stress, which can be related much better to mean velocity than to Reynolds numbers.

One of the first reported practical suggestions was to obtain a high shear stress at the wall by ensuring that the fluid velocity is high, a minimum of 1.5 ms^{-1} being recommended for effective cleaning. Jackson and Low (1982), with tomato juice, found a wall shear stress threshold of 0.8 Pa, above which there was a remarkable effect on cleaning. High shear stresses can be induced through turbulent flow conditions, where the shear stress is proportional to the mean velocity. On the other hand, Jennings, McKillop, and Luick (1957) postulated a minimum Reynolds number of 25,000, above which the mechanical effect is remarkable. It is noteworthy that at a constant volumetric flowrate, both the Reynolds number and the wall shear stress will decrease as the diameter increases. Therefore, any expansions or enlargements of the pipe system may not be so easy to clean. These matters should be considered at the design stage.

Increased cleaning rates have also been observed when the direction of flow is reversed every 5 or 15 seconds (Grasshoff, 1983). There are some claimed advantages of dispersing air bubbles into the cleaning solution (two-phase flow). Under appropriate conditions, annular flow takes place; in this case the rate of momentum transfer in the annular liquid layer becomes very intense and a strong cleaning action can be expected, with limited pumping expenses. Plett (1985) suggests that the amplitude and frequency of the shear stress peaks (arising from turbulent bursts) may play a more important role and be a more significant parameter than the average wall shear stress. On some equipment, it may be necessary to install an additional pump to obtain these cleaning velocities or to bypass the homogenizer, if this is limiting the flow.

Plett (1985) has reviewed kinetic models in more detail. Deviations from first-order reaction kinetics occur, which is not surprising, considering the overall complexity of the process. The ability of first-order kinetics to model the overall dynamic behavior has been attributed to the fact that the rate-limiting process that occurs at the end of the process is the mechanism of diffusion of cleaning reaction products.

It has often been noted that soil removal is more rapid than predicted at the beginning of the process and that it decreases toward the end. It is probable that this arises from the fact that at the beginning of the process the soil adheres to another layer of soil, whereas at the end of the operation the forces of adhesion between the soil and the support must be broken and that these are greater.

Thus as cleaning proceeds the reverse processes (to fouling) are taking place; that is, restoration of the overall heat transfer coefficient and reduction in the pressure drop over the fouled section. Thus, cleaning can be measured indirectly by restoring OHTC, or ensuring that a minimum pressure drop has been attained. Grasshoff (1983) and others have observed that pressure drop initially increases during cleaning due to significant swelling of the deposit. Also, the rate of cleaning initially is low, reaching a maximum after a short time, followed by a steady decline.

Without inspecting the surfaces, there is no way of easily establishing how effective the process has been. However, in practice, this is more difficult than is suggested.

8.7.3 Some Practical Observations

Simple deposits may be cleaned with hot water. Cheow and Jackson (1982a) studied water cleaning of a heat exchanger fouled with tomato juice at 20°, 70°, and 90°C. The most effective temperature was 70°C because some protein denaturation occurred at 90°C.

For most operations, detergents are required. In fact, when the same deposits were cleaned with 2% caustic soda, the rate of cleaning was faster at 90°C than at 70°C and no protein denaturation was seen (Cheow & Jackson, 1982b). They also found that when pressure was increased, the rate of cleaning decreased.

Fouling deposits from milk found on heated surfaces consist mainly of protein, minerals, and some fats. Alkali (caustic)-based detergents are effective in dissolving and removing fat (by hydrolysis) and protein, whereas acid-based detergents are effective in removing minerals; for specialized applications proteolytic enzymes may also be incorporated to increase efficiency although they are not so widely used on heat exchangers (more so with membrane equipment). The widely accepted traditional method to remove them is a two-stage, alkali-acid procedure. This involves at least five steps: prerinse, alkali circulation, intermediate rinse, acid circulation, and final rinse. This two-stage procedure was considered to be the most effective in removing milk deposits (Kane & Middlemiss, 1985).

A single-stage procedure has also been in use in the United Kingdom for many years, reducing the number of steps to three. Timperley and Smeulders (1987) have compared the single- and two-stage processes for bench-scale and pilot-scale tubular pasteurizers. The use of a single-stage detergent was shown to produce physically clean surfaces in half the time taken by the two-stage alkali-acid procedure, which did not remove calcium deposits completely. Similar findings were confirmed by Timperley et al. (1994) on pilot-scale and production-scale UHT sterilizers, leading to significant reductions in cleaning time (20% to 30%) in addition to lower water and energy consumption.

For many applications, alkaline-based detergent strengths between 1% and 2% are recommended. However, Bird (1994) reported that an 11% caustic solution cleaned glucose deposits in the shortest time, and a range from 4% to 12% gave acceptable performances. The use of higher concentrations led to longer cleaning times, and a solution of 20% caustic resulted in the same performance as a 1% solution. For a potato starch deposit, 14% caustic was found to be the optimum. Attention was drawn to the fact that these values are much higher than those needed for protein removal (0.5%). Grasshoff (1997) provides a recent review on cleaning of heat exchangers, which discusses the chemistry of detergents and the mechanics of cleaning, together with a summary of experimental studies on the effects of cleaning parameters on deposit removal. Still, one of the biggest problems is that of on-line monitoring of cleaning performance. Withers (1995) describes a collaborative CEC-FLAIR project related to fouling and cleaning, which includes the development of an ultrasonic cleaning device for use in tubular heat exchangers. Grasshoff (1997) points out that the main problem is with designing an oscillator that can be installed in a plate heat exchanger.

There is an increasing interest in biofilms. There may be some relevance to thermal processing operations, as improperly cleaned deposits may form breeding grounds for microbial growth, which may compromise product safety; there are also implications for corrosion. There have been a number of recent articles devoted to the removal of biofilms and to the inactivation of different microorganisms in biofilms. In thermal processing, the effective removal of biofilms (if they are present) would seem to be the most important priority. Biofilm formation and its elimination from food-processing equipment has been recently reviewed by Gibson et al. (1995) and Sjoeberg, Wirtanen, and Matilla-Sandholm (1995).

8.8 DISINFECTING AND STERILIZING

Disinfecting and sterilization of the equipment (as appropriate) should always be done immediately before the process commences and perhaps also after cleaning. Wherever possible, disinfecting and sterilizing by heat is the simplest and most effective procedure, and this is also the most commonplace for continuous heat treatment. For this purpose, steam or superheated water up to 140°C are used. One of the main advantages over chemical cleaners is that there is no residue to wash out and no risk of cross-contamination.

8.8.1 Heat

For pasteurized products, postpasteurization contamination is the most important determinant of keeping quality (see Chapter 5); this is reduced by ensuring that all surfaces that come in contact with the product downstream of the holding tube reach a temperature of 95°C for 30 minutes. For a UHT plant, it is crucial to sterilize the equipment downstream of the holding tube. To achieve this requires a temperature of 130°C for 30 minutes. It may also take some time to actually achieve this temperature. This is a very severe process, as can be determined by calculating the F_0 value, which would be about 230 (compare 3 to 18 for low-acid foods). It may be difficult to apply such temperatures to the cooling sections with direct-refrigeration sections.

As a general design principle, the aim to simplify plant pipework downstream of the holding tube; minimize pipe runs to ensure that they can be sterilized in a practicable time. Positioning the homogenizer downstream will increase the time for plant sterilization. Instruments based on thermal imaging for measuring surface temperatures are useful to check whether all pipework is adequately sterilized. Temperature-sensitive stick-on indicators provide a cheaper useful alternative for this purpose.

8.8.2 Sterilization by Chemicals

There may be some situations where chemical disinfecting or sterilizing agents are required when materials may be sensitive to heat. Some refrigeration systems may be difficult to sterilize by heat due to expansion of the refrigerant. One problem arising, which is more acute when a sterile surface is required, is that the chemical must be removed from the plant, raising the problem, what is used to do this without losing sterility? One possibility is sterile water, but this then requires a supply of sterile water. For plant sterilization, using heat is the simplest solution.

Note that caustic soda and nitric acid are themselves bactericides; some data are presented by the Society of Dairy Technology (SDT, 1990). Active chlorine is also used, usually in the form of sodium hypochlorite in the active chlorine concentration range of 10 to 200 ppm. It should be used cold, as it is corrosive at high temperature. Note that a concentration of 1,000 ppm of chlorine has no sporicidal effect if the pH is high. It should also be noted that the concentration may diminish with time due to chemical reaction; for example, chlorine reacts with oxidizable noncellular groups.

Hydrogen peroxide is used to sterilize surfaces and for aseptic packaging (Chapter 7). Others used are peracetic acid, peroxyacetic acid, and iodophores. If required, information on their chemistry and use can be found in Romney (1990) and Shapton and Shapton (1991).

REFERENCES

Adams H.W., Nelson, A.L., & Legault, R.R. (1955). Film deposition on heat exchanger coils. *Food Technology* **9**, 354–357.

Al-Roubaie, S.M.A. (1977). *Factors affecting deposit formation from heated milk on surfaces*. PhD thesis. Reading, UK: University of Reading.

Al-Roubaie, S.M.A., & Burton, H. (1979). Effect of free fatty acids on the amount of deposit formed from milk on heated surfaces. *Journal of Dairy Research* **46**, 463–471.

Barlow, I.E., Hardham, J.F., & Zadow, J.G. (1984). Stability of reconstituted whey protein concentrates to UHT processing. *Journal of Food Science* **49**, 32–33, 39.

Barton, K.P., Chapman, T.W., & Lund, D. (1985). Rate of precipitation of calcium phosphate on heated surfaces. *Biotechnology Progress* **1**, 39–45.

Belmer-Beiny, M.T., &Fryer, P.J. (1993). Preliminary stages of fouling from whey protein solutions. *Journal of Dairy Research* **60**, 467–483.

Bird, M.R. (1994). Cleaning agent concentration and temperature optima in the removal of food-based deposits: Continuous flow processing. *DTI Food Processing Sciences* LINK. Seminar, Campden and Charleywood Food Research Association, Chipping Campden, UK.

Burdett, M. (1974). The effect of phosphates in lowering the amount of deposit formation during the heat treatment of milk. *Journal of Dairy Research* **41**, 123–129.

Burton, H. (1964). The effect of forewarming on the formation of deposits from separated milk on a heated wire. XVII International Dairy Congress, B607–612.

Burton, H. (1965). A method for studying the factors in milk which influence the deposition of milk solids on a heated surface. *Journal of Dairy Research* **32**, 65–78.

Burton, H. (1966). A comparison between a hot-wire laboratory apparatus and a plate heat exchanger for determining the sensitivity of milk to deposit formation. *Journal of Dairy Research* **33**, 317–324.

Burton, H. (1968). Reviews of progress of dairy science, section G: Deposits from whole milk in heat treatment plant—a review and discussion. *Journal of Dairy Research* **35**, 317–330.

Burton, H. (1988). *UHT processing of milk and milk products*. London: Elsevier Applied Science.

Cheow, C.S., & Jackson, A.T. (1982a). Cleaning of plate heat exchanger fouled by tomato juice: 1. Cleaning with water. *Journal of Food Technology* **17**, 417–430.

Cheow, C.S., & Jackson, A.T. (1982b). Cleaning of plate heat exchanger fouled by tomato juice: 2. Cleaning with caustic soda solution. *Journal of Food Technology* **17**, 431–440.

Corrieu, G., Lalande, M., & Ferret. (1981). New monitoring equipment for the control and automation of milk pasteurization plants. In P. Linko, Y. Malkki, J. Olkku, & J. Larinkari (Eds.). *Food process engineering*, Vol. 1 (pp. 165–171). London: Food Processing Systems, Applied Science Publishers.

Daufin, G., Kerherve, F.L., & Quemerais, A. (1985). Electrochemical potential influence upon fouling of stainless steel and titanium heat surfaces by sweet whey. In D.B. Lund, E. Plett, & C. Sandu (Eds.), *Cleaning in food processing* (pp. 263–275). Madison, WI: Madison University Press.

Davies, T.J., Henstridge, C., Gillham, C.R., & Wilson, D.I. (1997). Investigation of whey protein deposit properties using heat flux sensors. *Trans Chem Eng* **75**, Part C, 106–110.

de Jong, P., Bouman, S., & van der Linden, H.J.L.J. (1992). Fouling of heat treatment equipment in relation to the denaturization of β-lactoglobulin. *Journal of the Society of Dairy Technology* **45**, 3–8.

Delplace, F., Leuliet, J.C., & Tissier, J.P. (1994). Fouling experiments of a plate heat exchanger by whey protein solutions. *Trans I Chem Eng* **72**, Part C, 163–169.

Donnelly, W.J., & Horne, D.S. (1986). Relationship between the ethanol stability of bovine milk and natural variations in milk composition. *Journal of Dairy Research* **53**, 22–33.

Environmental Concepts, Ltd. Case studies. Lakeview House, 12 Lakeside, Funtley, Fareham, Hants, PO17, 5EP.

Fox, P.F., & Morrisey, P.A. (1977). Reviews of the progress of dairy science: The heat stability of milk. *Journal of Dairy Research* **44**, 627–646.

Fryer, P. (1989). The uses of fouling models in the design of food process plant. *Journal of the Society of Dairy Technology* **42**, 23–28.

Fryer, P.J., Pritchard, A.M., Slater, N.K.H., & Laws, J.F. (1984). A new device for studying the effects of surface shear stress on fouling in the presence of heat transfer. In *Fouling in heat exchange equipment*. Symposium of the American Society of Mechanical Engineers (1984), 27–32.

Gallot-Lavallee, T., Lalande, M., & Corrieu, G. (1981). An optical method to study removing deposit by sodium hydroxide cleaning solution, 215–214. In B. Hallstrom, D.B. Lund, & C. Tragadh (Eds.), *Fundamentals and applications of surface phenomena associated with fouling and cleaning in food processing*, Proceedings, Tylosand, Sweden.

Gibson, H., Taylor, J.H., Hall, K.H., & Holah, J.T. (1995). Removal of bacterial biofilms, Report No. 2. Chipping Campden, UK: Campden and Chorleywood Food Research Association.

Gotham, S.M., Fryer, P.J., & Pritchard, A.M. (1992). β-lactoglobulin denaturation and aggregation reactions and fouling deposit formation: A DSC study. *International Journal of Food Science and Technology* **27**, 313–327.

Grandison, A.S. (1988). UHT processing of milk: Seasonal variation in deposit formation in heat exchangers. *Journal of the Society of Dairy Technology* **41**, 43–49.

Grandison, A.S. (1996). Foul play. *Dairy Industries International* **4**, 15–17.

Grasshoff, A. (1983). Cited in Plett (1985).

Grasshoff, A. (1997). *Cleaning of heat treatment equipment*, 32–44. In IDF Bulletin No. 328.

Gynnig, K., Thome, K.E., & Samuelsson, E.G. (1958). The burning of milk in plate pasteurizers. *Milchwissenschaft* **13**, 62–70.

Hiddink. J., Lalande, M., Maas, A.J.R., & Streuper, A. (1986). Heat treatment of whipping cream. I. Fouling of the pasteurisation equipment. *Milchwissenschaft* **41**, 542–546.

Horne, D.S. (1984). The ethanol stability of milk. Annual Report of Hannah Research Institute, 89–100.

Horne, D.S., & Parker, T.G. (1982). Some aspects of the ethanol stability of caprine milk. *Journal of Dairy Research* **49**, 459–468.

IDF Bulletin. (1981). *New monographs on UHT milk*, No. 133.

IDF Bulletin (1997). *Fouling and cleaning of heat treatment equipment*, No. 328.

Jackson, A.T., & Low, M.W. (1982). Circulating cleaning of a plate heat exchanger fouled by tomato juice, III. The effect of fluid flow rate on cleaning efficiency. *Journal of Food Technology* **17**, 745–752.

Jennings, W.G., McKillop, A.A., & Luick, J.R. (1957). Circulating cleaning. *Journal of Dairy Science* **42**, 1763.

Jones, A.D., Ward, N.J., & Fryer, P.J. (1994). The use of a heat flux sensor to monitor milk fluid fouling. In *Fouling and cleaning in food processing*, Jesus College, Cambridge, UK, 199–205.

Jowitt, R. (Ed.). (1980). *Hygienic design and operation of food plant.* Chichester, UK: Ellis Horwood.

Jungro-Yoon, & Lund, D.B. (1994). Magnetic treatment of milk and surface treatment of plate heat exchangers: Effects on milk fouling. *Journal of Food Science* **59**, 964–969, 980.

Kane, D.R., & Middlemiss, N.E. (1985). Cleaning chemicals—State of knowledge in 1985. In D. Lund, A. Plett, & C. Sandu (Eds.), *Fouling and cleaning in food processing* (pp. 312–335). Madison, WI: University of Wisconsin Extension Duplicating.

Kastanas, P. (1996). *Studies on the fouling of heat exchangers during the sterilisation of cows' and goats' milks.* PhD thesis, Reading, UK: University of Reading.

Kastanas, P., Lewis, M.J., & Grandison, A.S. (1995). Design and development of a miniature UHT plant for fouling studies and its evaluation using milk adjusted to different pH values, goats' milk and milk permeate. *Trans I Chem E*, **73**, Part C, 83–92.

Kastanas, P., Lewis, M.J., & Grandison, A.S. (1996). Comparison of heat exchanger performance for goat and cow milk. IDF/FIL Publication No. 9602, *Production and utilization of ewe and goat milk*, 221–230.

Kern, D.Q., & Seaton, R.E. (1959). A theoretical analysis of thermal surface fouling. *British Chemical Engineering* **4**, 258–262.

Lalande M., & Corrieu, G. (1979). *Investigation of fouling kinetics during heat treatment of milk in a plate heat exchanger.* International Union of Food Science and Technology [Food Engineering Symposium] Abstr. No. 1.2.19.

Lalande M., & Corrieu, G. (1981). Fouling of plate heat exchanger by milk, 279–288. In B. Hallstrom, D.B. Lund, & C. Tragadh (Eds.), *Fundamentals and application of surface phenomena associated with fouling and cleaning in food processing*. Proceedings, 6–9 April 1981, Tlosand, Sweden, 279–288.

Lalande, M., Tissier, J.P., & Corrieu, G. (1984). Fouling of plate heat exchanger used in ultra-high-temperature sterilisation of milk. *Journal of Dairy Research* **51**, 557–568.

Laxminarayana, H. & Dastur, N.N. (1968). Buffaloes' milk and milk products. *Dairy Science Abstracts* **30**, 231–241.

Ling, A.C., & Lund, D.B. (1978). Fouling of heat transfer surfaces by solution of egg albumin. *Journal of Food Protection* **41**, 187–194.

Loncin, M., & Merson, R.L. (1979). *Food engineering*. New York: Academic Press.

Lund, D.B., & Bixby, D. (1974). *Fouling of heat exchangers by biological fluids.* Proceedings of the IV International Congress of Food Science and Technology, Vol. IV, 434–442.

Lyster, R.L.J. (1965). The composition of milk deposits in an ultra-high temperature plant. *Journal of Dairy Research* **32**, 203–210.

Morgan, A.I., & Carlson, R.A. (1960). Fouling inside heat exchanger tubes. *Food Technology* **14**, 594–596.

Mottar J., & Moermans, R. (1988). Optimisation of the forewarming process with respect to deposit formation in indirect ultra high temperature plants and the quality of milk. *Journal of Dairy Research* **55**, 563–568.

Newstead, D. (1994). Observations on UHT plant fouling by recombined milk, AP279-1.

Paterson, W.R., & Fryer, P.J. (1988). A reaction engineering theory for the fouling of surfaces. *Chemical Engineering Science* **43**, 1714–1717.

Plett, E.A. (1985). Cleaning of fouled surfaces. In D. Lund, A. Plett, & C. Sandu (Eds.), *Fouling and cleaning in food processing*. Madison, WI: University of Wisconsin, Extension Duplicating.

Romney, A.J.D. (Ed.). (1990). *Society of Dairy Technology*. CIP: *Cleaning in place manual.*

Sandu, K., & Lund, D. (1985). Minimising fouling in heat exchanger design. *Biotechnology Progress* **1**, 10–17.

Schreier, P.J.R., Green, C.H., Hasting, A.P.M., Pritchard, A.M., & Fryer, P.J. (1994). Development of a fouling and cleaning test bed for milk fluids: Continuous flow processing. DTI Food Processing Sciences LINK.

Shapton, D.A., & Shapton, N.F. (1991). *Principles and practices for the safe processing of foods.* Oxford, UK: Butterworth-Heimemann. Reprinted 1988, Woodhead Publishing Ltd.

Shilton, N.C., Johnson, A., & Lewis, M.J. (1992). An investigation of a possible relationship between the ethanol stability of milk and the fouling of milk in an ultra-high-temperature process. *Journal of the Society of Dairy Technology* **45**, 9–10.

Sjoeberg, A.M. Wirtanen, G., & Matilla-Sandholm, T. (1995). Biofilm and residue investigations of detergents on surfaces of food processing equipment. *Food and Bioproducts Processing* **73**, (C1), 17–21.

Skudder, P.J. (1981). Effects of adding potassium iodate to milk before UHT treatment: 2. Iodate-induced proteolysis during subsequent aseptic storage. *Journal of Dairy Research* **48**, 115–122.

Skudder, P.J., & Bonsey, A.D. (1985). The effect of pH and citrate concentration on the formation of deposit during UHT processing, 226–234. In D. Lund, A. Plett, & C. Sandu (Eds.), *Fouling and cleaning in food processing*, Madison, WI: University of Wisconsin, Extension Duplicating.

Skudder, P.J., Brooker, B.E., Bonsey, A.D., & Alvarez-Guerrero, R. (1986). Effect of pH on the formation of deposit from milk on heated surfaces during ultra high temperature processing. *Journal of Dairy Research* **53**, 75–87.

Skudder, P.J., Thomas, E.L., Pavey, J.A., & Perkin, A.G. (1981). Effects of adding potassium iodate to milk before UHT treatment: I. Reduction in the amount of deposit on the heated surfaces. *Journal of Dairy Research* **48**, 99–113.

Society of Dairy Technology. (1990). *CIP, Cleaning in place manual.* Edited by A.J.D. Romney.

Swartzel. K.R. (1983). The role of heat exchanger fouling in the formation of sediment in aseptically processed and packaged milk. *Journal of Food Processing and Preservation* **7**, 247–257.

Timperley, D.A., Hasting, A.P.M., & De Goederen, G. (1994). Developments in the cleaning of dairy sterilization plant. *Journal of the Society of Dairy Technology* **47**, 44–49.

Timperley, D.A., & Smeulders, C.N.M. (1987). Cleaning of dairy HTST plate heat exchangers: Comparison of single-and two-stage procedures. *Journal of the Society of Dairy Technology* **40**, 4–7.

Tissier, J.P., Lalande, M., & Corrieu, G. (1983). A study of milk deposit on heat exchange surface during UHT treatment, 49–58. In B.M. McKenna (Ed.), *Engineering and food,* Vol. 1. London: Elsevier Applied Science.

Toyoda I., Schreier, P.J.R., & Fryer, P. (1994). A computational model for reaction fouling from whey protein solutions. In *Fouling and cleaning in food processing*, Jesus College, Cambridge, UK, 23–25 March 1994, 191–196.

Wadsworth, K.D., & Bassette, R. (1985). Laboratory scale system to process ultra-high-temperature milk. *Journal of Food Protection* **48**, 530–531.

Walstra P., & Jenness, R. (1984). *Dairy chemistry and physics*. New York: John Wiley.

Withers, P.M., Taylor, J.H., Richardson, P.S., & Holah, J.T. (1994). Ultrasonic detection and monitoring of fouling during UHT processing. In *Fouling and cleaning in food processing.* Jesus College, Cambridge, UK, 23–25 March, 181–190.

Withers, P.M. (1994). Ultrasonics in the detection of fouling. *Food Technology International—Europe*, 96–99.

Withers, P.M. (1995). Sanitation of food processing plant. *International Food Hygiene* **5**(6), 5, 7.

Zadow, J.G., & Hardham, J.F. (1978). The influence of milk powder, pH and heat treatment on the reflectance and stability of recombined milk sterilized by the ultra high temperature process. *Australian Journal of Dairy Technology* **33**(1), 6–10.

Zadow, J.G., Hardham, J.F., Kocak, H.R., & Mayes, J.J. (1983). The stability of goat's milk to UHT processing. *Australian Journal of Dairy Technology*, March, 20–23.

CHAPTER 9

Storage

9.1 INTRODUCTION: RAW MATERIALS/PROCESSING AND STORAGE

In general, the raw materials that are used as the starting materials or ingredients for heat-processed foods are diverse. In the context of this book (pumpable fluids), these include fruit, vegetables, and legumes, which may be converted to juices, purées, or soups; milk from cows, goats, sheep, buffaloes, and other animals; and coffee beans and tea leaves for beverage production. Many of these are perishable and should be processed as soon as possible after their production or harvesting, or if not, held at refrigerated temperatures until being processed. An obvious example is milk, where it is recommended that it be processed as soon as possible after the animal has been milked or either held refrigerated or thermized (Section 5.5) until it can be processed. Other materials (ingredients) supplied in dried form, such as milk-based powders, sugar, and cocoa, may be kept for longer periods, but their microbiological spore loading and their enzymatic activity should be low.

The ultrahigh temperature (UHT) process itself, including heating, holding, and cooling, may take from 1 to 6 minutes, during which time it will spend only a few seconds at the high temperature of approximately 140°C. After cooling, there may be some intermediate storage in an aseptic storage tank before the product is aseptically filled, or it may be filled directly into retail or bulk packaging.

In contrast to this, products that are commercially sterile may then be stored at ambient temperature for up to 6 months, which is the normal shelf life for such products. Ambient temperature itself may range from below 0°C to 55°C. Thus the quality of the product that is ultimately being appraised (evaluated) by the consumer will depend upon the culmination of the reactions that have taken place during raw material storage, processing and packaging operations and subsequent storage. The product may be consumed at any time up to the sell-by date (or even later). The storage period may in fact be several orders of magnitude longer than raw material storage and processing periods.

In terms of the effect on product quality, storage is also by far the most variable and therefore the most indeterminate of these processes. Thus, it is much easier for the food manufacturer to influence the raw material quality, and the processing and packaging, by means of the quality assurance systems that are put in place. However, as far as storage is concerned, there is much less scope for the food manufacturer to be able to control it. It is important to establish what the product will be like at the end of its shelf life, under the most adverse storage conditions.

It is the aim of this chapter to explore those changes taking place to food components during storage, especially those that may affect product quality. Also, the use of product incubation and accelerated storage trials are briefly discussed, as these can provide useful information about the stability of a product in a shorter time period. For UHT products where there has been a failure of heat treatment or postprocessing contamination, changes can occur quite rapidly and are accentuated by incubation at elevated temperatures. For example, such faulty milk products may go sour and/or produce gas within 24 to 72 hours at 30°C; this is easily detected by a fall in pH or dissolved oxygen; this would also be noticeable by smell and by coagulation. However, where processing conditions are being well controlled, the large majority of products will be commercially sterile (1 in 10^4), and will show no measurable increase in microbial activity. Nevertheless, they will still be subject to changes brought about by chemical reactions, residual enzyme activity, and physical processes. These may well reduce the nutritional value of the food as well as changing the sensory characteristics. It should be pointed out at this early stage that all chemical reactions are temperature dependent. Thus a sequence of reactions, for example, those involved in both flavor and color changes in UHT milk, will take place much more quickly at elevated temperatures. In general, foods will have a shorter shelf life (i.e., become unacceptable more quickly as the storage temperature increases). This is particularly noticeable above 30°C.

Also of importance in many of these changes is the concentration of dissolved oxygen in the food. Kessler (1989) emphasizes this by reviewing storage changes in heat-treated foods in two categories, those that are influenced by dissolved oxygen and those that are not. Flavor changes during storage and the rate at which some vitamins are lost are examples that are affected by the level of dissolved oxygen level, whereas browning and gelation are apparently not influenced by dissolved oxygen concentration.

Important changes that may affect the sensory characteristics and/or nutritional value are as follows:

- Color reactions, with browning being the predominant reaction here
- Flavor: cooked flavor and changes in sulfhydryl groups; the development of stale and oxidized flavors. These changes are caused by chemical reactions involving proteins and fats and may lead to the production of volatile components. The role of oxygen, light-induced changes, and changes induced by enzymes and their role in off-flavor defects, such as bitterness and lipolysed flavors.
- Texture changes such as sedimentation, thickening, or gelation
- Changes in nutritional value arising from loss of vitamins and changes in proteins, fats, and minerals

These changes are discussed in more detail in the following sections. Again, the main body of research is for milk and milk-based products, but many of the changes described will affect other foods, and so the principles involved will be applicable to them.

9.2 COLOR/BROWNING REACTION

The subject of browning is of great interest to the food industry. For UHT milk-based products, it will be shown that the amount of browning taking place during storage is much more significant, compared to the extent of browning produced by the UHT process itself, especially when storage temperatures are above 30°C. One would normally assume that heat treatment would increase browning, and this is in fact usually the case. However, experimental measurements have shown that UHT milk immediately after production in fact is whiter than raw milk. Although some browning will have occurred, this is overshadowed by a combination of other changes such as a reduction in fat globule size caused by homogenization and an increase in casein micelle size (see Section 6.4.9.2), both of which alter its light-scattering properties, making it whiter in appearance.

The browning reaction is better known as the Maillard reaction. It takes place between reducing sugars and amino groups, which are abundant in many foods. The chemistry is briefly summarized in Section 2.10. In milk this would be the lactose and primarily lysine of milk proteins. The reactions are extremely complex, even in model systems involving one amino acid and sugar, and proceed through a large number of intermediate products, with the end products being brown pigments. Labuza (1994) has reviewed the problems involved in interpreting the complexity of the kinetics of the Maillard reaction in a food context. One problem when analyzing the reaction is selecting the most appropriate compounds to be measured in order to monitor it. For example, there are many intermediate compounds formed on the way to these brown pigments, which have also been studied in order to understand the course of the reaction. One of the most commonly examined is hydroxymethylfurfural (HMF), which is covered in Section 10.5.7. However, it has not been easy to demonstrate convincingly any increases in such intermediates of the Maillard reaction during storage, except where storage temperatures are above 35°C (Burton, 1988; Kessler & Fink, 1986). This has been substantiated by Jimenez-Perez et al. (1992), who stored UHT milks for 90 days at five storage temperatures between 6° and 50°C. Their results indicated that increases in HMF and also lactulose took place in the temperature range 40° to 50°C.

Other compounds that have been analyzed during milk storage include lactulose lysine and fructose lysine (Moller, Andrews, & Cheeseman, 1977). Although these do change during storage, their complex methods of analysis make them too laborious to be used as a routine procedure (IDF, 1984). Furosine, which is a new amino acid, is liberated by the hydrolysis of lactulose lysine. This has also been used as an indicator of heat treatment (Section 10.5.7).

The Maillard reaction makes an important contribution to both the flavor and color of the product. In order to understand the reaction, these brown pigments may be measured by extraction and color measurement, or the color of the product can be measured directly. Methods of measuring color for milk products have been reviewed by Burton (1988) and more generally in foods by Hutchings (1994) and Hunt (1995).

Thus, although UHT milk is white immediately after production, it will develop a chocolaty brown color after 4 months' storage at 40°C. Products with more reducing sugars, for example, hydrolyzed milk, sweetened milk, or ice cream mix, or additional amino groups will go browner more rapidly than normal milk. Increasing the pH will also accelerate browning. There is no evidence to suggest that it is affected by the level of dissolved oxygen.

Based on classical D and *z*-value theory, *z*-values of 21.3°C (Burton, 1954) and 26.2°C (Horak & Kessler, 1981) have been reported. More significantly, according to the latter workers, browning can be described by one activation energy (107 kJ/mol) over the temperature range 25° to 140°C. If this is so, comparisons can be made for different situations; for example, storage conditions involving long time periods at relatively low temperatures can be directly compared to processing conditions involving short periods at high temperatures. Exhibit 9–1 shows such calculations for different storage and processing conditions and allows some interesting comparisons to be made. According to these Figures, the extent of browning caused by heating at 140°C for 4 seconds would be the same as about 1 week's storage at 25°C or about 1 day at 40°C. Browning caused by sterilization at 121°C for 3 minutes would be about 10 times more severe than 140°C for 4 seconds. Although this is what is predicted by the Arrhenius equation, it does not necessarily result in the products being noticeably browner. Fink and Kessler (1988) give 400 seconds at 121°C as the threshold for color detection. Note that this corresponds to an extent of browning reaction of $26{,}740 \times 10^{-16}$. In one factory investigation to find out why UHT milks were going excessively brown, the explanation was found to be a failure in the cooling system, with products being packed in excess of 50°C and then remaining at that temperature for a long time.

Although the extent of browning is low during an ideal UHT process (140°C/4 seconds), it should be also be noted that there may be significant contributions if the heating and cooling rates are extended (Section 6.4.1). It should also be remembered that browning will be accompanied by a significant fall in pH (e.g., 0.1 to 0.3 pH unit). For example, Adhikari and Singhal (1991) reported a change after 24 days of storage from 6.82 to 6.52 at 22°C and 6.45 at 37°C.

A further interesting point is how these calculated differences in kinetics translate to measured and perceived differences in color. One system now widely used for color measurement is the CIELAB system, which uses three parameters: L*, a*, and b*, where

Exhibit 9–1 The Extent of Browning at Different Processing and Storage Conditions

All values are 10^{-16}	
1 month at 25°C	4,675
4 onths at 25°C	18,700
1 month at 40°C	36,900
4 months at 40°C	147,600
Higher temperatures for shorter times (HTST) pasteurization	13
10 minutes at 95°C	3,960
Retort—3 minutes at 121°C	11,900
UHT direct—4 seconds at 140°C	1,190
Browning threshold (Fink & Kessler, 1988)	26,740

Note: Extent of browning = $\int e^{-E/RT} \cdot dt$, see Section 6.4.9.

L^* is a measure of lightness (0 black to 100 white)
a^* is a measure of red/green (positive +100 [red] to negative –80 [green])
b^* is a measure of yellow/blue (positive +70 [yellow] to negative –80 [blue])

Within these spaces, equal distances in space represent approximately equal color differences. As foods are heated and browning proceeds, the trends are for L^* to decrease and for a^* and b^* to increase.

In the CIELAB color space, the color difference between two samples (ΔE) is defined as the difference between the two points (1 and 2) representing the colors in the CIELAB space. It is calculated from

$$\Delta E = \{ (L_1^* - L_2^*)^2 + (a_1^* - a_2^*)^2 + (b_1^* - b_2^*)^2 \}^{0.5}$$

For color differences to be noticed (perceived), ΔE values should be between 2 and 5. Pagliarini, Vernille, and Peri (1990) showed that a minimum ΔE of 3.8 should be attained before there was a visual perception of milk browning. Color development during storage increases as the severity of the heat treatment increases. If browning is a problem with a specific product, and this may be likely at higher storage temperatures, there may be some potential for looking at strategies to reduce the heat treatment given in addition to reducing the storage temperature.

It is likely that the flavors produced by the Maillard reaction during storage will contribute to the decline in acceptability of UHT milk with prolonged storage (Burton, 1988).

9.2.1 Fruit Juice Browning

Fruit juices will be susceptible to considerable enzymatic browning during their production, and steps should be taken to control this. The enzymes responsible will be largely inactivated during pasteurization. Fruit juices are also susceptible to considerable nonenzymic browning during storage. Compounds involved in the browning reaction include ascorbic acid (particularly in its oxidized form), other sugars, and some of the nitrogenous compounds. Note that it is suggested that juices should be deaerated prior to heat treatment to reduce vitamin C oxidation.

Rassis and Saguy (1995) processed aseptically concentrated orange juice at three temperatures (84°, 87°, and 90°C for 72 seconds) and stored them for up to 7 weeks at 32°C and 15 weeks at 22°C. No differences were found in non–enzymic browning (NEB), vitamin C, sucrose, fructose, and glucose or HMF due to the different thermal treatments. Sensory analysis showed a nonsignificant preference for concentrated orange juice processed at the lower temperature. The dominating factors affecting vitamin C retention and NEB were storage time and temperature. A lag-time was observed in NEB formation, and its length depended on storage temperature.

Lee, Yoon, and Lee (1995) measured the browning index in stored canned orange juice and found an activation energy of 45.3 kcal/mole between 40° and 50°C. The Q_{10} for browning (the factor by which quality changes for a 10°C rise in temperature) was estimated to be 2.1.

The shelf life of canned orange juice was predicted to be 10, 6, and 3 months for storage at 10°, 20°, and 30° C, respectively. The loss of vitamin C was also highly temperature dependent: the total vitamin C content of canned and bottled orange juice stored at 40° to 50°C for 24 weeks fell from 100% to 13% and 4%, respectively. Storage of canned and bottled orange juice at 30°C resulted in total vitamin C contents of 76% and 50%, respectively, relative to levels before storage. Discoloration of the orange juice, as measured by browning index, was only slight in canned juices stored at 20° and 40°C; it was slightly more pronounced in bottled juices stored at these temperatures. Storage at 50°C caused large increases in browning index with time for both bottled and canned orange juice.

Naim et al. (1993) showed that addition of L-cysteine significantly reduced browning in orange juice, as well as helping to improve the flavor; also, as the cysteine concentration increased, the amount of ascorbic acid retained during storage increased. Wong and Stanton (1993) found that removal of amino acid residues (using adsorbent resins) reduced browning in kiwi fruit during storage; removal of phenolics had no effect. Wong, Stanton, and Burns (1992) also studied the effect of initial oxidation during juice preparation on NEB in four kiwi fruit juice concentrates. Model concentrates were also included to enable the contribution to browning of the remaining nonmeasured juice components to be assessed. Kiwi fruit juices and model solutions were aerated for varying lengths of time and concentrated and stored at 20°C for 20 weeks. Increased initial oxidation of ascorbic acid in the juices and model systems led to increased browning in the stored concentrates. The time course of browning in oxidized samples consisted of a sharp initial phase over the first 2 weeks of storage, followed by a slower and steady rate of browning. Loss of total ascorbic acid and total phenolics was correlated with the observed increases in browning and haze on storage. The juice concentrates exhibited greater browning than the model concentrates at a similar level of oxidation, suggesting that minor juice components make a major contribution to browning. Ibarz, Bellmunt, and Bota (1992) measured browning in concentrated apples during storage and found activation energies of 30.22 kcal/mol (126.3 KJ/mol) for browning and 24.77 kcal/mol (103.5 KJ/mol) for HMF formation.

It is suggested that a grapefruit juice of good color could be produced through careful selection of highly pigmented grapefruits, controlled blending, and further color enhancement; β-carotene and lycopene were considered to be important for the color of pigmented fruit (Lee, 1997). Results indicate that quantitation and location of the pigment in fruits and pigment solubility appear to have more influence on juice color than changes associated with browning under adverse conditions. Addition of sodium metabisulfite will also reduce browning, and vitamin C supplementation will help reduce any enzymatic browning due to polyphenol oxidases.

9.3 DESTABILIZATION/DEPOSIT FORMATION AND GELATION

Deposit formation and gelation are two problems that may be encountered with UHT milk, and are both likely to arise from some form of destabilization of the protein in milk, particularly the casein micelle.

9.3.1 Deposit Formation

Most UHT milks develop a slight sediment (about 1 part in 1,000), which is not normally sufficient to be a problem. Burton (1968) suggested that this sediment is produced by the same mechanism that is responsible for fouling and represents fouling material that has not deposited onto the wall of the heat exchanger. The quantity of sediment depends upon the type and quality of the raw milk and the severity of the heat treatment, and increases as the severity of the process increases. There was also more sediment found in directly heated milks compared to indirectly heated milks (for equal sterilization effects). The amount of sediment decreases with increasing homogenization pressure and shows how homogenization can be effective in reducing the chalky defect in UHT milk.

Sometimes, a very much more voluminous sediment appears. It is probably related to relatively poor raw milk quality in hot, dry, climatic conditions. Sediment volume increases sharply if the pH of the milk is below 6.6, and high-ionic calcium levels appear to play a part, as the problem can be reduced by adjusting the pH upward and adding compounds such as phosphates, both of which will reduce ionic calcium. This also fits in well with more sediment from poorer-quality raw milk. The alcohol stability test (Section 8.4.3) has been suggested to be useful in identifying raw milk that might produce high amounts of sediment.

Two further pieces of evidence for the involvement of ionic calcium come from processing of goat's milk; this has a high content of ionic calcium and it is noticeable that it is very susceptible to fouling and that a heavy deposit forms in the container within a few hours of processing. Deposit formation can be reduced by increasing the pH prior to heat treatment and by addition of phosphates (Section 8.4.5), both of which reduce ionic calcium (Zadow, Hardham, Kocak, & Mays, 1983). Some detailed measurements of ionic calcium in milks, creams, and concentrated milks, using a selective ion electrode, have been made by Ranjith (1995).

In addition to this, storing milk at a high ambient temperature will cause it to go brown quite quickly. This will be accompanied by a drop in pH and an increase in ionic calcium. Thus the browning reaction itself will further exacerbate the problem, and this will definitely occur at the hottest time of the year.

There appears to be no easy solution to eliminating deposit formation. Some practical suggestions to reduce the amounts are to try to reduce the storage temperature or to consider reducing the shelf life during the hot period, for example from 6 months to 3 months. It may also be possible to increase the pH or reduce ionic calcium of the milk prior to processing, as suggested by Burton (1988).

Sediment formation in buffalo's milk has been studied by Sharma and Prasad (1990). They showed that UHT treatment increased the viscosity of milk. Over the processing temperature range 120° to 150°C for 2 seconds, there was very little effect on sediment formation; sediment formation increased with increased storage time, but the amount of sediment was slightly higher at 15°C compared to 30°C. Sediment formation is a major problem when goat's milk is UHT processed (Section 6.4.11), and it may appear within hours of processing.

9.3.2 Gelation

Gelation is a problem that can occur during storage for UHT milk and milk-based products. Where gelation is found there may be several possible causes. The simplest explanation is that the problem is microbiological in origin, since gelation also often accompanies milk souring. Gelation and souring would tend to take place in a short time period. Also, if this were the case, it would usually be accompanied by a number of other changes that would make the milk unacceptable (i.e., a fall in pH, some separation, and the production of off-flavors and odors). Sharma and Prasad (1990) reported such observations for buffalo's milk that was heated at 120° and 125°C, whereas milk processed between 130° and 150°C remained normal. In some recent UHT trials (unpublished) where there was a very high level of postprocessing contamination, the dissolved oxygen fell quickly from high concentrations to almost zero. In fact, changes in dissolved oxygen were more rapid than changes in pH (personal observation).

If the cause is not microbiological, it could arise from either physical and chemical changes involving the casein micelle (more common in concentrated milks), or enzymically, due to reactions caused by heat-resistant proteases that survive the heat treatment or to a lesser extent by the survival of indigenous protease. This enzyme-induced gelation is more common in normal milks (unconcentrated) and may be accompanied by an increase in bitter flavor in the milk. Also, some articles dealing with gelation do not differentiate between the different mechanisms.

9.3.3 Age Gelation

UHT treatment leads in general to a slight increase in casein micelle size. In addition to that, there is also an increase in the number of small particles, which may be denatured whey protein complexes and/or casein that has dissociated from the micelle. Age gelation is used to describe the sudden progressive increase in the viscosity of a UHT liquid during storage, to be followed by the formation of a gel that cannot be redispersed. This follows a period where viscosity remains constant for a long period and maybe even slightly decreases. It is much more of a problem in UHT milk than in sterilized milk and more so in concentrated milks that are subject to UHT sterilization, the severity of the problem increasing as the concentration factor increases. The mechanisms are far from being fully understood. The physicochemical and biochemical processes involved in age gelation are summarized in Exhibit 9–2.

Walstra and Jenness (1984) conclude that several reactions may be involved, and, under different conditions and for different products, the mechanisms responsible may well be (partly) different. Some general observations are as follows.

In general, a more intense heat treatment will delay gelation. Age gelation occurs in the absence of whey proteins; blocking of thiol groups does not prevent age gelation. Nevertheless, it is fairly certain that redox reactions are involved. Maillard compounds have been held responsible and may either cause or delay gelation. Addition of sugar delays gelation, particularly when added after concentration. Age gelation was found to occur faster in ultrafiltered low-fat UHT milk (1.4% fat and 5.18% protein) than in normal UHT milk of the same total solids, possibly due to the increased protein content (Reddy, Nguyen, Kailasapathy, & Zadow, 1992).

According to Burton (1988), it is thought that the casein-serum protein complex produced during UHT processing dissociates during storage, and during dissociation the α-casein undergoes cleavage, which makes it sensitive to calcium ions. Muir (1984) considers that milk fat is involved and that the homogenization process is important through the production of fat-protein complexes. He discusses the contradiction that high homogenization pressures are required to prevent fat separation but tend to predispose the product to age gelation.

Muir (1984) suggested that adjustment of the milk to pH 7.4 before forewarming and concentrating and that the addition of carrageenan or phosphatides would postpone this type of age gelation. More recently, McMahon (1996) has proposed that aggregation involves a complex formed between β-lactoglobulin and κ-casein (βκ complex). During UHT processing the denatured β-lactoglobulin covalently bonds to the κ-casein to form large polymeric βκ complexes. During storage, this complex is gradually released from the micelle, due to weakening of the ionic bonds that anchor the κ-casein to the micelle. These βκ complexes accumulate in the serum phase but some may remain partially attached to the micelle. Once these reach a critical volume concentration, a gel network of cross-linked βκ complex is formed, with any attached casein micelles also being incorporated into the network. This cross-linking continues, until a semirigid gel is produced.

Exhibit 9–2 Processes Involved in Age Gelation

Physicochemical
- Dissociation of the casein/whey protein complexes
- Cross-linking due to Maillard reaction
- Removal or binding of calcium ions
- Conformational changes of casein molecules: breakdown of micelle structure; interaction of β-lactoglobulin and κ-casein; S—S exchange reactions; pH change; dephosphorylation of casein; and interaction of casein and carbohydrate

Biochemical
- Heat resistance and reactivation of natural and bacterial proteinases
- Survival of bacterial spores

Source: Reprinted from M.J. Lewis, Advances in the Heat Treatment of Milk, in *Modern Dairy Technology*, Vol. 1, 2nd ed., R.K. Robinson, ed., p. 44, © 1994, Aspen Publishers, Inc.

Release of the βκ complex is observed (by electron microscope) as protuberances and tendrils on the micelle surface. Involvement of the casein micelles in age gelation occurs through attached βκ complex appendages and not via direct contact between the micelle surfaces. McMahon (1996) proposes that the following observed phenomena related to age gelation can be explained on the basis of this theory:

- Increasing the heat intensity delays gelation.
- Storing at higher temperature delays gelation, but browning becomes a problem.
- Gelation increases as concentration increases.
- Orthophosphates accelerate gelation, whereas hexametaphosphate delays gelation.
- Both proteolysis by plasmin and proteases promote age-onset gelation.
- Increasing the pH above its natural level accelerates age-onset gelation, which contradicts the findings of Muir.

Walstra and Jenness (1984) reported that homogenized UHT cream is also susceptible to age thickening. McKenna and Singh (1991) reported that UHT-reconstituted concentrated skim milks made from high-heat powders had considerably longer gelation times than those made from medium- or low-heat powders. However, sediment formation was also greater. They concluded that both physicochemical and proteolytic processes play some part in the mechanism of gelation in this product.

Harawalka (1992) provides a review of age gelation of sterilized milks.

9.3.4 Enzyme-Induced Gelation

Enzyme-induced gelation is the most common form of gelation encountered for normal (unconcentrated) UHT milks. Originally it was thought that both gelation and a bitter flavor might be caused by protein changes, perhaps arising from the reactivation of indigenous proteases. Although this may sometimes be the case, it is now well established that proteolytic enzymes arising from pyschrotrophic bacteria may survive UHT treatment and cause gelation. Adams, Barach, and Speck (1976) concluded that the growth of pyschrotrophs in raw milk led to detectable proteolysis, particularly κ- and β-caseins, even after 2 days. Coagulation shortly after UHT treatment was observed and increased with increasing psychrotrophic count and decreasing severity of heat treatment. These findings were confirmed by Law et al. (1977), who found that the level of pyschrotrophic bacteria in raw milk determined whether UHT milk would gel during subsequent storage. Samples containing greater than 8×10^6 colony-forming units of *Pseudomonas fluorescens* AR 11 gelled between 10 and 12 days after production when stored at 20°C; below this count, no gelation was observed after 20 weeks of storage.

Many organisms that produce heat-resistant proteases have now been identified and the resistance of the enzymes determined (Section 6.4.10). Typically these bacterial enzymes break down κ-casein to *para*-κ-casein, in a way similar to the action of rennet. This destabilizes the casein micelle and leads to the formation of a gel, without any significant change in pH. In this sense, the term *sweet-curdling* may be used. However, extensive proteolysis will give rise to a bitter flavor. Collins, Bester, and McGill (1993) found that proteolytic activity in skim milk correlated well with bitterness, especially for milks stored at 30°C as opposed to 20°C. It appeared that proteinases played a more important role in loss of taste and acceptability of UHT skim milk than lipases.

Thus the main way of avoiding this type of gelation is to avoid the use of raw milk with high counts of pyschrotrophs and therefore high concentrations of protease enzyme. However, a simple pyschrotroph count on the mixed flora associated with raw milk is not well correlated with the level of bacterial enzymes and is not necessarily an indication of the probable sensitivity of the milk to age gelation following UHT processing. The enzyme level depends not only on the number of organisms, but also their type and their stage of growth (Driessen, 1989).

Where there is some uncertainty about protease activity in raw milk, extending the holding time rather than increasing the processing temperature (generally increasing the C* value) will be more effective in significantly reducing its activity, because of the high *z*-values for enzyme inactivation.

The reactions taking place during proteolysis lead to an increase in nonprotein nitrogen and the casein macropeptide (CMP) and hence one of its components, which is sialic acid (Section 6.4.10.2). Suggestions have been made to monitor CMP (Picard, Plard, & Collin, 1996) and sialic acid (Zalazar, Palma, & Candioti, 1996) as a means of monitoring and predicting whether gelation may occur during storage or, better still in my opinion, for screening raw milks prior to UHT processing. Other measurements involve the use of limulus assay (Section 6.4.10.3). Although it may be useful to monitor levels during storage, it is not so effective as a general control strategy that should aim to prevent it from happening in the first place.

UHT milks that are moving toward gelation become less stable to alcohol and calcium. Gelation effects are reported to be reduced in higher-fat products and at lower storage temperatures. Early-lactation milks were found to gel before late-lactation milks. Within each state of lactation, those with a higher somatic cell count tended to gel first, but this was not so significant an effect (Auldist et al., 1996). Al-Kanhal, Abu-Lehia, and Al-Saleh (1994) measured proteolysis in UHT milks made from fresh and recombined milks and found significant differences in proteolysis. Proteases surviving heat treatment in UHT custard do not cause age gelation, but rather a drop in viscosity, which was equally undesirable (Burton, 1988).

There has been further work on survival of indigenous enzymes (plasmin); plasmin has been shown to be able to cause gelation (note that the mechanisms are different), but there is some debate about whether it should be present in UHT milk; if so, it is more likely to be present in milk that is less severely heat-treated. The use of the limulus test should also help distinguish whether gelation was caused by bacterial proteases.

Note that the author has processed raw milk that was 8 days old and found that immediately after UHT processing it had formed a soft gel. At this stage, the gelled milk tasted neither bitter nor sour. On other occasions, raw milk that was about the same age formed a firm gel when placed in a test tube in a heated water bath at 45°C for 2 hours. Both milks must have had considerable proteolytic activity for such gelation to occur.

9.4 DISSOLVED OXYGEN

The oxygen content in good-quality raw milk is normally close to saturation, at about 10 to 11 mg/L; pumping and gentle agitation ensure that it remains close to saturation. Dissolved oxygen also has an inverse temperature solubility relationship, becoming less soluble as temperature increases (as does calcium phosphate). However, despite this negative temperature solubility effect, the dissolved oxygen in freshly produced indirectly heated UHT milk is often between 80% and 90% saturated. It is prevented from coming out of solution during UHT processing by the application of about 1 bar in excess of the saturated vapor pressure, at the corresponding processing temperature. Also, since the product is sealed (enclosed), if any does come out of solution in the high-temperature section, it may go back into solution in the cooling section. In contrast, the dissolved oxygen concentration is considerably lower in directly heated UHT milk; personal observations suggest between 40% and 50% saturation, although Burton (1988) quotes values of less than 1 mg/L (about 10% saturated).

The incorporation of a deaeration unit will also decrease dissolved oxygen concentration. This may be used in situations where considerable aeration may have taken place in the preparation stage, for example, in the mixing of dried ingredients, such as ice cream mix, fortified milk drinks, or drinks made by powder reconstitution.

The type of packaging will have a marked effect on dissolved oxygen levels throughout storage. If the packaging material is oxygen permeable, there will be an almost infinite supply of oxygen, for example, polyethylene. The change in dissolved oxygen during storage will depend upon the extent of those reactions that consume oxygen and the supply of oxygen through the packaging material. Otherwise, in "normal" UHT products, this dissolved oxygen is available for chemical reactions, the extent of some reactions being dependent upon the level of dissolved oxygen and the temperature. The most important oxidation reactions involve sulfhydryl groups, vitamin C, and fats. In adverse situations, such as those where spoilage is found, dissolved oxygen will disappear very quickly.

9.5 FLAVOR CHANGES

Flavor is a property detected by the senses, in particular taste and smell, and therefore requires taste panel work for its evaluation. Most flavor research involves a combination of taste panel work and chemical analysis. Flavor changes in food during storage arise because of changes in chemical constituents. One of the most extensively studied area is flavor changes during storage of UHT milk.

9.5.1 Cooked Milk Flavor

It is widely acknowledged that consumer acceptance of and preference for a certain type of milk is influenced more by its flavor than by any other attribute. The flavors of particular interest in heated milk are the "cooked" flavor and the bitter and oxidized flavors that develop during long-term storage of UHT and sterilized milk. The picture is further complicated because the cooked flavor that develops on heating changes rapidly during the early days of storage, and these changes are affected by temperature and dissolved oxygen levels.

The vocabulary used for describing the cooked flavor is also not straightforward, with terms such as *cooked*, *boiled*, *cabbagey*, *sulfury*, and *caramelized* being frequently used. More comprehensive information for heat-treated milks is given by Prasad (1989) and Prasad, Thomson, and Lewis (1990). It is this cooked flavor that is unpopular with most UK consumers and the main reason why UHT milk sales remain low in the United Kingdom. The cooked flavor is easily detected both in liquid milk and in many drinks such as tea and coffee, despite relatively small amounts of milk being used; it is less easy to identify in flavored milks, particularly those with strong flavors such as chocolate or toffee.

Josephson and Doan, as early as 1939, made some observations on the source and significance of the cooked flavor in milk heated to temperatures above 170°F (76.7°C). They concluded that sulfhydryl groups were wholly responsible for the cooked flavor. These sulfhydryl groups also reduce the oxidation-reduction potential of heated milk and act as antioxidants. They suggested that the lactalbumin fraction and some of the proteins associated with the fat globular membrane were the main sources of the sulfhydryl groups. Thus the role of sulfhydryl groups was recognized at an early stage. Since then later work has drawn attention to active and total sulfhydryl groups, disulfide groups, and low-molecular-weight volatiles.

The different approaches that have been taken to examine the source of the cooked flavor in heat-treated milk are as follows:

- Measurement of the amount of sulfur components in milk. These include active sulfhydryl (SH) groups; total—SH and disulfide (S—S) groups; total (—SH + S—S), expressed as —SH groups; sulfur-containing amino acids; and low-molecular-weight volatile components (e.g., H_2S and CH_3SH). The development of methodology and results obtained has been reviewed by Lewis (1994).
- Examining the denaturation levels of the proteins in milk, particularly β-lactoglobulin and α-lactalbumin, as these are usually implicated as the major source of sulfur-containing components. Whey protein denaturation is discussed in Section 6.4.7.1.

One of the best accounts of the flavor changes in milk on heating and during storage is given by Ashton (1965). He recognized two phases, each with a number of distinct stages. These are summarized as follows:

Primary phase

(a) Initial heated flavor, accompanied by a strong sulfhydryl or cabbagey smell
(b) Weaker sulfhydryl or cabbage odor with residual cooked flavor
(c) Residual cooked flavor with normal, acceptable, agreeable flavor

Secondary phase

(d) Normal, acceptable to agreeable, flat, acceptable flavor
(e) Flat, acceptable to mild oxidized flavor
(f) Incipient oxidized flavor (or rancidity) to pronounced rancidity

It can be seen that the flavor changes taking place are considerable. UHT milk is best consumed while in stages (c) and (d). However, this is difficult to ensure because the rate at which any sample of milk progresses through this sequence of events will depend upon the storage temperature and the level of dissolved oxygen.

These flavor changes can now be explained in terms of the major oxidation reactions taking place throughout storage. There is a well-defined oxidation-reduction sequence: first sulfhydryl groups are oxidized (those exposed by heat as well as compounds such as hydrogen sulfide), followed by ascorbic acid, and finally the lipid fraction. In fact, both sulfhydryl groups act as antioxidants toward vitamin C, which in turn acts as an antioxidant toward eventual oxidation of the fat.

Immediately after production, the milk produced by the indirect process has been found to have a very intense sulfhydryl or cabbagey smell, most likely due to a high concentration of hydrogen sulfide, as well as some other sulfur-containing volatiles. Some of these have been identified as carbonyl sulfide (COS), methanethiol (CH_3SH), carbon disulfide (CS_2), and dimethyl sulfide $(CH_s)_2S$ (Jaddou, Pavey, & Manning, 1978). These were thought to be more responsible than the reactive sulfhydryl groups for this early cooked flavor.

Gaafar (1987) made a detailed investigation into sulfur-containing components, in relation to the onset of cooked flavor. He concluded that the threshold of cooked flavor corresponds to a β-lactoglobulin denaturation of about 60%, a hydrogen sulfide concentration of 3.4 μg/L and a reactive sulfhydryl concentration of 0.037 mmol/L. The mechanisms for the formation of some of these sulfur volatiles have been discussed in more detail by Walstra and Jenness (1984).

Most of these volatile low-molecular-weight components disappear rapidly within a few days, at a rate that depends upon temperature and dissolved oxygen concentration, leaving a weak residual cooked flavor.

Once this and other reactive sulfhydryl groups (produced mainly from unfolding of β-lactoglobulin) have been oxidized, ascorbic acid will start to be oxidized to dehydroascorbic acid (with no change in nutritional value). Once this is complete, oxidation of the fat and development of the flavors described as oxidized, tallowy, stale, and cardboardy appear. Jeon, Thomas, and Reineccius (1978) analyzed volatiles produced in UHT milk heated at 145°C for 3 seconds, which were stored up to 150 days. Twenty-six compounds were identified, most of which were carbonyl compounds, and it was believed that aldehydes were the most important contributor to off-flavors in UHT milk; no mention was made of sulfur-containing components. Mehta (1980) reviewed the factors affecting the onset of the stale or oxidized flavor, which appears after the cooked flavor has disappeared, together with the volatile components responsible for it. Methyl ketones were the largest class of compounds isolated, although aldehydes were thought to make the most significant contribution to this off-flavor. The presence of high concentrations of dissolved oxygen in packaged milk and higher storage temperatures will accelerate all these relevant reactions (i.e., the disagreeable cooked flavor will disappear more quickly but the unacceptable oxidized note will appear more quickly). In contrast, milk produced by direct-steam injection has a much reduced cooked (cabbagey) smell and taste, resulting from some removal of the low-molecular-weight volatile sulfur components in the flash-cooling process. Badings (1977) reported that hydrogen sulfide in indirect UHT milk was 82.5 $\mu g\ kg^{-1}$; this was about eight times the level found in direct heated milk.

There is also a reduction in the dissolved oxygen content, quoted to be less than 1 mg/L. Thus, immediately after production, the milk is probably at stage (c) in Ashton's scheme.

Table 9–1 Hedonic Order of the Stored UHT Milk Samples at the Three Storage Temperatures

Hedonic Order	*Storage Temperature* 4°C	22°C	30°C
	Sample Age		
1	2 weeks	2 weeks	4 days
2	1 month	4 days	2 weeks
3	2 months	1 week	1 month
4	3 months	1 day	1 day
5	4 days	1 month	1 week
6	1 week	3 months	2 months
7	4 months	4 months	4 months
8	5 months	2 months	5 months
9	6 months	5 months	3 months
10	1 day	6 months	6 months

Source: Reprinted with permission from S.K. Prasad, D.M.H. Thomson, and M.K. Lewis, Trends in Food Product Development, in *Proceedings of the World Congress of Food Science and Technology (Singapore)*, T.C. Yan and C. Tan, eds., © 1990, International Union of Food Science and Technology.

However, it may progress through the scheme more slowly thereafter because of the lower level of dissolved oxygen. Nevertheless, it is often reported that it is difficult to distinguish direct UHT milk from pasteurized milk.

These complex changes will affect the acceptability of UHT during its storage period. Burton (1988) summarizes that it is generally accepted that the flavor of UHT milk begins to deteriorate after 2 to 3 weeks of storage at ambient temperature. The rate of deterioration depends upon temperature and is much reduced at refrigeration temperatures, although this would somewhat defeat the purpose of UHT processing. Some later work involved collecting UHT milk from a commercial dairy on its day of production and storing it at three different temperatures (4°, 22°, and 30°C) (Prasad et al., 1989). The milks were evaluated by a trained panel (20 assessors) over a period of 6 months to assess the milks. The results from the hedonic testing are given in Table 9–1.

At 4°C and 22°C, the milks most preferred were 2 weeks old, whereas at 30°C the milk that was 4 days old was most preferred. At all three storage temperatures, the general preference was for milks that were days or weeks old rather than months. At 22°C and 30°C (typical ambient temperatures), milks that were 2 months or older were preferred less, and milks that were 6 months old were least preferred. However, for milks stored at 4°C, milks that were 1 week old or less also scored badly, showing that the initial objectionable cooked flavor disappeared slowly at low temperature.

It would also appear that the flavors that develop on long-term storage at higher temperatures are more unacceptable than the pronounced initial cooked flavor. It also seems certain that Maillard-type reactions contribute to this decline in acceptability, and the UHT milks take on the sensory characteristics of sterilized milk. One solution is to consume the milks within 2 months of purchase, or within 4 months of the recorded end of shelf life. Some UHT milks produced in India were reported to be unsatisfactory after only 22 to 24 days at 37°C, again attributed to the Maillard reaction. The pH had dropped from 6.82 to 6.45 (Adhikari & Singhal, 1991).

9.5.2 Flavor Improvement

There have been a number of attempts to improve the cooked flavor of UHT milks. One approach is to add substances to milk before heat treatment to reduce the intensity of the cooked flavor. Badings (1977) reported that cooked flavor could be reduced by addition of L-cystine, to milk prior to heating. For indirect heating, addition of 30 and 70 mg of cystine per kilogram of milk prior to heating reduced the hydrogen sulfide concentration in the heated milk down from 82.5 to 9.5 and 1.7 $\mu g\ kg^{-1}$. Again, this hydrogen sulfide was found to disappear quickly during the first 24 hours of storage. There was also no inclination to oxidation or other flavor defects; it was later proposed that hydrogen sulfide was removed by L-cystine, with L-cysteine as the reaction product. On the other hand, addition of L-cysteine resulted in a massive increase in the amount of deposit as well as a most unacceptable cabbagey or sulfury flavor.

Skudder, Thomas, Pavey, and Perkin (1981) found that potassium iodate (10 to 20 ppm) also reduced cooked flavor, by causing the oxidation of any exposed sulfhydryl groups. Unfortunately, bitter components were formed about 14 days after processing at these levels of iodate addition. This was attributed to either increased instability of proteases within the milk or suppression of natural protease inhibitors.

Swaisgood (1977) patented a process to remove cooked flavor from milk by immobilized sulfhydryl oxidase attached to glass beads (Swaisgood, 1980). He suggested putting such a reactor downstream of the UHT holding tube. To my knowledge, this has not been commercialized. Another approach is to reduce the intensity of the heat treatment, for example, 120°C for 2 seconds to avoid the cooked flavor and combine this with the use of nisin (150 IU/mL) to prevent the growth of microbes that will survive the heat treatment. Note that nisin is very effective against gram-positive bacteria (Section 5.6.3.1). Control milks without nisin spoiled in 3 to 4 days at 30°C, whereas most of the milk samples with nisin were still sound after 30 days.

Most attention has been paid to cooked flavor, but Mehta (1980) has reviewed the factors affecting the onset of stale or oxidized flavors, which appear after the volatiles responsible for the cooked flavor have disappeared. Methyl ketones were the largest class of compounds isolated, although aldehydes were thought to make the most significant contribution to this off-flavor (Blanc & Odet, 1981). Hutchens and Hansen (1991) measured fat-soluble carbonyls in UHT cream, standardized to 10% fat and identified butanal, hexanal, heptanal, nonanal, and decanal. All were found to decrease during storage. The various types of flavor defects in milk have been reviewed by Shipe et al. (1978), Badings (1984), and more recently by McSweeney, Nursten, and Urbach (1997).

9.6 CHANGES TO OTHER COMPONENTS DURING STORAGE

UHT products are potentially more interesting than those in in-container sterilized products, since more of the active food components will survive the heat treatment. The role of storage temperature on acceptability should be again emphasized.

9.6.1 Proteins

Changes of the Maillard type occur during storage; these lead to covalent polymerization of caseins and an increased resistance to proteolysis. Henle, Schwarzenbolz, and Kostermeyer (1996) estimated that the content of polymerized casein increased from an initial value of 8.2% to 11.7%, 19.5%, and 53.9% in UHT skim milk stored for 6 months at 4°, 20°, and 37°C, respectively.

Apart from (reducible) disulfide bonds formed by thiol-disulfide reactions between protein-bound cysteine residues, protein oligimerization is mainly influenced by the formation of irreversible (nonreducible) covalent cross-links. Reactions attributed to this type of cross-linking are the formation of isopeptides, the formation of dehydroalanine-derived amino acids such as lysinoalanine, and the Maillard reaction.

Six months of storage at 30° to 37°C will give a loss of available lysine of about 10%, which is equivalent to that caused by in-container sterilization. This will lead to the formation of lactulosyl lysine and fructosyl lysine (Section 10.5.7).

9.6.2 Fats

There may be some development of free fatty acids during storage; this is more likely at higher temperatures and for higher fat contents and greater after direct than indirect processing. It is most likely to arise in situations where heat-resistant lipases have survived the UHT process, but problems arising from lipolysis are thought to be far fewer than those arising from proteolysis. Milk with an acid-degree value of greater than 2.0 is generally accepted to have an unacceptable "lipolyzed" flavor.

Choi and Jeon (1993) detected residual lipase activity in commercial UHT milks, the greatest activity being found in the cream fraction, but activity was also found in the aqueous supernatant and the casein precipitates. During storage, lipolytic activity was much higher at 35°C than at 23°C. Coconut-type flavors in stored buffalo's milk have been correlated with the presence of C10 and C12 delta lactones (Manju-Bansal & Sharma, 1995).

Note that most milks are homogenized, with a big increase in the surface area of the fat phase. Al-Kanhal et al. (1994) observed that fat separation increased significantly with increase in storage temperature, both in fresh and recombined UHT milks.

9.6.3 Vitamins

In the absence of light, the fat-soluble vitamins A, D, and E are stable for at least 3 months at ambient temperature. Vitamin A has been found to be stable for 8 months at 25°C. It is destroyed by light at room temperature (40% loss in diffuse daylight after 14 days, compared to none in dark).

Burton (1988) summarizes published data on the effects of storage on the water-soluble vitamins (except for ascorbic acid and folic acid) in Table 9–2. Some of the water-soluble vitamins are also stable on storage in the absence of light. However, the most marked loss of vitamins during storage of UHT products occurs with ascorbic acid, of which only the reduced form will survive UHT treatment, and folic acid. The losses of these vitamins are

Table 9–2 Losses of Some Water-Soluble Vitamins during Storage of UHT Milk in the Dark

	Loss (%)		
Vitamin	*A*	*B*	*C*
Thiamin	ns	10	ns
Riboflavin	ns	10	10
Nicotinic acid	ns	20	—
Vitamin B_6	50	35	—
Vitamin B_{12}	40	—	15
Pantothenic acid	ns	30	—
Biotin	ns	20	—

ns = not significant.
A. Ford et al. (1969), 3 months at 15° to 19°C.
B. Görner and Uherová (1980), 6 weeks at 20° to 25°C.
C. Thomas et al. (1975), 9 weeks at 23°C.

Source: Reprinted with permission from H. Burton, *Ultra-High-Temperature Processing of Milk and Milk Products*, p. 269, © 1988.

interlinked and depend upon the availability of oxygen (See Section 9.4) and hence the method of processing.

Ford et al. (1969) showed that the loss of both ascorbic acid and folic acid was dependent upon dissolved oxygen concentration and therefore took place at a much higher rate in indirect processed milk, or milk without any deaeration (Figure 9–1).

Where UHT milk contained a high level of dissolved oxygen (about 9 mg/L), there was a rapid destruction of ascorbic acid and also of folic acid. Ascorbic acid and folic acid disappear within about 14 days; ascorbic acid disappears slightly more quickly than folic acid. If

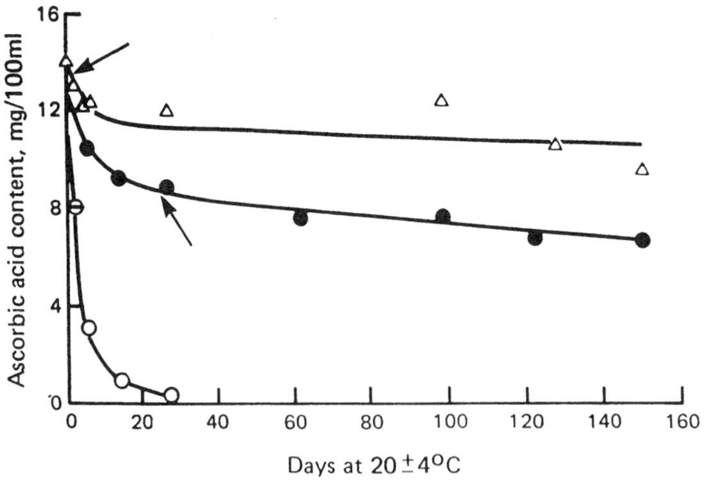

Figure 9–1 Loss of ascorbic acid in UHT milk with different oxygen levels, in impermeable containers in the dark. Initial O_2 levels: Δ 1.0mg/L; • 3.6 mg/L; 08.9 mg/L; ↑ point where O_2 level falls below 1mg/L (this is never reached at the highest initial level). *Source:* Reprinted with permission from H. Burton, *Ultra-High-Temperature Processing of Milk and Milk Products*, Fig. 9.4, © 1988.

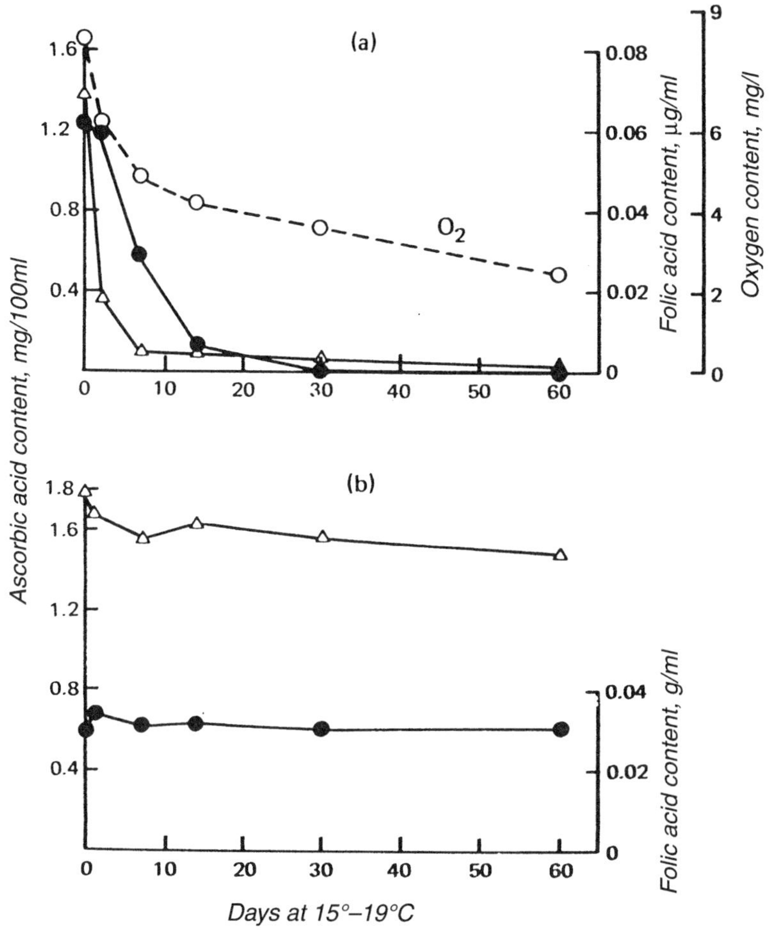

Figure 9–2 Effect of storage of UHT milk in airtight containers in the dark on the ascorbic acid (△) and folic acid (●) contents (derived from Ford et al., 1969). (a), Initial O_2 content 8.2 mg/L. (b), Initial O_2 content 1 mg/L. *Source:* Reprinted with permission from H. Burton, *Ultra-High-Temperature Processing of Milk and Milk Products*, Fig. 9.3, © 1988.

the milk is in an oxygen-impermeable container, the dissolved oxygen level will fall; it has been estimated that by the time that folic acid and vitamin C have completely disappeared, there is still calculated to be more than 50% of the original oxygen to be present. As the initial dissolved oxygen level fell, the rate of loss of ascorbic acid also fell (Figure 9–2) (Thomas, Burton, Ford, & Perkin, 1975). It was suggested that ascorbic acid losses fell sharply when the dissolved oxygen level fell below 2 mg/L, and at a level of 1 mg/L it was unable to sustain the oxidation reactions. Note that while ascorbic acid oxidation reactions are taking place, fat oxidation reactions are prevented; thus ascorbic acid acts as an antioxidant and helps reduce off-flavors produced by oxidation reactions. From a vitamin activity standpoint this loss of folic acid is very serious; ascorbic acid, however, is oxidized to dehydroascorbic acid, which still shows vitamin activity. However, it is much more heat labile than ascorbic acid, and this vitamin activity may be lost if the milk is subsequently severely heat treated, for example, when used in heated drinks.

It was also reported that vitamin B_{12} losses increased slightly at high dissolved oxygen levels, with a reduction from 20% to 10% loss over a 9-week storage period when dissolved oxygen was reduced from 8 to 9 mg/L to below 1 mg/L. Andersson and Oeste (1992) reported that vitamin B_{12} losses were dependent upon the concentration after processing, with losses ranging from 0 to almost 70% after 21 weeks. These variable results were thought to be due to the different proportions of the different forms of the vitamin in milk.

Container permeability will influence the course of these reactions. Polyethylene, polyethylene-paper laminates, and polypropylene are all highly permeable to oxygen. If milk low in oxygen is filled into such containers, the oxygen level will increase and may even approach saturation after a few days. Therefore, there is no protection against oxidation reactions in these containers, as the replacement of oxygen through the container is more rapid than its depletion. The headspace in the container may also be a significant source of oxygen, especially in oxygen-impermeable containers. It has been estimated that if the headspace volume is 15% and milk is saturated with oxygen at 10°C, the ratio of headspace oxygen to dissolved oxygen is about 5:1. This will not be significant in oxygen-permeable containers.

The stability of vitamins in stored UHT skim milk were studied by enriching it with the following vitamins: A, D_3, B_1, B_2, B_6, B_{12}, C, α-tocopherol, nicotinic acid, calcium pantothenate, and folic acid (Dolfini et al., 1991). Each vitamin was added at a concentration of at least twice the recommended daily allowance (RDA) and then subjected to a direct UHT process. The vitamin contents were in general still higher than the RDA values after 20 weeks of storage at 20°C. Contents of vitamin C were found to decrease considerably during storage. Addition of vitamin C (as a means of protecting other vitamins) to indirect UHT milk generally had little protective effect on the contents of other vitamins during storage, with the exceptions of vitamin A and folic acid.

Ideally, slight improvements in nutritional value may result if products are deaerated prior to UHT processing. Burton (1988) concluded that the effects on flavor may be of much greater commercial significance than changes in nutritional characteristics.

9.6.4 Light-Induced Changes

The majority of aseptic packaging systems are cartons, so product will be kept in the dark. However, for plastic bottles or glass containers, light-induced reactions are important.

Vitamin A was found to be destroyed by light at room temperature; there was no loss after 14 days at room temperature in the dark, but a 40% loss in diffuse daylight. The effects of light may explain some of the losses of this vitamin on storage. Exposure to light during storage can cause loss of water-soluble vitamins, where none would have otherwise occurred (e.g., riboflavin) or accelerate a loss (e.g., vitamin B_{12} and vitamin B_6) (Burton, 1988).

9.6.5 Enzyme Regeneration

The enzyme phosphatase has been found by some workers to regenerate in UHT milk. This regeneration was inhibited by the presence of dissolved oxygen and was less likely to occur in oxygen-permeable containers or in indirect UHT milk. In UK milk, it was also found that reactivation reached a sharp peak in midsummer, with a flat minimum in the winter months. Phosphatase reactivation is not considered to be a practical problem as it does not affect the quality of the milk. There is no evidence that other enzymes are reactivated after UHT processing (Burton, 1988).

9.7 SOME OTHER PRODUCTS

Most of the work is reported for cow's milk. This section briefly reviews other products where storage changes have been studied.

Al-Kanhal et al. (1994) compared changes in the quality of fresh and recombined UHT milks during storage at temperatures of 6°, 20°, and 35°C. The UHT recombined milks showed a decrease in sensory quality with increased storage temperature and the reduction was higher than for UHT fresh milk. Other factors measured that changed during storage included pH, acid degree value, viscosity, and sedimentation. Changes in buffalo's milk during storage have been reported by Singh and Patil (1989a, 1989b). In general, buffalo's milk showed a faster deterioration than cow's milk. The formation of lactones during their storage has been studied by Manju-Bansal and Sharma (1995).

Changes in ultrafiltered (UF) reduced-fat milks have been measured by Reddy et al. (1992). Sensory scores of UF reduced-fat milks were the same as whole milks after 10 weeks at 37°C; in fact, both milks were considered to be not suitable for consumption.

Changes in physicochemical and sensory characteristics of UHT flavored peanut beverages during storage have been studied by Rustom, Lopez-Leiva, and Nair (1995, 1996). Deterioration in the sensory qualities was highly correlated with an increase in sedimentation index and a decrease in pH and emulsion stability. The viscosity of strawberry-flavored beverage was constant, whereas chocolate-flavored beverages gelled after 19 weeks at all temperatures.

UHT soy beverages were stored at 5°, 25°, 35°, and 45°C and analyzed throughout; pH reflectance and sensory scores decreased during storage, whereas viscosity, proteolysis lipolysis, oxidation, and browning increased. These rates of reaction were highly temperature dependent. Proteolysis was the main reason for the decline in acceptability (Anantha-Narayanan, Kumer, & Patil, 1993).

9.8 ACCELERATED STORAGE

Samples of heat-sterilized foods are incubated at elevated temperatures to establish more quickly whether they are microbiologically sound. This is necessary because the counts will usually be too low to allow accurate detection within a short time period of production.

A typical storage combination is 30°C for 14 days, although other temperatures (i.e., 37°, 45°, and 55°C may be used (Shew, 1981; IDF 1981). General experience has shown that with any of these techniques, a positive result (arising from a failure of heat treatment or postprocessing contamination) is indicated by a massive microbial count, and a satisfactory product from a complete absence of growth. Our experience with UHT failures arising from postprocessing contamination bear this out, with milk samples coagulating within 2 to 3 days at 30°C. The presence of 1 or 2 colonies in a sample may be attributed to atmospheric contamination when setting up the test. Often, an incubation temperature of 55°C is stipulated, although Shew (1981) questions the necessity for use of this temperature. Our experience is that growth is much more prolific at 30°C than 55°C. Preincubation is further discussed in Section 10.5.2.

Such incubation procedures can also be used for accelerated storage testing, for giving a quicker indication about how the sensory characteristics may be adversely changed during storage, especially with regard to development of off-flavors, gelation, or unacceptable changes in color.

9.9 REFRIGERATED STORAGE

There is no doubt that refrigerated storage will maintain UHT products acceptable for a longer time period, since the chemical reactions that give rise to undesirable changes in the sensory characteristics of the foods will be retarded. However, if the product is microbiologically sound, there will be no further improvement in its microbiological status.

Pasteurized products are stored refrigerated; the effects of storage temperature on microbial growth rate and shelf life have been discussed in detail in Chapter 5.

Another approach is to make use of time-temperature indicators or integrators (TTIs) (Section 4.20). For refrigerated storage, TTIs can work either at a simple level, that is, as a critical temperature indicator, which indicates that a critical temperature has been exceeded (for freeze-thaw) or at a more complex level as a true TTI, which takes into account both these important factors. Three major measurement principles have been investigated for storage problems: (1) using a dye that diffuses, (2) using an enzyme-catalyzed reaction, and (3) using a solid polymerization substrate that gets darker.

The challenge is to match the kinetic response of the TTI in terms of both reaction rate and temperature dependence with the kinetics of the quality parameter being investigated. Unfortunately, it will not be able to account for the idiosyncrasies of processed foods, such as day-to-day variations in microbial load and other factors that either promote or retard microbial growth. My own personal view is that TTIs provide a very useful means for monitoring temperature abuse in those parts of the cold chain that cannot be easily controlled or that are out of control of the food-processing industry. It may also provide the added benefit of helping to make consumers more aware of their responsibility in ensuring a safe food supply.

TTIs will also have a role for products stored at ambient temperature to indicate when storage temperatures have been excessively high, to warn that undesirable changes may well have taken place.

REFERENCES

Adams, D.M., Barach, J.T, & Speck, M.L. (1976). Effect of psychrotrophic bacteria from raw milk on milk proteins and stability of milk proteins to ultrahigh-temperature treatment. *Journal of Dairy Science* **59**, 823–827.

Adhikari, A.K., & Singhal, O.P. (1991). Changes in the flavour profile of indirectly heated UHT milk during storage: Effect of Maillard browning and some factors affecting it. *Indian Journal of Dairy Science* **44**, 442–448.

Al-Kanhal, H.A., Abu-Lehia, I.H., & Al-Saleh, A.A. (1994). Changes in the quality of fresh and recombined ultra high temperature treated milk during storage. *International Dairy Journal* **4**, 221–236.

Anantha-Narayanan, K.R., Kumar, A., & Patil, G.R. (1993). Kinetics of various deteriorative changes during storage of UHT soy beverage and development of a shelf-life prediction model. *Lebensmittel Wissenschaft und Technologie* **26**, 191–197.

Andersson, I., & Oeste, R. (1992). Loss of ascorbic acid, folacin and vitamin B_{12}, and changes in oxygen content of UHT milk. II. Results and discussion. *Milchwissenschaft* **47**, 299–302.

Ashton, T.R. (1965). Practical experience: The processing and aseptic packaging of sterile milk in the United Kingdom. *Journal of Society of Dairy Technology* **18**, 65–85.

Auldist, M.J., Coats, S.J., Sutherland, B.J., Hardham, J.F., McDowell, G.H, & Rogers, G.L. (1996). Effect of somatic cell count and stage of lactation on the quality and storage life of ultra high temperature milk. *Journal of Dairy Research* **63**, 377–386.

Badings, H.T. (1977). A method of preventing the occurrence of cooked flavour in milk and milk products. *Nordeuropaesik, Mejeri-Tidsskrift* **43**, 379–384.

Badings, H.T. (1984). Flavours and off-flavours. In P. Walstra & R. Jenness (Eds.), *Dairy chemistry and physics*. New York: John Wiley & Sons.

Blanc, B., & Odet, G. (1981). Appearance, flavour and texture aspects: Recent developments. In IDF Bulletin No. 133, *New monograph on UHT milk.*

Burton, H. (1954). Colour changes in heated milks: 1. The browning of milk on heating. *Journal of Dairy Research* **21**, 194–203.

Burton, H. (1968). Reviews of the progress of dairy science: Section G, Deposits from whole milk in heat-treatment plant: A review and discussion. *Journal of Dairy Research* **35**, 317–330.

Burton, H. (1988). *UHT processing of milk and milk products*. London: Elsevier Applied Science.

Choi, I.W., & Jeon, I.J. (1993). Patterns of fatty acids released from milk fat by residual lipase during storage of ultra-high temperature processed milk. *Journal of Dairy Science* **76**, 78–85.

Collins, S.J., Bester, B.H., & McGill, A.E.J. (1993). Influence of psychrotrophic bacterial growth in raw milk on the sensory acceptance of UHT skim milk. *Journal of Food Protection* **56**, 418-425.

Dolfini, L., Kueni, R., Eberhard, P., Fuchs, D., Gallman, P.U., Strahm, W., & Sieber, R. (1991). Behaviour of supplemented vitamins during storage of UHT skim milk. Ueber das Verhalten von zugesetzten Vitaminen waehrend der Lagerung von UHT-Magermilch. *Mitteilungen aus dem Gebiete der Lebensmitteluntersuchung und Hygiene* **82**, 187–198.

Driessen, F.M. (1989). Inactivation of lipases and proteases, 71–93. In IDF Bulletin No. 238, *Heat-induced changes in milk.*

Fink, R., & Kessler, H.G. (1986). HMF values in heat-treated milk. *Milchwissenschaft* **41**, 638–641.

Fink, R., & Kessler, H.G. (1988). Comparison of methods for distinguishing between UHT treatment and sterilization of milk. *Milchwissenschaft* **43**, 5, 275–279.

Ford, J.E., Porter, J.W.G., Thompson, S.Y., Toothill, J., & Edwards-Webb, J. (1969). Effects of ultra-high-temperature (UHT) processing and subsequent stoarge on the vitamin content of milk. *Journal of Dairy Research* **36**, 447.

Gaafar, A.A.M. (1987). *Investigation into the cooked flavour in heat-treated milk.* PhD thesis, Reading, UK: University of Reading.

Gorner, F., & Uherova, R. (1980). Vitamin changes in UHT milk during storage. *Nahrung* **24**, 373–379.

Harawalkar, V.R. (1992). Age gelation of sterilised milks. In P.F. Fox (Eds.), *Advanced dairy chemistry-1, Proteins* (pp. 691–734). London: Elsevier Applied Science.

Henle, T., Schwarzenbolz, U., & Kostermeyer, H. (1996). Irreversible crosslinking of casein during storage of UHT-treated skim milk, 290–298. In IDF (1996) Bulletin No. 9602, *Heat treatment and alternative methods.*

Horak, F.P., & Kessler, H.G. (1981). Colour measurement as an indicator of heat-treatment of foods, as exemplified by milk products. *Zeitschrift fur Lebensmittel-Technologie und Verfahrenstechnik* **32**, 180–184.

Hunt, R.W.G. (1995). *Measured colour* (2nd ed.). London: Ellis Horwood.

Hutchens, R.K., & Hansen, A.P. (1991). The effect of various UHT processing parameters and storage conditions on the saturated aldehydes in half-and-half cream. *Journal of Food Protection* **54**, 109–112.

Hutchings, J.B. (1994). *Food colour and appearance*. London: Chapman Hall.

Ibarz, A., Bellmunt, S., & Bota, E. (1992). Non-enzymic browning during storage of apple concentrates. *Fluessiges-Obst* **59**(1), 9–11.

IDF Bulletin. (1981). *New monograph on UHT milk*, No. 133.

IDF Bulletin. (1989). *Heat-induced changes in milk*, No. 238.

IDF. (1996). *Heat treatments and alternative methods*, IDF/FIL No. 9602.

Jaddou, H.A., Pavey, J.A., & Manning, D.J. (1978). Chemical analysis of flavour volatiles in heat-treated milks. *Journal of Dairy Research* **45**, 391–403.

Jeon, I.J., Thomas, E.L., & Reineccius, G.A. (1978). Production of volatile sulphur compounds in ultrahigh-temperature processed milk during storage. *Journal of Agricultural and Food Chemistry* **26**, 1183–1188.

Jiminez-Perez, S., Corzo, N., Morales, F.J., Delgado, T., & Olano, A. (1992). Effect of storage temperature on lactulose and 5-hydroxymethyl-furfural formation in UHT milk. *Journal of Food Protection* **55**, 304–306.

Josephson, D.V., & Doan, F.J. (1939). Observations on the cooked flavour in milk: Its source and significance. *Milk Dealer* **29**, 35, 36, 54, 58–61.

Kessler, H.G. (1989). Effect of thermal processing on milk. In S. Thorne (Ed.), *Food preservation*, Vol. 5 (pp. 80–129). London: Elsevier Applied Science.

Kessler, H.G., & Fink, R. (1986). Changes in heated and stored milk, with an interpretation by reaction kinetics. *Journal Food Science* **51**, 1105–1111, 1155.

Labuza, T.P. (1994). Interpreting the complexity of the Maillard reaction. In T.P. Labuza, G.A. Reineccius, et al. (Eds.), *Maillard reactions in chemistry, food and health* (pp. 176–181). Cambridge, UK: Royal Society of Chemistry.

Lee, H.S. (1997). Issue of color in pigmented grapefruit juice. *Fruit-Processing* **7**, 132–135.

Lee, K., Yoon, Y., & Lee, R. (1995). Computation of Q_{10} values and shelf-life for canned and bottled orange juices. *Korean Journal of Food Science and Technology* **27**(5), 748–752.

Lewis, M.J. (1994). Advances in the heat treatment of milk. In R.K. Robinson (Ed.), *Modern dairy technology* (2nd ed.), Vol. 1. London: Chapman Hall.

Manju-Bansal, & Sharma, U.P. (1995). Formation of lactones during storage of UHT processed buffalo milk. *Journal of Dairy Research* **62**, 645–650.

McKenna, A.B., & Singh, H. (1991). Age gelation in UHT-processed reconstituted concentrated skim milk. *International Journal of Food Science and Technology* **26**, 27–38.

McMahon, D.J. (1996). Age-gelation of UHT milk: Changes that occur during storage: Their effect on shelf-life and the mechanism by which gelation occurs, 315–326. In IDF (1996) *Heat treatments and alternative methods*, IDF/FIL No. 9602.

McSweeney, P.L.H., Nursten, H.E., & Urbach, G. (1997). Flavours and off-flavours in milk and dairy products. In P.F. Fox (Ed.), *Advanced dairy chemistry* Vol. 3 (pp. 403–468) London: Chapman Hall.

Mehta, R.S. (1980). Milk processed at ultra-high-temperatures: A review. *Journal of Food Protection* **43**, 212–225.

Moller, A.B., Andrews, A.T., & Cheeseman, G.C. (1977). Chemical changes in ultra-heat treated milks during storage. *Journal of Dairy Research* **44**, 267–275.

Muir, D.D. (1984). UHT-sterilized milk concentrate: A review of practical methods of production. *Journal of Society of Dairy Technology* **37**, 135–141.

Naim, M., Wainish, S., Zehavi, U., Peleg, H., Rouseff, R.L., & Nagy, S. (1993). Inhibition by thiol compounds of off-flavor formation in stored orange juice. I. Effect of L-cysteine and N-acetyl-L-cysteine on 2,5-dimethyl-4-hydroxy-3(2H)-furanone formation. *Journal of Agricultural and Food Chemistry* **41**, 1355–1358.

Pagliarini, E., Vernille, M., & Peri, C. (1990). Kinetic study on colour changes in milk due to heat. *Journal of Food Science* **55**, 1766–1767.

Picard, C., Plard, I., & Collin, J.C. (1996). Application of the inhibition ELISA method to the study of proteolysis caused by heat-resistant *Pseudomonas* proteinases specific towards kappa-casein in heated milk. *Milchwissenschaft* **51**, 438–442.

Prasad, S.K. (1989). *The sensory characteristics of heat-treated milks, with special reference to UHT processing*. PhD thesis, Reading, UK: University of Reading.

Prasad, S.K., Thomson, D.M.H., & Lewis, M.J. (1990). Descriptive sensory analysis of milks. In T.C. Yan, & C. Tan (Eds.), *Trends in food product development* (pp. 191–203). Singapore: Proceedings of the World Congress of Food Science and Technology, 1987.

Ranjith, H.M.P. (1995). *Assessment of some properties of calcium-reduced milk and milk products from heat treatment and other processes*. PhD thesis, Reading, UK: University of Reading.

Rassis, D., & Saguy, I.S. (1995). Kinetics of aseptic concentrated orange juice quality changes during commercial processing and storage. *International Journal of Food Science and Technology* **30**, 191–198.

Reddy, K.K., Nguyen, M.H., Kailasapathy, K., & Zadow, J.G. (1992). Evaluation of reduced-fat ultrafiltered UHT milk. *ASEAN Food Journal* **7**, 152–156.

Rustom, I.Y.S., Lopez-Leiva, M.M., & Nair, B.M. (1996). UHT-sterilized peanut beverages: Kinetics of physicochemical changes during storage and shelf-life prediction modeling. *Journal of Food Science* **61**,195–203, 208.

Rustom, I.Y.S., Lopez-Leiva, M.H., & Nair, B.M. (1995). UHT-sterilized peanut beverages: Changes in physicochemical properties during storage. *Journal of Food Science* **60**, 378–383.

Sharma, D.K, & Prasad, D.N. (1990). Changes in the physical properties of ultra high temperature processed buffalo milk during storage. *Journal of Dairy Research* **57**, 187–196.

Shew, D.I. (1981). Technical aspects of quality assurance, 115–121. In IDF Bulletin. No. 133, *New monograph on UHT milk.*

Shipe, W.F., Bassette, R., Deane, D., Dunkley, W.L., Hammond, E.G., Harper, W.J., Kleyn, D.J., Morgan, M.E., Nelson, J.H., & Scanlan, R.A. (1978). Off flavours of milk: nomenclature, standards and bibliography. *Journal of Dairy Science* **61**, 855–869.

Singh, R.R.B., & Patil, G.R. (1989a). Storage stability of UHT buffalo milk. I. Change in physico-chemical properties. *Indian Journal of Dairy Science* **42**, 150–154.

Singh, R.R.B., & Patil, G.R. (1989b). Storage stability of UHT buffalo milk. II. Sensory changes and their relationship with physico-chemical properties. *Indian Journal of Dairy Science* **42**, 384–387.

Skudder, P.J., Thomas, E.L., Pavey, J.A., & Perkin, A.G. (1981). Effects of adding potassium iodate to milk before UHT treatment. *Journal of Dairy Research* **48**, 99–113.

Swaisgood, H.E. (1977). US Patent, 4.053 644.

Swaisgood, H.E. (1980). Sulphydryl oxidase: Properties and applications. *Enzyme and Microbial Technology* **2**, 265–272.

Thomas, E.L., Burton, H., Ford, J.E., & Perkin, A.G. (1975). The effect of oxygen content on flavour and chemical changes during aseptic storage of whole milk after ultra-high-temperature processing. *Journal of Dairy Research* **42**, 285–295.

Walstra, P., & Jenness, R. (1984). *Dairy chemistry and physics.* New York: John Wiley & Sons.

Wong, M., & Stanton, D.W. (1993). Effect of removal of amino acids and phenolic compounds on non-enzymic browning in stored kiwifruit juice concentrates. *Lebensmittel Wissenschaft und Technologie* **26**, 138–144.

Wong, M., Stanton, D.W., & Burns, D.J.W. (1992). Effect of initial oxidation on ascorbic acid browning in stored kiwifruit juice and model concentrates. *Lebensmittel Wissenschaft und Technologie* **25**, 574–578.

Zadow, J.G., Hardham, J.F., Kocak, H.R., & Mayes, J.J. (1983). The stability of goat's milk to UHT processing. *Australian Journal of Dairy Technology*, March, 20–23.

Zalazar, C., Palma, S., & Candioti, M. (1996). Increase of free sialic acid and gelation in UHT milk. *Australian Journal of Dairy Technology* **51**, 22–23.

CHAPTER 10

Quality Assurance

10.1 INTRODUCTION

Today, quality assurance (QA) and quality control (QC) are terms used commonly in the food industry. Harrigan and Park (1991) provide definitions, which are internationally agreed. *Quality assurance* is defined as "all those planned and systematic actions necessary to provide adequate confidence that a product or service will satisfy given requirements for quality." *Quality control* is defined as "the operational techniques and activities that are used to fulfil requirements for quality."

According to Hubert et al. (1996), QA refers to the corporate oversight function to clearly state the corporate product quality objectives and goals as well as to affirm that the QC program is functional and achieving these objectives and goals. QC, on the other hand, is an on-line or production function that establishes and administers the day-to-day policies, procedures, and programs at plant level. Thus there may be a companywide QA program outlining what was expected to ensure safety in all food products, whereas it may be up to each plant manager to ensure that in-plant QC activities in the production of its product were done satisfactorily to meet the overall corporate QA goal.

Quality control is therefore one component of an integrated system of quality assurance.

10.1.1 Safety and Quality Considerations

The main purpose of heat treatment is to reduce microbial activity to low levels, thereby making foods safe for consumption. At the same time, a wide range of chemical, biochemical, and physical reactions will take place in the food, which will affect the quality and overall acceptability (appearance, flavor, and texture) of the product (see Chapter 2). Therefore, the conflicting issues of *safety* and *acceptability* are two major issues in thermal processing. In this chapter quality assurance embraces both these important aspects.

Safety issues relate primarily to reducing the activity of food poisoning and food spoilage bacteria. The range of heat treatments is wide; pasteurization is widely practiced but it is a mild process and the product has a relatively short shelf life. In some situations, either making a conscious decision not to pasteurize your product or suffering a pasteurization failure could lead to the presence of vegetative pathogens in the final product. Prior to the widespread use of pasteurization, there were many documented cases of food poisoning attributed to raw milk, and some still occur where milk is not pasteurized. More recently there was a serious food poisoning outbreak in North America, attributed to *Escherichia coli* O157 (see Section 5.4.6) in unpasteurized apple juice. This may appear surprising, since it is an acidic product, where pathogenic bacteria would not be expected to grow; however, if they were present in the raw material, through contamination or other means, they appear to survive in an acid environment. Thermoduric organisms and heat-resistant spores will survive pasteurization (see Section 10.2.2) or the product may also become recontaminated after pasteurization (postprocessing contamination [PPC]), due to poor cleaning and hygiene procedures. Also, microbial counts in pasteurized products that are still considered to be both safe and acceptable can be quite high (e.g., 5×10^4/ml) (see Section 10.2.1), and in general there are no problems in the detection and enumeration of microorganisms in such foods. However, one of the main areas of activity has been concerned with the development of rapid microbiological methods, to provide quicker feedback of product quality, especially for the increasing number of products with shelf lives of 2 days or less.

On the other hand, monitoring sterilization processes provides some more interesting and challenging problems, since the aim is to achieve zero defective product items. In this case a defect refers usually to a spoiled product. This means that (in theory) large numbers of samples need to be examined to ensure that this is being achieved (see Section 10.3.3).

A serious food poisoning outbreak can cost a company millions, even if there is only one fatality. A reputable branded product can be ruined by such adverse publicity. Although rare in the commercial sector, there have recently been serious botulinum outbreaks due to canned salmon and hazelnut yogurt. In both these cases, the cause was attributed to PPC rather than a failure of heat treatment. In the case of the salmon, a pinhole leak led to the product's being recontaminated, and in the case of the yogurt, the source of the botulinum toxin was thought to be the hazelnut purée itself—perhaps it could be argued that this in itself was a failure of heat treatment, since the purée had been underprocessed. Where documented cases of botulinum food poisoning do occur, it is usually from home-heat-preserved vegetables that have been underprocessed. Again this illustrates the importance of understanding the principles affecting the safety of heat-preserved low-acid foods.

There have been serious salmonella outbreaks in pasteurized products that have led at least in one instance to closure of the plant. Failures of pasteurization and other sources of postprocessing contamination may also give rise to pathogen-related problems with dehydrated products, for example, in the production of powders and infant formulations by spray drying.

Other safety issues include the unwanted presence of veterinary residues (e.g., antibiotics and growth hormones) and environmental contaminants, such as heavy metals, pesticides, herbicides, and radionuclides. In all cases, preliminary processes should be employed to remove these. Radionuclides hit the headlines suddenly, with the Chernobyl disaster in 1986, which was followed by limits being placed on the levels of radionuclides of many internationally traded foods. Some radionuclides have a very short half-life (e.g., radioactive iodine), which usually contaminates milk has a half-life of about 11 days), but others may stay in the environment for much longer; for example, cobalt 60 has a half-life of 5.26 years. Some sheep meat in the United Kingdom still remains contaminated following Chernobyl. Common radionuclides monitored in the environment include strontium 90, cesium 132, cobalt 60, and potassium 40.

There may also be some natural toxins present; one example is trypsin inhibitor in soy beans; in this case heating has the benefit that it will inactivate it (Section 6.4.4). However, inactivation is not so temperature dependent as microbial inactivation, so ultrahigh temperature (UHT) processing will inactivate less compared to in-container sterilization processes. Eating only a few raw kidney beans can lead to severe stomach cramp and again the compounds responsible can be inactivated by heat. Some substances can cause strong allergic reactions; for example, recently serious allergies have been reported in some individuals to peanuts, sesame seeds, and shellfish, ingestion of which may cause anaphylactic shock. Many people also suffer food intolerances, for example, intolerance to lactose in milk products; this can be remedied by lactose removal by hydrolysis or fermentation.

Another aspect of concern may be additives. For liquid milk it is not usually permitted to use any additives, but for milk-based flavored drinks, some creams, and many other formulated, thermally processed foods, additives are permitted to improve their functional behavior or their sensory characteristics—labeling regulations require that these be declared. In general, the use of heat treatment reduces the requirement for preservatives for controlling microbial growth, although nisin has been found to be effective to control spore activity in some canned foods. The effectiveness of small nisin additions provides scope for improving the sensory characteristics by being able to reduce the severity of the heat process (Section 5.6.3.1).

The sensory characteristics of the food will contribute to the perceived quality and acceptability of the product. These include its appearance, flavor, and texture.

Other issues may also impinge on safety and quality and need consideration. These include energy utilization and conservation, water supply, effluent disposal, environmental issues, and the general well-being and welfare of the work force.

In the past, the concept of quality control was based mainly upon end-product testing for detecting faulty products, or worse still from consumer complaints. Once a fault was noted, investigations were undertaken to find the cause of the fault and provide corrective action. However, if it is not found quickly, then a large amount of product could be wasted. Over the past 25 years the approach has changed to a philosophy based on a more thorough understanding of the process, leading to implementing procedures to control the process more effectively, thereby trying to prevent problems from occurring in the first place. One widely used approach is based on hazard analysis, critical control points (HACCP), based on identification of critical control points and introducing procedures for their effective control (ICMSF, 1988). In essence this involves identification of the hazards involved, determination of the critical control points, specifying the control criteria, monitoring the process, and taking any necessary corrective action.

This approach has been very useful in improving both the safety and quality of heat-processed foods. Some examples of its application are discussed in Section 10.4.3. The general principles and practices for the safe processing of foods are covered in more detail by Shapton and Shapton (1991).

10.2 MICROBIOLOGICAL SPECIFICATIONS FOR FOODS

Microbiological standards or specifications for foods may by introduced for quality assurance purposes and in some cases may be incorporated into a country's food regulations. Harrigan and Park (1991) summarize what a correctly drawn up specification would include:

- A statement of the microorganisms or microbial toxins of concern
- A description of the sampling plan to be applied to obtain the samples to be examined
- The microbiological limits (number of microorganisms or concentration of toxin) appropriate to the food and its intended market
- The precise analytical method to be employed, including, for example, the methods of sub-sampling and preparation of dilutions, the medium to be employed, the source of the medium or constituents, the incubation temperature and the time

A recent publication (IFST, 1999) deals with the development and use of microbiological criteria for foods in greater detail. The role of rapid microbiological methods and use of chemical or physical measurements as indirect indicators of microbial activity are discussed in Section 10.3.

10.2.1 Short Shelf-Life Products

The analytical problems encountered with pasteurization are quite different from those for sterilized products, as it is expected that considerable microbial activity will be found in the sample, since pasteurization would not be expected to reduce microbial activity totally (Chapter 5). Sampling plans are becoming more commonplace as the basis for experimental design and for interpreting microbiological data for such products. The merits of 2-class and 3-class sampling plans are discussed by Harrigan and Park (1991) and more recently by the IFST Professional Food Microbiology Group (1997). In industry, 2-class sampling plans are applied mainly when testing for specific pathogens, using detection (presence/absence) tests. The 3-class sampling plans are applied when using enumeration tests.

Note also that some enzymes are also used to provide indirect information on microbial inactivation during processing; for example, a negative phosphatase activity and positive lactoperoxidase activity should be found in pasteurized milk and a negative amylase activity for pasteurized egg. Table 10–1 shows a typical specification for a cook-chilled recipe dish, which fall into this short shelf-life category.

One example of the use of such 2-class and 3-class sampling plans are in the Dairy Products Hygiene Regulations (SI, 1995) for pasteurized milk. This should meet the following microbiological standards in any random sampling checks carried out in the treatment establishment.

- 2-class: pathogenic microorganisms;
- Absence in 25 g, $n = 5$, $c = 0$
- 3-class: coliforms (per ml)

Table 10–1 Total Viable Count (TVC) and Typical Counts for Pathogens Found in Cook-Chilled Recipe Dishes

	Satisfactory	*Unacceptable*
TVC	$<10^4$	$>10^5$
Enterococci	$<10^3$	$>10^4$
Escherichia coli		$>10^2$
Bacillus cereus		10^3
Clostridium perfringens		$>10^3$
Staphylococcus aureus		$>10^2$
Listeria monocytogenes		$>10^2$
Salmonella		1 in 25 g

Source: Reprinted with permission from *Food Innovation Centre Newsletter*, Jan. 1997, © 1997, Food Innovation Centre.

- n = 5, c = 1, m = 0, M = 5 plate count at 21°C (per ml) after incubation at 6°C for 5 days
- n = 5, c = 1, m = 5×10^4
- M = 5×10^5

where: n = number of samples; c = number of samples allowed between the lower limit (m) and the upper limit (M).

Thus for pasteurized milk to comply with these regulations, five samples should be analyzed; ideally all samples should give counts below 5×10^4/ml. However, if one sample is higher than 5×10^4, but below 5×10^5, then the milk is also satisfactory, If any sample has a count higher than 5×10^5, then the milk would be deemed to have failed.

It is also recommended that such microbiological results are used to establish one's own control limits and for following daily trends in production behavior. This should allow the identification or anticipation of problem areas, and allow faults to be rectified before they become too serious. The role of PPC and its effects on keeping quality for these products cannot be overemphasized. Methods for measuring have been reviewed by the IDF (1993).

10.2.2 Bacillus Cereus

A brief review of *Bacillus cereus* is given, as it is a common foodborne pathogen that is not inactivated by pasteurization and may well grow at refrigerated storage. *Bacillus cereus* spores may be present in low numbers in raw milk. *B. cereus* is found only sporadically in bulk-tank milk. Milk churns are a possible source of contamination. A 90 % removal (1D) has been reported by bactofugation. It is also found in starch, sugar, spices, and milk powders. Therefore, these ingredients will need monitoring. It is also often found in rice and may be a problem with all dishes containing rice. Although often found in low numbers in powders, it can increase in numbers quickly when they are reconstituted, particularly if the foods are held at elevated temperatures. It is capable of growth during refrigerated storage. However, no growth of spores was reported in rice pudding at 4° and 7°C, even at high incubation levels. A regeneration time of 17.5 hours at 9°C has been reported. More toxin was produced in aerated milk. None was reported at 8°C.

Low numbers of *Bacillus cereus* in foods does not pose a health hazard. In fact, low numbers are often isolated from many foods. High counts (greater than 10^5 cfu/g) isolated from foods that have caused food poisoning have suggested that it may be responsible. *B. cereus* food poisoning fortunately is rather a mild form of food poisoning; symptoms have been described as abdominal pain, profuse watery diarrhea, and some nausea, about 10 hours after eating. Recent work has concentrated on isolating and characterizing the toxins that can produce either diarrhea symptoms or vomiting.

There is considerable difference in the thermal stability of these toxins. *B. cereus* food poisoning usually arises when foods have been kept warm for too long. Temperature control becomes a major control point. Conditions can be mistaken or confused with those provided from *Clostridium perfringens* and *Staphylococcus* outbreaks. Despite its occurrence in milk, food poisoning incidents attributed to milk have been low; one reason for this is that it tends to make the milk extremely bitter and hence undrinkable. *Bacillus cereus* spores are not inactivated by pasteurization. One survey reported it to be present in only 2% of pasteurized milk samples tested. Mean counts are from 4 to 300/L.

The heat resistance of bacteria is affected by water activity, fat content, and pH. The heat resistance of *Bacillus cereus* is greater than that of other mesophilic spore formers. The most comprehensive review of the heat resistance data is given by Bergere and Cerf (1992). There is a great deal of discrepancy in the data. Seventeen strains isolated from various sources showed D_{100} values of 0.6 to 27 minutes and z from 7.4° to 14.5°C. All that can be concluded with certainty is that for some of the more heat-resistant strains reported, a temperature of 116°C for several seconds would result in far less than 2 decimal reductions. For the less heat-resistant forms, a significant reduction may result. For example, heat treatment at 110°C for 15 seconds gave a 99.97% reduction (3 to 4 D).

Some high z-values have been recorded, for example, 35.9°C by Burton (1988). Rees and Bettison (1991) recorded decimal reduction times of 5.5 minutes at 100°C and 2.37 minutes at 121°C. Serotype 1 has been shown to be one of the most heat-resistant serotypes, with a D_{95} ranging between 22.4 and 36.2 minutes. Some values in custard were 3.7 minutes at 90°C, 3.0 minutes at 95°C, and 2.2 minutes at 100°C. Heat resistance decreased slightly as pH was reduced.

Bacillus cereus was reported to be the least sensitive of the spore-forming bacillus to inactivation by hydrogen peroxide. An irradiation dose of 43 to 50 kGy was required for 12 D, at –78°C, indicating a high resistance to irradiation. Heat resistance also increased in the presence of carbon dioxide. Growth has been reported to be inhibited by methanol acetone extracts, derived from *Lactobacillus acidophilus*, potassium sorbate, potassium nitrite, furanones, and sodium metabisulfite. Toxin production is highest when milk is constantly aerated. There is some evidence that it is inhibited by lactic acid and other organic acids. These treatments may be further combined with the use of additives, such as nisin, which is known to inhibit the growth of gram-positive spore-forming bacteria (see Sections 5.6.3 and 5.7).

10.3 COMMERCIALLY STERILE PRODUCTS

In sterilization processes, the aim is to produce products that are deemed to be commercially sterile. The inactivation of most heat-resistant spores follows first-order reaction kinetics. Therefore, the number of spores surviving the process will depend upon the initial population and the severity of the heat treatment. According to first-order reaction kinetics it is not possible to reduce the population to zero, and if sufficient of the sample is analyzed, some surviving spores may be found. However, providing that they do not proliferate or affect either the safety or the acceptability of the product, they are not considered to be a problem, although it is likely that a small percentage may cause spoilage. For this reason a distinction is made between absolute sterility and *commercial sterility*. Although the ideal situation is to achieve zero spoilage, most commercial producers of canned or UHT products would be happy to achieve a spoilage rate of 1 in 10,000 units and preferably 1 in 10^5. In practice, 1 in 5,000 may be acceptable, whereas more than 1 in 1,000 would be cause for serious concern.

Thus commercial sterility is concerned with inactivating the most heat-resistant spoilage bacterial spores and then ensuring that the product is not recontaminated. *Bacillus stearothermophilus* falls into the category of being very heat resistant; a process that would achieve 2D would produce at least 24D for *Clostridium botulinum*. Thus if commercial sterility is achieved, there should be no problems of food safety arising from *Clostridium botulinum*. This is borne out by the relatively few cases of botulinum produced by commercially processed heat-treated foods, when compared to the many millions of units processed each day.

One problem is verifying that a sterilization process is under control; this will involve analyzing a large number of samples when first commissioning a process. For example, for a spoilage rate of 1 in 10^4 samples, one would need to analyze between 22,500 and 46,000 samples to detect one spoiled unit, depending upon the probability chosen (i.e., 90% to 99% in this example). However, even analyzing this number of samples would not ensure absolute sterility and safety. Some of the problems related to sampling frequency with commercially sterile foods will be discussed in more detail. With these products, the role of postprocessing contamination on keeping quality cannot be overemphasized.

10.3.1 Sampling Theories and Probabilities

As mentioned, the ideal situation is to achieve zero defective items. The probability of finding no defective items in a sample of n items, drawn at random from a population whose fractional proportion of defectives is p (proportion satisfactory is q, where $q = 1 - p$) is given by the first term of the binomial expansion $(q + p)^n$, which is q^n. Example: If there is 1 defective item in a sample of 1,000, then $p = 0.001$ and $q = 0.999$. If a sample of 500 is analyzed, then the probability of finding zero defectives p (0) is $0.999^{500} = 0.606$ (i.e., there is a 60.6 % chance of finding zero defectives).

The chance of finding 1 or more defectives is given by 1 – p(0), which in this case would equal 1 – 0.606 = 0.394. Therefore, there is a 39.4% chance of finding one or more defective items. Thus the expression q^n can be used to evaluate the probability of finding zero defectives for combinations of different sample sizes and fraction defectives in the population.

Some results are summarized in Table 10–2, for sample sizes from 10 to 10,000 and for percentage defectives in the population ranging from a high level of 10% (1 in 10), which is considered to be a gross failure of sterilization, up to an acceptable level of 0.01% (1 in 10,000). This corresponds to values of ρ from 0.1 to 0.0001. Thus, if 10,000 samples are analyzed at this latter target level of spoilage, the chance of finding no defective items is 36.8%. There will be a 63.2% chance of finding one or more defective items in what is an acceptable product.

An alternative way of presenting these data is in the form of characteristic curves (Figure 10–1). These data outline some of the inherent problems that arise from this sampling theory. If only 10 samples were analyzed, for a population with a spoilage rate of 1 in 10, there would be only a 36.8% chance of finding a defective item and a 1% chance of finding a defective item at a population spoilage rate of 1 in 1,000. Therefore, in situations where sample sizes were small, if one or more defective sample was found, the main conclusion would be that spoilage rates were very high and that there was a major problem with the sterilization process. However, if no defective items were found, there would be no justification for complacency; it would not verify that the process was being well controlled.

If 50 samples were taken, then for a population spoilage rate of 1 in 10, there would be a 99% chance of detecting it and for a population spoilage rate of 1 in 1,000, there would be only a 5% chance of detecting it. However, there is a small possibility that an unacceptable batch would be accepted; that is, a 1% chance that a population with 1 in 10 defectives, or a 95% chance that a population with 1 in 1,000 defectives would be accepted.

Table 10–2 The Probability of Finding No Defective Items for Different Sample Sizes and Spoilage Rates

	Sample Size (n)								
	10	*50*	*100*	*200*	*500*	*1,000*	*2,000*	*5,000*	*10,000*
% Defective	*Probability of Finding No Defectives*								
0.01 (1 in 10^4)	0.999	0.995	0.990	0.980	0.951	0.904	0.819	0.607	0.368
0.05	0.995	0.975	0.951	0.905	0.779	0.607	0.368	0.082	0.007
0.1	0.990	0.951	0.905	0.819	0.607	0.368	0.135	0.007	
0.5	0.951	0779	0.607	0.368	0.082	0.007			
1	0.905	0.607	0.368	0.135	0.007				
5	0.607	0.082	0.007						
10	0.368	0.007							

Note: p = 0.0001 equates to 0.01% defectives (1 in 10^4); p = 0.1 equates to 10% defectives (1 in 10)

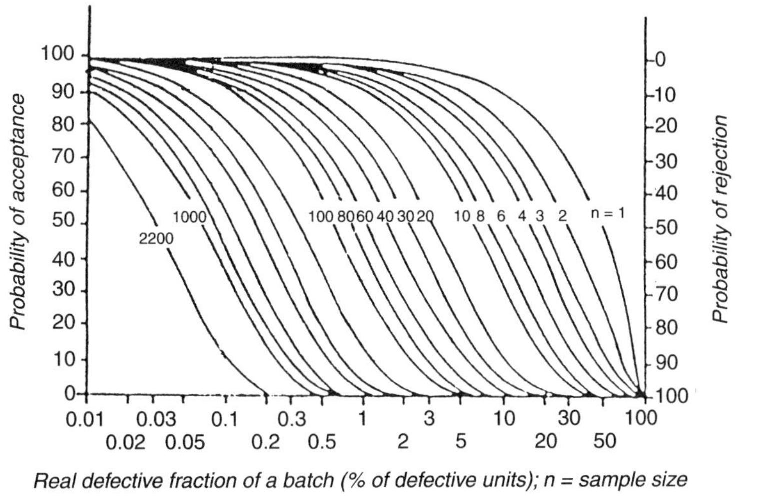

Figure 10–1 Examples for operating characteristic curves. *Source:* Reprinted from H. Reuter, *Aseptic Packaging of Food*, p. 252, with permission from Technomic Publishing Co., Inc., Copyright 1989.

This gives rise to the concept of *consumer risk*—the risk that an unacceptable batch would be accepted. If large numbers of samples are measured, the consumer risk decreases significantly. However, if 2,000 samples are analyzed and the process is well under control (population spoilage of 1 in 10^4), the chance of finding zero defectives is 81.9%. There is a 18.1% chance of finding one or more defectives (i.e., the chances of an acceptable batch being rejected becomes high). This gives rise to the concept of *producer risk* (an acceptable batch being rejected). Therefore, interpreting the results from sampling plans has to be done with care, and nothing can be ever 100% certain or uncertain. Increasing the number of samples analyzed will decrease the level of uncertainty, but will also increase the time, effort, and amount of wasted product.

10.3.2 Sampling for Verification of a Process

It can be seen that large numbers of samples need to be analyzed in order to verify that a process is under control. This procedure would be required when commissioning a new process or product. A further question that needs to be answered is if n samples are analyzed and no defective samples are found, what does this tell you about the probable spoilage rate in the population? This can be determined within 95% confidence (c = 0.95) by determining the fractional proportional satisfactory (q) that would result in a fractional probability of 0.05, i.e. (1 – c) or (5% chance) of finding one or more defective items, using the following equation:

$$q = (1 - c)^{1/n}$$

where q = fraction proportion satisfactory, c = confidence limit (expressed as a fraction), and n = sample number.

Thus if 500 samples were analyzed and no defective items are found for a confidence limit of 95% (c = 0.95), then q = 0.99402 and p (fractional proportion defective = 1 – 0.99402 = 0.0059). Thus, the best estimate of the number of defectives in the population is 0.0059 or 5.9 in every 1,000. It is important to appreciate that this is nowhere near the 1 in 10,000 target Figure, despite the large number of samples analyzed. Table 10–3 shows the most likely percentage level of defective samples in the population for different sample sizes (if zero defects are found).

In fact, the number of samples that need to be analyzed, without any defects being found, to verify a target spoilage rate of 1 in 10^4 is just under 30,000 (at the 95% confidence level). As mentioned earlier, increasing the sample size would increase the reliability of the determination: it would also increase the workload, as well as the amount of product wasted.

The ideal situation is to aim for zero spoilage. Although this is not realistic with food products, nevertheless the philosophy of wishing to achieve zero defects can be adopted by seeking to identify and control all hazards that exist in the food and food-production process (Harrigan & Park, 1991).

One neat solution would be to analyze all samples by nondestructive methods. There is a real need for such an on-line, noninvasive sampling system, which will check every item and in principle be able to identify defective items.

Table 10–3 can be used to determine the likely level of spoilage when no defects were found. It is also possible to determine the likely spoilage rate in the population if one or more defectives are found in the batch (Table 10–4), as may be the case in many practical situations.

10.3.3 Some Associated Practical Aspects

The method used for analysis is also important: if microbiological examination is involved, it is important to ask what will constitute a failure for a product that is required to be commercially sterile. For example, UHT milk heat-treatment regulations stipulate that the product will be deemed to have failed if the count is greater than 100/ml. Although this may

Table 10–3 Estimates for the Number of Defective Items in the Population for Different Sample Sizes (No defectives found in the sample)

Sample Size	*Percentage Defective Units in Population*	
100	2.95	29.5/1,000
500	0.6	6/1,000
1,000	0.3	3/1,000
2,000	0.15	1.5/1,000
10,000	0.03	3/10,000
50,000	0.006	0.6/10,000

Table 10–4 Estimates for the Percentage Maximum Number of Defective Items in the Population for Different Sample Sizes (One or more defectives found in the sample)

Sample Size	*Number of Defective Units in the Sample* 0	1	2	3	4	5
100	2.95	4.66	6.15	7.55	8.92	10.26
500	0.6	0.94	1.26	1.55	1.82	2.09
1,000	0.3	0.47	0.63	0.77	0.91	1.04
2,000	0.15	0.24	0.31	0.39	0.46	0.52
10,000	0.03	0.05	0.06	0.08	0.09	0.10

Source: Reprinted from H. Reuter, *Aseptic Packaging of Food*, p. 256, with permission from Technomic Publishing Co., Inc., copyright 1989.

appear to be a generous standard in some terms, a milk product with such a count (or even up to 2.5 orders of magnitude higher) would still most likely be safe to drink and is unlikely to be a health hazard, especially when considering what is an acceptable count in pasteurized and cooked/chilled foods (see Table 10–1). Similar questions about what changes can be tolerated may be asked if pH or titratable acidity is used for detecting defective items.

Another important aspect is to ascertain the cause of the microbiological failure. There are two main types of failure for heat-treated products, these being blown containers and acid production (flat-sour). Blown containers are in general caused by anaerobic bacteria, primarily *Clostridia*. In the majority of cases these are associated with some form of postprocessing contamination or packaging fault, perhaps a defective seal (e.g., pinholing). These are relatively easy to identify due to bulging of the pack, and defective items can be picked out before the product is dispatched.

The second cause is acid production, which may also cause coagulation or gelation with milk-based products; there is no gas production and for this reason the problem is known as flat-sour. The main possible causes are *Bacillus stearothermophilus* and other spore-forming bacteria. This is one of the most heat-resistant of the spoilage bacteria and will reduce the pH down to about 5.2. As there is no gas production or bulging, it is not easy to see and detect.

Discovering what microorganisms are responsible for spoilage is very important, as it will lead to a fuller understanding of the overall causes of failure and give further opportunity to prevent it from happening in the future. Williams (1996) concludes that problems from UHT plant failure are rare and easily monitored, more common are contamination of transfer lines or aseptic tanks, and most frequent are failures arising from aseptic filling (see Chapter 7).

Usually, excessive microbial growth will change the appearance and/or flavor of UHT milk. However, Burton (1988) reported that several organisms have been found that gave no change in appearance, smell, or taste, even at counts of 10^8/ml. More recently, attention has focused upon a very heat-resistant mesophilic spore former (HRS), which has been found and isolated from UHT milks in various parts of Europe and has caused them to fail the regulatory microbiological tests (Hammer et al., 1996). These have been characterized as

Gram-positive to Gram-labile rods, with terminal-produced endospores. Growth was observed at temperatures between 20° and 52°C, and it took 2 days to reach a count of 10^5/ml in UHT milk at 30°C; oxygen was required for growth. However, counts very rarely increased much above this level and there were no noticeable changes in the organoleptic characteristics (including gelation), except in cases where a pink coloration was observed when milk was stored in plastic bottles with a low oxygen barrier. Oxygen depletion may have limited its further growth. This organism has been recently classified as *Bacillus sporothermodurans* (Pettersson et al., 1996).

The heat resistance of these spores was investigated by several laboratories, and wide variations were observed (Table 10–5). In some cases they were found to be very heat resistant. One suggested reason for these differences between laboratories was the difference in heat resistance of HRS in naturally contaminated milk and laboratory-cultured strains. Toxicity tests have been conducted and at present there is no evidence to suggest any pathogenic or toxic properties of the HRS. The extreme heat resistances reported would suggest that these organisms may well be surviving UHT treatment.

Thus an interesting situation has arisen that UHT milks are being produced that fail the prescribed legal tests for such milk, but that may well be still safe for consumption. This again raises the question about exactly when a product such as UHT milk is deemed to be defective. Fortunately, milk has its own built-in safety system, with products being unfit to drink well before they become unsafe to drink. This may not necessarily be the case for other products. In general, it is important to note that thermal processes are designed such that the chances of a pathogen surviving the heat treatment are several orders of magnitude less than that of general spoilage bacteria surviving.

Another case recently reported is where products have gone sour, but the organisms have been very difficult to detect, as they grow very slowly. One such microorganism is considered to be a *Lactobacillus*; although it is present in products, it is not easy to detect. Its heat resistance has been reported to be high—and very variable.

As mentioned, in terms of problem solving or trouble-shooting, it is important to be able to identify the source of the contaminating organisms; large numbers of gram-negative or vegetative bacteria would suggest that the product has been subject to postprocessing contami-

Table 10–5 Heat Resistance of HRS

Institute	*Result*
Canning Research Institute, Campden (UK)	D_{100} = 5.09 min
Institute for Food Technology, Weihenstephan (D)	D_{121} = 8.3–34 s
Tetra Pak, Lund (S)	$F_0 < 10$ min (contaminated milk)
	$F_0 > 68$ min (pilot plant, production plant)
Tetra Pak Research, Stuttgart (D)	$D_{98} > 60$ min; D_{120} = 10 min
Netherlands Institute for Dairy Research, Ede (NL)	D_{126} = 1 min; D_{147} = 5 s

Source: Reprinted with permission from P. Hammer, F. Lembke, G. Suhren, and W. Heeschen, Characterisation of Heat Resistant Mesophilic *Bacillus* Species Affecting the Quality of UHT Milk: Heat Treatment and Alternative Methods, S.I. 9602, © 1996, International Dairy Federation.

nation. Heat-resistant spores would suggest that the product has been underprocessed or that the raw materials may have contained an unusually high spore count.

Until on-line methods become available that will allow for 100% testing—using nondestructive methods—it will not be possible to achieve 100% reduction of microorganisms, and a low level of spoilage is inevitable. However, it may be more difficult to convince the general public that a spoilage rate of 1 in 10^4 is an acceptable level of control. If 1 million units were produced every day, this would imply that 100 units would be unacceptable and may well be released onto the market. It may be difficult to argue convincingly in a court of law that this was acceptable, faced with the emotive situation of such spoilage giving rise to food poisoning and being opposed by an articulate prosecuting advocate and a hostile press.

10.4 QUALITY ASSURANCE/COMMERCIALLY STERILE PRODUCTS: THE CURRENT APPROACH

10.4.1 Introduction

It is the aim of this section to provide a fuller understanding of the factors affecting the safety and quality of UHT-treated products. As discussed in the introduction, the emphasis is on preventing problems from occurring, based on a fuller understanding of the process, which will then lead to the establishment of effective strategies for controlling the process, thereby further reducing the chances of serious problems occurring.

There are various publications that help provide the guidelines and framework for an effective control mechanism. These include Good Manufacturing Practice (GMP) (IFST, 1998), Guidelines for safe production of heat preserved foods (1994), HACCP publications (ICMSF, 1988), microbiological criteria for foods (IFST, 1999), and other texts dealing with the safe processing of food such as Shapton and Shapton (1991).

The guidelines encourage each manufacturer to have and to be able to demonstrate an effective system of quality management, which is appropriate to its individual circumstances, and which implements the underlying principles of the guidelines. Care needs to be paid to all products, but particular care is advised to specialized products, such as infant formulae and pharmaceutical products.

It is a strict adherence to these principles that will ensure that thermal processing is a safe form of processing. It is recognized that UHT processing is more complex than conventional thermal processing (IFST, 1998). The philosophy of UHT processing should be based upon preventing and reducing microbial spoilage by understanding and controlling the process, as briefly discussed earlier. One way of achieving this by using the principles of HACCP (ICMSF, 1988). The hazards of the process are identified and procedures adapted to control them. These are summarized for UHT processing in Figure 10–2. Critical control points are identified as raw material quality, processing conditions, and postprocessing contamination; where the homogenizer is placed downstream, this also becomes a major control point. It is also important to be able to provide documented records of all procedures involved.

An initial target spoilage rate of between 1 in 10^4 and 1 in 10^5 should be aimed for (see Section 10.3) Such low spoilage rates require very large numbers of samples to be taken to

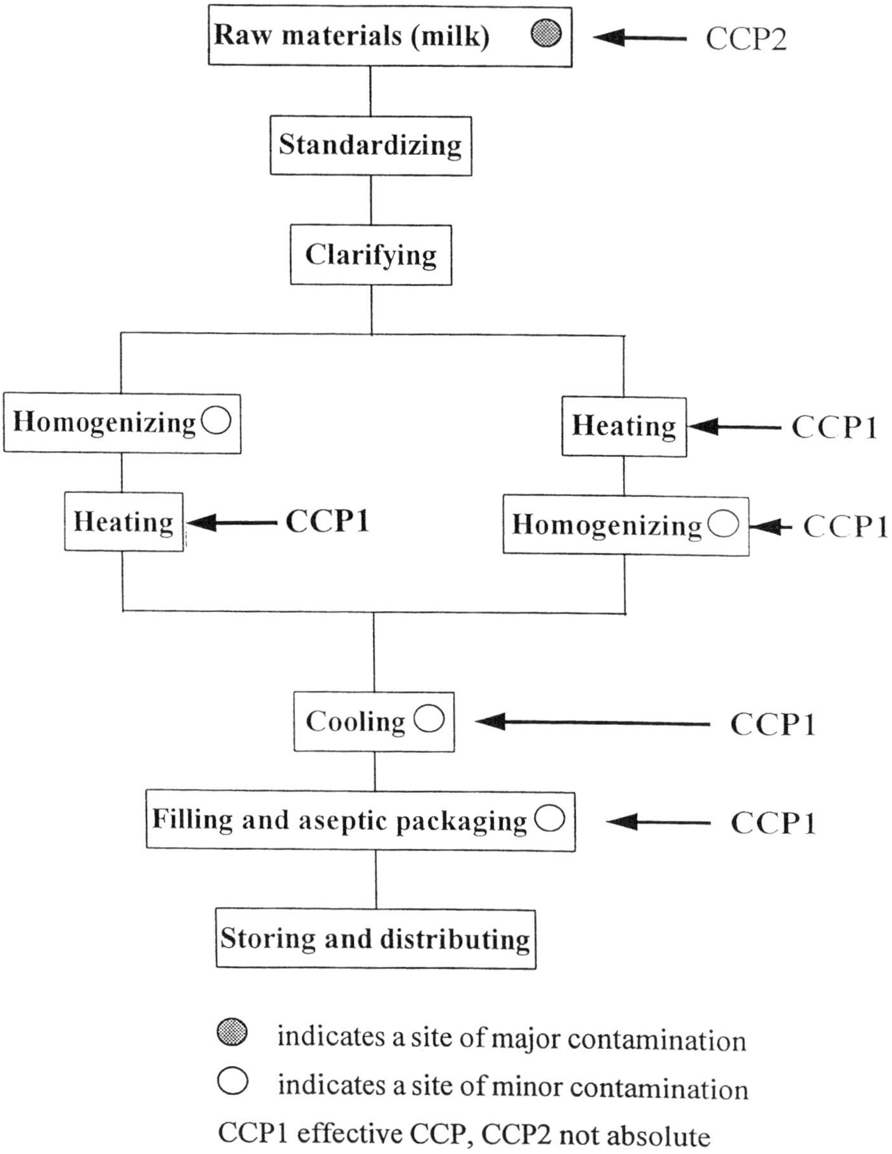

Figure 10–2 Flow diagram for production of UHT milk in aseptic retail packs. *Source:* Reprinted with permission from *Microorganisms in Foods 4*, © 1988, Blackwell Science Ltd.

verify that the process is being performed and controlled at the desired level (Section 10.3). When commissioning a new process, the sampling rate should be 100%, with at least 30,000 samples being taken. Once it is established that the process is under control, sampling frequency can be reduced and sampling plans can be designed to detect any spasmodic failures. More success will result from targeting high-risk occurrences, such as start-up, shut-down, and product changeovers (Williams, 1996). In terms of controlling the process, the following aspects merit some attention.

10.4.2 Raw Material Quality and Preparation Processes

All aspects of raw material deserves attention, from an understanding of the physical properties described earlier, through to spore loadings and chemical composition. For some products, there are microbiological standards for raw materials; those for milk intended for heat-treated drinks are as follows (1995):

cow's milk	< 100,000/ml
goat's, ewe's, or buffalo's milk	< 3,000,000/ml

A recent review of raw milk quality has been published by the IDF (1996).

Also of particular concern for UHT-processed products would be high levels of heat-resistant spores and enzymes in the raw materials, as these could lead to increased spoilage and stability problems during storage. Quality assurance programs must ensure that such poor-quality raw materials are avoided. Burton (1988) points out that attention should be paid to how spores are determined. For example, in dairying, spores are described as those microorganisms that will survive 80°C for 10 minutes, whereas other definitions use 100°C for 10 minutes. Such a difference may give rise to a 100-fold difference in the measured population (2 × z-values). In general, materials with spore counts of greater than 1,000/g should be avoided. Dried products such as milk, other dairy powders, cocoa powder, rice, and other functional powders need to be carefully monitored. Spices may contain high bacterial loads, and irradiation or other disinfecting processes may be appropriate to reduce them. In general, procedures should be put in place to monitor spore counts of all raw materials. Care should also be taken to avoid materials with high concentrations of heat-resistant enzymes, which may withstand UHT treatment, for example, lipases and proteases in raw milk or powdered milks. Pyschrotrophic counts will only give an indication that problems may be encountered. The author is unaware of any simple, direct test for detecting bacterial proteases in raw materials. Milk with a high acidity may also be susceptible to fouling and sedimentation, and a combination of the alcohol stability test and titratable acidity may help avoid the use of problematic milk. Milk should have an alcohol stability greater than 74% and a titratable acidity of less than 0.17% lactic acid (Burton, 1988).

The product formulation is also important: the nature and source of the principal ingredients; the levels of sugar, starch, and salt, as well as the pH of the mixture, particularly if there are appreciable amounts of protein. Some thought should be given to water quality, particularly its mineral content. Calcium and magnesium in water may reduce the heat stability of milk proteins and make it more susceptible to fouling (Chapter 8). The use of softened water may slightly increase the sodium content of the product, which may not be appropriate for some members of the community.

For products requiring mixing, reproducibility in metering and weighing ingredients is also important, as is ensuring that powdered materials are properly dissolved or dispersed and that there are no clumps; if clumps are present, it will take longer for heat to penetrate and these may protect any heat-resistant spores toward the interior of the clump. Improperly dissolved materials may block filters, plates, and homogenizer valves and generally interfere with the flow of the product through the process. Care should be taken to avoid too much air being entrained into the product during mixing; this may cause foaming and decrease the density of the fluid. Entrained air will later come out of solution and form bubbles or air locks in the heat exchanger. Excessive agitation during the mixing or blending period or excessive holding temperatures may cause some other unwanted reactions. With milk, it may damage the milk fat globular membrane, promoting lipolysis and the development of a pronounced soapy flavor. Other undesirable reactions that might take place during mixing include proteolysis and microbial activity. If mixing takes place at 40° to 50°C, which is commonplace, the amount of time it spends may be important, as that is the optimum temperature for many enzyme-catalyzed reactions. An intermediate cooling stage following mixing may be required if the product is not to be processed within a short time of mixing.

10.4.3 Processing Aspects: Process Characterization for Safety and Quality

To some extent, requirements for microbial safety and food acceptability and its perceived quality are conflicting, as a certain amount of chemical change will occur during adequate sterilization of the food. Therefore it is important to ask what is meant by quality and what is the scope for improving the quality.

One aspect of quality that has already been discussed is reducing microbial spoilage. A second important aspect is minimizing chemical damage and reducing nutrient loss. In this aspect, UHT processing offers some distinct advantages over in-container sterilization. Chemical reactions are less temperature sensitive, so the use of higher temperatures, combined with more rapid heating and cooling rates, helps to reduce the amount of chemical reaction. There is also a choice of indirect heat exchangers available, such as plates, tubular, and scraped surface, as well as direct-steam injection or infusion plants. Considerable differences arise in the heating and cooling rates and the shear conditions encountered for indirect processes and between the direct and indirect processes due to steam injection and flash cooling. Because of these differences similar products processed on different plants may well be different in quality. A more detailed discussion is given by Burton (1988).

A number of parameters can be used to evaluate the amounts of microbial inactivation or chemical reaction taking place during a heat treatment, including F_0, B*, and C*, as discussed in Sections 3.22 and 6.4.6. For a better understanding of the UHT process, it is required to know the temperature/time profile for the product. Some examples of such profiles are shown for a number of different UHT process plants (see Figure 6–3). Calculations of B* and C* values for plants with different temperature/time profiles are given in Sections 3.22 and 6.4.6. It is a worthwhile exercise for any UHT processor to calculate these parameters for their particular plant configuration, as they will give an indication of the severity of the process being used. This then can be related to other product quality parameters.

Chemical changes to the product could be further reduced by using temperatures in excess of 145°C. The best solution would be the direct process, with its accompanying rapid heating and cooling, whereby heating and cooling between 75° and 140°C are almost instantaneous (Sections 3.11 and 6.2.2). Care should be taken to ensure that the steam does not contaminate the product, and precautions must be taken for the removal of impurities from the steam, such as water droplets, oil, and rust. The flash-cooling process also removes volatiles and oxygen. Advantages of process are reduced chemical damage and a less intense cooked flavor for many products. Claims that products produced by direct UHT cannot be distinguished from pasteurized products may be slightly exaggerated, but the cooked flavor intensity is low compared to indirect UHT milk. One problem arises from the very short holding times required, and the control of such short holding times. In theory it should be possible to obtain products with very high B* and low C* values at holding times of about 1 second. For indirect processes, the use of higher temperatures may be limited by fouling considerations, and it is important to ensure that the heat stability of the formulation is optimized; it may be worthwhile developing simple tests to assess heat stability. The alcohol stability test (Section 8.4.3.1) has proven useful for milk products. Generally, direct systems give longer processing runs than indirect processes.

Thus holding time and temperature are perhaps the two most critical parameters. Recording thermometers should be checked and calibrated regularly, and accurate flow control is crucial (as for pasteurization).

In continuous processes there is a distribution of residence times. Two important physical properties are density and viscosity. These properties, combined with flowrate and pipe dimensions, will also determine whether the flow is streamline or turbulent. This in turn will influence heat transfer rates and the distribution of residence times within the holding tube and also the rest of the plant. The main concern is for viscous fluids, where the flow in the holding tube is likely to be streamline, and there will be a wide distribution of residence times. For Newtonian fluids the minimum residence time will be half the average residence time. Turbulent flow will result in a narrower distribution of residence times, with a minimum residence time of 0.83 times the average residence time. In both cases, the minimum residence time should be greater than the stipulated residence time, to avoid underprocessing. Residence time distributions and their implications for UHT processing are discussed in more detail in Section 3.23 onward.

Sufficient pressure must also be applied in order to achieve the required temperature. A working pressure in the holding tube in excess of 1 bar over the saturated vapor pressure, corresponding to the UHT temperature, has been suggested. This will prevent temperatures falling below the set-point temperature.

Homogenization conditions may be important; is it necessary to homogenize and if so at what pressures? Should the homogenizer be positioned upstream or downstream of the holding tube? Will two-stage homogenization offer any advantages? Homogenization upstream offers the advantage of breaking down any particulate matter to facilitate heat transfer, as well as avoiding the need to keep the homogenizer sterile during processing. All of these aspects will influence both the safety and the quality of the products. It will also take much longer to sterilize a plant with the homogenizer downstream, since the entire homogenizer block must be heated to above 130°C by the circulating hot water.

UHT processes, like canned products, will also be susceptible to postprocessing contamination. This will not usually give rise to a public health problem, although contamination with pathogens cannot be ruled out. However, high levels of spoiled product will not improve its quality image, particularly if not detected before it is released for sale. Contamination may arise from the product's being reinfected in the cooling section of the plant, or in the pipelines leading to the aseptic holding or buffer tank or the aseptic fillers.

10.4.3.1 The Sterilization and Cleaning Procedures

Sterilization and cleaning procedures will need attention: Are they adequate and have they been properly accomplished? The plant should be sterilized downstream of the holding tube at 130°C for 30 minutes. Cleaning should be adequate (detergent concentrations and temperatures) to remove accumulated deposits, and the extent of fouling should be monitored if possible. Steam barriers should be incorporated if some parts of the equipment are to be maintained sterile while other parts are being cleaned (see Chapter 7).

10.4.3.2 Aseptic Filling Procedures

Sterilization procedures should be verified. The seal integrity of the package should be monitored as well as the microbial quality of package. The packaging material may have defects, or the seals may not be airtight, or packaging may be damaged during subsequent handling. All these could result in an increase in spoilage rate. Further detail is provided in Chapter 7.

10.4.3.3 Other Factors

Recording and reporting all the important experimental parameters is an important procedure. This can help ensure that any peculiarities can be properly investigated. There should be regular inspection and maintenance of equipment, particularly to eliminate leaks. Staff education programs are important; all staff involved with the process should be educated in order to understand the principles and be encouraged to be diligent and observant with respect to changes from the norm. This should be combined with monitoring plant and processing conditions. Keeping records of critical processing conditions and temperature measurement and recording are fundamental, but pressure readings can also be good indicators, especially on steam lines, to give advanced warning of fouling; flowrates should also be monitored.

The factory layout is also important as far as ensuring that there is no cross-contamination between raw materials and finished products, even when there is apparently only a remote possibility, as in the case of heat-sealed foods. It must not be forgotten that some of the most serious food poisoning outbreaks in heat-treated foods have occurred as a result of postprocess contamination.

With experience, further hazards will become apparent and methods for controlling them introduced. The overall aim should be to reduce spoilage rates further and to improve the quality of the product.

10.5 THE ROLE OF ANALYTICAL TESTING

10.5.1 Introduction

Analysis of the raw materials and final products is crucial to understanding and thereby controlling the factors affecting the safety and quality of the product. Thus keeping accurate laboratory records will be a valuable aid in terms of ensuring that quality criteria are being met and standards maintained. They are also useful for detecting deviations and problems at an early stage, thereby nipping them in the bud before they become too serious. Overall, this will reduce the need for trouble-shooting and having to solve problems after they have happened. Ideally, the best control is achieved by anticipating problems in advance and thereby preventing them from occurring. In a factory situation, the analytical methods used should be as quick and simple as possible; it may also help if they are fully automated (at a cost).

10.5.2 Product Testing—Microbiological and Other Tests

It has been shown that it is difficult to verify by routine sampling that a UHT process is under control, as large numbers of samples are required. Although this is the case, there is no doubt that microbiological sampling does play an important role in quality assurance. It will certainly detect any gross failures of heat treatment, as can be seen from Table 10–2. Williams (1996) records that statistically to detect 0.5% spoilage with 99% confidence, 919 units must be tested. More usually, 1 in 350 units is tested per run.

The emphasis in the discussion is on defective items, where it is implied that the defect is caused by microbiological spoilage. This is the main problem that will arise in UHT processing. Burton (1988) states that it is important to establish the most reliable method of detecting spoilage in an individual pack and most economic in terms of time and equipment. Most methods for detecting spoilage involve the destruction of the sealed pack and therefore the loss of salable product.

In general, microbiological examination is time consuming and tedious, and results are not normally available for 2 to 3 days. Care should especially be taken to ensure that any microbiological analyses do not lead to cross-contamination from the laboratory to the processing plant.

Burton (1988) reviews some of the recommendations on sampling plans and the literature on determining optimum preincubation conditions. *Sample preincubation* is required, as the low microbial counts that might be present, even in spoiled packs immediately after processing, will make them difficult to detect immediately after production. Therefore, the products are usually preincubated prior to analysis. It appears that the optimum preincubation conditions are 5 days at 30° to 35°C. If time is at a premium, it could be reduced to 3 days at the cost of missing some potential spoilage. In most cases there seems to be little point in preincubation at more than one temperature; generally, little more information is obtained, and fewer packs are likely to be taken at any one temperature, thus reducing the accuracy of estimation of spoilage levels.

In fact the preincubation times and temperatures may be specified in the recommended test methods and also in some legislation. In general, tests are based on preincubation at 30°C for 14 days or 55°C for 7 days. For most conditions the former preincubation periods are more appropriate, as the latter conditions will rarely be encountered in practice and will only allow the growth of thermophilic bacteria. In the United Kingdom, the microbiological specifications for sterilized and UHT milk are that after 14 days in a sealed container at 30°C, the sample should have a microbiological count at 30°C of less than 100/ml. In addition, it should be organoleptically normal (whatever that means), and not show any sign of deterioration. Some browning will inevitably occur at the higher temperatures.

Microbiological analysis can be used for evaluating the quality of raw materials. Important are the numbers of spores and pyschrotrophic bacteria, specifically for enzymes, proteases, and lipases. Some raw materials will be purchased against specifications, so periodically these need to be analyzed to ensure that they comply with the specifications. They can also be used for checking the processing efficiency and the effectiveness of cleaning and sterilizing regimens, as well as the general level of factory hygiene.

Where processes or procedures are in doubt, challenge testing (using specific heat-resistant spores) can be used, both of process and packaging efficiencies. However, these are best done when establishing and commissioning a process and should if possible be avoided in the midst of general production operations. Use of temperature-time integrators may provide an alternative method for process validation (Section 4.21). Often the specific methods used would reflect the type of product being produced (i.e., the safety issues involved and the likely quality defects to be found).

10.5.3 Rapid Methods

Waites (1997) estimates that about 90% of the tests carried out for the detection of microorganisms of interest to the food microbiologist involve the use of classical techniques and involve agar-plating techniques or most probable number counts. However, there has been some development in rapid microbiological methods in order to be able to monitor processes more quickly, especially for the detection and enumeration of pathogens in pasteurized and cook-chill products. Attention is paid in this section to those that may be useful for UHT-treated products.

Five types of procedure have been developed (Harrigan & Park, 1991), based on various measurement principles:

1. Those that produce colony counts more rapidly than traditional methods
2. Those that rely on direct detection of microbes by microscopy
3. Those that assess the amount of some particular component of microbes and assume that its concentration is proportional to the microbial count
4. Those that exploit some physiological property of the organisms
5. Those that assay products of metabolism

The rapid tests available are reviewed in Table 10–6. The most direct tests involve direct microscopic examination, using simple staining methods. However, it is only possible to make direct counts if microbial counts are in excess of 3×10^5/ml and to get accurate results many fields have to be counted. The direct epifluorescent technique (DEFT) is a refinement, which involves filtering the sample and concentrating the microorganisms on the surface of a membrane, followed by the use of a fluorescent dye to stain and make visible the microbes. It is capable of detecting lower levels than direct microscopy, with a range of 6,000 to 10^7/ml and can distinguish between live and dead bacteria. As can be seen, it would not be suitable for UHT milk without prior incubation, but it may be useful for gauging the quality of raw products to be used for UHT processing (IDF, 1991). Perhaps the most appropriate tests for

Table 10–6 Advantages and Disadvantages of New Detection Methods

Detection Method	*Advantages*	*Disadvantages*
ATP (for use with cleaning)	Very fast (2 min)	None
ATP (for microbial count)	Very fast On-site, cheap	Problems with ATP from plants and animals Detects dead cells
Impedance	Fast (often 24 h) Organisms can still be typed Good for sterility testing Detects viable cells only	Calibration required Depends on selective media Expensive
Antibody linked probes	Very fast (1 h) Immunomagnetic Separation possible	Not sensitive (10^5–10^6 cells) Detects dead cells also Expensive
Phage-based assays	Fast (4 h) Sensitive (1 cell) Detects only viable cells	Not yet commercial
DNA/RNA probes/ polymerase chain reaction	Fast (4 h) Sensitive (1 cell) Very specific	Detects dead cells Food components can interfere Complete systems relatively expensive

Source: Reprinted with permission from W.M. Waites, Principles of Rapid Testing and a Look to the Future, *Journal of the Society of Dairy Technology*, Vol. 50, No. 2, p. 59, © 1997, Society of Dairy Technology.

UHT products are based upon methods that provide a suitable environment for growth and detect physiological changes that take place as a result of growth. The sample is incubated with a suitable growth medium, and in principle a single colony will grow and any metabolic reaction products will be detected. The most common detection methods are based upon conductance or resistance measurement and involve measuring the changes in electrical properties of the medium as the microbes grow. Such commercial machines are the Bactometer, Malthus, and Rabit. Modern instruments allow many samples to be measured simultaneously, making it an ideal method for UHT products, since large numbers of negative samples can be checked quickly (often within 24 hours for initial counts of less than 100/ml). Other systems have evolved that make use of color indicators, to give a visual indication of defective products. Samples can be tested without any preincubation period. It is prone to contamination and it fails to detect some organisms.

Rapid methods based on ATP measurement are widely available. Two problems that arise are that they do not distinguish (easily) between live and dead cells, and there can be considerable interference from nonmicrobial DNA. Bacterial spores contain no ATP, and sublethally stressed cells contain less. The test gives an almost immediate response, but it is expensive. A number of commercial kits are available based on this method, for hygiene testing and counting purposes. It is particularly useful in kit form for testing the microbiological counts of food processing surfaces and for monitoring hygiene procedures.

There is a also a role for rapid methods for enzyme analysis; for example, phosphatase activity can be measured very quickly and sensitively, using the rapid Fluorophos method (Section 5.4.1). This use of more sensitive tests can further help to improve quality and allow for earlier detection of changes in product character. There is need for a simple and rapid test for detecting excessive protease and lipase activity.

10.5.4 Noninvasive Methods

The methods normally employed for analyzing foods are destructive. A number of methods have been evaluated for noninvasive sampling, for detecting spoiled samples. The obvious advantages are that there would be no sample waste and it would be possible to achieve 100% sampling. In practical UHT situations, it is difficult to detect any changes where microbial growth is low, so an incubation period is also required to make the test more effective. Most of the test methods involve incubating UHT products (packs) with spoilage organisms to test whether the method may have practical applications.

One of the earliest was Electester for viscoelastic changes caused by changes in viscosity and gelation as a result of spoilage. It is only suitable for milk and packages of a certain type. Equipment has been evaluated for detecting spoiled products (by whatever cause), based upon ultrasonics. An ultrasound beam is transmitted into the product, inducing acoustic streaming.

Ahvenainen, Wirtanen, and Manninen (1989) found that ultrasound imaging was an effective noninvasive method for monitoring microbial growth. They suggested that bacterial loads in UHT-processed foods, such as soft ice cream and processed vanilla sauce, could be detected at levels of 10^5 cfu/g. Various bacteria were used, including *Bacillus cereus*, *Staphylococcus aureus*, *Clostridium perfringens*, and *Escherichia coli*. It was also shown to be effective in detecting gelation caused by proteolytic enzymes.

Wirtanen, Ahvenainen, and Mattila-Sandholm (1992) examined factors such as frequency, probe area, and other technical factors for the sensitivity of ultrasound imaging with respect to UHT milk and UHT base material for soft ice cream packed in Tetra-Brik cartons. Frequency (3.75 to 7.5 MHz), dynamic range, echo enhancement, and gamma compensation had significant effects in ultrasound imaging on accuracy and rapidity of the methods. Gestrelius, Mattila-Sandholm, and Ahvenainen (1994) observed a decrease in the streaming velocity was measured as spoilage occurred, induced by addition of *Staphylococcus epidermis* and *Bacillus subtilis*, suggesting that the ultrasound method detected infected packs.

The drawbacks are that these methods will not detect low levels of spoilage and will only be applicable after the sample has been incubated. Thus it becomes difficult to apply them in a QC role immediately after production; they could be applied after the normal period of incubation at the factory, prior to release.

10.5.5 Indirect Methods

10.5.5.1 Nonmicrobiological Indicators of Spoilage

Even after such preincubation periods, it may be preferable or more convenient to use some nonmicrobiological measurements as indices of microbiological spoilage. For milk and milk-based products, microbial growth in most cases results in a fall in pH and an increase in titratable acidity. These two measurements will often suffice to provide an initial rapid indication of whether spoilage has occurred; where spoilage is excessive, no measurement might be required, as it can be detected by visual observation (i.e., coagulation and smell). The level of ionic calcium also increases as pH falls, due to dissociation of calcium phosphate from the casein micelle.

One other test that has been investigated is the alcohol stability. In the milk industry the alcohol stability test can be used as an indirect indicator of microbial activity and susceptibility to fouling (see Section 8.2.7). As milk sours its alcohol stability falls; thus tested milk should be stable in 68% alcohol and be organoleptically acceptable (Burton, 1988).

Also for some products, dissolved oxygen, and redox potential are useful. Dissolved oxygen disappears very quickly when there is excessive microbial activity, and the redox potential will also fall. Dye-reduction tests, such as the resazurin test for raw milk quality and the methylene blue test for the keeping quality of pasteurized milk or cream, are still used occasionally, although the interpretation of the results is subject to debate (Burton, 1988).

10.5.6 Time-Temperature Indicators

Time-temperature indicators (see Section 4.21) are defined as small, inexpensive devices that show a time-temperature dependent change that can be easily measured and that is designed to mimic the change in a target parameter in a food that is subject to temperature change. The two obvious examples are to validate thermal processes, usually from a microbiological standpoint or to monitor microbiological changes taking place during refrigerated storage to help predict the end of shelf life and for chemical changes in products stored at ambient temperature.

For thermal process validation, some indicator that is safer and more convenient to use than *Clostridium botulinum* is needed. It should have a similar *z*-value (activation energy) and it must not undergo too much reaction in the temperature range being investigated, so that it can be measured in the final product. It should also be small in size, so as not to interfere with the heat transfer process; easy to recover; and give an accurate readout. In process validation, the TTI may be a component in the food or it may be added to the food system. The use of TTIs in heat treatment is being investigated using enzymes to monitor that important pathogens are inactivated during pasteurization and sterilization (see Section 4.21).

The ultimate extension of these processes is to have an indicator that can be incorporated into the packaging to indicate that the food has been adequately sterilized or has not been subject to high temperatures during ambient storage.

10.5.7 Chemical and Physical Tests—Quality Evaluation

Other important tests for UHT products involve assessing the severity of heat damage from a chemical or an enzymatic standpoint, for example, cooking values (Section 2.10) and C* values (Section 3.22). If the same processing conditions are used each day, the amount of chemical change should remain the same. Thus any deviations would indicate that processing conditions were not being controlled as closely as required. Any deviations from the norm may provide an early warning of more serious problems.

A variety of tests have been used to assess the amount of chemical damage. The turbidity test is still widely used for milk and is simple to perform. It involves mixing 20 ml of milk with 4 g of ammonium sulfate; at this concentration, ammonium sulfate will cause precipitation of the casein and denatured whey protein; the undenatured whey protein will remain in solution. The mixture is filtered, and any undenatured whey protein remains in solution. When the filtrate is boiled it will go turbid, the amount of turbidity being proportional to the amount of undenatured whey protein. In the United Kingdom, sterilized milk must give a negative turbidity, whereas most UHT milks will give positive turbidity, indicating some residual undenatured whey protein. The turbidity test is related to whey protein denaturation (Section 6.4.1).

Burton (1988) reviewed the literature for distinguishing between sterilized and UHT milk. A great deal of effort went into this in the 1970s and 1980s, and tests were based on the premise that a greater amount of chemical damage would be caused by the in-container sterilization process. It was suggested that the turbidity test would be useful. However, some sterilized milks were found to give a slight turbidity, whereas some UHT milks, especially those where a preholding stage was included to reduce fouling, gave a negative turbidity. The author has measured many UHT milks that have also given a negative turbidity, most likely arising from indirect plants with high regeneration efficiencies.

Some of the subjectivity was removed by measuring the turbidity using a nephelometer, but it was not still found possible to distinguish between the milks in every case (Moermans & Mottar, 1984). One suggestion for this was the complex nature of the denaturation process, and the break in reaction kinetic conditions at 95°C (Section 6.4.7.1). At best, the turbidity test will give a qualitative picture of the extent of the whey protein denaturation process and identify milks where the process has been severe, from a chemical standpoint.

Of late, less emphasis has been placed on distinguishing between UHT and sterilized milks. All these tests discussed are useful in that they will measure the severity of heat treatment. Owing to the wide range of conditions used, it may not always be possible to avoid an overlap between sterilized and UHT products.

The level of hydroxymethylfurfural (HMF) was also measured in heat-treated milks for the same purpose. Again, the values in sterilized milks were considerably higher than those in UHT milks, but there was some overlap between the methods that arose from assumptions made about the value of the blank used for the unheated milk, which varied from one sample to another. The distinctions were better when the blank could be measured and accounted for, but this would not be possible for routine quality control. One additional problem that was identified was a reduction in HMF in sterilized milk after 1 week of storage at 20°C, bringing it down into the range for UHT milk.

A further component of interest is lactulose, since its concentration in unheated milk is negligible, thus eliminating the problem of access to the original sample (see Section 2.10). When Andrews (1984) first investigated the factors affecting lactulose formation, the analytical procedure was long (24 h) and complex and involved using several enzymes. Burton's review (1988) describes work done up to that time, mainly using this complicated assay method. Andrews (1984) summarized the distribution of lactulose contents of 82 commercial UHT and in-container sterilized milks (Figure 10–3). It can be seen that there were three main groups, with no overlap between UHT direct and indirect, but some overlap between indirect UHT and sterilized milk. He suggested that 71.5 mg/100 ml should be used to distinguish between the two types. Later work by Andrews and Morant (1987) suggested that a limit of 68.2 mg/100 ml. Discrimination was further improved when lactulose data were combined with color measurements.

As well as the interest in lactulose for differentiating between the types of heat treatment, there is also interest in its absolute value. This is of particular interest as there was a good correlation between lactulose level and consumer reaction to product flavor (Andrews & Morant, 1987). Increases in lactulose in UHT milks were minimal during storage, whereas increases in sterilized milks were noticed, at a rate that depends upon the level immediately after processing. Lactulose can now be measured much more quickly using high-pressure liquid chromatography (HPLC). This has made it much more convenient to measure, and it is

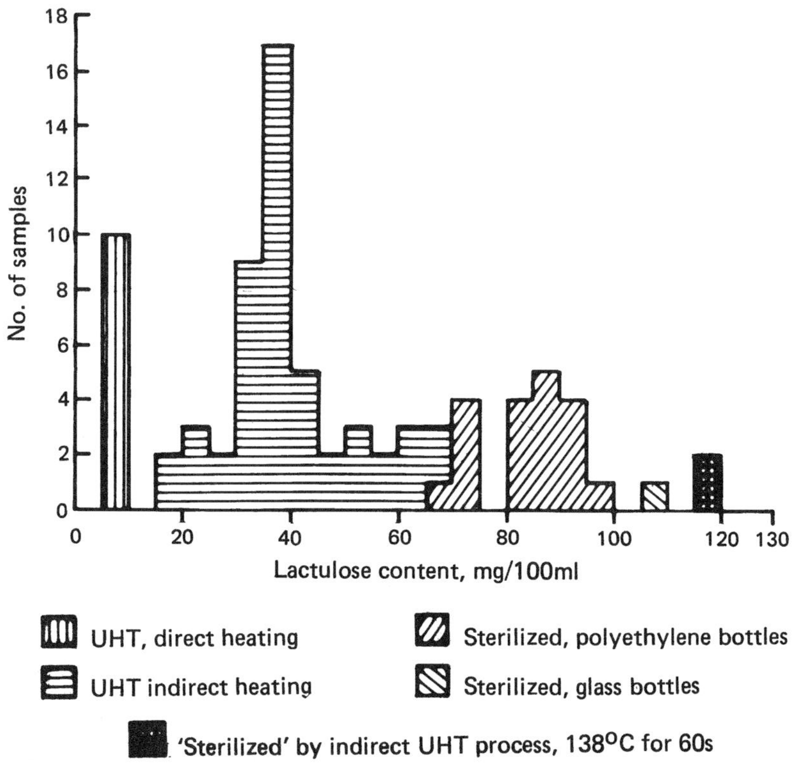

Figure 10–3 Lactulose contents of commercial UHT and in-container sterilized milks. *Source:* Adapted with permission from G.R. Andrews, Distinguishing Pasteurized, UHT, and Sterilized Milks by Their Lactulose Content, *Journal of the Society of Dairy Technology*, Vol. 37, p. 92, © 1984, Society of Dairy Technology.

now much more widely measured to indicate the amount of heat damage occurring immediately after processing. It is still not a simple method for routine laboratory use, however. A recent survey showed that 50% of indirect UHT milks had lactulose levels higher than 40 mg/100 g (a level that corresponds to the HRS inactivation).

Color changes are also interesting. It is possible to distinguish between sterilized products and UHT products immediately after processing; however, browning reactions will continue during storage, and UHT milks stored at high temperatures will soon approach sterilized milks in their color. Color data alone therefore will not provide definitive information about the severity of processing, but it may be useful in conjunction with other information, such as lactulose and/or whey protein denaturation, to determine whether there has been temperature abuse during storage.

Other chemical tests that may be performed on the product are the measurement of protein, fat, and carbohydrate, as these will all contribute to the energy value (calorific value) through the Atwater factors: carbohydrate, 16 MJ/kg; protein, 17 MJ/kg; fat, 37 MJ/kg; and alcohol, 29 MJ/kg). This is important for nutritional labeling. For products that contain organic acids, this also has to be determined.

In nutritional terms, the mineral content may be important, with specific minerals for specific products (e.g., calcium enrichment of milk). The role of ionic calcium in milk stability has been discussed (see Section 8.4.2). There is also interest in magnesium supplementation. This is also a divalent cation and will tend to destabilize heat-treated milk in a fashion similar to calcium.

Vitamin determination and vitamin fortification are important aspects of the nutritional profile. Other measurements might be done on free fat and free fatty acids: rancidity and nonprotein nitrogen (NPN) or protein fractions to test for hydrolysis or denaturation. Others that have been measured include available lysine. Again, many of the conventional methods have been superceded by rapid analyses; for example, near infrared analysis is widely used for the rapid measurement of protein, fat, and lactose in dairy products.

Physical measurements such as viscosity are important for milks, creams, and desserts; for low-viscosity fluids such as milks, capillary flow viscometers are useful and very accurate, whereas for more viscous products and those exhibiting non-Newtonian behavior, rotational viscometers should be used. Penetrometer readings and force-deformation relationships may be useful for when products gel or for measuring the clotting ability, which itself may be an indirect means of assessing heat damage. Alternatively, controlled stress or strain rheometers can be used to measure the viscous and elastic components of the material.

The fat globule size and distribution will give an indication of the stability of the fat phase in a heat-treated emulsion. Some differences between unhomogenized and homogenized milks are shown in Figure 10–4. Clumping in cream can be also identified by an increase in fat globule size.

Freezing point depression is another rapid method that permits the detection of added water. It is useful for steam injection/infusion processes for ensuring that the correct amount of water has been removed. It may also help detect whether plates or valves are leaking or whether extraneous water is added due to short changeover periods.

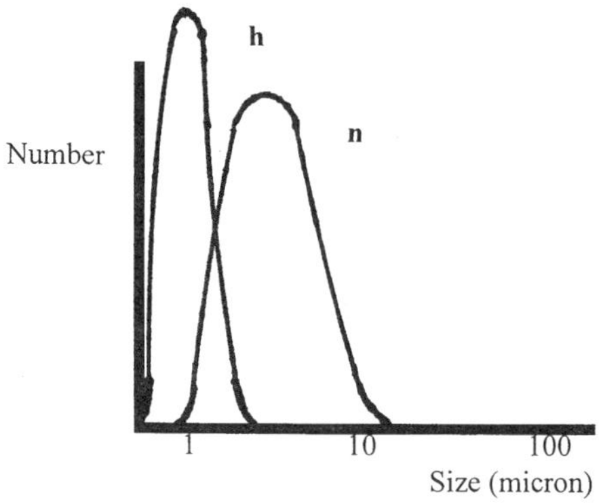

Figure 10–4 Differences between unhomogenized (n) and homogenized milks (h)

Measuring the amount of sediment may indicate whether poor-quality raw milk was used. Also, the limulus assay for assessing lipopolysaccharide (LPS) arising from the cell walls of gram-negative bacteria can be used for this purpose.

However, all these experimental measurements will at best provide answers only to what has happened in the past. The key is to prevent it happening again in the future! The ideal scenario is to build up a comprehensive profile for each process and product: this will contribute to a fuller understanding of the interaction between the product and the process, which in turn will help achieve the twin objectives of safety and quality.

10.6 SENSORY CHARACTERISTICS

The sensory characteristics of the final product are extremely important and are affected by changes that take during processing (Chapters 5 and 6), and especially throughout storage (Chapter 7), where an overview of changes in sensory characteristics has been given. The major sensory characteristics are as follows:

- The presence or absence of visual defects, whether it has visibly separated or coagulated or changed its color
- The aroma and taste, which together contribute to the flavor
- The texture and mouthfeel. Defects for milk may be described as watery, thin, coagulated, sandy, gritty, or separated.

Sensory evaluation techniques may employ one or more of the following three methodologies:

1. Discrimination testing for finding out whether there is a perceived difference between two products.

2. Acceptance or affective testing to determine the degree of acceptance or preference of a product, often using a standard product as a control. Hedonic scales may be set up typically with five, seven, or nine points.
3. Descriptive analysis, which is a much more detailed approach, giving a total sensory description of a product, and which usually involves training a panel.

Prasad (1989) and Prasad, Thomson, and Lewis (1990) describe the development of a vocabulary for heat-treated milks, which involved the use of 38 descriptive terms, covering its appearance, odor, flavor, mouthfeel, and aftertaste. A list of terms used is given in Exhibit 10–1, together with their descriptions.

In most canneries, it is a standard practice for samples of heat-treated products from every batch to be tasted by as many people as possible who work at the factory. These people will soon become experts in those specific products and in principle should be able to detect any sensory quality defects and prevent substandard product from being released for sale. However, this is not so straightforward with some UHT products, especially milk, as the products have a high cooked flavor intensity immediately after production, which is not generally liked. For reasons described, samples should be tasted both shortly after processing and also throughout storage, preferably after 1 to 2 weeks of incubation, as changes in sensory characteristics will occur throughout.

It is also argued that for milk, most microbiological problems can be picked up much more quickly from changes in sensory characteristics (e.g., gelation or off-flavor development). In support of organoleptic testing for quality control purposes, Ashton (1965) analyzed samples from 17,856 organoleptically normal cartons microscopically, and none showed the presence of microorganisms. In contrast, several pure cultures have been found that could grow in milk to higher counts without any organoleptic changes, and 25% of nonsterile UHT packs showed no organoleptic change (IDF, 1981). Hersom and Hulland (1980) in their excellent book on canned foods, provide a very interesting chapter on the microbiology of sound canned foods.

10.7 SOME LEGAL ASPECTS

There are statutory heat treatment requirements for some products that lay down conditions under which heat treatment should be performed. These may include temperature/time combinations; statutory testing, which may occasionally be microbiological; and other aspects, such as flow control, flow diversion, and specific cooling rates. Some legislative aspects for UHT products have been discussed by Staal (1981).

Food legislation is always in a state of flux. For example, Table 10–7 shows the current standards for the various heat treatments in the European Community, together with tentative proposals for discussion, involving the measurements of both β-lactoglobulin and lactulose in a range of heat-treated milks (Wilbey, 1996). These stipulate that lactulose levels in UHT milk should exceed 100 mg/l and in sterilized milks it should exceed 600 mg/l, thus distinguishing between UHT and sterilized milk.

Exhibit 10–1 Adjectives Selected for Describing the Sensory Characteristics of Milks and Their Corresponding Definitions

TERM	DEFINITION
Visual characteristics (by eyes)	
(1) Degree of creamy/yellowness (none–intense)	A continuum exemplified by the lack of creamy/yellow color at one end to butter-like yellow at the other
(2) Degree of brownness (none–intense)	A continuum exemplified by canned or in-bottle sterilized milks as the maximum
(3) Degree of grey/greenness (none–intense)	A continuum exemplified by milk permeate or whey (maximum) and full cream milk (minimum)
(4) Transmittance (translucent–opaque)	Passage of light through milk. Skim milk with virtually no fat shows maximum transmittance while homogenized full cream milk exemplifies the minimum
(5) Phase separation (none–extreme)	A continuum exemplified by homogenized and skim milks as minimum and pasteurized full cream milk as maximum
(6) Visual consistency (thin/watery–thick/creamy	The resistance observed when the sample is swirled. The maximum and minimum of the continuum exemplified by pasteurized high-fat milks and skim milks, respectively
Olfactory characteristics (by nose)	
(7) Odor intensity (nil–strong)	Overall odor intensity
(8) Characteristic milkiness (nil–strong)	Characteristic milk aroma exemplified by pasteurized full-cream milks
(9) Sweet smell (nil–strong)	An odor sensation that is analogous to the sweet taste
(10) Degree of freshness (not fresh–very fresh)	Odor characteristic normally associated with fresh, pasteurized, full-cream milk. (Assessors could not define freshness without further associations.)
(11) Cooked odor (nil–strong)	A continuum exemplified by pasteurized milk at the lower end and in-bottle sterilized milk at the other extreme
(12) Stale/old (nil–strong)	A negative odor attribute that is different from "not fresh." Again the assessors could not define staleness by simple association.
(13) Cardboard/oxidized (nil–strong)	Exemplified by a cardboard smell or oxidized fat in milk
(14) Sourness (nil–strong)	An odor sensation that is analogous to sour taste
(15) Cheesiness (nil–strong)	Exemplified by mature Cheddar cheese
(16) Synthetic/artificial (nil–strong)	A foreign odor not normally associated with milk
(17) Antiseptic/phenolic (nil–strong)	Exemplified by TCP (a commercial preparation of trichlorophenol)
(18) Metallic (nil–strong)	An odor associated with metal, particularly copper and iron
(19) Nutty (nil–strong)	Fatty, burnt, roasted, or sulfury odor note, characteristic of certain nuts
Oral characteristics (by mouth)	
Flavor	
(20) Flavor intensity (nil–strong)	Overall flavor intensity
(21) Characteristic milkiness (nil–strong)	Characteristic milk flavor exemplified by fresh, good, pasteurized full-cream milk

continues

Exhibit 10–1 continued

TERM	DEFINITION
(22) Sweetness (nil–strong)	A taste similar to sucrose. Perceived mainly at the front of the tongue
(23) Degree of freshness (not fresh–very fresh)	Flavor characteristic normally associated with fresh, pasteurized, full-cream milk. (Assessors could not define freshness without further associations.)
(24) Cooked flavor (nil–strong)	A continuum exemplified by pasteurized milk at the lower end and in-bottle sterilized milk at the other extreme
(25) Stale/old (nil–strong)	A negative flavor attribute that is different from "not fresh." Again the assessors could not define staleness by simple association.
(26) Cardboard/oxidized (nil–strong)	Exemplified by a cardboard smell or oxidized fat in milk
(27) Sourness (nil–strong)	A taste sensation that is analogous to sour milk
(28) Cheesiness (nil–strong)	Exemplified by mature Cheddar cheese
(29) Synthetic/artificial (nil–strong)	A foreign flavor not normally associated with milk
(30) Antiseptic/phenolic (nil–strong)	Exemplified by TCP (a commercial preparation of trichlorophenol)
(31) Metallic (nil–strong)	A flavor associated with metals, characterized by the sensation of sucking a copper alloy
(32) Nutty (nil–strong)	Fatty, burnt, roasted, or sulfury odor note, characteristic of certain nuts
(33) Bitter (nil–strong)	A basic taste sensation characterized by a solution of caffeine or quinine sulfate
Mouthfeel	
(34) Oral consistency (thin/watery–thick/creamy)	A textural continuum exemplified by skim milk at the lower end and high-fat milk at the other extreme
(35) Greasiness (not greasy–very greasy)	A sensation associated with a film or particles of fat on the tongue and or teeth. Characterized by high-fat milks
(36) Graininess (smooth–grainy)	The sensation of fine particles or granules. Characterized by a suspension of ground chalk
Aftertaste	
(37) Persistent aftertaste (nil–very persistent)	Duration for which the flavor remains after sample is consumed
(38) Drying effect (nil–very dry)	The feeling of shrivelling or puckering on the tongue or lining of the mouth

Source: Reprinted with permission from S.K. Prasad, D.M.H. Thomson, and M.K. Lewis, Trends in Food Product Development, in *Proceedings of the World Congress of Food Science and Technology (Singapore)*, T.C. Yan and C. Tan, eds., © 1990, International Union of Food Science and Technology.

Table 10–7 Current Standards for the Various Heat Treatments and Tentative Proposals for Discussion (in italics)

Assay	*Pasteurized*	*High Temperature Pasteurized*	*UHT*	*Sterilized*
Alkaline phosphatase	–ve	–ve	–ve	–ve
Lactoperoxidase	+	–ve	–ve	–ve
Turbidity	(+)	(+)	(±)	–ve
Lactulose	*not detected*	*< 50 mg/L*	*> 100 mg/L*	*> 600 mg/L*
β-lactoglobulin	*> 2600 mg/L*	*≥ 2000 mg/L*	*> 50 mg/L*	*< 50 mg/L*

Source: Reprinted with permission from R.A. Wilbey, Estimating the Degree of Heat Treatment Given to Milk, *Journal of the Society of Dairy Technology*, Vol. 49, No. 4, pp. 109–112, © 1996, Society of Dairy Technology.

REFERENCES

Ahvenainen, R., Wirtanen, G., & Manninen, M. (1989). Ultrasound imaging: A non-destructive method for control of the microbiological quality of aseptically packed foodstuffs. *Lebensmittel Wissenschaft und Technologie* **22**, 273–278.

Ahvenainen, R., Wirtanen, G., & Mattila-Sandholm, T. (1991). Ultrasound imaging: A non-destructive method for monitoring changes caused by microbial enzymes in aseptically-packed milk and soft ice-cream base material. *Lebensmittel Wissenschaft und Technologie* **24**, 397–403.

Andrews, G.R. (1984). Distinguishing pasteurized, UHT and sterilized milks by their lactulose content. *Journal of the Society of Dairy Technology* **37**, 92.

Andrews, G.R., & Morant, S.V. (1987). Lactulose content and the organoleptic assessment of ultra heat treated and sterilized milks. *Journal of Dairy Research* **54**, 493–507.

Ashton, T.R. (1965). Practical experience: The processing and aseptic packaging of sterile milk in the United Kingdom. *Journal of the Society of Dairy Technology* **18**, 65–85.

Bergere, J.L., & Cerf, O. (1992). Heat resistance of *Bacillus cereus* spores, 23–25. In IDF Bulletin No. 275, *Bacillus cereus* in milk and milk products.

Burton, H. (1988). *UHT processing of milk and milk products*. London: Elsevier Applied Science.

Cerf, O. (1989). Statistical control of UHT milk. In H. Reuter (Ed.), *Aseptic packaging of foods* (p. 244). Lancaster, Basel: Technomics.

Food Innovation Centre Newsletter, January, 1997.

Gestrelius, H., Mattila-Sandholm, T., & Ahvenainen, R. (1994). Methods for non-invasive sterility control in aseptically packaged foods. *Trends in Food Science and Technology* **5**, 379–383.

Guidelines for the safe production of heat preserved foods. (1994). Department of Health, London: HMSO.

Hammer, P., Lembke, F., Suhren, G., & Heeschen, W. (1996). Characterisation of heat resistant mesophilic *Bacillus* species affecting the quality of UHT milk. IDF (1996). *Heat treatment and alternative methods*, SI 9602.

Harrigan, W.F., & Park, R.W.A. (1991). *Making safe food*. London: Academic Press.

Hersom, A.C, & Hulland, E.D. (1980). *Canned foods*. Edinburgh, Scotland: Churchill Livingstone.

Hubert, W.T., Hagstad, H.V., Spangler, E., Hinton, M.H., & Hughes, K.L. (1996). *Food safety and quality assurance* (2nd ed.). Ames, IA: Iowa State University Press.

ICMSF. (1988). *Micro-organisms in foods*, Vol. 4, *Application of the hazard analysis critical control point (HACCP) system to ensure microbiological safety*. Oxford, UK: Blackwell Scientific Publications.

IDF Bulletin. (1981). *New monograph on UHT milk*, No. 133.

IDF Bulletin. (1991). *Methods for assessing the bacterial quality of raw milk from the farm*, No. 256.

IDF Bulletin. (1992). Bacillus cereus *in milk and milk products*, No. 275.

IDF Bulletin. (1993). *Catalogue of tests for the detection of PPC of milk*, No. 281.

IDF. (1996a). *Bacteriological quality of raw milk*, Ref. SI 9601.

IDF. (1996b). *Heat treatment and alternative methods*, Ref. SI 9602.

IFST. (1997). Development and use of microbial criteria for foods. *Food Science and Technology Today* **11**, 137–177.

IFST. (1998). *Food and drink—Good manufacturing practice: A guide to its responsible management* (3rd ed.). London: Author.

IFST. (1999). *Development and use of microbiological criteria for foods.* London: Author.

Moermans, R., & Mottar, J. (1984). The turbidity test and the quality of UHT treated milk. *Milchwissenschaft* **39**, 94–95.

Pettersson, B., Lembke, F., Hammer, P., Stackebrandt, E., & Priest, F.G. (1996). *Bacillus sporothermodurans*, a new species producing highly heat-resistant endospores. *International Journal of Systematic Bacteriology* **46**, 759–764.

Prasad, S.K. (1989). *The sensory characteristics of heat-treated milks, with special reference to UHT processing*. PhD thesis, Reading, UK: University of Reading.

Prasad, S.K., Thomson D.M.H., & Lewis, M.J. (1990). Descriptive sensory analysis of milks. In T.C. Yan, & C. Tan (Eds.), Singapore: Proceedings of the World Congress of Food Science and Technology.

Rees, J.A.G., & Bettison, J. (1991). *Processing and packaging of heat preserved foods.* Glasgow, Scotland: Blackie.

Shapton, D.A., & Shapton, N.F. (1991). *Principles and practices for the safe processing of foods*. Oxford, UK: Butterworth-Heinemann, Reprinted by Woodhead Publishing Limited (1998).

Staal, P.F.J. (1981). Legislative aspects. In IDF Bulletin, *A new monograph on UHT milk*, No. 133.

Statutory Instruments (SI). (1995). *Food milk and dairies: The dairy products (hygiene) regulations*, No. 1086. London: HMSO.

Waites, W.M. (1997). Principles of rapid testing and a look to the future. *International Journal of Dairy Technology* **50**, 57–60.

Wilbey, R.A. (1996). Estimating the degree of heat treatment given to milk. *Journal of the Society of Dairy Technology* **49**, 109–112.

Williams, T.W. (1996). Storage and quality aspects of UHT products. In *Training Course Notes, Principles of UHT Processing*. Leatherhead, UK: Leatherhead Food Research Association.

Wirtanen, G., Ahvenainen, R., & Mattila-Sandholm, T. (1992). Non-destructive detection of spoilage of aseptically-packed milk products: Effect of frequency and imaging parameters on the sensitivity of ultrasound imaging. *Lebensmittel Wissenschaft und Technologie* **25**, 126–132.

Appendix A

Mathematical Model for Heat Transfer to a Sphere

The mathematical model is based on dividing a sphere into 20 spherical elements of equal radial increment and performing a heat balance on each one. Equations representing the boundary conditions for the innermost and outermost elements are also written, and the equations successively integrated numerically, using a fourth-order Runge Kutta algorithm (Figure A–1). The temperature of each element was taken to be at the center of that element (e.g., for element i, the temperature was taken to be that at a radius midway between r_i and r_{i-1}).

Performing a heat balance on element i,

Heat in from element $i+1=$

$$k \cdot 4 \cdot \pi \frac{r_i^2}{\Delta r}(\theta_{i+1} - \theta_i)$$

where Δr is the difference between successive radii, i.e., $r_i - r_{i-1}$ and k is the thermal conductivity of the solid.

Heat out to element $i-1=$

$$k \cdot 4 \cdot \pi \frac{r_{i-1}^2}{\Delta r}(\theta_i - \theta_{i-1})$$

Heat accumulation in element $i=$

$$\frac{4}{3} \cdot \pi \cdot (r_i^3 - r_{i-1}^3) \cdot \rho \cdot c_p \cdot \frac{d\theta}{dt}$$

From the law of conservation of energy,

heat accumulation = heat input – heat output

Therefore, equating the three terms above and rearranging, for i = 2 to (n – 1):

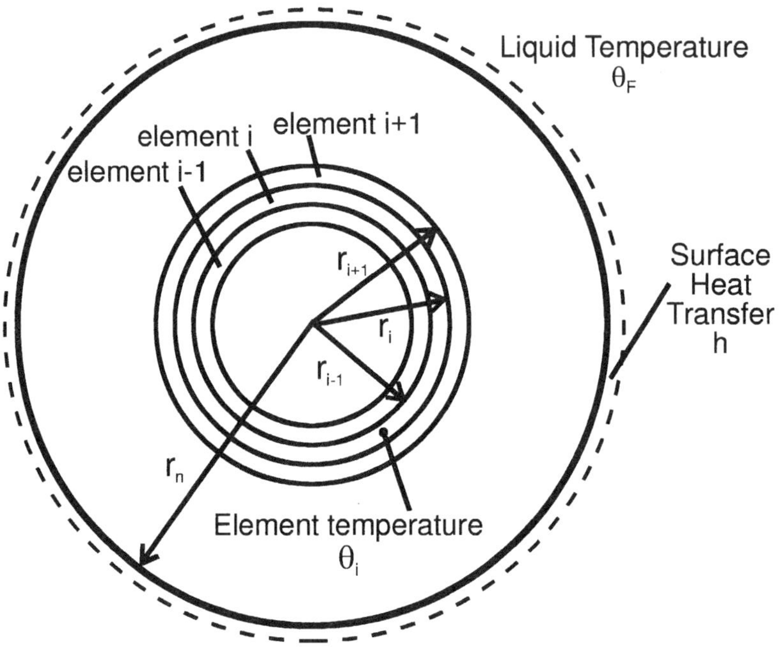

Figure A–1 Fourth-order Runge Kutta algorithm

$$\frac{d\theta}{dt} = 3 \cdot \frac{k}{\rho c_p} \cdot \frac{[r_i^2(\theta_{i+1} - \theta_{i-1}) - r_{i-1}^2(\theta_i - \theta_i - 1)]}{\Delta r(r_i^3 - r_{i-1}^3)}$$

and note that the thermal diffusivity, α, is defined as $(k/\rho\ c_p)$

The boundary conditions are:

(a) for the innermost element, element 1, heat enters and leaves only by the next innermost element, i.e., element i – 1 does not exist. Therefore:

$$\frac{d\theta}{dt} = \frac{3 \cdot \alpha}{r_i \Delta r} \cdot (\theta_2 - \theta_1)$$

(b) for the outermost element, element n, the thickness from the surface to the center of the outermost element is half the radius. Taking the temperature of the surrounding liquid as θ_f:

Heat in by conduction from surroundings =

$$k \cdot 4 \cdot \pi \frac{r_n^2}{(\Delta r / 2)}(\theta_F - \theta_n)$$

If a convective surface heat transfer coefficient, h, exists, then:

$$\text{Heat in by convection from surroundings} = h \cdot (4 \cdot \pi \cdot r_n^2)(\theta_f - \theta_n)$$

The total heat flow into the element from the surroundings can then be calculated by adding the conductances inversely:

$$\text{Total heat in from surroundings} = \frac{2 \cdot h \cdot k}{(2k + \Delta rh)} \cdot 4\pi r_n^2 (\theta_f - \theta_n)$$

Therefore, for the outer element:

$$\frac{d\theta}{dt} = \frac{3 \cdot \alpha}{(r_n^3 - r_{n-1}^3)} \cdot \left[\frac{2h}{(2k + \Delta rh)} \cdot r_n^2 (\theta_F - \theta_n) - r_{n-1}^2 (\theta_n - \theta_{n-1}) \right]$$

A fourth-order Runge Kutta algorithm with fixed integration-step is used for the numerical integration. The model is given below, written in FORTRAN, together with typical data (DAT.IN) and temperature profile (TPROF.IN) input files. The program reads physical and thermal data in file "DAT.IN" and the time-temperature-heat transfer coefficient profile from "TPROF.IN" and gives the model output both on the screen and in the file "MODEL.OUT", which can be renamed and deleted before the next run. The model may be verified by assuming an instantaneous temperature rise in the liquid and comparing the temperature of each element to the result obtained using the solution to the Fourier equation (Equation 4.34) or to the charts from Gurney and Lurie (1923). The integration step length has been set at 0.00005 s in file "DAT.IN", which has been found to give the balance between accuracy and computer run time.

Modifications are easily made. Nonuniform initial conditions can be accommodated in the INITIALIZATION section by setting TEMP(I) for elements I = 1 to 20. If nonuniform thermal properties are required, a 20-element array for CP, RHO, or TC (specific heat, density, or thermal conductivity, respectively) can be set up and incorporated in the derivative section just before line 69. A nonlinear temperature profile can be accommodated by incorporating another algorithm for TEMP(21) in lines 67–68 or giving more time-temperature data points. The model may be modified to calculate the proportion of surviving spores in the sphere as a whole, rather than F_0 value of the center. At each integration step, the D value of the spores in the element can be calculated from its temperature and the proportion of surviving spores from the current number and Equation 2.5. The surviving spores in the whole sphere can be totaled at each communication interval in the program.

```
*****************************************************************
  PROGRAM
* -WITH UHT PLANT TEMP PROFILE (ST LINE HEATING)
* -DOUBLE PRECISION - VARIABLE HTC VALUE
  COMMON/VALS/ALPHA,DELTAR,HTC,TC//TEMP(21),R(21)
  DOUBLE PRECISION F0(21),A1,A2,A3,A4,ALPHA,BDIAM
  DOUBLE PRECISION CP,DELT,DELTAR,F0FL,FOUR,HTC,R,RHO,TC
  DOUBLE PRECISION TEMP,TIME,TSTP,VAR,ZZ
  DOUBLE PRECISION TIMD(20),TEMPD(20),H(20),BIOT(20)
   LOGICAL FLAG
   CHARACTER*8 RUNNO
   FLAG=.FALSE.
  OPEN(10,FILE='DAT.IN',STATUS='OLD')
  OPEN(11,FILE='MODEL.OUT',STATUS='NEW')
  OPEN(12,FILE='TPROF.IN',STATUS='OLD')
  READ(10,1050) RUNNO
  READ(10,*) TC
  READ(10,*) RHO
  READ(10,*) CP
  READ(10,*) HTC
  READ(10,*) BDIAM
  READ(10,*) DELT
  READ(10,*) INTOP
  CLOSE(10)
* READ IN TIME-TEMPERATURE-HTC PROFILE
   NVALS=0
   DO 200 L=1,20
   READ(12,*,END=210)TIMD(L),TEMPD(L), H(L)
   NVALS=NVALS+1
200 CONTINUE
   NVALS=NVALS-1
210 TSTP=TIMD(NVALS)
   CLOSE(12)
* RADIUS CALCULATION'
    R(20)=BDIAM/2.0
   DO 20 I=1,20
   R(I)=(I*R(20)/20.0)
   TEMP(I)=TEMPD(1)
   F0(I)=0.0
20 CONTINUE
  F0FL=0.0
   DELTAR=R(20)-R(19)
   TEMP(21)=TEMPD(1)
```

```
   F0(21)=0.0
   NDERC=INT(INTOP/DELT)
   ALPHA=TC/(RHO*CP)
* INITIALIZATION ********
   WRITE(*,1005)ALPHA/10000.0
   WRITE(*,1030)NDERC*DELT
* DYNAMIC SECTION **********
   J=1
   HTC=H(J)
   BIOT(J)=(R(20)*HTC)/TC
   DO 100 IDYC=1,4000
   IF (TIME.GT.TSTP) GOTO 500
   FOUR=ALPHA*TIME/(R(20)*R(20))
   WRITE(11,1010)TIME,TEMP(1),TEMP(21),F0(1),F0FL
   WRITE(*,1020)IDYC,TIME,TEMP(1),TEMP(21),F0(1),F0FL, HTC
* DERIVATIVE SECTION ********
   DO 120 IDERC=1,NDERC
    TIME=TIME+DELT
230 IF (TIME.GT.TIMD(J+1)) THEN
   J=J+1
   HTC=H(J)
   BIOT(J)=(R(20)*HTC)/TC
   GOTO 230
        ELSE
   TEMP(21)=TEMPD(J)+((TEMPD(J+1)-TEMPD(J))*(TIME-TIMD(J))
  &/(TIMD(J+1)-TIMD(J)))
   END IF
* INTEGRATIONS
   VAR=TEMP(1)
   CALL FUNC3(VAR,ZZ)
    A1=DELT*ZZ
   VAR=VAR+A1/2.0
   CALL FUNC3(VAR,ZZ)
    A2=DELT*ZZ
   VAR=VAR+A2/2.0
   CALL FUNC3(VAR,ZZ)
    A3=ZZ*DELT
   VAR=TEMP(1)+A3
   CALL FUNC3(VAR,ZZ)
    A4=DELT*ZZ
   TEMP(1)=TEMP(1)+(A1+2.0*A2+2.0*A3+A4)/6.0
   DO 110 K=2,19
   VAR=TEMP(K)
```

```
      CALL FUNC1(K,VAR,ZZ)
       A1=DELT*ZZ
      VAR=VAR+A1/2.0
      CALL FUNC1(K,VAR,ZZ)
       A2=DELT*ZZ
      VAR=VAR+A2/2.0
      CALL FUNC1(K,VAR,ZZ)
       A3=ZZ*DELT
      VAR=TEMP(K)+A3
      CALL FUNC1(K,VAR,ZZ)
       A4=DELT*ZZ
      TEMP(K)=TEMP(K)+(A1+2.0*A2+2.0*A3+A4)/6.0
  110 CONTINUE
      VAR=TEMP(20)
      CALL FUNC2(VAR,ZZ)
       A1=DELT*ZZ
      VAR=VAR+A1/2.0
      CALL FUNC2(VAR,ZZ)
       A2=DELT*ZZ
      VAR=VAR+A2/2.0
      CALL FUNC2(VAR,ZZ)
       A3=ZZ*DELT
      VAR=TEMP(20)+A3
      CALL FUNC2(VAR,ZZ)
       A4=DELT*ZZ
      TEMP(20)=TEMP(20)+(A1+2.0*A2+2.0*A3+A4)/6.0
* END OF DERIVATIVE LOOP ****
      F0(1)=F0(1)+(DELT/60.0)*(10.0**((TEMP(1)-121.1)/10.0))
      F0FL=F0FL+(DELT/60.0)*(10.0**((TEMP(21)-121.1)/10.0))
  130 CONTINUE
  120 CONTINUE
* END OF DYNAMIC LOOP *****
  100 CONTINUE
* TERMINAL SECTION *******
  500 WRITE(*,1045)
      WRITE(11,1050) RUNNO
      WRITE(11,1005)ALPHA/10000.0
      WRITE(11,1055) TC, RHO, CP, BDIAM
      WRITE(11,1070)
      DO 300 I=1,NVALS
      WRITE(11,1060)TIMD(I),TEMPD(I),H(I),BIOT(I)
  300 CONTINUE
      CLOSE(11)
```

```
1000 FORMAT(' BIOT=',F6.2,' ALPHA=',E9.3,'m2/s')
1005 FORMAT(' ALPHA=',E9.3,'m2/s')
1010 FORMAT(1X,F7.2,1X,2F7.2,1X,2F7.2)
1020 FORMAT(1X,I4,1X,F7.2,1X,2F7.2,1X,F6.2,F6.2, 1X, F8.4)
1030 FORMAT(' MODEL OUTPUT INTERVAL (s) IS ',F6.3)
1040 FORMAT(' **  END OF HEATING PERIOD ')
1045 FORMAT(' **  END OF SIMULATION ')
1050 FORMAT(A15)
1055 FORMAT(1X,' TC=',F5.3,' RHO=',F5.3,' CP=',F5.3,'
BDIAM=',F5.2)
1060 FORMAT(1X,F5.1,3X,F5.1,3X,F8.4, 3X,F10.4)
1070 FORMAT(' TEMP   TEMP    HTC BIOT  TEMPERATURE  PROFILE')
       STOP
       END
       SUBROUTINE FUNC1(J,VAR,ZZ)
       COMMON /VALS/ALPHA,DELTAR,HTC,TC//TEMP(21),R(21)
       DOUBLE PRECISION ALPHA, DELTAR,HTC,R,TC,TEMP,VAR,ZZ
       ZZ=(((TEMP(J+1)-VAR)*R(J)*R(J))-(VAR-TEMP(J-1))
       &*(R(J-1)*R(J-1)))*3.0*ALPHA/(DELTAR*(R(J)*R(J)*R(J)
       &-R(J-1)*R(J-1)*R(J-1)))
       RETURN
       END
       SUBROUTINE FUNC2(VAR,ZZ)
       COMMON /VALS/ALPHA,DELTAR,HTC,TC//TEMP(21),R(21)
       DOUBLE PRECISION ALPHA,DELTAR,HTC,R,TC,TEMP,VAR,ZZ
       ZZ=3.0*ALPHA*((2.0*HTC*R(20)*R(20)*(TEMP(21)-VAR)
       +/(2.0*TC+DELTAR*HTC))-(R(19)*R(19)*(VAR-TEMP(19))
       +/DELTAR))/(R(20)*R(20)*R(20)-R(19)*R(19)*R(19))
       RETURN
       END
       SUBROUTINE FUNC3(VAR,ZZ)
       COMMON /VALS/ALPHA,DELTAR,HTC,TC//TEMP(21),R(21)
       DOUBLE PRECISION ALPHA,DELTAR,HTC,R,TC,TEMP,VAR,ZZ
       ZZ=(3.0*ALPHA*(TEMP(2)-VAR)*R(1)*R(1))
       (DELTAR*R(1)*R(1)*R(1))
       RETURN
       END
```

```
******************************************************************
    File DAT.IN
    *******************************
    title
    0.62      TC (thermal conductivity, W/mK)
```

```
    1.05       RHO (density, kg/m³ x 10⁻³)
    2.8        CP (specific heat, J/gK)
    0.1240     HTC (liquid-particulate heat transfer coefficient,
                 W/m²K x 10⁻³)
    10.0       BDIAM (Diameter of particulate in mm)
    0.00005    DELT (integration step length, s)
    1          OUTPUT INTERVAL in seconds
    ***********************************************************
    File TPROF.IN
    Time (s), Temperature (°C), heat transfer coefficient (for next
      section) (W/m²K x 10⁻³)
***********************************************************
0.0,80.0,1248.0
26.1,116.4,0.1248
30.0,116.4,1248.0
60.0,130.0,0.1248
120.0,130.0,1248.0
150.0,64.5,1248.0
180.0,64.5,0.1248
```

Index

A

Acidic food, 7–10
Acidity, 7–10
Aerosol, 321–322
Age gelation, 376–377
Air/liquid interface, 28–29
Alcohol stability test, fouling, 349–350
Ambient temperature, extended shelf life, 231
Analytical testing
  chemical test, 417–421
  indirect methods, 416–417
  noninvasive methods, 408
  nonmicrobiological indicators of spoilage, 416–417
  physical test, 417–421
  quality assurance, 407–412
  rapid methods, 414–416
  time-temperature indicator, 417
Anthocyanin, 54
Arrhenius equation model, 37–38
Aseptic packaging, 295–327
  chemical sterilant, 298–300
  dry heat, 297–298
  ethylene oxide, 299–300
  hydrogen peroxide, 298–299
    ultraviolet irradiation combined, 301–302
  ionizing radiation, 302
  methods of container sterilization, 296–302
  peracetic acid, 300
  quality assurance, 407
  requirements, 295–296
  saturated steam, 296–297
  types of filling systems, 302–327
  ultraviolet irradiation, 300

B

Bacillus cereus
  extended shelf life, 227–229
  quality assurance, 391–393
Bacillus stearothermophilus, 129
Bacillus subtilis, 129
Bacterial proteinase, 268–269
Bacteriocin
  extended shelf life, 229–230
  pasteurization, 229–230
Bactofugation, 225
Batch processing
  flow diagram, 3
  history, 2
Batter, fouling, 356
Beer, pasteurization, 223
Biochemical change
  kinetics, 49–50
  temperature, biochemical component changes at high temperatures, 49
Biofilm, 332
Biot number, 23–24, 25
Blanching, 12
Blow-molded plastic bottle, 312
Bottle
  blow-molded plastic bottle, 312

blown, filled, and sealed at single station, 314–315
glass bottle, 320–321
nonsterile bottles, 312–313
sterile blown bottles, 313, 314
Bovine spongiform encephalopathy, 5
Browning reaction, 52, 371
fruit juice, 373–374
processing, 371–374
storage, 371–374
Bulk filling system, 322–323

C

Calcium dipicolinate, thermal death kinetics, 45
Can, 302–303
Canners' (constant *z* value) model, 36–37, 38
Carbon dioxide, raw material, 226–227
Carotenoid, 54
Carton, 303, 304
Chemical, sterilization of equipment, 364–365
Chemical marker, 181
Chemical sterilant, aseptic packaging, 298–300
Chemical test, quality assurance, 417–421
Chlorophyll, 54
Cleaning, 358–363
dynamics of cleaning, 361–362
kinetics, 361–362
measuring cleaning effectiveness, 359
quality assurance, 407
Clostridium botulinum, 237
acceptable level, 35
thermal process criterion, 41, 42
Coffee, ultrahigh temperature processing, 280
Color, 30, 264–265
degradation, 54
loss of pigments, 54
processing, 369–370
storage, 369–370
Complex colloidal system, 28–29
Compressibility
gas, 12, 13
vapor, 12, 13
Concentric tube heat exchanger, 72–75
Conduction, 19–21
resistance, 20–21
Conduction pack, 10
Constant lethal temperature, microorganism death, 33–36
Consumer risk, 395
Continuous processing, 4
advantages, 4
Continuous-flow thermal processing
legal aspects, 414–417
packaging, 294
particulate liquid, 143–188
chemical marker, 181
difficulties, 143–144
enzyme marker, 180–181
mathematical modeling, 174–178
microbiological marker, 180
overall process validation, 178–181
prediction of sterilization effect, 178–181
time-temperature integrator, 181
residence time distribution, 139–141
practical evaluation, 141
viscous liquid product
equipment selection, 134
modification of velocity profile for Hershel-Bulkley fluids, 139, 140
non-Newtonian viscosity effect on laminar flow velocity profile, 137–139
non-Newtonian viscosity effect on Reynolds number, 136–137
residence time distribution by section shape, 139–140
Continuous-flow thermal processing plant, 57–130
acceptable pack failure rates, 105–107
capital cost of process, 57–58
direct heating system, 84–101
culinary steam supply, 99–101
indirect heating system compared, 101–103
product concentration or dilution control, 95–98
direct use of electricity, 103–104
flow control, 59–60
friction heating, 103–104
heat exchange equipment, 62
heat recovery, 58
heating energy source, 58–59
homogenization, 60–62

indirect heat exchanger comparison, 81–82
indirect heating system, 63–84
infusion system, 91–94
injection system compared, 94–95
microwave, 104
Ohmic heating, 104
operating costs, 57–58
pasteurization, 65–66
plant performance evaluation, 105–113
plate heat exchanger, 64–65
prediction of process performance
from time-temperature profile, 105–113
using measured plant data, 107
preprocessing heat treatment, 62–64
process equipment, 59–62
product quality, 57
pumping, 59–60
purpose, 57
residence time distribution, 113–130
chemical marker, 130
direct process evaluation, 128
dispersion model, 120–122, 123
effect on quality of foods, 127
errors in prediction, 123–124
low-viscosity Newtonian liquid velocity profiles, 114–115
marker organism, 128–130
measurement, 117–119
plant performance effect, 113–114
plant performance evaluation, 127–130
practical residence time distribution curves analysis, 123
residence time distribution data models, 119–122
residence time distribution for any input signal, 118–119
residence time distribution theory, 115–117
sterilization efficiency effect, 124–127
tanks-in-series model, 120
scraped-surface heat exchanger, 79–81, 82
steam injection system, 85–91
infusion system compared, 94–95
sterilization
advanced plate-type indirect heating plant, 70–71
heat exchanger specification, 67–68
low-viscosity liquid, 66–68, 68–72
temperature
chemical marker, 130
direct process evaluation, 128
marker organism, 128–130
plant performance evaluation, 127–130
time-temperature profile
chemical quality criteria, 112
determination, 108, 109
microbiological criteria, 111
performance calculation, 109–113
performance prediction, 107–113
tubular heat exchanger, 72–79
concentric tube heat exchanger, 72–75
shell-and-tube tubular heat exchanger, 75–79
types, 58–59
Convection, 22
Convection pack, 10
Cooking, characterized, 50–54
Cream
fat content, 217
fouling, 354
mouthfeel, 217–218
pasteurization, 216–218
types, 217
ultrahigh temperature processing, 271–273
viscosity, 217–218
Crystallization, enthalpy, 19
Custard, ultrahigh temperature processing, 277–280

D

Decimal reduction time
defined, 36
temperature, 36–39
Arrhenius equation model, 37–38
canners' (constant *z* value) model, 36–37, 38
models compared, 38–39
Density, 11–13
defined, 11
fluid, air, 12
liquid, 12
scales, 12

specific gravity, 11
temperature, 12
Deposit formation
fouling, 335
storage, 375
Detergent raw materials, properties, 360
Dielectric constant, 27
values, 27, 28
Dielectric loss factor, 27
values, 27, 28
Diffusion properties, 27–28
Dilatant behavior, 14
Dimensionless analysis, 24
Dimensionless group, 24–25
Dimensionless temperature ratio, 194
Direct contact heating, resistance, 23–24
Direct heating, sterilization, 245–248
Direct heating system, continuous-flow thermal processing plant, 84–101
culinary steam supply, 99–101
indirect heating system compared, 101–103
product concentration or dilution control, 95–98
Disinfection, heat, 364
Dispersion model, 120–122, 123
Dissolved air, fouling, 351
Dissolved oxygen, storage, 372
Double processing, extended shelf life, 227–229
Dry heat, aseptic packaging, 297–298

E

Egg
fouling, 355–356
pasteurization, 219–220
minimum requirements, 215
Electrical conductivity, 25–26
food materials at 19°C, 26
temperature, 26
Electrical properties, 25–27
Enthalpy, 18–19
crystallization, 19
defined, 18
differential scanning calorimetry, 19
fruit juice, 20
vegetable juice, 20
Enzyme inactivation, 1, 51, 214–216, 266–269, 270
Enzyme marker, 180–181
Enzyme regeneration, storage, 388–389
Enzyme-induced gelation, 378–379
Equipment sterilization, 364–365, 412
Ethylene oxide, aseptic packaging, 299–300
Evaporated milk
sterilization, 273–274
ultrahigh temperature processing, 273–274
Extended shelf life
ambient temperature, 231
Bacillus cereus, 227–229
bacteriocin, 229–230
double processing, 227–229
extended heat treatment, 227–229
heat-resistant spore, 227–229
higher temperature for shorter time, 227–229
refrigerated product, 224–230
extended heat treatment, 227–229
raw material quality, 224–227
thermoduric bacteria, 227–229

F

F value, sterilization, 41
Fat, 263
storage, 377
Flavor, 264
storage
cooked milk flavor, 380–384
flavor changes, 380–384
flavor improvement, 384
Flip-flop batch vat pasteurization, flow diagram, 3
Flow, thermal processing, 30
Flowrate, fouling, 342
Fluid, density, air, 12
Food
diffusion properties, 27–28
electrical properties, 25–27
optical properties, 29–30
physical properties, 10–15

sensory characteristics, 29–30
shape, 10–11
size, 10–11
surface areas, 10–11
surface properties, 28–29
thermal properties, 15–19
Food composition, 7–10
tables, 8
Food quality
residence time distribution, 127
thermal processing, 5
Food safety
process characterization, 410–412
quality assurance, 395–397
thermal processing, 4–5
Forewarming, fouling, 350
Form-fill-seal system, 318–320
Fouling
additives, 351
aging raw milk, 350
alcohol stability test, 349–350
batter, 356
components, 359
construction material, 341–342
cream, 353–354
defined, 331
deposit formation, 335
dissolved air, 350–351
economic implications, 331–332
egg, 356
factors affecting, 341–344
flowrate, 342
forewarming, 350
fouling period, 333
fouling rigs, 338, 340
fouling sensor, 338–341
goat's milk, 352–353
indirect indicators, 341
induction period, 333
measurement, 335–341
milk, 345–354
mechanisms, 351–352
methods to reduce fouling, 348–351
milk permeate, 352–353
raw milk quality, 346–348
reconstituted milk, 354
models, 344
overall heat transfer coefficient, 331
decrease, 335, 338
plant construction, design, and operation, 341
pressure monitoring, 337–338, 339, 340
resistances, 333
surface finish, 341–342
temperature, 342, 343
temperature measurement, 336, 337
terms, 333–335
tomato, 355
whey protein, 354–355
Fourier number, 25
Friction heating, continuous-flow thermal processing plant, 103–104
Froude number, 25
Fruit drink, pasteurization, 220–222
Fruit juice
browning, 365–366
enthalpy, 20
pasteurization, 220–222

G

Gas, compressibility, 12, 13
Gelation, storage, 376–379
age gelation, 376–378
enzyme-induced gelation, 378–379
Glass bottle, 320–321
Goat's milk, fouling, 352–353
Good Manufacturing Practice, 407

H

Heat
disinfection, 364
sterilization of equipment, 364
Heat activation, 46
thermal death kinetics, 44–45
Heat exchanger
fouling, 31
particulate liquid, 159–160

viscous liquid product
heat transfer coefficients, 141–143
non-Newtonian liquid, 142–143
Heat film coefficient, 22
Heat recovery, continuous-flow thermal processing plant, 58
Heat resistance
heat-resistant spore, 227–229, 237–239, 240, 398–399
vegetative bacteria, 204–209
Heat transfer, 19–25, 47
multiple resistances, 22–23
sphere, mathematical model, 164, 174–178, 427–434
Heat transfer coefficient, particulate liquid, 159–160
experimental measurement, 170–171, 172–173
experimental methods for determination, 166–168
between liquids and solid bodies, 164–166
minimum value, 165–166
Heating food
history, 2
range of products, 2
reasons for, 1–2
severity of process, 1
Heat-resistant spore
extended shelf life, 227–229
heat resistance, 398–399
sterilization, 237–239, 240
Hershel-Bulkley liquid, velocity profile, 139, 140
Higher temperature for shorter time, extended shelf life, 227–229
High-intensity pulsed electric field, pasteurization, 230
Homogenization
continuous-flow thermal processing plant, 60–62
sterilization, 248–250
Hot filling, 294–295
Hydrogen peroxide
aseptic packaging, 298–299
ultraviolet irradiation combined, 301–302
raw material, 226

I

Ice cream, pasteurization, 218–219
Ice cream mix, ultrahigh temperature processing, 274
In-container sterilization, 240
Indirect heat exchanger, continuous-flow thermal processing plant, exchanger comparison, 81–82
Indirect heating system
continuous-flow thermal processing plant, 63–84
sterilization, 243–245
Infant formulation, ultrahigh temperature processing, 280
Infusion system, continuous-flow thermal processing plant, 91–94
injection system compared, 94–95
Interfacial tension, 29
Ionizing radiation, aseptic packaging, 302
Irradiation, pasteurization, 230

K

Keeping quality
pasteurization, 201–211
pathogen inactivation, 204–209
process characterization, 207–209
raw material considerations, 203
storage conditions, 210–211
time/temperature combinations, 203–207, 208
temperature, 210–211

L

Lactulose, 53, 263–264
Laminar flow, non-Newtonian liquid, modification of velocity profile for power law fluids, 137–138, 139
Latent heat, 18
Lethality table, sterilization, 237, 238
Lipase, 266–267
Liquid, density, 12

Log-logistic theory of microbial death, 48–49
Low-acid food, 10
Low-viscosity liquid, 66–68, 68–72
Lysine, 263

M

Maillard reaction, 52–53
Marker organism, 128–130
Mechanistic theory of microbial death, 36–39, 47–48
Microbial inactivation, 4
Microbiological marker, 180
Microbiological testing, quality assurance, 413–414
Microfiltration, 225
Microorganism, 1
  clumping, 46
Microorganism death
  constant lethal temperature, 33–36
  criteria, 41–42
  kinetics, 33–49
    methods for determination, 43–44
  measurement problems, 44–45
  temperature, overall effect of changing, 40
Microwave heating
  continuous-flow thermal processing plant, 104
  pasteurization, 201
Milk, 28–29
  fouling, 345–354
    mechanisms, 351–352
    methods to reduce fouling, 348–351
    milk permeate, 352–353
    raw milk quality, 346–348
    reconstituted milk, 354
  pasteurization, 211–216
    changes during, 212–214
    critical control points in raw milk handling, 204
    enzyme inactivation, 205, 214–216
    hazards involved, 204
    residual enzyme activity, 205
  pasteurized, sales, 6
  sterilization, 250–252
    sales, 6
  thermization, 223–224
  ultrahigh temperature processing, 250–251, 252–256
    bacterial proteinase, 268–269
    changes taking place during, 255–256
    color, 264–265
    effects on (milk) food components, 256–263
    enzyme inactivation, 266–269, 270
    fat, 263
    flavor, 264
    flow diagram, 408
    lactulose, 263–264
    lipase, 266–267
    lysine, 263
    mineral, 259
    non-cow's milk, 269–271
    protease, 266, 267
    protein, 256–259, 260
    sales, 6
    sensory characteristics, 264–266
    texture, 265–266
    thiamine, 263
    vitamin, 260–262, 263
Milk analogue, ultrahigh temperature processing, 274–277, 278, 279
Milk-based drink, pasteurization, 216
Mineral, 259
Mouthfeel, cream, 217–218
Mycobacterium tuberculosis, 212

N

New variant Creutzfeldt-Jakob disease, 5
Non-cow's milk, 269–271
Non-Newtonian liquid
  laminar flow, modification of velocity profile for power law fluids, 137–138, 139
  residence time distribution, different-shaped sections, 139–140
  Reynolds number, non-Newtonian viscosity, 136–137
  rheogram, 16
  viscosity, 14
Nusselt number, 24, 165, 166, 167

Nutritional drink, ultrahigh temperature processing, 280

O

Ohmic heating
continuous-flow thermal processing plant, 104
particulate liquid, 183–188
continuous-flow thermal processing, 185–186
design, 186–187
flow diagram, 185
modeling, 186–187
verification, 187–188
Overall heat transfer coefficient, 22–23
fouling, 331
decrease, 335, 338

P

Packaging, 285–327. *See also* Aseptic packaging
aseptic packaging, 295–327
connection of packaging equipment to process plant, 285–290
continuous flow thermal processing, 294
direct connection of filler to process plant, 286–287
hot filling, 294–295
mechanical principles, 293–294
sterile balance tank, 287–291
cleaning, 290–291
construction, 289–290
integral with filler, 287
large, 287–290
small, 287
sterilization, 290–291
sterile barrier, 292
Paperboard carton
formed from reel, 303–307
preformed, assembled from blanks on filler, 307–309
Particulate liquid
continuous-flow thermal processing, 143–188
chemical marker, 181
difficulties, 143–144
enzyme marker, 180–181
mathematical modeling, 174–178
microbiological marker, 180
overall process validation, 178–181
prediction of sterilization effect, 178–181
time-temperature integrator, 181
heat exchanger, 159–160
heat transfer coefficient, 159–160
experimental measurement, 170–171, 172–173
experimental methods for determination, 166–168
between liquids and solid bodies, 164–166
minimum value, 165–166
Ohmic heating, 183–188
continuous-flow thermal processing, 185–186
design, 186–187
flow diagram, 185
modeling, 186–187
verification, 187–188
processing equipment, 144–145
separate processing of liquid and solid phases, 144–145
pumping, 145–148
residence time distribution
effect of different parameters, 152–159
principles, 158–159
Stork Rota-Hold process, 181–183
surface heat transfer
analytical solutions, 162–163
graphical solutions, 161–162
mathematical models, 163–164
rate of heating on particulate solid, 160
time-temperature integrator, 166, 168–170
transport of particulate solids by liquids, 148–152
critical velocity, 151–152
physical factors, 149–152
Pasteurization, 193–231
advantages of batch system, 194

bacteriocin, 229–230
beer, 223
continuous processes, 194
continuous-flow thermal processing plant, 65–66
cream, 216–218
defined, 193
dimensionless temperature ratio, 194
egg, 219–220
minimum requirements, 215
fruit drink, 220–222
fruit juice, 220–222
high(er) temperatures for short(er) times, 195–201
advantages, 195–196
control, 197–201
engineering, 197–201
layout, 197, 198
high-intensity pulsed electric field, 230
historical developments, 194
objections, 194, 195
ice cream, 218–219
irradiation, 230
keeping quality, 201–211
pathogen inactivation, 204–209
process characterization, 207–209
raw material considerations, 203
storage conditions, 210–211
time/temperature combinations, 203–207, 208
microwave heating, 201
milk, 211–216
changes during, 212–214
critical control points in raw milk handling, 204
enzyme inactivation, 205, 214–216
hazards involved, 204
residual enzyme activity, 205
sales, 6
milk-based drink, 216
novel techniques, 230
postpasteurization contamination, 193, 209–210
pulsed light, 230
tomato, 223
UHP processing, 230
vitamin, 213–214
whey protein denaturation, 213
Peracetic acid, aseptic packaging, 300
pH, 7–10
values, 9
Physical test, quality assurance, 409–413
Plastic cup, preformed, 315–318
Plastic pouch, 310–312
Plate heat exchanger
continuous-flow thermal processing plant, 64–65
sterilization, 243–245
Postpasteurization contamination, 209–210
Power law, viscosity, 137–138, 139
Prandtl number, 24
Preparation, quality assurance, 401–402
Preprocessing heat treatment, continuous-flow thermal processing plant, 62–64
Pressure
fouling, 337–338, 339, 340
thermal processing, 30
Probability, quality assurance, 401–403
Process characterization
food safety, 410–412
quality assurance, 410–412
Processing
browning reaction, 363–366
color, 363–366
raw material, 361–362
Producer risk, 403
Product range, thermal processing, 5–6
market sectors, 6
Product testing, quality assurance, 405–406
Protease, 266, 267
Protein, storage, 377
Pulsed light, pasteurization, 230
Pumping, particulate liquid, 145–148

Q

Quality assurance, 395
analytical testing, 412–421
aseptic packaging, 412
Bacillus cereus, 399–401

chemical test, 417–421
cleaning, 404
commercially sterile products, 401–407
current approach, 407–411
defined, 395
food safety, 395–397
legal aspects, 422–425
microbiological specifications for foods, 398–401
short shelf-life products, 398–399
microbiological testing, 413–414
physical test, 417–421
practical aspects, 404–407
preparation, 409–410
probability, 401–403
process characterization, 410–412
product testing, 413–414
raw material, quality, 409–410
sampling for verification of process, 403–404, 405
sampling theory, 401–403
sensory characteristics, 421
sterilization of equipment, 412
time-temperature indicator, 417
Quality control, defined, 395

R

Radiation, 19
Raw material, 224–227
attributes, 225–227
carbon dioxide, 226–227
hydrogen peroxide, 226
natural antimicrobial systems, 225–226
processing, 369–370
quality assurance, quality, 409–410
storage, 369–370
Refractive index, 30
Refrigerated product, extended shelf life, 224–230
extended heat treatment, 227–229
raw material quality, 224–227
Refrigerated storage, 382
Regeneration, 1
Regeneration efficiency, 196
Residence time distribution
continuous-flow thermal processing, 139–141
practical evaluation, 141
in process equipment, 140–141
continuous-flow thermal processing plant, 113–130
chemical marker, 130
direct process evaluation, 128
dispersion model, 120–122, 123
effect on quality of foods, 127
errors in prediction, 123–124
low-viscosity Newtonian liquid velocity profiles, 114–115
marker organism, 128–130
measurement, 117–119
plant performance effect, 113–114
plant performance evaluation, 127–130
practical residence time distribution curves analysis, 123
residence time distribution data models, 119–122
residence time distribution for any input signal, 118–119
residence time distribution theory, 115–117
sterilization efficiency effect, 124–127
tanks-in-series model, 120
non-Newtonian liquid, different-shaped sections, 139–140
particulate liquid
effect of different parameters, 152–159
principles, 158–159
sterilization, 124–127
Resistance, 25–26
conduction, 20–21
direct contact heating, 23–24
Reynolds number, 24
non-Newtonian viscosity, 136–137
Rinsing, 356–358
water/product changeover, 357–358

S

Sampling theory, quality assurance, 401–403

Saturated steam
aseptic packaging, 296–297
properties, 13
Scraped-surface heat exchanger,
continuous-flow thermal processing plant, 79–81, 82
Shear-thickening behavior, 14
Shelf life. *See* Extended shelf life
Shell-and-tube tubular heat exchanger, 75–79
Soup, ultrahigh temperature processing, 277–280
Soy milk
storage, 381
ultrahigh temperature processing, 274–277, 278, 279
Specific enthalpy, 18–19
Specific gravity, density, 11
Specific heat, 15–18
values, 17
water, 17
Sphere, heat transfer, mathematical model, 164, 174–178, 427–434
Spray pasteurizer, 201, 202
Starch-based product, ultrahigh temperature processing, 277–280
Steam injection system, continuous-flow thermal processing plant, 85–91
infusion system compared, 94–95
Sterile balance tank, sterilization, 290–291
Sterile barrier, packaging, 292
Sterilization, 237–280
canning process, 241–242
continuous systems, 242–248
continuous-flow thermal processing plant
advanced plate-type indirect heating plant, 70–71
heat exchanger specification, 67–68
low-viscosity liquid, 66–68, 68–72
criteria, 237–242
direct heating, 245–248
evaporated milk, 273–274
F value, 41
heat-resistant spore, 237–239, 240
history, 2
homogenization, 248–250
in-container sterilization, 240
indirect heating, 243–245
lethality table, 237, 238
milk, 250–252
plate heat exchanger, 243–245
reaction kinetic data, 253
reaction kinetic parameters, 240
residence time distribution, 124–127
time-temperature sterilization, 243
tubular heat exchanger, 243–245
ultra-high temperature (ultrahigh temperature) product, 242–248
Sterilization of equipment
chemical, 364–365
heat, 364
quality assurance, 395
Storage, 369–393
accelerated storage, 389–390
browning reaction, 371–374
color, 371–374
deposit formation, 375
destabilization, 374–379
dissolved oxygen, 380
enzyme regeneration, 388–389
fat, 385
flavor
cooked milk flavor, 380–384
flavor changes, 380–384
flavor improvement, 384
gelation, 376–379
age gelation, 376–378
enzyme-induced gelation, 378–379
light-induced changes, 388
protein, 385
raw material, 369–370
refrigerated storage, 390
soy milk, 389
vitamin, 385–388
Stork Rota-Hold process, particulate liquid, 181–183
Surface area to volume ratio, 10–11
Surface heat transfer, particulate liquid
analytical solutions, 162–163
graphical solutions, 161–162
mathematical models, 163–164
rate of heating on particulate solid, 160
Surface properties, 28–29

Surface tension, 29
  values, 29

T

Tanks-in-series model, 120
Tea, ultrahigh temperature processing, 280
Temperature
  biochemical component changes at high temperatures, 49
  continuous-flow thermal processing plant
    chemical marker, 130
    direct process evaluation, 128
    marker organism, 128–130
    plant performance evaluation, 127–130
  decimal reduction time, 36–39
    Arrhenius equation model, 37–38
    canners' (constantz value) model, 36–37, 38
    models compared, 38–39
  density, 12
  electrical conductivity, 26
  fouling, 342, 343
    measurement, 336, 337
  microorganism death, 40
  thermal conductivity, 20
  thermal processing, 30
  viscosity, 14, 15
Texture, 265–266
Thermal conductivity, 19–21
  defined, 19
  models for predicting, 19–20
  temperature, 20
  values, 21
Thermal death kinetics
  activation, 45
  aging phenomenon, 45
  calcium dipicolinate, 45
  heat activation, 44–45
  nonlinear data, 45–47
  problems, 44–45
Thermal diffusivity, 21
Thermal processing
  competing techniques, 31
  criteria, 41–42
  flow, 30
  food quality, 5
  food safety, 4–5
  general cooking changes, 50–54
  levels of ingredients, 30
  measurements variables, 30
  pressure, 30
  product range, 5–6
    market sectors, 6
  temperature, 30
Thermization, milk, 223–224
Thermochromic paint, 167–168
Thermoduric bacteria, extended shelf life, 227–229
Thermoresistometer, 43
Thiamin, 52, 263
Thixotropy, 14
Time-temperature indicator
  analytical testing, 417
  quality assurance, 417
Time-temperature integrator, 181
  particulate liquid, 166, 168–170
Time-temperature profile, continuous-flow thermal processing plant
  chemical quality criteria, 112
  determination, 108, 109
  microbiological criteria, 111
  performance calculation, 109–113
  performance prediction, 107–113
Time-temperature sterilization, 243
Titratable acidity, 10
Tomato
  fouling, 355
  pasteurization, 223
Tubular heat exchanger
  continuous-flow thermal processing plant, 72–79
    concentric tube heat exchanger, 72–75
    shell-and-tube tubular heat exchanger, 75–79
  sterilization, 243–245
Tunnel pasteurizer, 201, 202

U

Ultrahigh temperature processing
  coffee, 280

cream, 271–273
custard, 277–280
evaporated milk, 273–274
ice cream mix, 274
infant formulation, 280
milk, 250–251, 252–256
bacterial proteinase, 268–269
changes taking place during, 255–256
color, 264–265
effects on (milk) food components, 256–263
enzyme inactivation, 266–269, 270
fat, 263
flavor, 264
flow diagram, 400
lactulose, 263–264
lipase, 266–267
lysine, 263
mineral, 259
non-cow's milk, 269–271
protease, 266, 267
protein, 256–259, 260
sales, 6
sensory characteristics, 264–266
texture, 265–266
thiamine, 263
vitamin, 260–262, 263
milk analogue, 274–277, 278, 279
nutritional drink, 280
pasteurization, 230
soup, 277–280
soy milk, 274–277, 278, 279
specialized products, 280
starch-based product, 277–280
sterilization, 242–248
tea, 280
vitamin C, 262, 263
whey protein denaturation, 257–259, 260
Ultraviolet irradiation, aseptic packaging, 300
hydrogen peroxide combined, 301–302

V

Vapor, compressibility, 12, 13
Vegetable juice, enthalpy, 20
Vegetative bacteria, heat resistance, 204–209
Velocity profile, Hershel-Bulkley liquid, 139, 140
Viscosity, 13–15
characterization, 134–136
cream, 217–218
non-Newtonian fluid, 14
power law, 137–138, 139
temperature, 14, 15
values, 14
water, 14
Viscous liquid product, 133–143
continuous-flow thermal processing
equipment selection, 134
modification of velocity profile for Hershel-Bulkley fluids, 139, 140
non-Newtonian viscosity effect on laminar flow velocity profile, 137–139
non-Newtonian viscosity effect on Reynolds number, 136–137
residence time distribution by section shape, 139–140
defined, 133
heat exchanger
heat transfer coefficients, 141–143
non-Newtonian liquid, 142–143
Vitalistic theory, 48
Vitamin, 260–262, 263
degradation, 51–52
pasteurization, 213–214
storage, 377–380
Vitamin C, 51–52
ultrahigh temperature processing, 262, 263

W

Water
specific heat, 17
viscosity, 14
Water activity, 25
values, 26
Whey protein, fouling, 354–355
Whey protein denaturation
pasteurization, 213
ultrahigh temperature processing, 257–259, 260